MW00598296

GEOPHYSICAL WAVES AND FLOWS

Waves and flows are pervasive on and within Earth. This book presents a unified physical and mathematical approach to waves and flows in the atmosphere, oceans, rivers, volcanoes and the mantle, emphasizing the common physical principles and mathematical methods that apply to a variety of phenomena and disciplines. It is organized into seven parts: introductory material; kinematics, dynamics and rheology; waves in non-rotating fluids; waves in rotating fluids; non-rotating flows; rotating flows; and silicate flows. The chapters are supplemented by 47 'fundaments', containing knowledge that is fundamental to the material presented in the main text, organized into seven appendices: mathematics; dimensions and units; kinematics; dynamics; thermodynamics; waves; and flows. This book is a valuable reference for graduate students and researchers seeking an introduction to the mathematics of waves and flows in the Earth system, and can serve as a supplementary textbook for a number of courses in geophysical fluid dynamics.

DAVID E. LOPER is an Emeritus Professor at Florida State University. After completing his PhD in Mechanical Engineering at Case Institute of Technology, and after a brief stint in the aerospace industry and a post-doctoral appointment at the National Center for Atmospheric Research in Boulder CO, Professor Loper took a joint position in Applied Mathematics and the Geophysical Fluid Dynamics Institute at Florida State University, where he was named Distinguished Research Professor and awarded a Named Professorship. He was a co-founder and secretary of SEDI (Study of the Earth's Deep Interior, a Committee of the International Union of Geodesy and Geophysics) and is Fellow of the American Geophysical Union.

GEOPHYSICAL WAVES AND FLOWS

Theory and Applications in the Atmosphere, Hydrosphere and Geosphere

DAVID E. LOPER

Florida State University

CAMBRIDGE
UNIVERSITY PRESS

University Printing House, Cambridge CB2 8BS, United Kingdom

One Liberty Plaza, 20th Floor, New York, NY 10006, USA

477 Williamstown Road, Port Melbourne, VIC 3207, Australia

314–321, 3rd Floor, Plot 3, Splendor Forum, Jasola District Centre, New Delhi – 110025, India

79 Anson Road, #06-04/06, Singapore 079906

Cambridge University Press is part of the University of Cambridge.

It furthers the University's mission by disseminating knowledge in the pursuit of education, learning, and research at the highest international levels of excellence.

www.cambridge.org
Information on this title: www.cambridge.org/9781107186194
DOI: 10.1017/9781316888858

First published 2017

Printed in the United States of America by Sheridan Books, Inc.

A catalogue record for this publication is available from the British Library.

Library of Congress Cataloging-in-Publication Data
Names: Loper, David E., author.
Title: Geophysical waves and flows : theory and applications in the atmosphere, hydrosphere and geosphere / David E. Loper, professor emeritus, Florida State University.
Other titles: Waves and flows
Description: Cambridge : Cambridge University Press, 2017. |
Includes bibliographical references and index.
Identifiers: LCCN 2017032905 | ISBN 9781107186194 (hardback : alk. paper)
Subjects: LCSH: Waves. | Geophysics. | Flows (Differentiable dynamical systems) | Elastic waves. | Kinematics.
Classification: LCC QC157 .L67 2017 | DDC 531/.1133–dc23
LC record available at https://lccn.loc.gov/2017032905

ISBN 978-1-107-18619-4 Hardback

Contents

Preface

This monograph is the outgrowth of a set of notes that were prepared some years ago for a course that in fact never was taught. The notes lay fallow until the occasion of my 75th birthday, which inspired me to try to clean up my computer files. Upon discovering these notes, I consulted David Furbish regarding using these as the basis of a monograph on geophysical waves and flows. His encouraging response gave me the confidence to attempt to do so. This is the outcome of that endeavor.

The goal of that course was – and of the current monograph is – to present a unified approach to geophysical waves and flows, starting from the simplest case and progressively adding complicating factors in a systematic manner. Simplest are sound waves that occur in air and water. Similar, but somewhat more complicated, are compressive body waves (P waves), transverse waves (S waves) and edge waves (Rayleigh and Love waves) in Earth's mantle. Seemingly similar, but dynamically distinct, is the fluid edge wave (think of ocean waves), occurring in both deep and shallow water. It is a short step from shallow-water waves to flows in a horizontal channel. Next, if the channel is sloping, we encounter gravitationally driven flow and this leads naturally to the study of turbulent flows. These topics are found in Chapters 9–13 and 19–23 of this monograph.

In order to approach the analysis of waves and flows properly with a sound theoretical basis, we need to begin from square one, quantifying the manner in which a continuous body can move (kinematics), the nature of the forces that make it do so (dynamics) and the form of its response (rheology). These essential topics are covered in Chapters 3–7. Of course, rotation of Earth affects many types of waves and flows, so this subject is introduced in Chapter 8, with waves affected by rotation investigated in Chapters 15–18 and flows affected by rotation investigated in Chapters 25–28. It is interesting to compare and contrast the readily visible flows within the atmosphere and oceans to silicate flows occurring within Earth's mantle and in volcanoes; these latter flows are investigated in Chapters 29–35.

The book attempts to present a wide range of waves and flows fairly simply, while retaining mathematical rigor. A minimum prerequisite of prior knowledge is an understanding of multivariate calculus and ordinary differential equations; a previous knowledge of partial differential equations, complex variables and linear algebra is desirable but not necessary. The book is written in an informal manner; my guide in this approach has been to imagine writing for a precocious grandchild: Noah Bliss. Finally, I wish to acknowledge the significant assistance provided by Paul Roberts, who read the manuscript with an eagle eye and provided me with many suggestions for improvement.

General Description of Contents

The main part of this book consists of 35 chapters organized into seven parts, consisting of I: introductory material, II: kinematics, dynamics and rheology, III: waves in non-rotating fluids, IV: waves in rotating fluids, V: one-dimensional flows, VI: rotating flows and VII: silicate flows. The chapters are supplemented by 47 "fundaments," containing knowledge that is fundamental to the material presented in the main text, organized into seven appendices: A: mathematics, B: dimensions and units, C: kinematics, D: dynamics, E: thermodynamics, F: waves and G: flows. These fundaments are intended to aid the reader who may have some gaps in background or to serve as a resource. The 35 chapters and 47 fundaments may be thought of as the warp and woof of a grand tapestry, with related topics joined together by several hundred footnotes.

Part I consists of an introductory chapter discussing a number of broad issues, including mathematical modeling, continuum mechanics, energy and planetary cooling and the concept of a continuous body, and a chapter containing a number of more detailed preliminaries, including establishment of a reference coordinate system. Part II deals with kinematics, dynamics and rheology, beginning in Chapter 3 with an analysis of the kinematics of deformation and flow, and followed in Chapter 4 with an investigation of dynamics and the stress tensor. Next, in Chapter 5 we are introduced to some thermodynamics relevant to geophysical waves and flows and in Chapter 6 to the fundamentals of shear rheology. Chapter 7 contains an introduction to the concept of a static reference state and the equations governing perturbations of this state. This part concludes with an introduction to rotating fluids in Chapter 8.

We begin our study of waves in Part III, focusing on waves unaffected by rotation. Chapter 9 contains an introduction to waves, with an investigation of sound waves in § 9.3. The various types of waves that can occur in elastic bodies are studied in Chapter 10, with seismic waves in Earth's mantle discussed in § 10.5. Deep-water waves are considered in Chapter 11, which includes a discussion of the nature of ocean waves in § 11.5. Following this, we investigate linear shallow-water waves in Chapter 12 and nonlinear shallow-water waves in Chapter 13. Our survey of non-rotating waves concludes in Chapter 14 with analyses of capillary, interfacial and internal-gravity waves.

Our study of rotating waves in Part IV begins in Chapter 15 with geostrophic, inertial and Rossby waves. In Chapter 16 we investigate how rotation affects surface, interfacial and internal-gravity waves. The last two chapters of this part deal with ocean waves that rely solely on rotation for their existence: equatorial Kelvin and Rossby waves studied in Chapter 17 and coastal Kelvin and topographic Rossby waves studied in Chapter 18.

Part V contains six chapters devoted to simple one-dimensional flows, beginning in Chapter 19 with an orientation to the topic and a review of the relevant governing equations.

Steady and unsteady flows in a uniform horizontal channel are investigated in Chapters 20 and 21, respectively. We introduce gravitational forcing of channel flows, by allowing the channel to tilt downward, in Chapter 22. This leads us to the topic of turbulence, which is investigated in some detail in Chapter 23, leading to a simple model of turbulent diffusion. This model is applied to some simple turbulent flows (katabatic winds, avalanches and cumulonimbus clouds) in Chapter 24.

Flows in rotating fluids are investigated in Part VI, beginning with a detailed study of laminar and turbulent Ekman layers in Chapter 25. Atmospheric flows, including the general circulation, thermal winds and the jet stream, are investigated in Chapter 26, and oceanic currents, including the Sverdrup balance, western-boundary currents and thermo-haline circulation, are investigated in Chapter 27. This part concludes in Chapter 28 with a survey of vortices and brief analyses of the structure and dynamics of tornadoes and hurricanes.

In Part VII our attention returns to the "solid" Earth with an investigation of various silicate flows. We begin in Chapter 29 with a survey of the equations governing silicate flows and estimates of parameter values within the mantle. Earth's mantle is convecting in an effort to cool the mantle and core. The nature of this flow is reviewed in Chapter 30, then the modes of mantle convection involved in cooling the mantle and core are investigated in Chapters 31 and 32. In Chapter 33 we survey the various types of volcanic flows, then in Chapter 34 investigate the nature of flow in volcanic conduits and conclude in Chapter 35 with a study of lava sheet flow on the surface.

The fundaments are organized into seven appendices. The first, Appendix A, deals with mathematical issues, including vector algebra, vector calculus, curvilinear coordinates, Taylor series, Fourier series and integrals, classification of partial differential equations, a listing of the Greek alphabet and introductions to scalar and vector potentials and the stream function. Appendix B deals with dimensions and units, including introductions to dimensional analysis and the SI system of units, and contains several tables giving values of parameters relevant to geophysical waves and flows. Kinematic topics surveyed in Appendix C include non-inertial frames of reference and virtual forces, the material derivative, finite deformation and flow lines and points. Appendix D deals with dynamics and surveys viscoelastic behavior, silicate rheology, the constants of elasticity, surface tension, the general conservation law, the Euler and Bernoulli equations, kinetic and internal energies, thin-layer equations and variables, the Proudman–Taylor theorem and the vector vorticity equation. Perhaps the most important appendix is the fifth, Appendix E, surveying thermodynamics, including storage and transfers of energy, the mole, the first law of thermodynamics, thermodynamic potentials, variables and parameters, equations of state for density and entropy, ideal mixtures, the energy equation, ideal gases, thermodynamics of the atmosphere, phase equilibrium and thermodynamic efficiency. Appendix F contains fundaments on waves, including waves in three dimensions, Fourier representation of waves, Stokes waves, Kelvin's ship waves, energy of deep-water waves, Laplace's tidal equations and inertial waves. Finally, Appendix G contains three fundaments of flows: shear-flow instability, boundary-layer theory and a critique of models of open-channel flows.

Part I

Introductory Material

In this first part, we get oriented to the general subject matter of the book and the style of approach. The orientation in Chapter 1 includes a discussion of broader aspects, such as mathematical modeling, continuum mechanics, the role of energy in the universe and the apparent need of astronomical bodies to cool as rapidly as possible. Central to our study is the concept of a continuous body, as discussed in § 1.2.

The second chapter provides a bit more introductory detail, including the reference coordinate system, a definition of waves and flows, a summary of the scope of the book and brief introductions to a number of related issues.

1
Introduction

This monograph is a survey of – and introduction to – the ways in which the atmosphere, oceans and various parts of the so-called solid Earth can deform and move. The most common type of deformation is a *wave* and the most common type of motion is *flow*. Our goals are to quantify and understand these deformations and motions. In order to achieve these goals, we will need to develop an appropriate set of procedures and tools.

An appropriate procedure is *mathematical modeling*, which consists of three steps:

- *Formulation* of a mathematical model of a physical system. This is the crucial element in the procedure. If the model does not mimic the relevant physical processes, it will fail to yield useful results.
- *Solution* of the mathematical problem. This is the part stressed in most mathematics courses, but in fact is the most routine aspect of the procedure.
- *Interpretation* of the solution. This is the payoff; solution of a well-conceived model should yield new physical insight and make testable predictions.

We won't be following these steps formally, instead they will come naturally as we investigate each type of wave and flow.

Our mathematical models will treat the physical system (that is, the portion of Earth under consideration) as a continuous body, using the mathematical apparatus of *continuum mechanics*. The concept of a continuous body is discussed in § 1.2. Continuum mechanics consists of three fundamental elements:

- *Kinematics* describes how bodies can move and deform. For example, it limits the types of deformation which leave a body competent (i.e., unfractured).
- *Dynamics* explains why bodies move and deform. They do so in response to external and internal forces, within the context of Newton's second law.
- *Rheology* quantifies the kinematic response to dynamic forces, taking into account the material properties of the body under investigation. Two identically shaped bodies, composed of water and of steel, respond quite differently to the same applied forces. Rheology encompasses the change of volume of a body induced by a change in pressure (*compressive rheology*) and the change of shape caused by a change in deviatoric stresses (*shear rheology*).

The nature of the physical system under consideration is determined primarily by its shear rheology. While continuous bodies can exhibit a wide range of possible rheological behaviors, we will consider only a few relatively simple types: linear elastic solid and three types of fluid:[1] inviscid, Newtonian and power-law.[2] Other exotic states of matter, such as plasmas (found in the Van Allen belts) and thixotropic solids (such as clays) having history-dependent rheology, will not be included in our survey.

The emphasis throughout this monograph will be on the fundamentals of the processes being modeled with mathematical complexity minimized where possible. In that spirit, although we live in a three-dimensional world, many physical processes are effectively two or even one dimensional. The fundamental concepts will be introduced in general form, but will be applied only in simple situations.

At various points in the text concepts are introduced which may be unfamiliar to some readers. As an aid to these readers, many of these concepts are explained in various fundaments, found in the appendices.

1.1 A Broader Perspective

Before delving into details of geophysical waves and flows in the following chapters, it may be helpful to place them within a broader cosmological context by briefly discussing the related concepts of energy and cooling.

1.1.1 Energy

Energy is the essence of the universe, with much of it in a frozen phase called mass. It is possible that the total energy of the universe is zero, with the (negative) gravitational energy equaling all other (positive) forms of energy. The study of energy is concerned with both its storage within a body and its transfer between bodies. Broadly speaking, there are four forms of energy:

- *Energy associated with the short-range nuclear forces.* This energy is quantified by the rest mass, M, which is related to the energy E by Einstein's celebrated equation: $M = E/c^2$, where c is the speed of light. Barring nuclear reactions, this energy is in a "frozen" state and is considered separately from all other forms of energy. Kinematics is the study of the behavior of mass within a continuous body; see Chapter 3 and Appendix C. The deformation of a continuous body is limited by the equation of conservation of mass, also known as the continuity equation; see § 3.3.1 and § 3.4.1.
- *Electromagnetic energy.* The study of electromagnetism is called *electrodynamics*. Relatively little energy is stored as electromagnetic energy, but electromagnetic photons very efficiently transfer energy between bodies separated in space, a process referred to

[1] We learned in grade school that there are three types of matter: solid, liquid and gas. However, liquids and gases are both fluids, distinguished primarily by their compressibility.
[2] The rate of flow of power-law fluid is proportional to the deviatoric stress raised to an exponent or power.

as *radiative transfer*. However, we will be concerned with this mode only peripherally, with Earth receiving heat in the form of solar electromagnetic radiation, primarily in and near the visible range of frequencies, and casting heat off to outer space, primarily in the infrared. Electromagnetic forces affect waves and flows within Earth's core and in the ionosphere, but these regions are not included in the scope of this book.

- *Energy stored in disorganized motions of the atoms comprising a body. Thermodynamics* is the study of this form of energy; see Appendix E. Commonly the term "energy" is taken to mean this form, but properly it should be called thermodynamic energy.
- *Energy stored in the organized positions and motions of macroscopic bodies.* This includes kinetic and potential energy (including gravitational potential energy). These energies are relative quantities, dependent on the position and velocity of the observer. Since they are not intrinsic quantities, kinetic and potential energy are studied separately from thermodynamics, in a subject area called *dynamics*; see Chapter 4. The equation of conservation of momentum[3] is a primary focus of dynamics; see § 4.6.

1.1.2 Cooling

The visible Universe is composed of stars, planets and a myriad of smaller objects each of which appears to be attempting to *cool as rapidly as possible*. The predominant mode of cooling for celestial bodies is electromagnetic radiation, and Earth is no exception; it cools by radiation to outer space. A small fraction of the radiation emitted by our Sun is absorbed by Earth. This solar insolation is a dominant factor in Earth's heat balance. Earth in turn re-radiates this to space, along with a much smaller amount of heat from the interior, as the bulk of Earth slowly cools.

Portions of Earth are in motion because these flows aid in the cooling of Earth. For example, most of the solar insolation strikes the tropics and this heat is re-radiated to space more efficiently by the poleward transport of heat via atmospheric and oceanic circulations. In addition, the bulk of Earth (that is, the mantle and core) cool more rapidly via convection within the mantle than by conduction. All the wave motions and flows on Earth are caused either primarily or incidentally by the need for Earth to cool as rapidly as possible.

1.2 The Concept of a Continuous Body

Perhaps the first step in defining a continuous body is to define what is meant by a body. A physical body is a group of atoms or molecules held together by inter-atomic forces, typically electro-chemical in origin, or confined by impenetrable walls. In considering the collection of atoms as a body, we abandon the possibility of describing the behavior of each individual atom, and consider only the collective behavior of relatively large numbers of atoms. Given that the number of atoms in a body of hand-specimen size is on the order of 10^{26}, whereas the largest computers currently available (in 2017) can model fewer than

[3] Momentum is the gradient of energy.

10^{12} particles, this is an eminently reasonable approach. Even if we could model the behavior of each atom, we must question the motivation for doing so: what sense could we make of that large amount of information? As the old saying goes, in such a case, "we couldn't see the forest for the trees".

We wish to describe and predict the behavior of bodies as they move and deform at a level of accuracy sufficient for scientific curiosity and practical need. With this in mind, we shall treat the body as a macroscopic continuum, ignoring its microscopic atomic nature. In this approach, the smallest volume we shall consider, called a *particle*, contains a large number of atoms.[4] This number is sufficiently large that atomic fluctuations in position and velocity are not evident. Given the large value of Avogadro's number (the number of atoms in a mole of substance[5]), we can do this with some confidence. For example, suppose we wish to model the large-scale dynamics of the oceans. If we chose our particle to be the size of a drinking glass (which is far smaller than necessary) the number of molecules of water in each particle would be greater than the number of such particles in all the oceans.

In the context of macroscopic continuum mechanics, the particles that comprise a continuous body are of infinitesimal size and the locations of the particles are quantified by the positions of a continuum of points as a function of time, e.g., $\mathbf{x}(t)$, where $\mathbf{x}$ is the position vector, measured using some reference coordinate system, and t is time; see §2.1.

In order to quantify deformation, we must consider localized groups of particles, called *parcels*. A parcel consists of all the particles within the neighborhood[6] of a given particle. Each particle is a member of a large number of parcels. Particles have no discernible size or shape, while parcels and bodies do.

To summarize:

- a *continuous body* is composed of a vast number of particles;
- a *particle* is composed of a vast number of atoms or molecules and is identified by its reference position or current location;
- a *point* is a position in three-dimensional space used to identify or locate particles;[7] and
- a *parcel* is a tiny body composed of all particles in the neighborhood of a given point or particle.

Displacements of the particles comprising a body relative to a reference state are constrained by kinematics. The displacement of a body consists of *rigid-body motion* plus *deformation* and *flow*. Rigid body motion consists of *translation* and *rotation*, while deformation and flow consist of change of volume and change of shape, as illustrated in Figure 1.1.

[4] Except when considering the behavior of an ideal gas, when a particle is an atom or a molecule; see Appendix E.8.4.

[5] See Appendix E.2.

[6] A *neighborhood* is a small volume surrounding a specified point.

[7] See § 2.1.

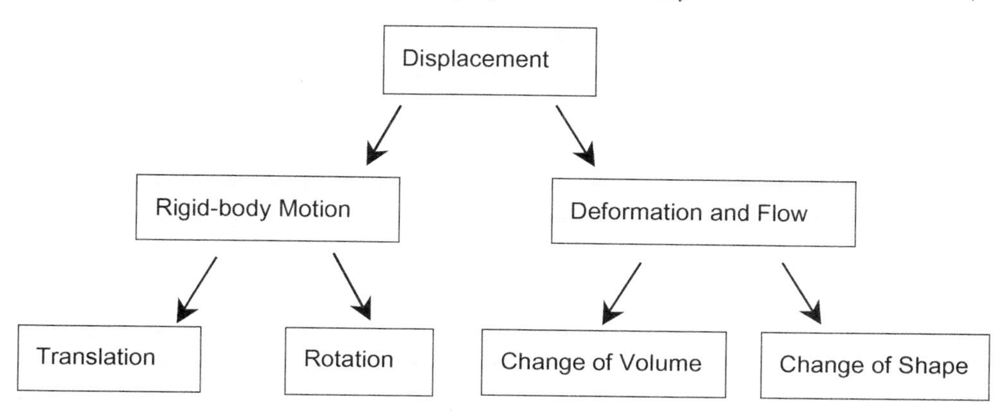

Figure 1.1 Components of displacement of a continuous body.

The motions of bodies are quantified by use of a reference coordinate system. We will employ a Cartesian reference system.[8] Large-scale translation and rotation (that is, rigid-body motion) may be removed from the kinematic description by using a coordinate system which moves with the body under investigation. While this simplifies kinematics, it complicates dynamics. In particular, the use of a non-inertial coordinate system introduces so-called virtual forces into the force balance.[9]

1.2.1 Composition of a Continuous Body

A particle may consist of various types of atoms and molecules in differing phases, with each type or phase being called a *constituent*.[10] The relative magnitudes of the constituents will be represented as mass fractions, treated as functions of position and time. Barring change of phase, the composition of a particle can change only by diffusion. Diffusion in solid bodies is an exceedingly slow process (as can be verified by the banding seen in geodes) and can be neglected. If the body is a fluid (a liquid or gas), adjoining particles can exchange atoms and molecules much more readily and diffusion can be an important process.

A body may be characterized by the chemical composition and state of matter at a point and by the physical structure in the neighborhood of a point. If the composition and state are the same at all points in a body, it is said to be *homogeneous*. If the structure of the body is the same in all directions, the body is said to be *isotropic*.

Transition

In this first chapter, we have discussed the broad goals of this book, introduced the concept of a continuous body and discussed the possible displacements of its constituent particles.

[8] Other possible reference systems are briefly discussed in Appendix A.3.
[9] See Appendix C.1.
[10] The chemically distinct materials comprising a body are called *components*. The phases of a component are differing constituents.

These displacements will be analyzed in Chapter 3, which deals with kinematics – how a continuous body can move and deform. Before delving into the study of kinematics, we need to establish a basis for quantification of deformation and flow, using a *reference coordinate system*. This is described in Chapter 2. This chapter also provides an orientation to the main topic of this book, waves and flows, explains the format of the book and briefly introduces a number of ancillary topics.

2
Getting Started

The first things we need to know are: where are we and where are we going? Assuming one is speaking of position and direction in space, this question is answered by constructing and using a reference coordinate system, as described in the following section. Assuming one is speaking of position and topic within this book, this question is answered in § 2.2 and § 2.3. This chapter concludes with brief presentations § 2.4 of a number of tools that will be useful as we proceed with our study of waves and flows.

2.1 Reference Coordinate System

A principal goal of continuum mechanics is to quantify the deformation and flow of continuous bodies. In order to do so, we must have a way to locate the particles of the body. This is accomplished using a *reference coordinate system*, consisting of an *origin*, i.e., a point of beginning, (labeled O in Figure 2.1 and referred to in the text as point O) together with a means of identifying positions relative to the origin.[1] The position of a particle, currently at point P, is represented symbolically by the vector **x**, extending from O to P:[2]

$$\mathbf{x} = \overrightarrow{\mathrm{OP}}.$$

Positions may be identified by distance traveled in three orthogonal (mutually perpendicular) directions. For example, the directions could be along straight axes, as in Cartesian coordinates shown in Figure 2.1, or could be up, south and east, as in spherical coordinates. In the diagram the dotted arrows denote unit vectors along the three independent directions of a Cartesian reference coordinate system. (These unit vectors could be placed anywhere, but commonly are placed with their tails at O.) These unit vectors[3] are named $\mathbf{1}_1$, $\mathbf{1}_2$ and $\mathbf{1}_3$, and are arranged in a right-hand configuration.[4] Any one of these will be designated by $\mathbf{1}_i$, with $i = 1$, 2 and/or 3, as the situation requires. In subsequent diagrams (such as

[1] See Appendix A.3.
[2] Vector notation is explained in Appendix A.1.1.
[3] In the following, a unit vector is denoted by the symbol **1** with a subscript identifying the direction.
[4] Such that $\mathbf{1}_1 \times \mathbf{1}_2 = \mathbf{1}_3$, where the cross product is described in Appendix A.1.3.

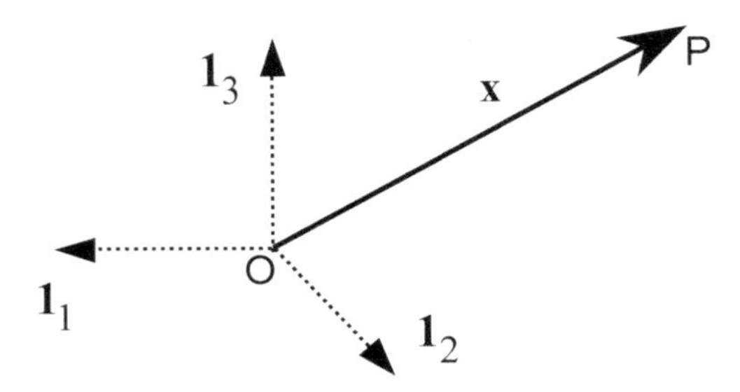

Figure 2.1 Points O (marking the coordinate origin) and P (marking a general point), relative to the unit vectors of the reference coordinate system.

Figure 3.1), the unit vectors will be omitted. A more detailed discussion of coordinates and coordinates systems is found in Appendix A.3.

By the additive properties of vectors, any vector $\mathbf{x}$ may be represented as a sum of vectors along the directions of the axes:[5]

$$\mathbf{x} = x_1 \mathbf{1}_1 + x_2 \mathbf{1}_2 + x_3 \mathbf{1}_3 = x_i \mathbf{1}_i .$$

That is, to get from point O to point P, we need to know how far to travel in three independent directions. If the point P is moving, the coordinates x_i (i = 1, 2, 3) and the vector $\mathbf{x}$ are functions of time.

Often the coordinates are labeled x, y and z rather than 1, 2, and 3 and the unit vectors are denoted by $\mathbf{1}_x$, $\mathbf{1}_y$ and $\mathbf{1}_z$ rather than $\mathbf{1}_1$, $\mathbf{1}_2$ and $\mathbf{1}_3$. We will use both conventions. The theoretical development in Part II is facilitated by the use of the numerical labeling, while the $\{x, y, z\}$ convention is preferable for the applications found in subsequent parts.

2.1.1 Reference Position and Reference State

Each particle of the body is identified by its *Lagrangian reference position*, represented by $\mathbf{X}$, as measured using the reference coordinate system. (Think of $\mathbf{X}$ as a label on each particle.) For simplicity, we will assume that initially (prior to displacement) all particles are in their reference positions and the body as a whole is in its *reference state*. That is, a particle of the body is identified by the vector $\mathbf{X}$ (and is initially located at $\mathbf{X}$); at some later time t the particle is located at $\mathbf{x}(\mathbf{X}, t)$ with, of course, $\mathbf{x}(\mathbf{X}, 0) = \mathbf{X}$. The particle having reference position $\mathbf{X} = \mathbf{0}$ is referred to as particle O. This particle is initially at the origin of the coordinate system, but at a later time is at position $\mathbf{x}_{\mathrm{O}}(t) = \mathbf{x}(\mathbf{0}, t)$. Similarly, the particle originally at point P will be referred to as particle P.

In the study of kinematics, we will take advantage of the freedom in placement of the reference coordinate system to simplify the analysis. That is, without loss of generality, we

[5] For an explanation of vectorial notation and the summation convention, see Appendix A.1, especially Appendix A.1.2.

may assume that point O is initially located within the body. And the particle initially at point O may be *any* particle of the body, so that, when quantifying deformation, we can focus on the neighborhood of the origin.

2.1.2 Lagrangian and Eulerian Frames of Reference

In our studies, we will need to look at deformation and flow from two points of view, called *Lagrangian* and *Eulerian*, distinguished by the manner in which we identify particles. We already have been introduced to the Lagrangian point of view, in which each particle has an identifying position $\mathbf{X}$, and we keep track of individual particles as they move about. The other way to keep track of particles is to sit at a point, identified by position $\mathbf{x}$, and note what passes by. Generally the Lagrangian point of view is used for deformation of elastic solids that have a definite reference position and typically experience small displacements, while the Eulerian point of view is commonly used when investigating the flow of fluids. The Eulerian description of motion is developed in Appendix C.2.

For clarity we will append a subscript $\mathbf{X}$ to the nabla symbol representing the del operator when employing the Lagrangian system.[6] If the nabla symbol lacks this subscript, we are using the Eulerian description.

2.1.3 Rotating Frames

We commonly will use a reference coordinate system fixed to – and rotating with – Earth. Let's take a few moments to understand positions and motions seen by two observers: observer O_f using a coordinate system moving with Earth, but not rotating (so that the distant stars appear stationary) and observer O_Ω using a coordinate system fixed to and rotating with Earth. To be specific, let's suppose that both observers are using cylindrical reference frames[7] $\{\varpi, \phi, z\}$ with common origin and common z axis that coincides with Earth's axis of rotation. With increasing ϕ being eastward,

$$\phi_f = \phi_\Omega + \Omega t,$$

where $\{\phi_f, \phi_\Omega\}$ is the angle measured by observer $\{O_f, O_\Omega\}$ and Ω is the rotation rate of Earth. A particle at rest on Earth a distance $\varpi = R$ from the axis of rotation is seen by observer O_Ω to be not moving (with $\phi_\Omega = 0$, say), while observer O_f sees the particle moving in a circle of radius R with $\phi_f = \Omega t$.

Turning this argument around, a particle that is not rotating ($\phi_f = 0$) is seen by observer O_Ω to be moving in a circle of radius R with $\phi_\Omega = -\Omega t$. This apparent motion, which is due entirely to the rotation of observer O_Ω's coordinate system, is the basis for virtual forces and for inertial and geostrophic waves.[8]

[6] See § 3.3.
[7] Cylindrical coordinates are explained in Appendix A.3.4.
[8] See § 15.1 and appendices C.1.3 and F.7.

2.2 Waves and Flows

Before getting too far into details, let's take a moment to get our bearings, by clarifying what we mean by waves and flows.

Waves are one of the most ubiquitous forms of natural motion. They occur in continuous media such as the atmosphere, oceans and Earth's interior. They also can occur in discrete media, for example in traffic flow. Then there is "the wave" performed at football games, which represents a cooperative behavior of many thousands of people. It is difficult to formulate a comprehensive definition of a wave, as waves occur in a variety of circumstances and take a variety of forms. One possible definition of a wave is a coherent deformation or signal which travels in or on a medium. In this monograph, we will be concerned only with mechanical waves, involving the deformation of a continuous body. Characteristics of mechanical waves include

- the presence of a restoring force, tending to drive the body back to an equilibrium configuration;
- time-dependence – the equations governing mechanical waves involve time derivatives; and
- no net displacement.

A variety of geophysical waves are possible. In the introductory comments to Part III, we will discuss the categorization of these waves.

Flows are motions of a continuous body that result in permanent displacement. Often flows are confined within topographic depressions (such as river beds), but confined flow may arise due to dynamic conditions. Examples of such flows include the jet stream in the atmosphere and the gulf stream in the North Atlantic Ocean.

The distinction between waves and flows is not sharp and the two blend in many situations. For example, deep water waves do not involve net displacement, but one readily sees large displacements of water in the swash zone on a beach. Also, the tides, which are waves, can induce bores that travel up rivers for considerable distances. However, it is helpful to have the ideal cases of waves and flows in mind when one encounters these more complicated cases.

2.2.1 Energies and Conservation Laws

All of our analyses of waves and flows are built on conservation laws. These laws describe how some quantity, such as density or velocity, varies with position and time. Broadly speaking, there is one fundamental conservation law for the Universe: conservation of energy. Most of the energy we experience is in a frozen form called mass. Since exchanges of mass and energy are rare (confined to the atomic processes of fission and fusion) and the weight of the energies we experience in daily life is very small ($M = E/c^2$, where $c \approx 3 \times 10^8$ m s^{-1} is the speed of light), we commonly treat mass separately from energy and consider conservation laws for each.

Simple systems are composed of only one type of mass (e.g., air, water or silicate) in a single phase, called a *constituent*,[9] and we need to consider only the equation of conservation of total mass; we will refer to this as the equation of conservation of mass or the continuity equation. The continuity equation constrains the variations with position and time of the density within a continuous body. Complicated systems are composed of two or more types of mass (such as the ocean, which is composed of water and salt) and we will need an extra conservation equation for each of these added constituents. Also, if a type of mass (such as water) changes phase, we will need to consider each phase as a separate constituent, and develop conservation equations for each. Of course, these equations will need to contain terms quantifying the rates of change of phase.

Just as there are different types of mass, there are different types of energy. Apart from mass, energy is broadly divided into two types: heat energy (energy stored in random motions of particles) and work energy (energy stored in organized deformations or motions of atoms). Since there are many forms of energy, the energy equation can become rather complicated; see Appendix E.7. Waves and flows are driven by variations in work energy and are affected by heat energy through the dependency of certain parameters (such as density and viscosity) on temperature. In many cases, this dependency is unimportant and we need not be concerned with the energy equation.

Work energy may be broadly separated into two types: kinetic and potential. Kinetic energy is stored in motion of macroscopic assemblages of particles, while potential energy is stored in the positions of particles. Dynamics is concerned with changes and transfers of kinetic and potential energies. Since these changes are accomplished by the action of forces, the study of dynamics focuses on the balance of forces, with conservation of momentum playing a dominant role, alongside conservation of mass, with conservation of energy typically in a supporting role. Although there are two types of momentum, linear and angular, in our investigation of waves and flows, we will need to consider only linear momentum.[10] Quantification of momentum is complicated by the fact that it is a vector, whereas mass and energy are scalars. The momentum $\mathrm{d}\mathbf{p}$ of a particle is the product of its mass $\mathrm{d}m$ and velocity $\mathbf{v}$: $\mathrm{d}\mathbf{p} = \mathbf{v}\mathrm{d}m$. The associated kinetic energy of the particle $\mathrm{d}E_K$ is the product of its momentum and velocity: $\mathrm{d}E_K = \mathbf{v} \bullet \mathrm{d}\mathbf{p}$. In very simple systems, conservation of momentum and energy are equivalent; see Appendix D.6. Since positions and velocities are relative to an observer, so are potential and kinetic energies.

An important term in the equation of conservation of momentum is the divergence of the pressure, which serves to ensure that the spatial distribution of displacements or velocities is such that the body remains continuous. If the density of the body becomes zero or negative anywhere, the assumption of continuity is not satisfied and the body no longer is continuous. When surfaces of constant density and pressure coincide everywhere within the body, then the deformation or flow is *barotropic* and may be determined without

[9] A type of material is called a component. Differing phases of a material are different constituents.
[10] Angular momentum is important in the dynamics of rigid bodies.

recourse to the energy equation. On the other hand, if these surfaces do not coincide, the deformation or flow is *baroclinic* and the energy equation is needed.

In addition to the fundamental quantities of mass, momentum and energy, we will encounter variants that are quantified by auxiliary conservation laws derived from the fundamental conservation laws, such as the vorticity vector; see § 2.2.1 and Appendix D.10.

2.3 Scope and Organization

2.3.1 Scope

We will investigate waves and flows in four major geophysical systems: atmosphere, rivers, oceans and Earth's mantle, with brief attention paid to glaciers and magmas. Our investigation will have much in common with a TV miniseries, with a wide range of characters and many episodes. The star of our show is Newton's second law, principally in the form of the Navier–Stokes equation, and the co-star is the equation of conservation of mass, known familiarly as the continuity equation. A prominent character actor in the series is the first law of thermodynamics, also known as the energy equation. Beyond these main characters, there is a wide range of bit players that appear from time to time, including:

- atmosphere
  - strong vertical variation of density
  - rotation of Earth
  - phase change of water
  - variation of density with temperature
  - turbulent drag at the surface
- rivers
  - gravitational forcing
  - turbulent flow
  - variable depth
  - hydraulic shocks
  - bores
  - flash floods
- oceans
  - stable stratification
  - rotation of Earth
  - variations in salt concentration
  - bottom topography
  - forcing by the atmospheric winds
  - influence of the equator and coastlines
- mantle
  - moderate variation of density with depth
  - strong variation of viscosity with pressure, temperature and composition
  - non-Newtonian shear rheology

- magmas
  - strong variation of viscosity
  - exsolution of volatiles
  - fragmentation
  - supersonic flow

2.3.2 Organization

This monograph is organized into seven parts, including this introductory part. In Part II we develop the equations that govern the motion of continuous bodies. Waves unaffected by rotation are studied in Part III, while those affected by – or due solely to – rotation are studied in Part IV. Similarly, flows unaffected by rotation are considered in Part V, while rotating flows are studied in Part VI. Finally, in Part VII we turn our attention to the flow of silicates, including mantle convection, volcanic eruptions and flow of magma.

This text attempts to balance simplicity and generality. Two dangers in this approach are that clarity may be lost and a proper depth of understanding may not be attained. These dangers are mitigated in two ways. First, detailed expositions of a number of key concepts are given in a series of fundaments found in the appendices. Second, words introducing important concepts are italicized, in order to focus the reader's attention and also to provide key words that can be used in an internet search for more information on selected topics.

As a guide to the reader, each chapter ends with a paragraph entitled "Transition" that provides a very brief summary of the chapter just completed and a quick look at the upcoming chapter.

2.4 Bits and Bobs

We shall complete this introductory part with brief discussions of several subjects that will help orient our investigations.

2.4.1 Notation

In the following, we will be dealing primarily with scalars, vectors and tensors. A scalar will be represented by a Roman or Greek letter in italic type, a vector by a Roman or Greek letter in bold type and a tensor by a Roman letter in double-struck type. As a general rule, variables having dimensions and units will be represented by Roman letters, while dimensionless quantities and variables will be represented by Greek letters. As a guide to the reader, the Greek alphabet is reproduced in Appendix A.7.

2.4.2 Relevance and Quantification

Different disciplines of science and mathematics employ differing approaches to the study of our physical world, with differing relevance and precision, as summarized in Figure 2.2.

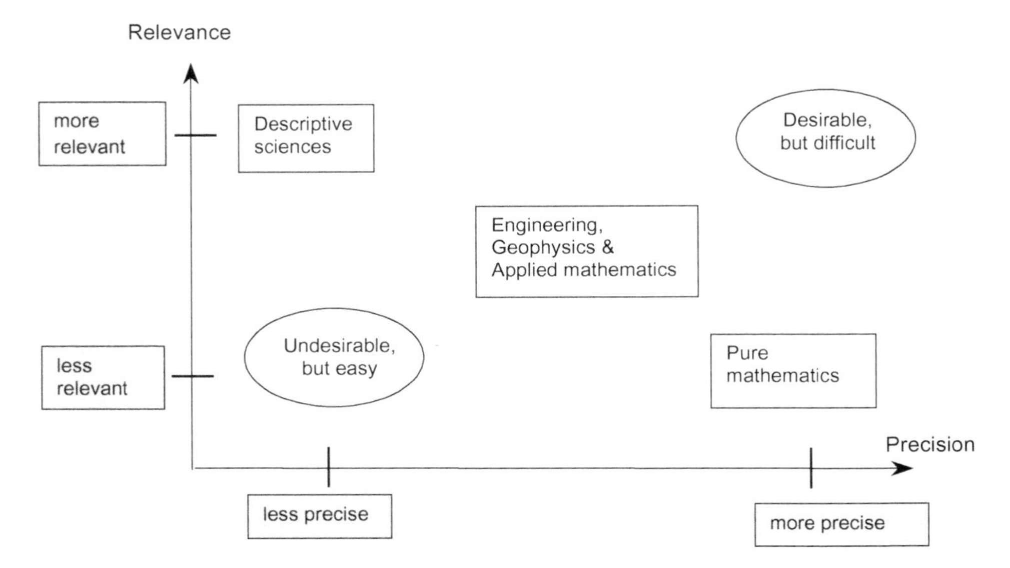

Figure 2.2 Relevance and precision.

Table 2.1. *Levels of quantification. The symbol O, called "big oh," is commonly employed in asymptotic analysis.*

Level	Mathematics used	Science used	Errors	Symbol
Description	None	None	—	—
Dimensional analysis	Algebra & geometry	Relevant quantities & dimensions	100%	$\sim$ and O
Basic balances	Integral calculus	Principles of physics & thermodynamics	10%	$\approx$
Detailed modeling	Partial differential equations	Advanced physics & thermodynamics	1%	$\doteq$
Analytic solutions & proof	Pure mathematics	Advanced physics & thermodynamics	0%	$=$ and $\equiv$

In the following, we will occupy the middle ground of engineering, geophysics and applied mathematics. In a similar vein, there are a number of levels of quantification, as summarized in Table 2.1. For the most part, we will employ either basic physical balances, with integral calculus as our primary mathematical tool, or detailed modeling, with partial differential equations as our mathematical tool. The symbol $=$ is commonly used in mathematical modeling, but it really should be replaced by $\doteq$ or even $\approx$.

2.4.3 Dimensions and Units

Scalars representing physical quantities, including the components of vectors and elements of tensors, have two attributes: magnitude and dimension. In this context the word "dimension" does not refer to the spatial extent of an object. Rather it refers to the physical nature of the quantity. In mechanical systems, all dimensions can be related to three base dimensions: mass, length and time. These dimensions are made manifest using a system of units. The most commonly used system is the meter-kilogram-second (MKS) system, supplemented by the radian. Alternative systems are the centimeter-gram-second system (cgs) and the English foot-pound-second system, supplemented by the degree. Most of the world employs the MKS system, but the English system is still commonly employed in the United States. Thermo-mechanical systems involve a fourth fundamental dimension: temperature. The MKS (or MKST) system is a subset of the International System (SI) of units; see Appendix B.2.

The equations of mathematical physics and their solutions, a few of which are investigated in this text, describe the behavior of physical systems. Typically the dependent variables, such as displacement or velocity, are functions of the independent variables position and time, plus a number of dimensional parameters describing the properties of the material, the shape of the system, the nature of the forcing, etc. Although the magnitude of a particular parameter depends on the system of units employed, the behavior of the physical system does not. This seemingly paradoxical situation is explained by the fact that the system behavior depends on a number of *dimensionless* parameters, whose numerical values are independent of the system of units being employed. The theoretical basis for assertion, called the *Buckingham Pi theorem*, is presented in the following subsection.

2.4.4 Buckingham Pi Theorem

The principal technique of dimensional analysis is that of non-dimensionalization. This technique consists of forming dimensionless groups of variables and parameters. By convention, these dimensionless groups are denoted by the capital Greek letter Pi (Π), often with a suitable subscript.[11] This process reduces the number of parameters in the problem to a minimum. Suppose a problem contains N_p dimensional parameters (including dependent and independent variables) that are expressed in N_d base dimensions. The minimum number N_Π of (dimensionless) parameters is determined by the Buckingham Pi theorem which states that

$$N_\Pi = N_p - N_d\,.$$

For more detail, see Appendix B.1.

[11] For reference, a listing of the Greek symbols is given in Appendix A.7.

2.4.5 Scale Analysis

Scale analysis (also called *scaling analysis*) may be thought of as enhanced dimensional analysis. The essence of scale analysis is to replace derivatives by their algebraic equivalent. For example, consider the derivative $\partial q/\partial x$, where q represents any dependent variable and x represents any independent variable. Scale analysis involves the replacement:

$$\frac{\partial q}{\partial x} \rightarrow \frac{Q}{X}$$

within a differential equation, where Q and X are constants representing typical magnitudes of q and x. When all derivatives are converted in this manner, the governing equations reduce to a set of algebraic scaling equations, which preserve the structure of the equations. In this process numerical, factors such as 2 and π are lost in the shuffle. To indicate this loss of precision, equals signs should be replaced by the similarity symbol $\sim$.

These algebraic scaling equations can be simplified using dimensional analysis, then the reduced algebraic equations can be analyzed to identify relations among the dimensionless variables and parameters. In particular, this procedure can identify terms that are small and can be neglected, as is done, for example, in § 8.6.1.

2.4.6 Similarity

Closely related to dimensional analysis and scale analysis is the concept of *similarity*. Two models, experiments or situations are dynamically similar when each is described by the same set of dimensionless parameters and the numerical values of these are identical. This property allows laboratory simulation of geological processes, such as mantle convection, that occur slowly and on large spatial scales. Dynamic similarity is typically identified through scale analysis.[12]

In addition, if a given flow problem lacks a characteristic length scale, this indicates that the problem possesses mathematical similarity, and that the governing partial differential equations can be reduced to ordinary differential equations. This makes the equations more amenable to analytic solution, analysis and understanding, aspects which retain value even though present-day computing power permits solution of many complicated problems.

2.4.7 Consistency of Units

The formal procedure of non-dimensionalization described in § 2.4.4 is particularly useful in simplifying a complicated mathematical problem. Often, if the problem is relatively simple, we don't need to resort to that formalism. However, the related concept of *consistency of units* provides a useful guide and check.

[12] See the previous section.

Consistency of units requires that each and every term in a given equation have the same units. In checking for consistency of units in mechanical systems, we must consider not only the "big three" (kilograms, meters and seconds), but also, additional units such as radians and moles. Often these additional units are overlooked, but they are potentially important; the magnitude of a term can be much different if degrees are used instead of radians to measure plane angles and if kilogram-moles are used instead of gram-moles.[13]

Transition

This marks the end of Part I, which has provided us with some orientation to the book: the study of waves and flows within continuous bodies. In the next part, Part II, we will develop the equations governing the behavior of waves and flows within continuous bodies.

[13] The "standard" mole is the gram-mole (see Appendix E.2), but we will be using the kilogram-mole in this book.

Part II
Kinematics, Dynamics and Rheology

In this part we develop the mathematical apparatus necessary for the quantification of waves and flows. As we noted in the introduction, this apparatus consists of three components. The first, kinematics, places a constraint on the possible motions of the particles comprising a continuous body in order that the body indeed remains continuous; this constraint is developed in Chapter 3. The second component, dynamics, essentially applies Newton's second law to each particle in the body. As we see in Chapter 4, this gives us a vector equation, called the momentum equation, relating the acceleration of a fluid particle to the forces acting on it. The third component of our apparatus, rheology, is a set of equations describing how the material of a body deforms or flows in response to forces. Our rheological equations include the change of volume induced by a change in pressure, as explained in Chapter 5, and the change of shape induced by changes in deviatoric stresses, as explained in Chapter 6.

The primary mathematical quantities of interest are the deformation (or its time derivative the velocity), the strain tensor (or its time derivative the rate-of-strain tensor) and the stress tensor. While stress and strain are used fairly interchangeably in colloquial usage, they have precise and distinct meanings in the present context. Stress quantifies the applied force and strain (or rate of strain) quantifies the deformation (or flow) that occurs in response to that force. More specifically, the stress tensor describes the structure of the forces acting within a body, while the strain (and rate-of-strain) tensor describes the structure of the internal deformation.

Waves and flows occur as changes to the static state of a body, so before tackling waves and flows, we need to determine the static state and develop the equations governing small deviations from this state; this is accomplished in Chapter 7. The statics and dynamics of rotating fluids are of particular importance in geophysical fluid dynamics; the basic concepts of – and equations governing – rotating fluids are considered in Chapter 8.

3

Kinematics of Deformation and Flow

We now turn our attention to the deformation and flow of a continuous body and in this chapter develop the concepts and mathematical expressions necessary for their quantification. We will be considering two types of body, *elastic* and *fluid*, depending on its behavior when subject to a non-hydrostatic or deviatoric stress (which we haven't defined yet). The displacement of an elastic body remains finite, while the displacement of a fluid body increases without bound. Whether a body behaves elastically and fluidly depends on several factors, including its composition, temperature and the state of stress. As a material is heated it tends to behave more like a fluid. And if the applied stresses are sufficiently great, a body will either begin to flow or break. Once a body has broken, it no longer is a continuous body and we would need a different mathematical formalism to describe its behavior.

Deformation is used primarily in reference to elastic bodies, while *flow* typically refers to fluid bodies, but they are closely related. Deformation is quantified by the relative change of position, i.e., *strain*, while flow is quantified by the time rate of change of position, i.e., the *rate of strain*. Equations governing strain and rate of strain are developed in the following sections.

3.1 Orientation to Kinematics

The form of the kinematic constraint on deformation and flow depends whether the body is elastic or fluid. An elastic body is one that can resist applied forces without continuing displacement (that is, $\mathbf{x}$ can be independent of t), while a fluid body experiences continuous displacement unless the internal (contact) forces are of a particular form (hydrostatic). In either case, these bodies must deform such that the amount of mass is conserved, and no voids form. The constraints on the deformation of elastic bodies and the flow of fluid bodies are developed in § 3.3 and § 3.4, respectively, leading to separate versions of the equation of *conservation of mass* for elastic and fluid bodies.

Two versions of the equation of conservation of mass arise in part because elastic and fluid bodies are observed differently. Deformation of an elastic body is quantified by following the motion of a specific particle, identified by its reference position $\mathbf{X}$, while deformation of a fluid body is quantified by viewing the motion of the body from a chosen frame of reference, with attention focused on the particle that happens to be currently at the position **x**. The former point of view is called Lagrangian and the latter is called Eulerian. The distinction between these two points of view is explained in Appendix C.2.

The deformation of a body can be decomposed into a change of volume without change of shape and a change of shape without change of volume, as illustrated in Figure 1.1:

- Deformation = Change of volume + Change of shape.

Equations governing the rheological behaviors associated with changes of volume and shape are developed in Chapters 5 and 6, respectively.

3.1.1 A Few Words About Tensors

In Chapters 3 and 4 we will encounter a number of tensors of rank two (having nine elements), including three symmetric tensors that play leading roles in mechanics: the strain tensor, its sibling the rate-of-strain tensor and their dynamic counterpart the stress tensor. These symmetric tensors are representations of the physical state of a body at a particular point and their elements may be thought of as potentials. These potentials are resolved into actual strains and stresses at a point P in the body by considering an imaginary planar surface S containing point P. Strains and stresses associated with S at P are vectors that depend on the orientation of S. The vector components perpendicular to the surface are termed *normal* and those parallel are termed *tangential* or *shear*. Often the elements of rank-two tensors are displayed in a 3 by 3 matrix array; e.g., see § 3.3.3 and § 3.4.3. The values of the elements of a tensor, as well as the description of S, depend on the orientation of the reference coordinate system. That is, the elements of a tensor and the formula describing S look quite different to two observers, A and B, who are using reference coordinate axes that point in different directions. However, stresses and strains are objective quantities and both observers will agree on the state of strain or stress on surface S.

It is explained in Appendix A.1.4 that a symmetric tensor has six independent elements and in Appendix C.3.1 that a rotation is specified by three scalars. This means that observer B, by cleverly orienting his coordinates relative to A's, can force three values of the symmetric tensor to take on chosen values. If he chooses to make the off-diagonal elements zero, then the tensor is said to be *diagonal*, and B's coordinate axes are called the *principal axes* of the tensor. With this point of view, strains and stresses on surfaces perpendicular to B's coordinate axes are oriented parallel to these axes; the tangential components are zero. The values of these normal strains and stresses are *invariants* of the tensors; all observers, not just B, can agree on their values.

3.2 Lagrangian Representation of Deformation and Flow

In order to separate displacement into its component parts as shown in Figure 1.1, we need to quantify the displacement of the particles comprising the parcel associated with particle P. Let's identify these particles by their reference position, given by $\mathbf{X} + \mathbf{Y}$, where $\mathbf{X}$ is

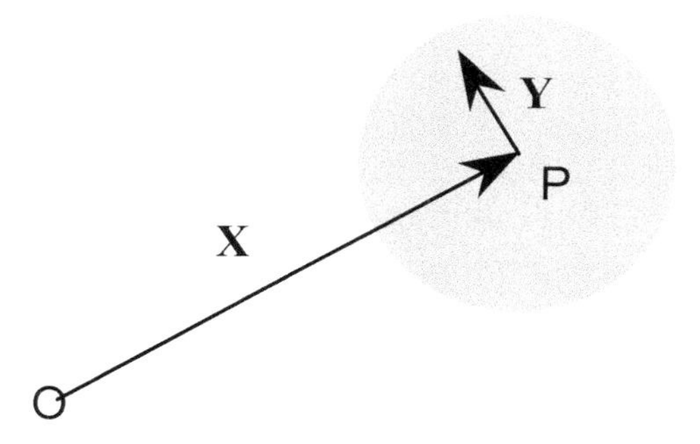

Figure 3.1 The parcel associated with particle P is illustrated as a gray circle. The reference position of a neighboring particle, relative to P, is quantified by $\mathbf{Y}$.

the reference position of particle P and $\mathbf{Y}$ is the reference position of a particle within the parcel, relative to P; see Figure 3.1. The displacement, $\mathbf{u}$, of a particle is given by

$$\mathbf{u} = \mathbf{x}(\mathbf{X}+\mathbf{Y},t) - \mathbf{X} - \mathbf{Y},$$

where $\mathbf{x}(\mathbf{X}+\mathbf{Y},t)$ is the current position of that particle. If the body is to remain continuous (i.e., no fractures or voids occur) at location P, $\mathbf{u}$ must vary smoothly with $\mathbf{Y}$; more precisely, it must be an *analytic function* having a Taylor series[1] about $\mathbf{Y} = \mathbf{0}$. With $\|\mathbf{Y}\|$ suitably small we need to keep only the first-derivative terms of the series. Since the vector $\mathbf{u}$ contains three components and each is a function of the three components of $\mathbf{Y}$, the series has nine first derivatives: $u_1 \approx k_{11}Y_1 + k_{12}Y_2 + k_{13}Y_3$, etc.; that is,[2]

$$\mathbf{u} \approx \mathbb{K} \bullet \mathbf{Y} \qquad \text{or} \qquad u_i \approx k_{ij}Y_j,$$

where $\mathbb{K}$ is the *tensor of deformation gradients*,[3] having elements given by[4]

$$k_{ij} = \left.\frac{\partial u_i}{\partial Y_j}\right|_{\mathbf{Y}=\mathbf{0}} = \frac{\partial x_i}{\partial X_j} - \delta_{ij},$$

where δ_{ij} is the *Kronecker delta*.[5] $\mathbb{K}$ is independent of time if the body has rotated or deformed and now has no relative motion, but is a function of time if rotation and/or deformation is ongoing.

If the deformation of the body is homogeneous (that is, the same at every point) and linear, then the above analysis is valid everywhere, and point P is located at

$$\mathbf{x}(\mathbf{X},t) = \mathbf{x}_{\mathrm{O}}(t) + \mathbf{X} + \mathbb{K} \bullet \mathbf{X},$$

[1] See Appendix A.4.
[2] Here and in the following, tensor equations are often given in both symbolic and subscript forms, with the subscript form following the Einstein summation rule; see Appendix A.1.2.
[3] Speaking more mathematically, $\mathbb{K}$ is the Jacobian matrix for the transformation from $\mathbf{X}$ to $\mathbf{x}$.
[4] Since $\mathbf{x}$ is a function of $X+Y$, $\partial x_i/\partial Y_j = \partial x_i/\partial X_j$.
[5] See Appendix A.1.4.

where $\mathbf{x}_O$ is the current position of point O. If the body experiences no displacement, then point P is located at $\mathbf{x} = \mathbf{X}$ and $\mathbb{K} = \emptyset$, where $\emptyset$ is the zero tensor[5]. If the body is translating, each point in a body has the same displacement, and point P is located at $\mathbf{x}(t) = \mathbf{x}_O(t) + \mathbf{X}$ and again $\mathbb{K} = \emptyset$. Translation is given by the additive term $\mathbf{x}_O(t)$; this is easily separated from the other components of displacement. Rigid-body rotation and deformation are determined by decomposition of $\mathbb{K}$; these components of displacement are found in the following section.

3.3 Rotation and Strain Tensors

The tensor $\mathbb{K}$ consists of rotation plus deformation or equivalently, rotation plus strain.[6] In this section we shall separate these, assuming that the associated displacements are infinitesimal in magnitude; that is $|k_{ij} - \delta_{ij}| \ll 1$ for all i and j. Rotation and deformation will be quantified by the *rotation tensor* $\mathbb{R}$ and the *strain tensor* $\mathbb{E}$, respectively. That is,

$$\mathbb{K} = \mathbb{R} + \mathbb{E} \qquad \text{or} \qquad k_{ij} = r_{ij} + \epsilon_{ij},$$

with[7]

$$\mathbb{R} = \tfrac{1}{2}\left(\mathbb{K} - \mathbb{K}^{\mathrm{T}}\right) + \mathbb{I} \qquad \text{or} \qquad r_{ij} = \frac{1}{2}\left(\frac{\partial x_i}{\partial X_j} - \frac{\partial x_j}{\partial X_i}\right) + \delta_{ij},$$

where $\mathbb{I}$ is the identity tensor,[8] the superscript T denotes the transpose of the tensor[9] and

$$\mathbb{E} = \tfrac{1}{2}\left(\mathbb{K} + \mathbb{K}^{\mathrm{T}}\right) - \mathbb{I} \qquad \text{or} \qquad \epsilon_{ij} = \frac{1}{2}\left(\frac{\partial x_i}{\partial X_j} + \frac{\partial x_j}{\partial X_i}\right) - \delta_{ij}.$$

Note that $\mathbb{E}$ is symmetric and the off-diagonal elements of $\mathbb{R}$ are antisymmetric.

The rotation tensor may be written as

$$\mathbb{R} = \mathbb{I} + \phi\mathbb{A} \qquad \text{or} \qquad r_{ij} = \delta_{ij} + \phi A_{ij},$$

where the antisymmetric tensor $\mathbb{A}$, written in display form, is

$$\mathbb{A} = \begin{bmatrix} 0 & -a_3 & a_2 \\ a_3 & 0 & -a_1 \\ -a_2 & a_1 & 0 \end{bmatrix}.$$

The non-zero elements of $\mathbb{A}$ form the components of a unit vector pointing along an axis of rotation $\mathbf{1}_\Omega = a_i \mathbf{1}_i$ (with $a_i a_i = 1$) and ϕ is the (small) angle of rotation about that axis. Rotation can be expressed as

$$\mathbb{R}\bullet\mathbf{x} = \mathbf{x} + \mathbf{\Phi} \times \mathbf{x}, \qquad \text{where} \qquad \mathbf{\Phi} = \phi\mathbf{1}_\Omega$$

[6] Strain is relative displacement.
[7] Generalizations of these expressions for finite deformation are found in Appendix C.3.
[8] See Appendix A.1.4.
[9] See Appendix A.1.5.

has the appearance of a vector, but does not behave as a vector if ϕ is not infinitesimal.[10]

The elements of the strain tensor may be expressed in terms of the components of the displacement vector $\mathbf{u}(\mathbf{X})$ as

$$\epsilon_{ij} = \frac{1}{2}\left(\frac{\partial u_i}{\partial X_j} + \frac{\partial u_j}{\partial X_i}\right).$$

Note that the divergence of the displacement vector is equal to the trace of the strain tensor:[11]

$$\nabla_{\mathbf{X}} \bullet \mathbf{u} = \text{trace}[\mathbb{E}] = \epsilon_{ii}, \qquad \text{where} \qquad \nabla_{\mathbf{X}} \equiv \frac{\partial}{\partial X_i}\mathbf{1}_i$$

is the del operator[12] acting with Lagrangian coordinates, X_i, rather than the usual Eulerian coordinates x_i.

The strain tensor may be split into isotropic and deviatoric parts

$$\mathbb{E} = \tfrac{1}{3}(\nabla_{\mathbf{X}} \bullet \mathbf{u})\mathbb{I} + \mathbb{E}' \qquad \text{or} \qquad \epsilon_{ij} = \tfrac{1}{3}(\nabla_{\mathbf{X}} \bullet \mathbf{u})\delta_{ij} + \epsilon'_{ij}.$$

The orientation of the reference coordinate system is arbitrary. It is possible to orient this system such that the strain tensor at a particular point of interest is *diagonal* (i.e., all off-diagonal elements are zero). The directions of this specially oriented coordinate system are called the *principal axes* of the strain tensor.[13] From this point of view, strain is an expansion or contraction of the body along three mutually perpendicular axes, with the fractional amount of change being given by the *eigenvalues* of the tensor. Expressing this another way, the strain tensor $\mathbb{E}$ changes a spherical parcel into an ellipsoid whose axes are the principal axes of $\mathbb{E}$ and the fractional changes of distance along each of these axes are the eigenvalues of $\mathbb{E}$.

3.3.1 Volume Change and Conservation of Mass

Deformation consists of change of volume (without change of shape) and change of shape (at constant volume). Change of volume is isotropic: the same in all directions. If the deformation is small, the volume V of a parcel is given by $V \approx V_0(1 + \nabla_{\mathbf{X}} \bullet \mathbf{u})$, where V_0 is the initial volume of the parcel.[14] During deformation, the mass of the parcel remains the same; that is, $\rho V = \rho_0 V_0$, where ρ is the mass density and ρ_0 is the initial density of the parcel. Eliminating V between these two equations and realizing that $|\nabla_{\mathbf{X}} \bullet \mathbf{u}| \ll 1$, we have $\rho \approx \rho_0(1 - \nabla_{\mathbf{X}} \bullet \mathbf{u})$. We have transformed the statement of conservation of volume from extensive to *intensive*. Now, the initial density of parcels may vary with location. To

[10] See Appendix C.3.1.
[11] The divergence of a vector is explained in Appendix A.2.
[12] See Appendix A.2.
[13] See Appendix A.1.4.
[14] See Appendix C.3.3.

emphasize this distinction, we may write

$$\rho \approx \rho_r (1 - \nabla_{\mathbf{X}} \bullet \mathbf{u}) ,$$

where a subscript r denotes a reference value that may depend on position. If the body is *incompressible*, then $\rho = \rho_r$ and $\nabla_{\mathbf{X}} \bullet \mathbf{u} = 0$.

The divergence of the strain tensor is an important quantity that may be expressed as[15]

$$\begin{aligned} \nabla_{\mathbf{X}} \bullet \mathbb{E} &= \frac{\partial \epsilon_{ij}}{\partial X_i} \mathbf{1}_j = \frac{1}{2} \frac{\partial}{\partial X_i} \left(\frac{\partial u_i}{\partial X_j} + \frac{\partial u_j}{\partial X_i} \right) \mathbf{1}_j \\ &= \frac{1}{2} \frac{\partial^2 u_j}{\partial X_i \partial X_i} \mathbf{1}_j + \frac{1}{2} \frac{\partial}{\partial X_j} \left(\frac{\partial u_i}{\partial X_i} \right) \mathbf{1}_j = \tfrac{1}{2} \nabla_{\mathbf{X}}^2 \mathbf{u} + \tfrac{1}{2} \nabla_{\mathbf{X}} (\nabla_{\mathbf{X}} \bullet \mathbf{u}) . \end{aligned}$$

3.3.2 Isotropic displacement

If an elastic material is expanded or compressed uniformly, then

$$\mathbb{E} = (\nabla_{\mathbf{X}} \bullet \mathbf{u}) \, \mathbb{I} \qquad \text{and} \qquad \nabla_{\mathbf{X}} \bullet \mathbb{E} = \nabla_{\mathbf{X}} (\nabla_{\mathbf{X}} \bullet \mathbf{u}) .$$

3.3.3 Uniform 2-D (plane) strain

Suppose the body deforms in the X_1 direction and the amount of deformation is linear in the X_2 direction; i.e., $\mathbf{x} = \mathbf{X} + 2\epsilon X_2 \mathbf{1}_1$. The tensor of deformation gradients for this case may be expressed in display notation as

$$\mathbb{K} = \begin{bmatrix} 1 & 2\epsilon & 0 \\ 0 & 1 & 0 \\ 0 & 0 & 1 \end{bmatrix}$$

and the rotation and strain tensors are

$$\mathbb{R} = \begin{bmatrix} 1 & \epsilon & 0 \\ -\epsilon & 1 & 0 \\ 0 & 0 & 1 \end{bmatrix} \qquad \text{and} \qquad \mathbb{E} = \begin{bmatrix} 0 & \epsilon & 0 \\ \epsilon & 0 & 0 \\ 0 & 0 & 0 \end{bmatrix} .$$

The displacement $\mathbb{K}$ is the sum of a rotation $\mathbb{R}$ and a change of shape $\mathbb{E}$, as illustrated in Figure 3.2.

3.4 Fluid Deformation and Rate of Strain

Fluid deformation refers to the time rate of change of size and shape of a body. Since fluid bodies have no reference configuration, the appropriate independent variable is the current

[15] The Laplacian ∇^2 is explained in Appendix A.2.5.

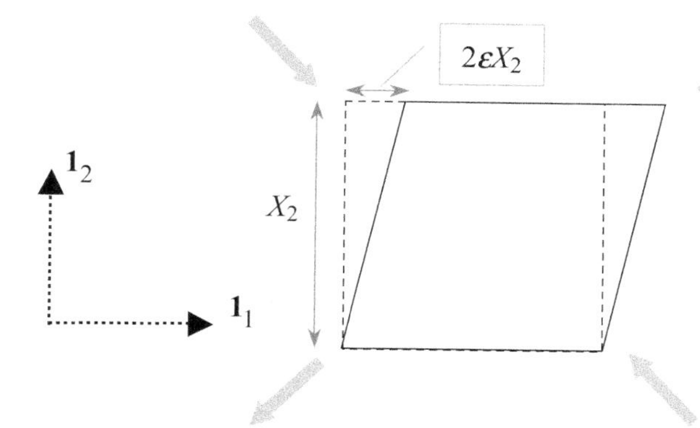

Figure 3.2 Illustration of plane strain. Displacement in the "1" direction is the sum of a clockwise rotation plus a change of shape indicated by the gray arrows.

position, $\mathbf{x}$, with the deformation being determined by the velocity, $\mathbf{v}$, of the body at point $\mathbf{x}$. Here $\mathbf{v}$ is the velocity experienced by a particle

$$\mathbf{v} = \left.\frac{\partial \mathbf{x}}{\partial t}\right|_{\mathbf{X}},$$

where the subscript $\mathbf{X}$ denotes a time derivative measured by an observer moving with the particle. This is often called the *material derivative*. The time derivative is more complicated if the observer does not move with the material.[16]

In this chapter we will represent $\mathbf{v}$ in component notation as[17]

$$\mathbf{v} = v_i \mathbf{1}_i \qquad (\text{sum over } i),$$

where $\mathbf{1}_i = \{\mathbf{1}_1, \mathbf{1}_2, \mathbf{1}_3\}$ are the unit vectors of our reference coordinate system.

As with elastic strain, in order to determine the fluid rate-of-strain we must investigate the deformation of a parcel associated with a specific particle P located at position $\mathbf{x}$. Let the positions of neighboring particles (within the parcel) be given by $\mathbf{x} + \mathbf{y}$, and let the velocity vary with $\mathbf{y}$ and t while $\mathbf{x}$ is held fixed. Assuming that the velocity of the body varies smoothly[18] in the neighborhood of P, we have that

$$\mathbf{v}(\mathbf{y}, t; \mathbf{x}) = \mathbf{v}(\mathbf{0}, t; \mathbf{x}) + \dot{\mathbb{K}}(\mathbf{0}, t; \mathbf{x}) \bullet \mathbf{y}$$

$$\text{or} \qquad v_i(\mathbf{y}, t; \mathbf{x}) = v_i(\mathbf{0}, t; \mathbf{x}) + \dot{k}_{ij}(\mathbf{0}, t; \mathbf{x})\, y_j,$$

where $\dot{\mathbb{K}}$ is the *tensor of velocity gradients* with elements given by $\dot{k}_{ij} = \partial v_i / \partial y_j$. These elements are functions of $\mathbf{x}$ and t and, since $\partial v_i / \partial x_j$ yields the same functions as $\partial v_i / \partial y_j$,

[16] See Appendix C.2.
[17] In later chapters, we will write $\mathbf{v} = u\mathbf{1}_x + v\mathbf{1}_y + w\mathbf{1}_z$.
[18] Recall that this is the essence of a continuous body; see § 1.2.

they may be determined by

$$\dot{k}_{ij} = \partial v_i / \partial x_j \, .$$

Note that these tensor elements have dimensions of inverse time. Deformation is uniform if the components of the velocity are linear functions of the position coordinates.

As with the deformation tensor, this tensor may be split into rate of rotation and rate of strain tensors:

$$\dot{\mathbb{K}} = \dot{\mathbb{R}} + \dot{\mathbb{E}},$$

with these tensors being the time derivatives of their steady counterparts:[19]

$$\dot{\mathbb{R}} = \frac{1}{2}\left(\dot{\mathbb{K}} - \dot{\mathbb{K}}^{\mathrm{T}}\right) \qquad \text{or} \qquad \dot{r}_{ij} = \frac{1}{2}\left(\frac{\partial v_i}{\partial x_j} - \frac{\partial v_j}{\partial x_i}\right)$$

and

$$\dot{\mathbb{E}} = \frac{1}{2}\left(\dot{\mathbb{K}} + \dot{\mathbb{K}}^{\mathrm{T}}\right) \qquad \text{or} \qquad \dot{\epsilon}_{ij} = \frac{1}{2}\left(\frac{\partial v_i}{\partial x_j} + \frac{\partial v_j}{\partial x_i}\right).$$

The rate-of-rotation tensor contains three independent elements and the rate-of-strain tensor contains six.

Note that rotation about a fixed axis is given by $\dot{\mathbb{R}} = \Omega \mathbb{A}$, where $\Omega = \dot{\phi}$ is the rate of rotation and the antisymmetric tensor $\mathbb{A}$ is given in § 3.3. This rotation is commonly represented as a vector: $\boldsymbol{\Omega} \equiv \Omega \mathbf{1}_{\Omega}$, where $\mathbf{1}_{\Omega} = a_i \mathbf{1}_i$ (such that $\dot{\mathbb{R}} \bullet \mathbf{x} = \boldsymbol{\Omega} \times \mathbf{x}$), though like rotation, rate of rotation is not a true vector.

The divergence of the rate-of-strain tensor is

$$\begin{aligned}
\boldsymbol{\nabla} \bullet \dot{\mathbb{E}} &= \frac{\partial \dot{\epsilon}_{ij}}{\partial x_i} \mathbf{1}_j = \frac{1}{2} \frac{\partial}{\partial x_i}\left(\frac{\partial v_i}{\partial x_j} + \frac{\partial v_j}{\partial x_i}\right) \mathbf{1}_j \\
&= \frac{1}{2} \frac{\partial^2 v_j}{\partial x_i \partial x_i} \mathbf{1}_j + \frac{1}{2} \frac{\partial}{\partial x_j}\left(\frac{\partial v_i}{\partial x_i}\right) \mathbf{1}_j = \tfrac{1}{2} \nabla^2 \mathbf{v} + \tfrac{1}{2} \boldsymbol{\nabla}(\boldsymbol{\nabla} \bullet \mathbf{v}) .
\end{aligned}$$

3.4.1 Conservation of Mass

Consider a parcel of mass M undergoing fluid deformation. Conservation of mass requires that[20] $\mathrm{d}M/\mathrm{d}t = \mathrm{d}(\rho V)/\mathrm{d}t = 0$. This implies that

$$\dot{V}/V + \dot{\rho}/\rho = 0,$$

[19] Presented in § 3.3, noting that $\dot{\mathbb{I}} = \emptyset$.

[20] Here time derivatives are taken in the Lagrangian reference frame, following the parcel.

where the dot denotes a time derivative. The rate of change, $\dot{V}$, of volume experienced by a body during fluid deformation is given by

$$\dot{V}/V = \nabla \bullet \mathbf{v}$$

giving

$$\dot{\rho} + \rho(\nabla \bullet \mathbf{v}) = 0 .$$

The Lagrangian time derivative is given by the material derivative:[21]

$$\dot{\rho} = \frac{\partial \rho}{\partial t}|_{\mathbf{x}} + \mathbf{v} \bullet \nabla \rho .$$

Finally, with $\partial \rho / \partial t|_{\mathbf{x}}$ written simply as $\partial \rho / \partial t$, the equation of conservation of mass becomes

$$\frac{\partial \rho}{\partial t} + \nabla \bullet (\rho \mathbf{v}) = 0 .$$

This is the fluid version of the equation of conservation of mass; it is also called the *continuity equation*. This is one of the principal equations governing fluid motion.

Anelastic Approximation

The secular term $\partial \rho / \partial t$ is important in the study of compressive waves,[22] but in the study of other types of waves and flows, we can safely adopt the *anelastic approximation* and ignore this term. Now the continuity equation is simply

$$\nabla \bullet (\rho \mathbf{v}) = 0 .$$

This equation can be "solved" by the introduction of a stream function:[23]

$$\rho \mathbf{v} = \nabla \times \boldsymbol{\Psi} .$$

The stream function is particularly useful if the flow has symmetry such that the reduced continuity equation has only two terms; for example, see § 15.3, § 16.2.1, § 27.1 and § 28.2.

3.4.2 Isotropic Motion and Deviatoric Rate of Strain

If the fluid is expanding or contracting isotropically, then the rate of strain is given by

$$\dot{\mathbb{E}} = (\nabla \bullet \mathbf{v})\, \mathbb{I} \qquad \text{or} \qquad \dot{\epsilon}_{ij} = \frac{\partial v_k}{\partial x_k} \delta_{ij}$$

and the divergence of the rate-of-strain tensor is

$$\nabla \bullet \dot{\mathbb{E}} = \nabla (\nabla \bullet \mathbf{v}) \qquad \text{or} \qquad \frac{\partial \dot{\epsilon}_{ij}}{\partial x_j} = \frac{\partial}{\partial x_j} \left(\frac{\partial v_k}{\partial x_k} \right) \delta_{ij} .$$

[21] See Appendix C.2.
[22] See § 9.3.
[23] See Appendix C.4.

We may separate this uniform expansion from the general rate of strain by writing

$$\dot{\mathbb{E}} = (\boldsymbol{\nabla} \bullet \mathbf{v})\,\mathbb{I} + \dot{\mathbb{E}}' \qquad \text{or} \qquad \dot{\epsilon}_{ij} = \frac{\partial v_k}{\partial x_k}\delta_{ij} + \dot{\epsilon}'_{ij},$$

where $\dot{\mathbb{E}}'$ is the deviatoric rate of strain having elements

$$\dot{\epsilon}'_{ij} = \frac{1}{2}\left(\frac{\partial v_i}{\partial x_j} + \frac{\partial v_j}{\partial x_i}\right) - \frac{\partial v_k}{\partial x_k}\delta_{ij},$$

or in display form,

$$\dot{\mathbb{E}}' = \frac{1}{2}\begin{bmatrix} -2\dfrac{\partial v_2}{\partial x_2} - 2\dfrac{\partial v_3}{\partial x_3} & \dfrac{\partial v_1}{\partial x_2} + \dfrac{\partial v_2}{\partial x_1} & \dfrac{\partial v_1}{\partial x_3} + \dfrac{\partial v_3}{\partial x_1} \\ \dfrac{\partial v_1}{\partial x_2} + \dfrac{\partial v_2}{\partial x_1} & -2\dfrac{\partial v_3}{\partial x_3} - 2\dfrac{\partial v_1}{\partial x_1} & \dfrac{\partial v_2}{\partial x_3} + \dfrac{\partial v_3}{\partial x_2} \\ \dfrac{\partial v_1}{\partial x_3} + \dfrac{\partial v_3}{\partial x_1} & \dfrac{\partial v_2}{\partial x_3} + \dfrac{\partial v_3}{\partial x_2} & -2\dfrac{\partial v_1}{\partial x_1} - 2\dfrac{\partial v_2}{\partial x_2} \end{bmatrix}.$$

Note that $\dot{\epsilon}'_{ii} = 0$ and[24]

$$\begin{aligned} 2\boldsymbol{\nabla} \bullet \dot{\mathbb{E}}' &= \nabla^2 \mathbf{v} - \boldsymbol{\nabla}(\boldsymbol{\nabla} \bullet \mathbf{v}) \\ &= -\boldsymbol{\nabla} \times (\boldsymbol{\nabla} \times \mathbf{v}). \end{aligned}$$

3.4.3 Shear Rate of Strain

Suppose the body deforms in the x_1 direction and the amount of deformation is linear in x_2, i.e., $\mathbf{v} = 2\dot{\epsilon}x_2\mathbf{1}_1$. The tensor of velocity gradients for this case may be expressed in display notation as

$$\dot{\mathbb{K}} = \begin{bmatrix} 0 & 2\dot{\epsilon} & 0 \\ 0 & 0 & 0 \\ 0 & 0 & 0 \end{bmatrix}$$

and the rotation and rate-of-strain tensors are

$$\dot{\mathbb{R}} = \begin{bmatrix} 0 & \dot{\epsilon} & 0 \\ -\dot{\epsilon} & 0 & 0 \\ 0 & 0 & 0 \end{bmatrix} \qquad \text{and} \qquad \dot{\mathbb{E}} = \begin{bmatrix} 0 & \dot{\epsilon} & 0 \\ \dot{\epsilon} & 0 & 0 \\ 0 & 0 & 0 \end{bmatrix}.$$

The rate of strain $\dot{\mathbb{K}}$ is the sum of a rotation $\dot{\mathbb{R}}$ and a change of shape $\dot{\mathbb{E}}$, as illustrated in Figure 3.2.

24 The alternate form has been obtained using the last formula of Appendix A.2.6.

3.5 Material Surface

A *material surface* is an important kinematic concept, with flows in the atmosphere and ocean commonly vertically averaged from one such surface to another; see § 8.4. Suppose a material surface is described by the equation $z - h(\mathbf{x}_H, t) = 0$, where $\mathbf{x}_H$ is the horizontal position vector and z is the upward coordinate. As we move along this surface, the function $z - h(\mathbf{x}_H, t)$ remains zero. That is, its material derivative[25] is zero:

$$\frac{\mathrm{D}}{\mathrm{D}t}(z-h) = \frac{\partial z}{\partial t} - \mathbf{v}_H \bullet \nabla_H h - \frac{\partial h}{\partial t} = 0,$$

where $\mathbf{v}_H$ is the horizontal velocity and ∇_H is the horizontal gradient. Noting that $\partial z/\partial t$ is equal to the vertical speed w, this equation may be written as

$$\text{at} \quad z = h(\mathbf{x}_H, t): \qquad w = \frac{\partial h}{\partial t} + \mathbf{v}_H \bullet \nabla_H h.$$

Often in applications the nonlinear term $\mathbf{v}_H \bullet \nabla_H h$ is negligibly small and the variables are evaluated at the equilibrium position of the surface.

3.6 Comments

Often in geophysical materials, such as Earth's crust, deformation is not smoothly distributed throughout the body, but is concentrated along discrete slip planes. This presents us with a knotty problem of how to describe the deformation and at what resolution. One possibility is to treat the problem by the boundary-layer formalism,[26] resolving the detailed structure of deformation across each slip plane. Another possibility is to treat the material between a pair of adjacent slip planes as a continuous body and quantify the relative motion of adjoining continuous bodies using some heuristic model, such as dry sliding.[27] This level of detail is necessary if, for example, we want to predict earthquakes. Unfortunately, earthquake prediction is presently impossible because it requires detailed information, obtainable only at great cost. On the other hand, long-term displacements that are the result of the cumulative effect of a great number of earthquakes are much easier to quantify and predict, using the tools we will be developing in the following chapters. [This is reminiscent of the situation in weather forecasting, where average behavior (climate) is easier to predict than daily weather.]

Transition

Section 3.1 has provided us with a brief orientation to kinematics and discussed the two ways of quantifying deformation: Lagrangian and Eulerian. The mathematical foundation for the study of kinematics – the deformation of elastic bodies and the flow of fluid

[25] See Appendix C.2.
[26] See Appendix G.2.
[27] See § 22.2.1.

bodies – is developed in § 3.2. Central to this development are the strain tensor (for elastic bodies) and the rate-of-strain tensor (for fluid bodies). In order that the body remain continuous, it must satisfy an equation of conservation of mass; this equation occurs in two versions, one for elastic bodies and one for fluid. The equations governing the dynamics of continuous bodies are developed in the next chapter, with emphasis on the stress tensor and the equation of conservation of momentum.

4
Dynamics and the Stress Tensor

In the previous chapter we investigated the ways in which a continuous body can deform and move. These deformations and motions are induced by three types of forces acting on the body, as explained in § 4.1. This chapter focuses on the quantification of the internal contact force, $\mathbf{F}_C$, and its relation to the *stress tensor*, denoted by $\mathbb{S}$. The stress tensor describes the state of stress within a continuous body as explained in § 4.2; it is the dynamic counterpart to the strain and rate-of-strain tensors of kinematics.

In the study of kinematics and dynamics, we have many choices. For example, in kinematics we are free to choose the location and orientation of our reference coordinate system. This system may be inertial or non-inertial; it may be arbitrarily oriented or aligned with the principal axes of the strain or rate-of-strain tensor. Similarly, in dynamics we are free to define what is meant by a body. Most of the time, we think of a body as a complete physical entity, with a physically defined boundary. However, we are free to define as a "body" any arbitrary portion of a physical body, such as the parcel introduced in § 1.2. We will use this freedom to our advantage in what follows. In the process of defining an arbitrary portion of a physical body as a "body," we must of necessity define its surfaces. The stresses on these defined surfaces are manifestations of the internal state of stress of the body at that location.

4.1 Forces

The deformations and motions of a continuous body are induced by three types of forces acting on the body:

- *Body forces*, denoted by $\mathbf{F}_B$, are applied directly to each internal particle of the body. Examples of body forces are the gravitational and electromagnetic (Lorentz) forces. The body force due to gravity is a known constant when studying the atmosphere or oceans and is a prescribed function of radius when studying mantle convection.
- *Virtual forces*, denoted by $\mathbf{F}_V$, arise from the use of a non-inertial frame of reference that is either accelerating, rotating or both. Examples of virtual forces are the centrifugal force due to the use of a rotating coordinate system and the Coriolis force, due to motion observed in a rotating frame of reference.[1]

[1] See Appendix C.1.

- *Contact forces*, denoted by $\mathbf{F}_C$, act on a body due to its contact with adjoining bodies. These may be external or internal. An external contact force is applied to an external surface of a body, typically in the form of a prescribed stress, i.e., force per unit area. An example of an external contact force is the force applied by the wind on the ocean at its surface. Alternatively, if the body under consideration is a small portion (a parcel) of a larger body, then internal contact forces act on the parcel as a result of the internal state of stress of the body, as quantified by the stress tensor. Internal contact forces consist of pressure and tangential (shear) stresses. Quantification of the internal contact forces is a major focus of dynamics.

Altogether the total force $\mathbf{F}_T$ acting on a body is the sum of these three forces:

$$\mathbf{F}_T = \mathbf{F}_B + \mathbf{F}_V + \mathbf{F}_C .$$

Dynamics are quantified by *Newton's second law of motion*, which states that

$$M\mathbf{a} = \mathbf{F}_T = \mathbf{F}_B + \mathbf{F}_V + \mathbf{F}_C ,$$

where $\mathbf{a}$ is the acceleration and M is the mass of the body. This is an extensive equation, with the magnitudes of its terms depending on the size or mass of the body. We shall find it preferable to employ an intensive version of this equation, by dividing by either mass or volume. The per-unit-mass version of Newton's second law is simply

$$\mathbf{a} = \mathbf{f}_B + \mathbf{f}_V + \mathbf{f}_C ,$$

where $\mathbf{f} = \mathbf{F}/M$ is the force per unit mass. Alternatively, we could use the per-unit-volume form

$$\rho\mathbf{a} = \rho(\mathbf{f}_B + \mathbf{f}_V + \mathbf{f}_C) ,$$

where $\rho = M/V$ is is the mass density, with V being the volume.

Newton's second law of motion is to be satisfied by every particle within a body; that is, at every internal point. This point-wise version of Newton's second law is called the equation of *conservation of momentum*. Versions of this equation for both elastic and fluid bodies are developed in the following sections.

4.2 Introduction to the Stress Tensor

To clarify the ideas discussed above and to introduce the concept of the stress tensor, let's consider a small internal disk-shaped "body" (a parcel) having volume Ab, where A is the area of its circular faces and b is its thickness. The surface of the body consists of the areas of the circular faces, represented by A_+ and A_-, and a thin cylindrical edge of width b. The

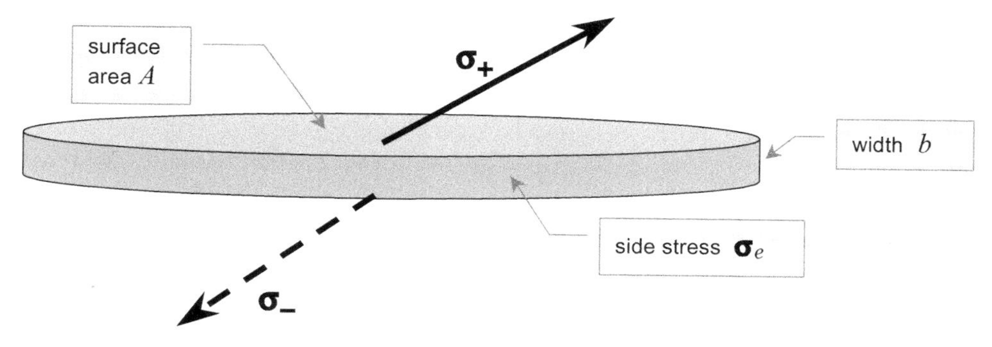

Figure 4.1 A thin disk-shaped body and its associated stresses.

total force $\mathbf{F}$ acting on the body[2] due to the stresses illustrated in Figure 4.1 is given by

$$\mathbf{F} = A\left(\boldsymbol{\sigma}_+ + \boldsymbol{\sigma}_-\right) + b\bar{\boldsymbol{\sigma}}_e,$$

where $\boldsymbol{\sigma}_+$ and $\boldsymbol{\sigma}_-$ are the stresses on the two surfaces and $\bar{\boldsymbol{\sigma}}_e$ represents the integrated effect of the stresses acting on the edge. The mass of the body is

$$M = \rho A b,$$

where ρ is the density of the material making up the body, and the force per unit mass (i.e., acceleration) is

$$\mathbf{a} = \frac{\mathbf{F}}{M} = \frac{1}{\rho}\left(\frac{\boldsymbol{\sigma}_+ + \boldsymbol{\sigma}_-}{b} + \frac{\bar{\boldsymbol{\sigma}}_e}{A}\right).$$

In order that the magnitude of the acceleration remain bounded as b becomes vanishingly small, we must have that

$$\boldsymbol{\sigma}_+ = -\boldsymbol{\sigma}_-.$$

That is, *the stresses on oppositely oriented surfaces at a point in a body must be equal and opposite*. This is a form of Newton's third law of motion: for every action, there is an equal and opposite reaction.

If we allow b to be small but finite, Newton's second law of motion requires that $\boldsymbol{\sigma}_+ = -\boldsymbol{\sigma}_- + b\left(\rho\mathbf{a} - \bar{\boldsymbol{\sigma}}_e/A\right)$ or[3] $\boldsymbol{\sigma}_+ = -\boldsymbol{\sigma}_- + O(b)$. From this we see that *the difference in stresses on opposite faces of a body is proportional to its thickness*. If the body is tiny (i.e., if b is small) then the stresses are nearly equal and opposite. This difference in stresses is important, as it controls the acceleration of the tiny body. We will quantify this difference below.

[2] For simplicity, we are omitting the body and virtual forces, which do not affect the following line of reasoning.
[3] The symbol O is explained in Appendix A.4.

Next consider a parcel in the shape of a rectangular parallelepiped (let's call it a "box") with faces aligned with our orthogonal reference coordinate system; see Figure 4.2. We may imagine this to be one of many such boxes created by cutting up a large body. (Think of dicing a potato.) This box has six faces, defined by the orientations of the unit vectors normal to them, pointing outward from the box, called the *external normals*. The three faces having external normals in the negative coordinate direction (dotted in the figure) are called back faces, while the three with external normals in the positive coordinate directions (solid in the figure) are front faces. We know from the previous argument that, in the limit the size of the box shrinks to zero, the stresses on the back faces must be equal and opposite to those on the front faces of the adjoining boxes, so it is sufficient to consider only stresses acting on the fronts. Each of the stress vectors acting on the three front faces has three components. That is, the stress $\boldsymbol{\sigma}$ associated with the front face having normal in the i direction is given by

$$\boldsymbol{\sigma}_i = \sigma_{ij}\mathbf{1}_j,$$

where σ_{ij} are the elements of the *stress tensor* $\mathbb{S}$. In this formula i is a free index, while the j's are summed.

4.3 Symmetry of the Stress Tensor

We have assured that a parcel does not experience rapid translational acceleration by requiring that the stresses on surfaces with oppositely directed normals be (nearly) equal

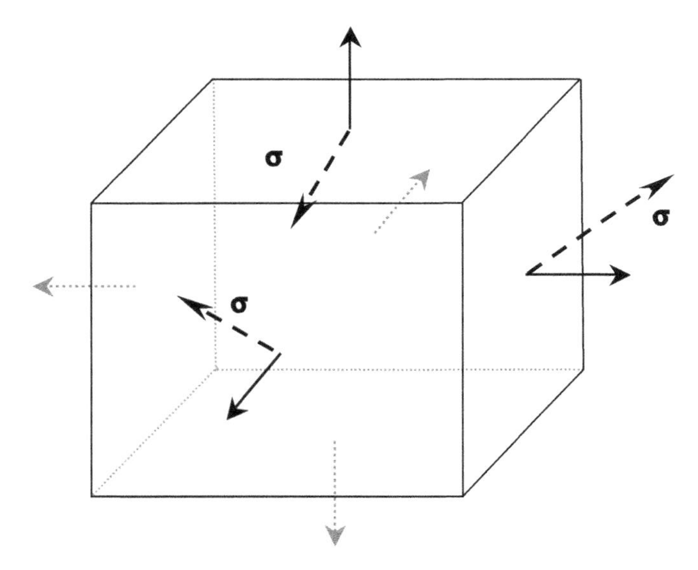

Figure 4.2 A small box extracted from a continuous body showing vectors normal to the front and back faces as solid black and dotted gray arrows, respectively, and the stresses on the front faces as dashed arrows.

and opposite. In this section, we develop the condition on the elements of the stress tensor which prevents a parcel from experiencing rapid angular acceleration. Consider a parcel in the shape of a box with spatial dimensions l_i along the three perpendicular directions as shown in Figure 4.3. The net torque acting on the box about an edge parallel to $\mathbf{1}_3$ is

$$\mathbf{T} = l_1 l_2 l_3 (\sigma_{12} - \sigma_{21}) \mathbf{1}_3 .$$

The moment of inertia of the body about that axis is

$$I_3 = \tfrac{1}{3} \rho l_1 l_2 l_3 (l_1^2 + l_2^2) ,$$

so that the angular acceleration about that axis has a magnitude

$$\alpha_3 = \frac{3(\sigma_{12} - \sigma_{21})}{\rho (l_1^2 + l_2^2)} .$$

Again we are free to choose the size of the box. If we let l_1 and l_2 become arbitrarily small, the acceleration will grow without bound unless $\sigma_{12} = \sigma_{21}$; we must impose this constraint on the stress tensor. We may repeat this procedure for axes aligned with the $\mathbf{1}_1$ and $\mathbf{1}_2$ coordinate directions and obtain that $\sigma_{23} = \sigma_{32}$ and $\sigma_{13} = \sigma_{31}$. That is, the stress tensor must be symmetric:

$$\mathbb{S} = \mathbb{S}^{\mathrm{T}} \qquad \text{or} \qquad \sigma_{ij} = \sigma_{ji} ,$$

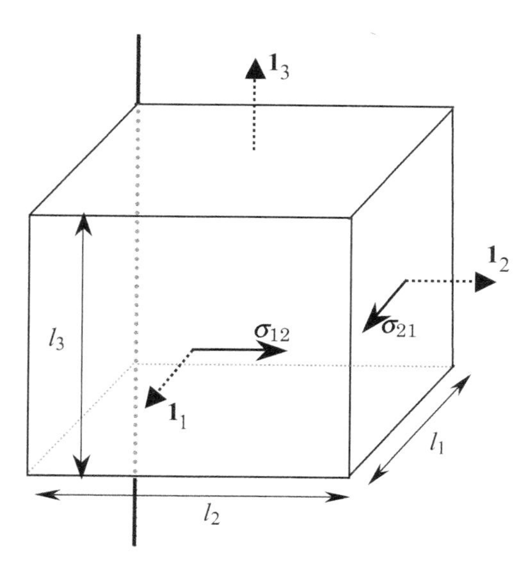

Figure 4.3 Illustration of the torque balance for the small box about the "3" axis. The relevant shear stresses are σ_{12} acting on the front "1" face and σ_{21} acting on the front "2" face.

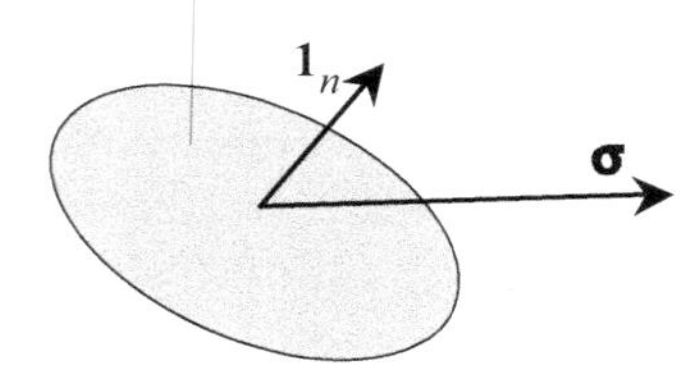

Figure 4.4 Manifestation of stress as a surface force on an exposed surface having normal $\mathbf{1}_n$.

where $\mathbb{S}^{\mathrm{T}}$ is the transpose[4] of $\mathbb{S}$. As with the strain and rate-of-strain tensors, the stress tensor is in general a function of position ($\mathbf{X}$ or $\mathbf{x}$) within the physical body under consideration and of time, t. Normal stresses are commonly represented by σ and shear (tangential) stresses by τ.

4.4 Stress and Force

The stress tensor at a point within the body is resolved into a stress (a force per unit area) by the introduction of a planar surface S through that point, with that portion of the body on one side (side A) of the surface being removed. The stress on side A of surface S is that necessary to maintain static equilibrium at the point. The surface is characterized by the orientation of a vector normal to side A. That is, if the unit normal vector is $\mathbf{1}_n = n_i \mathbf{1}_i$, then the stress on side A is given by[5]

$$\boldsymbol{\sigma} = \sigma_{ij} n_i \mathbf{1}_j ;$$

see Figure 4.4. For example, if $\mathbf{1}_n = \mathbf{1}_1$, then $\boldsymbol{\sigma} = \sigma_{1j}\mathbf{1}_j$. If the direction of the normal is reversed, i.e., if $\mathbf{1}_n = -\mathbf{1}_1$, then so is the direction of the stress: $\boldsymbol{\sigma} = -\sigma_{1j}\mathbf{1}_j$, in agreement with the conclusion of the previous section that the stresses on oppositely oriented surfaces at a point in a body must be the negative of each other.

If only the 12 and 21 elements of the stress tensor are non-zero, the body experiences a state of plane shear stress. This stress state is illustrated in Figure 4.5, where $\sigma_{12} = \sigma_{21} = \tau$. Note that a positive stress acting on a face with normal directed opposite to the unit vector represents a force acting in the negative coordinate direction.

Now let's calculate the net force acting on the tiny box shown in Figure 4.2. If the state of stress is non-uniform, then the stresses on the six faces of this box do not quite add to zero and result in a force on the box that we are calling the contact force, $\mathbf{F}_C$:

$$\mathbf{F}_C = (\mathbf{F}_{1+} + \mathbf{F}_{1-}) + (\mathbf{F}_{2+} + \mathbf{F}_{2-}) + (\mathbf{F}_{3+} + \mathbf{F}_{3-}),$$

where $\mathbf{F}_{i+}$ is the force exerted on the face with normal in the positive i direction and $\mathbf{F}_{i-}$ is the force exerted on the face with normal in the negative i direction.

4 See Appendix A.1.5.
5 With sums over i and j, the right-hand side of this equation represents nine terms; see Appendix A.1.2.

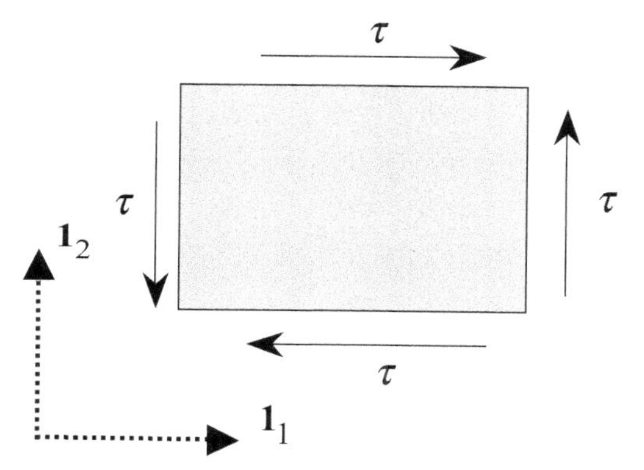

Figure 4.5 Illustration of plane shear stress.

We know from §4.2 that the pairs of forces in the three sets of parentheses are nearly equal and opposite. We need to quantify those small differences. Noting that force is the result of a stress acting on an area, we have that

$$\mathbf{F}_{1+} = (\sigma_{11}\mathbf{1}_1 + \sigma_{12}\mathbf{1}_2 + \sigma_{13}\mathbf{1}_3)\, l_2 l_3 = \sigma_{1j} l_2 l_3\, \mathbf{1}_j$$

and

$$\mathbf{F}_{1-} = -(\sigma_{11}\mathbf{1}_1 + \sigma_{12}\mathbf{1}_2 + \sigma_{13}\mathbf{1}_3)\, l_2 l_3 = -\sigma_{1j} l_2 l_3\, \mathbf{1}_j\,.$$

It is important to realize that these forces are applied at different locations, with $\mathbf{F}_{1-}$ being applied at position x_1 and $\mathbf{F}_{1+}$ at $x_1 + l_1$, and that the stress depends on position: $\sigma_{ij}(x_1, x_2, x_3)$. Rewriting the above equations more carefully, we have

$$\mathbf{F}_{1+} = \sigma_{1j}(x_1 + l_1, x_2, x_3)\, l_2 l_3\, \mathbf{1}_j \qquad \text{and} \qquad \mathbf{F}_{1-} = -\sigma_{1j}(x_1, x_2, x_3)\, l_2 l_3\, \mathbf{1}_j\,.$$

We are at liberty to assume that l_1 is sufficiently small that we may expand the stresses on face 1+ in Taylor series about the position of face 1− and keep only the linear term; that is

$$\sigma_{1j}(x_1 + l_1, x_2, x_3) = \sigma_{1j}(x_1, x_2, x_3) + \frac{\partial \sigma_{1j}}{\partial x_1}(x_1, x_2, x_3) l_1 + O(l_1^2)\,.$$

Using this we find that

$$(\mathbf{F}_{1+} + \mathbf{F}_{1-}) \approx V \frac{\partial \sigma_{1j}}{\partial x_1} \mathbf{1}_j\,,$$

where $V = l_1 l_2 l_3$ is the volume of the box and the derivatives are evaluated at (x_1, x_2, x_3). In the limit $l_1 \to 0$ the approximately equality becomes a precise equality. Representations of the other two pairs of forces follow by replacing 1 by 2 and 3 in this formula. Altogether we have

$$\mathbf{F}_C = V\left(\frac{\partial \sigma_{1j}}{\partial x_1} + \frac{\partial \sigma_{2j}}{\partial x_2} + \frac{\partial \sigma_{3j}}{\partial x_3}\right)\mathbf{1}_j = V\nabla \bullet \mathbb{S},$$

where

$$\nabla \bullet \mathbb{S} = \frac{\partial \sigma_{ij}}{\partial x_i}\mathbf{1}_j$$

is the divergence of the stress tensor. The per-unit-volume version of the internal force balance is

$$\rho \mathbf{f}_C = \nabla \bullet \mathbb{S},$$

where $\mathbf{f}_C = \mathbf{F}_C/M$ is the internal contact force per unit mass or specific contact force. Note that specific forces have the same dimensions as acceleration.

4.5 Pressure and Deviatoric Stress

The pressure is the mean of the normal stresses applied to the body:

$$p = -\tfrac{1}{3}\text{Trace}[\mathbb{S}] = -\tfrac{1}{3}\sigma_{ii}.$$

A minus sign appears in this equation because positive normal stress is tensile, while a positive pressure is compressive. Since the trace of a matrix is a scalar invariant,[6] the magnitude of p is independent of the orientation of the reference coordinate system. Structural materials, such as steel and wood, can sustain tensile stresses for an indefinite period of time, but most geological materials (and some structural materials such as concrete) cannot. As a general rule, if a geological material is subject to a negative pressure or a tensile stress, it will fail (i.e., break).

If a body experiences a hydrostatic state of stress, the stress tensor is simply

$$\mathbb{S} = -p\mathbb{I} \qquad \text{or} \qquad \sigma_{ij} = -p\delta_{ij}.$$

In many situations it is convenient to split the stress tensor into symmetric and deviatoric parts:

$$\mathbb{S} = -p\mathbb{I} + \mathbb{S}' \qquad \text{or} \qquad \sigma_{ij} = -p\delta_{ij} + \tau_{ij},$$

where $\mathbb{S}'$ is the deviatoric stress tensor having elements τ_{ij}. The divergence of the stress tensor now may be written as

$$\nabla \bullet \mathbb{S} = -\nabla p + \nabla \bullet \mathbb{S}' \qquad \text{or} \qquad \frac{\partial \sigma_{ij}}{\partial x_i}\mathbf{1}_j = -\nabla p + \frac{\partial \tau_{ij}}{\partial x_i}\mathbf{1}_j$$

[6] See Appendix A.1.4.

and

$$\mathbf{f}_C = -\frac{1}{\rho}\nabla p + \mathbf{f}_D, \qquad \text{where} \qquad \mathbf{f}_D = \frac{1}{\rho}\nabla \bullet \mathbb{S}'.$$

The subscript D stands for "deviatoric," "dissipative" or "drag". Note that Trace[$\mathbb{S}'$] $= \tau_{ii} = 0$.

4.5.1 Work of Deformation

As an elastic body deforms in response to an externally applied stress an amount of work

$$w = \epsilon_{ij}\sigma_{ij}$$

per unit volume is performed on the body. If the body is perfectly elastic, this energy is stored in the deformation and available to be recovered. This is the concept behind the springs that used to keep clocks and watches running. Splitting the strain and stress tensors into isotropic and deviatoric parts,[7] the work may be expressed as

$$w = -p(\nabla \bullet \mathbf{u}) + \epsilon'_{ij}\tau_{ij}.$$

As a fluid body moves in response to an externally applied stress an amount of power

$$\dot{w} = \dot{\epsilon}_{ij}\sigma_{ij} = -p(\nabla \bullet \mathbf{v}) + \dot{\epsilon}'_{ij}\tau_{ij}$$

per unit volume is performed on the body. The portion associated with change of volume is recoverable, while that involving the deviatoric rate of strain and stress is not; the power $\dot{\epsilon}'_{ij}\tau_{ij}$ goes into heating the system, showing up as a heating term in the energy equation; see § 5.4 and Appendix E.7.

4.6 Conservation of Momentum

Let's return to the per-unit-mass form of equation of Newton's second law given in § 4.1:

$$\mathbf{a} = \mathbf{f}_B + \mathbf{f}_V + \mathbf{f}_C.$$

We wish to develop two versions of this equation, relevant to elastic and fluid bodies. The major difference between the two is in the treatment of acceleration. The acceleration of a particle in an elastic body is expressed as

$$\mathbf{a} = \frac{\mathrm{d}^2\mathbf{u}}{\mathrm{d}t^2},$$

[7] Using $\epsilon_{ij} = (\nabla \bullet \mathbf{u})\delta_{ij}/3 + \epsilon'_{ij}$ and $\sigma_{ij} = -p\delta_{ij} + \tau_{ij}$.

where $\mathbf{u}$ is the displacement vector, while in a fluid body it is[8]

$$\mathbf{a} = \frac{\mathrm{D}\mathbf{v}}{\mathrm{D}t} \equiv \frac{\partial \mathbf{v}}{\partial t} + (\mathbf{v} \bullet \nabla)\,\mathbf{v}\,.$$

The factor $\mathbf{v} \bullet \nabla$ is a directional derivative.[9]

Now let's consider the forces. We found in § 4.4 that the contact force is variously expressed as

$$\mathbf{f}_C = \frac{1}{\rho}\nabla \bullet \mathbb{S} = -\frac{1}{\rho}\nabla p + \frac{1}{\rho}\nabla \bullet \mathbb{S}' = -\frac{1}{\rho}\nabla p + \mathbf{f}_D\,.$$

The body forces are gravity $\mathbf{g}$ and (possibly) Lorentz $\mathbf{f}_L$:

$$\mathbf{f}_B = \mathbf{g} + \mathbf{f}_L\,,$$

while the relevant virtual forces[10] are the centrifugal and Coriolis forces:

$$\mathbf{f}_V = -\boldsymbol{\Omega} \times (\boldsymbol{\Omega} \times \mathbf{x}) - 2\,\boldsymbol{\Omega} \times \mathbf{v}\,,$$

where $\boldsymbol{\Omega}$ is the rotation vector, describing the rotation of the coordinate frame, relative to the fixed stars. Typically the Lorentz and virtual forces are zero or negligibly small in an elastic body and we have the *Cauchy Momentum equation*

$$\frac{\mathrm{d}^2\mathbf{u}}{\mathrm{d}t^2} = \mathbf{g} + \frac{1}{\rho}\nabla \bullet \mathbb{S}\,,$$

while for a fluid body, we have

$$\frac{\mathrm{D}\mathbf{v}}{\mathrm{D}t} + 2\,\boldsymbol{\Omega} \times \mathbf{v} = \mathbf{g} + \mathbf{f}_L - \boldsymbol{\Omega} \times (\boldsymbol{\Omega} \times \mathbf{x}) - \frac{1}{\rho}\nabla p + \mathbf{f}_D\,.$$

The study of motions in Earth's core or in the ionized regions of the upper atmosphere are beyond the scope of this book. Consequently, the Lorentz force $\mathbf{f}_L$ will be ignored from now on.

4.6.1 Hiding the Centrifugal Force

The specific gravitational force, denoted by $\mathbf{g}_p$, experienced on Earth is the combined result of attraction of mass within Earth plus the centrifugal force due to the rotation of Earth:

$$\mathbf{g}_p = \mathbf{g} - \boldsymbol{\Omega} \times (\boldsymbol{\Omega} \times \mathbf{x})\,.$$

Using this, the momentum equation becomes

$$\frac{\mathrm{D}\mathbf{v}}{\mathrm{D}t} + 2\,\boldsymbol{\Omega} \times \mathbf{v} = \mathbf{g}_p - \frac{1}{\rho}\nabla p + \mathbf{f}_D\,.$$

[8] See Appendix C.2.
[9] See Appendix A.2.
[10] See Appendix C.1.

The direction "down" is parallel to $\mathbf{g}_p$ rather than $\mathbf{g}$.

The acceleration $\mathbf{g}_p$, like $\mathbf{g}$, is the gradient of a scalar. To see this, we may write $\mathbf{g}_p = -\nabla\psi_p$ and $\mathbf{g} = -\nabla\psi$. Noting that $\boldsymbol{\Omega}\times(\boldsymbol{\Omega}\times\mathbf{x}) = -\frac{1}{2}\nabla(\boldsymbol{\Omega}\times\mathbf{x})^2$ is the gradient of a scalar, we have

$$-\nabla\psi_p = -\nabla\psi + \tfrac{1}{2}\nabla(\boldsymbol{\Omega}\times\mathbf{x})^2 \qquad \text{or simply} \qquad \psi_p = \psi - \tfrac{1}{2}(\boldsymbol{\Omega}\times\mathbf{x})^2 .$$

On Earth, the centrifugal force is small compared with the gravitational force, so that the distinction between $\mathbf{g}$ and $\mathbf{g}_p$ is small. In order to simplify the presentation, this distinction will be ignored in the remainder of this text, with $\mathbf{g}$ appearing rather than the more-precise $\mathbf{g}_p$.

4.6.2 Natural Coordinates

The material derivative quantifies the change in the velocity vector with time, as experienced by an observer going with the flow. If we write $\mathbf{v} = v\mathbf{1}_v$, then

$$\frac{\mathrm{D}\mathbf{v}}{\mathrm{D}t} = \frac{\mathrm{D}v}{\mathrm{D}t}\mathbf{1}_v + v\frac{\mathrm{D}\mathbf{1}_v}{\mathrm{D}t} .$$

As we move in the direction of flow,

$$\frac{\mathrm{D}\mathbf{1}_v}{\mathrm{D}t} = \frac{\mathrm{d}l}{\mathrm{d}t}\frac{\mathrm{d}\mathbf{1}_v}{\mathrm{d}l} ,$$

where $\mathrm{d}l$ is a small length in the direction of motion. It is readily seen that $\mathrm{d}l/\mathrm{d}t = v$. The factor $\mathrm{d}\mathbf{1}_v/\mathrm{d}l$ may be calculated by imagining the unit vector is (instantaneously) moving around a circle of radius R: $\mathrm{d}\mathbf{1}_v/\mathrm{d}l = \mathbf{1}_n/R$, where $\mathbf{1}_n$ is a unit vector pointing to the center of the circle. Altogether, we have the natural form of the material derivative

$$\frac{\mathrm{D}\mathbf{v}}{\mathrm{D}t} = \frac{\mathrm{D}v}{\mathrm{D}t}\mathbf{1}_v + \frac{v^2}{R}\mathbf{1}_n .$$

$\mathrm{D}v/\mathrm{D}t$ is the Poincaré acceleration and v^2/R is the *centripetal acceleration*. Note that R and $\mathbf{1}_n$ may vary with position and time.

4.6.3 Alternate Forms

The form of the fluid momentum equation given immediately above is preferred for most applications, but several other forms are useful from time to time. Noting that[11]

$$(\mathbf{v}\cdot\nabla)\,\mathbf{v} = \tfrac{1}{2}\nabla(v^2) + \boldsymbol{\omega}\times\mathbf{v}, \qquad \text{where} \qquad \boldsymbol{\omega} = \nabla\times\mathbf{v}$$

[11] Using the third vector identity found in Appendix A.2.6.

is the *vorticity vector* and $v = \|\mathbf{v}\|$, and writing $\mathbf{g} = -\nabla\psi$, the momentum equation is

$$\frac{\partial \mathbf{v}}{\partial t} + \nabla\left(\tfrac{1}{2}v^2 + \psi\right) + (2\boldsymbol{\Omega} + \boldsymbol{\omega}) \times \mathbf{v} = -\frac{1}{\rho}\nabla p + \mathbf{f}_D .$$

If the fluid is barotropic, we can introduce a function P such that $\rho\,\mathrm{d}P = \mathrm{d}p$ and write the momentum equation as

$$\frac{\partial \mathbf{v}}{\partial t} + \nabla\left(\tfrac{1}{2}v^2 + \psi + P\right) + (2\boldsymbol{\Omega} + \boldsymbol{\omega}) \times \mathbf{v} = \mathbf{f}_D .$$

The terms in the parenthesis may be expressed as gH, where H is the head; see Appendix D.6.2.

Alternatively, we may use the natural form of the material derivative[12] and write the momentum equation as

$$\frac{\mathrm{D}v}{\mathrm{D}t}\mathbf{1}_v + \frac{v^2}{R}\mathbf{1}_n + 2\,\boldsymbol{\Omega} \times \mathbf{v} = -\nabla\psi - \frac{1}{\rho}\nabla p + \mathbf{f}_D .$$

If we take the dot product of this equation with a small step in the direction of flow, $\mathbf{dl} = \mathbf{v}\mathrm{d}t$, and note that the centripetal and Coriolis accelerations are normal to this path, we have

$$v\mathrm{d}v + \mathrm{d}\psi = -(1/\rho)\mathrm{d}p + \mathbf{f}_D \bullet \mathbf{dl} .$$

Note that, in multiplying the momentum equation by displacement, it has been changed from a force balance to an energy balance. (If the force term $\mathbf{f}_D$ is negligibly small, this is the stream-wise component of Euler equation; see Appendix D.6.1.) This equation states that changes in the kinetic and potential energy are impelled by the pressure force and retarded by the drag force. If the pressure decreases in the direction of flow, $\mathrm{d}p$ is negative and the kinetic and/or potential energy increase. In natural flow systems, $\mathbf{f}_D \bullet \mathbf{dl}$ is always negative; the drag force degrades mechanical energy. This term is equivalent to the dissipation term $\dot{\epsilon}'_{ij}\tau_{ij}$ introduced in § 4.5.1.

4.7 Summary

In the previous sections we have developed equations expressing the conservation of mass and momentum. Two versions of the each arise depending whether the material is elastic or fluid. In summary we have

	Conservation of Mass	**Conservation of Momentum**
Elastic	$\rho + \rho_r(\nabla \bullet \mathbf{u}) = \rho_r$	$\mathrm{d}^2\mathbf{u}/\mathrm{d}t^2 = \mathbf{g} + (1/\rho)\nabla \bullet \mathbb{S}$
Fluid	$\mathrm{D}\rho/\mathrm{D}t + \rho\nabla \bullet \mathbf{v} = 0$	$\mathrm{D}\mathbf{v}/\mathrm{D}t + 2\,\boldsymbol{\Omega} \times \mathbf{v} = \mathbf{g}$ $-(1/\rho)\nabla p + \mathbf{f}_D ,$

[12] From § 4.6.2.

where ρ is the mass density, ρ_r is the density of the reference state,[13] $\mathbf{u}$ is the displacement vector, $\mathbf{g}$ is the local acceleration of gravity, $\mathbb{S}$ is the stress tensor, $\mathbf{v}$ is the fluid velocity, $\boldsymbol{\Omega}$ is the coordinate rotation vector,

$$\frac{\mathrm{D}}{\mathrm{D}t} = \frac{\partial}{\partial t} + \mathbf{v} \bullet \nabla$$

is the material derivative and

$$\mathbf{f}_D = (1/\rho)\nabla \bullet \mathbb{S}'$$

is the deviatoric internal contact force.

Note that the fluid continuity equation may be written as

$$\frac{\partial \rho}{\partial t} + \nabla \bullet (\rho \mathbf{v}) = 0$$

and an alternate form[14] for the inertia term is

$$\frac{\mathrm{D}\mathbf{v}}{\mathrm{D}t} = \frac{\partial \mathbf{v}}{\partial t} + \tfrac{1}{2}\nabla(v^2) + \boldsymbol{\omega} \times \mathbf{v},$$

where $\boldsymbol{\omega} = \nabla \times \mathbf{v}$ is the vorticity of the fluid and $v = \sqrt{\mathbf{v} \bullet \mathbf{v}}$ is the speed.

4.7.1 Hiding Gravity

The acceleration of gravity may be expressed as the gradient of a potential: $\mathbf{g} = -\nabla \psi$. If the fluid is isentropic,[15] the density is a function of pressure: $\mathrm{d}\rho = \rho \mathrm{d}p/K(p)$, where K is the adiabatic incompressibility. In this case there exists a pressure function $P(p)$ such that $\mathrm{d}P = \mathrm{d}p/\rho$ and

$$\mathbf{g} - (1/\rho)\nabla p = -\nabla(\psi + P).$$

This shows that gravity and pressure act identically and the effect of gravity may be incorporated in the pressure without loss of physical generality. In particular, this tells us that if the fluid has a uniform density, then we can set $\mathbf{g} = \mathbf{0}$.

Transition

In this chapter we have developed the equations governing the distribution of stresses and forces within a continuous body. Central to this process are the stress tensor and the momentum equation. There is only one version of the stress tensor, applicable to both elastic and fluid bodies, but there are two versions of the momentum equation, for elastic and fluid bodies. In order to complete this set of governing equations, we need to

- develop an equation governing the variation of density; and
- relate the stress tensor (or deviatoric stress tensor) to the deformation or velocity.

These tasks are addressed in Chapters 5 and 6, respectively.

[13] Which may vary with position.
[14] Using the fifth formula in Appendix A.2.6.
[15] Or equivalently barotropic; see § 5.5.

5
Some Thermodynamics

We have investigated how a body may move and deform; this is kinematics, codified in the equation of conservation of mass. We also have investigated the distribution of stresses within a body which may cause it to deform; this is dynamics, codified in the equation of conservation of momentum. Versions of these conservation equations for elastic and fluid bodies are presented in § 4.7. We now turn our attention to the manner in which a body responds to the stresses; this is rheology. Commonly, rheology is taken to mean a change of body shape (called shear rheology), but it also encompasses the change of volume or density (without change of shape) induced by a change of pressure (called compression rheology). This latter response is determined by the *equation of state*,[1] giving the density ρ in terms of appropriate independent thermodynamic variables. Generally, the equation of state is presented in differential form (as in § 5.2), but for an ideal gas, it is a simple algebraic equation.[2]

As explained in Appendix E.5, in a system composed of one constituent[3] the density is a function of two independent thermodynamic variables, with some freedom in the choice of those variables. Two useful choices of independent variables are temperature T and pressure p, or pressure and specific entropy s. It should be noted that the density and pressure are unique variables, in that they play leading roles in both thermodynamics and dynamics. The differential form of the equation of state for density is presented in § 5.2, following a discussion in the next section of some new adjectives that we encounter while delving into the dynamics and thermodynamics of geophysical fluids. The thermodynamic presentation in this chapter is rather abbreviated; for more detail, refer to Appendix E.

5.1 New Adjectives

In our study of geophysical fluids, we will encounter a number of new adjectives. It may be helpful to define these clearly at this point, before plunging into the dynamic and thermodynamic details. These adjectives include:

[1] More correctly this should be called the density equation of state, to distinguish it from the equation of state for entropy; see Appendix E.5.

[2] See § 5.6.

[3] That is, one component in one phase.

- *adiabatic* – describes a thermodynamic process occurring without gain or loss of heat: $\mathrm{d}Q = 0$;
- *baroclinic* – describes a fluid having density a function of temperature and/or composition (as well as pressure): $\rho(p, T, \xi)$, where ξ is a constituent mass fraction;
- *barotropic* – describes a fluid having density a function of pressure only: $\rho(p)$;
- *homogeneous* – describes a fluid having properties that are the same at all points;
- *incompressible* – describes a fluid having density independent of pressure; the density may still vary with temperature and/or composition: $\rho(T, \xi)$;
- *isentropic* – describes a thermodynamic process occurring adiabatically, reversibly and without gain or loss of material: $\mathrm{d}S = 0$;
- *isotropic* – describes a material having properties that are the same in all directions;
- *isothermal* – describes a process occurring at constant temperature: $\mathrm{d}T = 0$; and
- *reversible* – describes an idealized process that occurs with the fluid remaining in thermodynamic equilibrium.

The adjectives homogeneous, incompressible and isotropic describe the properties of the fluid, baroclinic and barotropic describe the present state of the fluid,[4] while the remaining adjectives characterize processes experienced by the fluid.

The most common process in geophysics is convection; convective motions in the atmosphere and oceans are driven predominantly by solar heating at Earth's surface, while convection in the mantle is driven by both cooling at the top and heating at the bottom. Convection tends to drive these fluid bodies toward a *well-mixed* state having uniform composition and entropy.[5] More details on the structures in the atmosphere, oceans and mantle related to convection are given in § 7.1.

5.2 Differential Equation of State

The equation of state for density is best expressed in differential form and using p and T as independent thermodynamic variables; this is[6]

$$\mathrm{d}\rho = -\rho\alpha\,\mathrm{d}T + \rho\beta_T\,\mathrm{d}p + \rho_\xi\,\mathrm{d}\xi\,,$$

where[7]

$$\alpha = -\frac{1}{\rho}\frac{\partial\rho}{\partial T}\bigg|_{p,\xi} \qquad \text{and} \qquad \beta_T = \frac{1}{\rho}\frac{\partial\rho}{\partial p}\bigg|_{T,\xi}$$

[4] A fluid is barotropic if isotherms are coincident with isopycnals (surfaces of constant density) and baroclinic if they are not.
[5] The vertical thermal structure of a well-mixed body of fluid is called the *adiabat*.
[6] From Appendix E.5.
[7] See Appendix E.4.2.

are the coefficient of thermal expansion and isothermal compressibility, respectively, ξ is the mass fraction[8] of a second constituent (e.g., water vapor in air or salt in sea water) and

$$\rho_\xi = \left.\frac{\partial \rho}{\partial \xi}\right|_{p,T}.$$

If the system has more than two constituents, the factor involving $d\xi$ becomes a summation or vector; see Appendix E.5.

This equation of state introduces two thermodynamic variables, the temperature T and the composition ξ, each of which requires a governing equation. The variation of T is governed by the energy equation, given in § 5.4, and the variation of ξ is governed by a conservation equation, given in the following section.

5.3 Conservation of Material

The variation of a constituent (such as unsaturated water vapor in air or salt in seawater) is governed by a conservation equation of the form[9]

$$\frac{\partial \xi}{\partial t} + \nabla \bullet (\xi \mathbf{v}) = \nabla \bullet \left(\kappa_\xi \nabla \xi\right),$$

where ξ is the mass fraction of the constituent, $\mathbf{v}$ is the (mass averaged) velocity of the fluid and κ_ξ is the diffusion coefficient. The laminar diffusivity of material is very small, but – using Reynolds analogy[10] – the turbulent diffusivity is equal to the turbulent diffusivity of momentum, ν_T. A more general version this equation includes a source term, but this term is equal to zero for salt in the oceans and unsaturated water vapor in air.

When the water vapor in air is saturated, then the source term is non-zero and given by $s_\xi = \partial \xi^*/\partial t + \nabla \bullet (\xi^* \mathbf{v}) - \nabla \bullet \left(\kappa_\xi \nabla \xi^*\right)$, with the saturation vapor content, ξ^*, prescribed by[11]

$$\xi^* \approx \frac{p_0}{0.622 p} \exp\left(\frac{T_L}{T_B} - \frac{T_L}{T}\right),$$

where p is the pressure, p_0 is atmospheric pressure,[12] T is the temperature, $T_B = 373\,\mathrm{K}$ is the boiling temperature of water at 1 atmosphere and $T_L \approx 5400$ K.

5.4 Energy Equation

The energy equation[13] quantifies the change of temperature of a parcel:

$$\rho c_p \frac{\mathrm{D}T}{\mathrm{D}t} = \nabla \bullet (k \nabla T) + \alpha T \rho\, \mathbf{g} \bullet \mathbf{v} + \rho \Psi_R + \dot{\epsilon}'_{ij} \tau_{ij},$$

[8] See § 6.4.1.
[9] See Appendix D.5.
[10] See § 23.8.
[11] See Appendix E.9.4.
[12] 1 atmosphere = 101,325 Pa.
[13] Developed in Appendix E.7.

where c_p is the specific heat at constant pressure,[14] k is the thermal conductivity and Ψ_R is the radioactive heating (with dimensions of power per unit mass; SI units W/kg). Changes of temperature (represented by the terms on the right-hand side of this equation) are induced by diffusion of heat, compression induced by change of elevation, radioactive heating and dissipation of energy.[15]

5.5 The Isentropic State

The isentropic state of a fluid body is described by a pair of equations giving ρ and T as functions of p. The function $\rho(p)$ is determined from the density equation of state[16]

$$\frac{\mathrm{d}\rho}{\mathrm{d}p} = \frac{\rho}{K} = \frac{1}{U_s^2},$$

where K is the *bulk modulus*[17] and, as we shall see in § 9.3, U_s is the speed of sound. Note that

- a fluid obeying this equation is *barotropic* (if instead ρ depends on temperature as well as pressure, the fluid is *baroclinic*);
- the material becomes incompressible and the density becomes independent of pressure in the limit $K \to \infty$; and
- K must be positive. If it were not, a positive increase in pressure would cause an decrease in density; such a material would be unstable.

The incompressibility K is itself a thermodynamic function of p and s, where s is the specific entropy,[18] but if the motions are isentropic, then K is a function only of p. In this case, it is possible – in theory – to find a function $P(p)$ such that $\mathrm{d}p = \rho\,\mathrm{d}P$. This is useful in representing the pressure term in the momentum equation as the gradient of a scalar; see appendices D.6.1 and D.6.2.

The function $T(p)$ is determined from the differential equation of state for ρ given in § 5.2. Using the isentropic relation, this becomes

$$\left.\frac{\mathrm{d}T}{\mathrm{d}p}\right|_{s,\xi} = \frac{\beta_T}{\alpha} - \frac{1}{\alpha K}.$$

Using the identity[19]

$$\frac{1}{K} = \beta_T - \frac{\alpha^2 T}{\rho c_p}, \qquad \text{where} \qquad c_p = T\left.\frac{\partial s}{\partial T}\right|_p$$

[14] See Appendix E.4.2.
[15] The effects of changes of kinetic energy and composition have been ignored: see Appendix E.7.
[16] This is actually a definition: $K \equiv \rho\partial p/\partial\rho|_{s,\xi}$; see Appendix E.4.2.
[17] K is sometimes called the *adiabatic incompressibility* and is the inverse of the adiabatic compressibility β_s: $K\beta_s = 1$; see Appendix E.4.2.
[18] See Appendix E.3.3.
[19] See Appendix E.4.2.

is the specific heat at constant pressure, this is may be expressed as

$$\frac{\mathrm{d}T}{\mathrm{d}p} = \frac{\alpha T}{\rho c_p}.$$

This equation quantifies the increase of temperature that accompanies an increase in the pressure acting on an insulated body. If you squeeze something it gets warm, and if you let it expand, it gets cool. We commonly experience the latter when using a spray can; the can gets cold as its contents are sprayed out. Mechanical refrigeration is based on this principle. In a refrigerator, a working fluid that has been heated due to compression is allowed to cast off some of its heat. Then as the fluid expands, it gets colder, gaining the capacity to cool nearby objects by absorbing some of their heat.

The differential equations for ρ and T may be combined to give

$$\frac{\mathrm{d}T}{\mathrm{d}\rho} = \gamma_G \frac{T}{\rho}, \qquad \text{where} \qquad \gamma_G = \frac{\alpha K}{\rho c_p}$$

is the Grüneisen parameter.[20] If γ_G is constant, this yields $T \propto \rho^{\gamma_G}$.

If the body is in hydrostatic equilibrium, then $\mathrm{d}p = -\rho g \,\mathrm{d}r$, where g is the magnitude of the acceleration of gravity and r is the upward coordinate, and the adiabatic temperature gradient, often called the *adiabat*, is given by

$$\frac{\mathrm{d}T}{\mathrm{d}r} = -\frac{T}{r_A}, \qquad \text{where} \qquad r_A \equiv \frac{c_p}{\alpha g}$$

is the adiabatic scale height.

- For the atmosphere with $\alpha = 1/T \approx 1/288\ \mathrm{K}^{-1}$ and $c_p \approx 1005\ \mathrm{J{\cdot}kg^{-1}{\cdot}K^{-1}}$, $r_A \approx 3 \times 10^4$ m.[21]
- For the oceans with $\alpha \approx 2 \times 10^{-4}\ \mathrm{K}^{-1}$ and $c_p \approx 4184\ \mathrm{J{\cdot}kg^{-1}{\cdot}K^{-1}}$, $r_A \approx 2 \times 10^6$ m.
- For the mantle with $g \approx 10\ \mathrm{m{\cdot}s^{-1}}$, $\alpha \approx 3 \times 10^{-5}\ \mathrm{K}^{-1}$ and $c_p \approx 1250\ \mathrm{J{\cdot}kg^{-1}{\cdot}K^{-1}}$, $r_A \approx 4.2 \times 10^6$ m.

5.6 Ideal Gas

An *ideal gas* is composed of atoms having negligibly small volume (compared with the volume of the container) that interact only by bouncing off each other elastically. The equation of state $\rho(p, T)$ for an ideal gas is a simple algebraic relation:[22]

$$\rho = \frac{p}{R_s T}, \qquad \text{where} \qquad R_s = \frac{R}{M_A}$$

is the specific gas constant, $R = 8314\ \mathrm{m^2{\cdot}s^{-2}{\cdot}kmol^{-1}{\cdot}K^{-1}}$ is the ideal gas constant and M_A is the molar mass of the substance (measured in $\mathrm{kg{\cdot}kmol^{-1}}$, and numerically equal to the atomic number of the substance).

[20] See § 29.1.
[21] See Appendix E.8.5 and Table B.5.
[22] See Appendix E.8.

Noting that for an ideal gas[23] $\alpha T = 1$ and

$$K = \gamma p, \qquad \text{where} \qquad \gamma = c_p / c_v$$

is the ratio of specific heats, and using the ideal gas law, the differential equations for $\rho(p)$ and $T(p)$ developed in the previous section simplify to

$$\frac{d\rho}{dp} = \frac{\rho}{\gamma p} \qquad \text{and} \qquad \frac{dT}{dp} = \frac{R_s T}{c_p p}.$$

Since the coefficients c_p, R_s and γ are constants, these equations are readily integrated:

$$\rho = \rho_0 \, (p/p_0)^{1/\gamma} \qquad \text{and} \qquad T = \Theta \, (p/p_0)^{R_s/c_p},$$

where ρ_0, p_0 and Θ are reference constant values. Θ is called the *potential temperature* and $\Pi = (p/p_0)^{R_s/c_p}$ is the *Exner function*. When applied to the atmosphere, p_0 is usually set equal to atmospheric pressure and ρ_0 is the density of air at Earth's surface. The exponents R_s/c_p and γ are known constants:[24]

- for a monatomic gas, $R_s/c_p = 2/5$ and $\gamma = 5/3$;
- for a diatomic gas (e.g., the atmosphere), $R_s/c_p = 2/7$ and $\gamma = 7/5$; and
- for a tri-atomic gas (such as H_2O or CO_2), $R_s/c_p = 1/4$ and $\gamma = 4/3$.

Note that the speed of sound in an ideal gas is given by

$$U_s = \sqrt{\gamma R_s T}$$

and the adiabat may be expressed

$$\frac{dT}{dr} = -\frac{g}{c_p},$$

where r is the radial coordinate.

Transition

In this chapter we have developed the equation of state, which relates variations in density to variations in temperature and pressure. An interesting and useful version of the equation of state is that for an ideal gas. In many cases, motions occur adiabatically and density and temperature depend only on pressure. If the body is hydrostatic, pressure, density and temperature can be related to height; the variation of temperature with height is called the adiabat. These equations allow us to predict the structure of Earth's atmosphere.

The equation of state relates the isotropic part of deformation, specifically the density, to the pressure and temperature. The relation between deviatoric parts of deformation and stress, called shear rheology, is developed in the next chapter.

[23] See Appendix E.8.5.
[24] See Appendix E.8.5.

6
Shear Rheology

Shear rheology, commonly referred to simply as rheology, is the scientific study of the deformational response (strain and rate of strain) of a body to an applied force (stress), often called the *stress–strain relation*, expressed symbolically as $f(\mathbb{S}, \mathbb{E}, \dot{\mathbb{E}}) = \emptyset$. The response depends principally on the composition of the body, although the temperature and pressure can be important. (For example, the rheology of butter depends rather strongly on the temperature.) If the stress is independent of the rate of strain, the body is elastic and the symbolic relation simplifies to $f(\mathbb{S}, \mathbb{E}) = \emptyset$. Most elastic materials deform only slightly in response to an applied stress and the strain response to stress is adequately described by a linear relationship, obeying *Hooke's law*; this relationship is developed in § 6.1.

If the stress is independent of the strain, the body is fluid and the symbolic relation simplifies to $f(\mathbb{S}, \dot{\mathbb{E}}) = \emptyset$. If this relationship is linear, the fluid is said to be *Newtonian*. The two most commonly encountered fluids, liquid water and air, are Newtonian. However, for many geological materials the relation between stress and flow is nonlinear and the material is *non-Newtonian*. In general, there are two types of nonlinear rheology: *shear thinning* and *shear thickening*. A shear-thinning fluid deforms more readily as the shearing force is increased. Familiar examples of such material are latex paint and quicksand. On the other hand, a shear thickening fluid becomes more resistant to flow as the shearing stress is increased. Many geological materials, including ice (in glaciers), lavas and Earth's silicate mantle behave as shear-thinning fluids. This behavior is investigated in § 6.2.

The classification of materials as either elastic or fluid is simplistic, and somewhat misleading. For *viscoelastic* materials the stress depends on both strain and rate of strain. In reality a material may exhibit elastic, fluid or even more complex behavior depending on its thermodynamic state, the magnitude of the stresses applied and the time scale of interest. Most solid materials, as they are heated toward their melting point, develop more fluid-like behavior.[1] Many materials (such as emulsions like mayonnaise and whipped cream) behave elastically when the applied stress is small, but change their rheological character to fluid-like or viscoelastic at higher stresses. Some materials, notably the silicates of the

[1] A notable exception to this tendency is water ice; it is the only material to be brittle right up to the melting point.

mantle, behave elastically on short time scales (i.e., those associated with seismic waves and internal oscillations) but act like fluids on long time scales (i.e., those associated with post-glacial rebound and sea-floor spreading). Silly putty has a similar rheology. Often geological bodies (portions of Earth's crust) are not homogeneous; their response is often sensitive to volumetrically insignificant features, such as thin cracks or weak zones. The net result is that large geological bodies often have fluid-like character, especially on long time scales.[2] Permanent deformation that occurs slowly over a period of time is called *creep*. Then there are *thixotropic* materials (such as clays), the rheology of which depends on the recent history of stresses. In particular, a sudden change of stress can cause clays to transform from behaving as elastic solids to fluids, with disastrous results. This is why cities (such as Mexico City) that have been built on sediments are so vulnerable to earthquakes. With this in mind, the equations developed in the following sections should be employed with the realization that the rheological nature of the material may change drastically, depending on the circumstances.

With all these caveats, we now turn our attention to the rheology of elastic solids in § 6.1 and of fluids in § 6.2.

6.1 Elastic Solid

The rheology of an elastic solid is quantified by a functional relation between the stress and strain tensors. We will limit ourselves to the simplest possible form: a linear relation. But there is a complication we need to bear in mind. The complication is that, while strain can be defined locally in terms of displacements, **u**, from a reference state, stress cannot be defined entirely locally. That is, the stress tensor quantifies both the local effect of adjoining particles and the effect of the gravitational force, which represents the gravitational pull exerted by all other particles in the entire universe. So unless the body is in free fall, we must take into account the possibility that, when the body is in its reference state (that is, having zero strain), its state of stress is not zero. The simplest state of stress consistent with the existence of a gravitational force is a hydrostatic state, with the stress being isotropic; let's suppose that the reference stress tensor, denoted by $\mathbb{S}_r$, is given by

$$\mathbb{S}_r = -p_r \mathbb{I} \qquad \text{or} \qquad \sigma_{ijr} = -p_r \delta_{ij} ,$$

where p_r is the hydrostatic pressure associated with the reference state. The unit tensor $\mathbb{I}$ ensures that the force on any exposed surface will be normal to that surface and the minus sign indicates that the force will be compressive (with $p_r > 0$). Note that the reference-state pressure varies with position [that is, $p_r(\mathbf{x})$], but is independent of time.

As we noted previously, the simplest model of elastic deformation is that having a linear relation between stress and strain. Material with this behavior is said to obey Hooke's law. Hooke's law introduces two constants of elasticity. In geophysics[3] the relation between

[2] A number of simple forms of rheological behavior are briefly presented in Appendix D.1.
[3] Alternative versions of this relation are found in Appendix D.3.

the stress tensor $\mathbb{S}$ and the strain tensor $\mathbb{E}$ uses the bulk modulus K and rigidity or *shear modulus* μ:

$$\mathbb{S} + p_r\mathbb{I} = \left(K - \tfrac{2}{3}\mu\right)(\nabla_{\mathbf{X}} \bullet \mathbf{u})\mathbb{I} + 2\mu\mathbb{E} \qquad \text{or equivalently}$$

$$\sigma_{ij} + p_r\delta_{ij} = \left(K - \tfrac{2}{3}\mu\right)(\nabla_{\mathbf{X}} \bullet \mathbf{u})\delta_{ij} + 2\mu\epsilon_{ij},$$

where $\mathbf{u}$ is the displacement vector. Note that K and μ have the same units as pressure: Pa = kg·m^{-1}·s^{-2}. Recalling[4] that $\mathbb{E} = (\nabla_{\mathbf{X}} \bullet \mathbf{u})\mathbb{I} + \mathbb{E}'$ and $\mathbb{S} = -p\mathbb{I} + \mathbb{S}'$, this equation may be divided into two, describing isotropic and shearing behavior:

$$p = p_r - K\nabla_{\mathbf{X}} \bullet \mathbf{u} \qquad \text{and} \qquad \mathbb{S}' = 2\mu\mathbb{E}'.$$

In this form it is readily seen that K and μ determine the stresses needed to change the volume and shape of an elastic body, respectively. It is reassuring to note that, since $\nabla_{\mathbf{X}} \bullet \mathbf{u} \approx 1 - \rho/\rho_r$, this pressure-density relation is the same as the isentropic equation of state presented in § 5.5.

Using the expression for $\nabla_{\mathbf{X}} \bullet \mathbb{E}$ presented at the end of § 3.3.1 we see that

$$\nabla_{\mathbf{X}} \bullet \mathbb{S} = -\left(1 + \frac{\mu}{3K}\right)\nabla_{\mathbf{X}} p + \mu\nabla^2_{\mathbf{X}}\mathbf{u}.$$

Using this and the expression relating p and $\mathbf{u}$ above, the momentum equation for elastic bodies presented in § 4.7 becomes the *Navier equation*:

$$\rho\frac{\partial^2\mathbf{u}}{\partial t^2} = \left(K + \tfrac{1}{3}\mu\right)\nabla_{\mathbf{X}}(\nabla_{\mathbf{X}} \bullet \mathbf{u}) + \mu\nabla^2_{\mathbf{X}}\mathbf{u}.$$

6.2 Fluid Rheology

Nearly all geophysical fluid flows occur with a small rate of change of volume; the predominant mode of motion is shearing. Consequently, in the rheological development presented in this section, we will neglect terms containing the factor $\nabla \bullet \mathbf{v}$.[5] Note that, with the neglect of $\nabla \bullet \mathbf{v}$, $\dot{\mathbb{E}}' = \dot{\mathbb{E}}$. Changes of volume will, however, remain important in conservation of mass, through the subtle variations of ρ that drive convective motions and in the energy balance of the mantle (e.g., see § 29.2).

The goal of this section is to express the deviatoric body force, $\mathbf{f}_D$, in terms of velocity gradients, using[6]

$$\mathbf{f}_D = (\nabla \bullet \mathbb{S}')/\rho \qquad \text{and} \qquad \nabla \bullet \dot{\mathbb{E}}' = \tfrac{1}{2}\nabla^2\mathbf{v}.$$

[4] See § 4.5.
[5] This precludes consideration of the bulk viscosity of a fluid.
[6] The first of these was introduced in § 4.7 and the second was developed in § 3.4.2.

This requires the development of a relationship between the rate-of strain tensor, $\dot{\mathbb{E}}'$, and the deviatoric stress tensor, $\mathbb{S}'$, which is related to the stress ($\mathbb{S}$) tensor by $\mathbb{S} = -p\mathbb{I} + \mathbb{S}'$.[7] We begin this task by considering the stress-rate of strain relation in the form $\dot{\mathbb{E}}' = f(\mathbb{S}')$. Noting that $\dot{\mathbb{E}}' = \emptyset$ when $\mathbb{S}' = \emptyset$, this expression may be expanded in a tensorial Taylor series starting with the linear term

$$\dot{\mathbb{E}}' = \sum_{j=1}^{\infty} A_j \mathbb{S}'^j ,$$

where the power of a tensor is defined by the recurrence relation[8] $\mathbb{X}^j = \mathbb{X} \bullet \mathbb{X}^{j-1}$ together with the starting relations $\mathbb{X}^1 = \mathbb{X}$ and $\mathbb{X}^0 = \mathbb{I}$. A remarkable theorem[9] in tensor analysis in essence states that the series may be terminated at the quadratic term; without loss of generality we may write

$$\dot{\mathbb{E}}' = A_1 \mathbb{S}' + A_2 \mathbb{S}'^2 .$$

If $A_2 \neq 0$, this constitutive relation gives rise to secondary flows. Observations of glaciers and lavas do not reveal any of these secondary flows, so an adequate constitutive relation appears to be

$$\dot{\mathbb{E}}' = A_1 \mathbb{S}' \qquad \text{or equivalently} \qquad \dot{\epsilon}'_{ij} = A_1 \tau_{ij} ,$$

where τ_{ij} are the elements of $\mathbb{S}'$. This expression states that the structure of the rate-of-strain tensor, $\dot{\mathbb{E}}'$, is identical to that of the deviatoric stress tensor, $\mathbb{S}'$. Note that the coefficient A_1 has SI units $\text{Pa}^{-1}\cdot\text{s}^{-1}$ – the inverse of the dynamic viscosity. The elements of the rate-of-strain tensor are related to the velocity components by

$$\dot{\epsilon}'_{ij} = \frac{1}{2}\left(\frac{\partial v_i}{\partial x_j} + \frac{\partial v_j}{\partial x_i}\right) - \frac{\partial v_k}{\partial x_k}\delta_{ij} .$$

The remaining issue is parameterization of the coefficient A_1; this is a function of the two non-zero scalar invariants of $\mathbb{S}'$,[10] plus the relevant thermodynamic variables, p and T. If A_1 depends on the third scalar invariant (the determinant) of $\mathbb{S}'$, the constitutive relation gives rise to secondary flows that are not observed in geophysical flows. Consequently, we shall assume that A_1 depends only on the second scalar invariant of $\mathbb{S}'$,[11] defined here as

$$\tau^2 \equiv \tau_{ij}\tau_{ij} .$$

In the following three subsections we quantify the rheological behavior of inviscid, Newtonian and non-Newtonian fluids.

[7] See § 4.5.
[8] The dot product of two tensors is defined in Appendix A.1.
[9] Called the *Cayley–Hamilton theorem*.
[10] The first scalar invariant is zero; See Appendix A.1.4.
[11] This is an alternate form that is more convenient to represent symbolically; see Appendix A.1.4.

6.2.1 Inviscid Fluid

An inviscid fluid is able to deform freely, without any accompanying stress. This fluid is characterized by $A_1 = \infty$, so that $\mathbb{S}' = \emptyset$ and the fluid momentum equation presented in § 4.7 becomes *Euler's equation*:

$$\frac{\mathrm{D}\mathbf{v}}{\mathrm{D}t} + 2\,\boldsymbol{\Omega} \times \mathbf{v} = \mathbf{g} - \frac{1}{\rho}\nabla p\,.$$

6.2.2 Newtonian Fluid

Setting $A_1 = 1/2\eta$, the stress to rate-of-strain relation becomes

$$\mathbb{S}' = 2\eta\dot{\mathbb{E}}',$$

where the molecular *dynamic viscosity* (also called the *shear viscosity*) η is a thermodynamic function of p and T. With η being variable and recalling[12] that $2\nabla \bullet \dot{\mathbb{E}}' = \nabla^2\mathbf{v}$,

$$\begin{aligned}\nabla \bullet \mathbb{S}' &= \eta\nabla^2\mathbf{v} + 2\nabla\eta \bullet \dot{\mathbb{E}}' \\ &= [\nabla \bullet (\eta\nabla)]\,\mathbf{v} + \frac{\partial\eta}{\partial x_j}\frac{\partial v_j}{\partial x_i}\mathbf{1}_i\,.\end{aligned}$$

The viscous term is important in thin shearing boundary layers having flow predominantly parallel to the layer with large spatial gradients in the normal (z) direction. With the relevant component of stress denoted by τ and the rate of strain expressed as $(1/2)\partial u/\partial z$, the stress – rate-of-strain relation is

$$\tau = \eta\frac{\partial u}{\partial z}\,.$$

For shearing flow, the last term in the expression for $\nabla \bullet \mathbb{S}'$ is small and we can safely ignore it. Further, the density does not vary significantly across thin layers. Now the fluid momentum equation presented in § 4.7 becomes the *Navier–Stokes equation*:

$$\frac{\mathrm{D}\mathbf{v}}{\mathrm{D}t} + 2\,\boldsymbol{\Omega} \times \mathbf{v} = \mathbf{g} - \frac{1}{\rho}\nabla p + [\nabla \bullet (\nu\nabla)]\,\mathbf{v}\,,$$

where $\mathbf{g}$ includes the centrifugal force and $\nu = \eta/\rho$ is the *kinematic viscosity*. Note that the kinematic viscosity has dimensions of length^2/time.

The Navier–Stokes equation, coupled with the continuity equation

$$\frac{\partial\rho}{\partial t} + \nabla \bullet (\rho\mathbf{v}) = 0\,,$$

[12] See § 3.4.2.

are the principal equations of *fluid dynamics*. As written above, the terms in the Navier–Stokes equation are accelerations. By multiplying this equation by ρ, each term becomes a force per unit volume. However, for either version of this equation, it is common practice to refer to its terms as forces.

When considering various waves and flows, each of the forces in the Navier–Stokes equation may be neglected or modified, as follows:

- The inertial term. This term may be linearized when considering small deviations from a known velocity. For example, writing $\mathbf{v} = \mathbf{v}_r + \mathbf{v}'$, where $\mathbf{v}_r$ is known and the perturbation velocity $\mathbf{v}'$ satisfies $\|\mathbf{v}'\| \ll \|\mathbf{v}_r\|$, then

$$\frac{\mathrm{D}\mathbf{v}}{\mathrm{D}t} \approx \frac{\partial \mathbf{v}'}{\partial t} + (\mathbf{v}_r \bullet \nabla)\mathbf{v}'.$$

  Often in the study of wave motions $\mathbf{v}_r$ is set equal to zero.
- The Coriolis term. The Coriolis force is unimportant for flows having a time scale shorter than a day, such as tornadoes, or flows that are laterally confined, such as rivers. Also, since the atmosphere and oceans are thin layers, only the vertical component of the rotation vector is dynamically important, and the magnitude of this component is not constant, but varies with latitude; see Chapter 8 and Part IV.
- The gravitational term. The gravitational vector may be expressed as the gradient of a scalar: $\mathbf{g} = -\nabla\psi$. In the study of barotropic motions, gravity plays a passive role, while in the study of baroclinic motions, this is an important term.
- The pressure term. In the study of barotropic motions, the pressure term may be expressed as the gradient of a scalar:

$$(1/\rho)\nabla p = \nabla P.$$

  and combined with gravity. However if ρ is a function of temperature T, then the spatial variations of ρ, in conjunction with the gravity and pressure terms, can play an important dynamic role.
- The viscous term.
  - The molecular kinematic viscosities of air and water are negligibly small. However, as we shall see in Chapter 23, the turbulent viscosity, parameterizing the effect of small-scale flows, is significantly larger and dynamically important.
  - If viscosity is constant, the viscous term simplifies to $\nu\nabla^2\mathbf{v}$.
  - The turbulent viscosity varies with position, as does the molecular viscosity of silicates.[13] Spatial variation of viscosity is not commonly included in the formulation of the Navier–Stokes equation but it is important in determining the structure of the turbulent Ekman layer (studied in Chapter 25), atmospheric boundary layers (§ 24.1 and Chapter 28), the flow of silicates in mantle convection (Chapter 32) and lava flows (§ 35.1).

[13] See § 7.2.5.

For creeping flows the inertial and Coriolis terms may be ignored, giving the *Stokes equation*:

$$\frac{\partial \mathbf{v}}{\partial t} = \mathbf{g} - \left(\frac{1}{\rho}\right) \nabla p + [\nabla \bullet (\nu \nabla)] \mathbf{v}.$$

The temporal term on the left-hand side is important for Stokes waves,[14] but otherwise may be ignored.

Representation of Turbulent Flow

Nearly all natural flows of air and water are turbulent. Turbulent flow is chaotic, with motions and eddies occurring on all spatial scales. These motions may be separated into a large-scale part (having non-zero mean flow) and a set of deviations having zero mean. These two parts interact dynamically, with the deviations exerting a force (per unit mass), denoted by $\mathbf{f}_T$, on the mean motions. We will see in § 23.5.2 that the turbulent body force can be crudely approximated by a form similar to the laminar term: $\mathbf{f}_T = [\nabla \bullet (\nu_T \nabla)] \mathbf{v}$, where the turbulent viscosity ν_T parameterizes the effect of small-scale isotropic turbulent motions. This stress supplants the viscous term in the Navier–Stokes equation, giving

$$\frac{\mathrm{D}\mathbf{v}}{\mathrm{D}t} + 2\boldsymbol{\Omega} \times \mathbf{v} = \mathbf{g} - \frac{1}{\rho} \nabla p + [\nabla \bullet (\nu_T \nabla)] \mathbf{v}.$$

The major challenge in quantifying turbulence is to develop an accurate parameterization of ν_T. This challenge is addressed in § 23.5.

6.2.3 Non-Newtonian Rheology

Two important geological materials, glacial ice and silicates, have rather complex rheologies, behaving as (nonlinear) Maxwell materials[15] – elastic on short time scales and fluid on long time scales. This behavior permits large masses of ice to flow slowly, even though ice is brittle, and similarly permits Earth's mantle to transmit short-period waves elastically and to deform plastically on geological time scales. Since the characteristic time scales of seismic waves and mantle convection are widely separated and elastic waves in glaciers are unimportant, it is not necessary to develop a rheological model that combines both behaviors. That is, elastic waves in the mantle are adequately described by the linear elastic model presented in § 6.1. In this section we will focus on *sub-solidus* deformation and flow.[16]

Viscoelastic models of glaciers and silicates are characterized by two parameters: the shear modulus μ and the viscosity η. The shear modulus is equivalent to the elasticity of the spring in the simple rheological models investigated in Appendix D.1. Note that the SI

[14] Investigated in Appendix F.3.
[15] See Appendix D.1.2.
[16] Sub-solidus means that there no liquid phase is present.

units of μ and η are Pa and Pa·s, respectively, and the ratio η/μ has dimensions of time. The shear modulus for the mantle is known from seismic inversion; see Figure 7.2. As we shall see, silicate viscosity is a strong function of temperature.

Guided by the linear and nonlinear Maxwell models presented in appendices D.1.2 and D.1.2, we may write $A_1 = \alpha/2\eta$ and express the constitutive relation ($\dot{\epsilon}'_{ij} = A_1 \tau_{ij}$) developed earlier in this section as

$$\dot{\epsilon}'_{ij} = \frac{\alpha}{2\eta}\tau_{ij} \qquad \text{or simply} \qquad \dot{\epsilon} = \frac{\alpha}{2\eta}\tau\,,$$

where α is a dimensionless function of τ^2/μ^2. If $\alpha = 1$, the fluid is Newtonian; if it is variable, the fluid is non-Newtonian.

Non-Newtonian fluids are broadly separated into two classes depending whether the rate of strain increases faster or slower than linear with increasing stress, that is, whether the graph of rate of strain versus stress is concave up or down, as illustrated in Figure 6.1. Materials for which $\dot{\epsilon}$ increases slower than linear with τ are said to be *shear thickening*; these are illustrated schematically in Figure 6.1 by the dotted curve. Liquids containing long-chain polymers behave in this manner. Materials for which $\dot{\epsilon}$ increases faster than linear with τ are said to be *shear thinning*; these are illustrated schematically in Figure 6.1 by the solid curve. Crustal material behaves in this way, with fault zones being particularly weak. Shear-thinning materials commonly have a history-dependent rheology. Examples

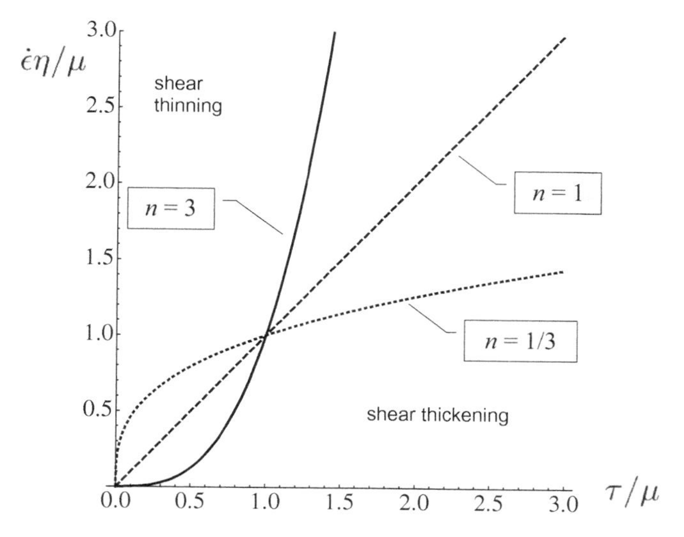

Figure 6.1 Illustration of the dimensionless rate of strain $\dot{\epsilon}\eta/\mu$ versus dimensionless stress τ/μ for Newtonian (dashed), shear thinning (solid) and shear thickening (dotted) fluid rheologies. Ice and silicates are shear thinning materials.

include gelatin, latex paint, clays, glacial ice, mantle silicates and magmas. Note that shear-thinning materials deform less than Newtonian for low levels of stress.

The function α must be analytic. As noted above, its simplest possible form is $\alpha = 1$ and the material is a Newtonian fluid. If we expand α in powers of τ^2/μ^2 and truncate at the linear term, the constitutive relation is

$$\dot{\epsilon}'_{ij} = \frac{\tau_{ij}}{2\eta}\left(1 + \alpha_3 \frac{\tau^2}{\mu^2}\right).$$

This is a shear-thinning form of *power-law creep*. Often the linear term negligibly small and creep is modeled by a general power:

$$\dot{\epsilon}'_{ij} = \frac{1}{2\eta}\left(\frac{\tau}{\mu}\right)^{n-1}\tau_{ij} \qquad \text{or simply} \qquad \dot{\epsilon} = \frac{\tau^n}{2\eta\mu^{n-1}},$$

where n is a positive number. This is referred to as *Glen's flow law* in glaciology and *Norton's law* in metallurgy. For glaciers estimates for n lie in the range $1.9 < n < 4.5$, but we have seen that ideally n should be an odd integer.

We have derived an expression for the rate of strain as a nonlinear function of stress. This will be more useful if it were inverted to give stress as a nonlinear function of rate of strain. In order to invert this, we may introduce the mean rate of strain:

$$\dot{\epsilon} \equiv \sqrt{\dot{\epsilon}'_{ij}\dot{\epsilon}'_{ij}} = \sqrt{\frac{1}{4\eta^2}\left(\frac{\tau}{\mu}\right)^{2n-2}\tau_{ij}\tau_{ij}} = \frac{\mu}{2\eta}\left(\frac{\tau}{\mu}\right)^n.$$

Using this we have

$$\tau_{ij} = 2\eta\,(2\eta\dot{\epsilon}/\mu)^{(1-n)/n}\,\dot{\epsilon}'_{ij}.$$

If the motion is simple shear then[17]

$$\dot{\epsilon}'_{ij} = \frac{1}{2}\frac{\partial v_i}{\partial x_j}$$

and

$$\tau_{ij} = \mu\left(\frac{\eta}{\mu}\frac{\partial v_i}{\partial x_j}\right)^{1/n}.$$

In taking the root of the rate of strain, it is important to select the solution having the same sign as the stress.

[17] See § 3.4.2.

6.2.4 Thermal Activation of Creep

Many diffusive processes, including creep, are *thermally activated*, with the viscosity η governed by the *Arrhenius equation*:[18]

$$\eta(p,T) = \eta_0 e^{-\beta/\upsilon_H}, \qquad \text{where} \qquad \upsilon_H = T/T_m(p)$$

is the *homologous temperature*, $T_m(p)$ is the melting temperature,[19] β is a dimensionless constant[20] and a subscript 0 denotes a constant value. The magnitude of β is a bit uncertain; a plausible range of estimates is from 25 to 30. The large value of β means that creep is strongly dependent on temperature. For example, with $\beta \approx 27$, $T \approx 3700$ K and $T_m \approx 4100$ K, say, the viscosity of silicate material changes by a factor of e if the temperature changes by ≈ 125 K.

If T has a maximum temperature T_0 at location $x = 0$, and the spatial variation of temperature near this location is given by $T = T_0(1 - (x/x_0)^\gamma)$, say (with $\gamma = 1$ or 2), then the viscosity increases exponentially with x for $|x| \ll x_0$: $\eta(p,T) \approx \eta_1 \exp(\beta_1 (x/x_0)^\gamma)$ where $\beta_1 = \beta T_0/T_m$ and $\eta_1 = \eta_0 \exp(-\beta T_0/T_m)$.

6.3 Comments on the Parameters and Forces

Each of the constitutive parameters occurring in elastic and fluid stress–strain relations is a thermodynamic parameter that – in general – varies with temperature and pressure. In most applications, these parameters are treated as constant. The second law of thermodynamics requires these parameters to be non-negative.

The forces associated with the elastic constitutive parameters are conservative; they permit the propagation of waves without dissipative loss. However, the fluid constitutive parameter (viscosity) is inherently dissipative and leads to damping of waves.

As noted in § 2.2, waves require a restoring force that pulls the displaced particles back toward equilibrium. In elastic bodies, this force is provided by pressure or the elastic stresses. In fluid bodies, there is a much larger variety of possible restoring forces. The most obvious fluid restoring force is the pressure (fluids lack elastic stresses), producing sound waves.

6.4 Mixtures

While most geophysical systems have a single phase, there are several important and interesting systems that are two-phase mixtures of gas and liquid (e.g., clouds, fog and rain), gas and solid (e.g., snow, hail, avalanches, pyroclastic flows and sand storms) or liquid and solid (e.g., turbidity currents). On the small scale (smaller than a parcel), the two phases of a mixture are distinct and separate. (This is in contrast to a solution, in which the

[18] See appendices D.2.2 and D.2.3.
[19] See Appendix E.10.2.
[20] Unrelated to the compressibility introduced in § 5.2.

two constituents are intermingled down to the molecular level.) However, on a large scale (parcel size or larger), a mixture may be treated as a continuum composed of two phases.

Mixtures may be in differing physical configurations, depending on the geometry of the phases. The simplest configuration consists of small isolated (fluid or solid) particles of phase p (p for particle) embedded in a fluid continuum of phase f (f for fluid); this is a *slurry*. Unless the particles are very small or the densities of the phases are nearly identical, our model of the mixture will need to account for the sedimentation (vertical drift) of the particles. Sedimenting particles aggregate at a boundary, increasing the concentration of p. If p is fluid, the particles coalesce (rain into puddles), but if it is solid, they accumulate as particles (snow on the ground). If the concentration of particles is sufficiently large that they touch, then they no longer drift vertically, but now act as a *fluidized bed* of particles of p immersed in a continuum of f, provided that they do not *sinter* (spontaneously weld together). If the particles do sinter, then the mixture consists of a rigid matrix of p with f filling the voids: a *porous matrix*, dynamically similar to a ground-water aquifer. These different configurations require differing mathematical models and the domain of the models may change with time as the system evolves.

The composition of a mixture varies with position $\mathbf{x}$ and time t due to sedimentation and also due to change of phase. That is, phase p might have condensed from f (think of cloud formation), with the amount of that phase depending on value of the thermodynamic variable pressure and temperature.[21] If the solid particles of the fluidized bed or porous medium are allowed to slowly deform over time, the mixture may undergo *compaction*, with a slow separation of the phases of the mixture.

6.4.1 Mass and Volume Fractions

The relative amounts of phases f and p in a binary mixture may be quantified using either volume fractions, denoted by Φ, or mass fractions, denoted by ξ. That is, suppose we have a parcel having volume $V = V_f + V_p$ and mass $M = M_f + M_p$, where V_f is the volume occupied by phase f, etc. The volume and mass fractions of particles are defined by

$$\Phi \equiv V_p/V \qquad \text{and} \qquad \xi \equiv M_p/M\,,$$

with $V_f = (1-\Phi)V$ and $M_f = (1-\xi)M$. Denoting intrinsic densities by ρ with suitable subscripts, we have $M_f = \rho_f V_f$ and $M_p = \rho_p V_p$. Putting these formulas together, the density of the mixture is given by

$$\frac{1}{\rho} = \frac{1-\xi}{\rho_f} + \frac{\xi}{\rho_p} \qquad \text{or} \qquad \rho = (1-\Phi)\rho_f + \Phi\rho_p\,.$$

Eliminating ρ between these two formulas, we have ξ in terms of Φ and vice versa:

$$\xi = \frac{\rho_p \Phi}{(1-\Phi)\rho_f + \Phi\rho_p} \qquad \text{and} \qquad \Phi = \frac{\rho_f \xi}{(1-\xi)\rho_p + \xi\rho_f}\,.$$

[21] See Appendix E.10.1.

The *reduced densities* of the phases (denoted by a tilde) are defined by

$$\tilde{\rho}_f \equiv (1-\xi)\rho_f \qquad \text{and} \qquad \tilde{\rho}_p \equiv \xi\rho_p .$$

Using these, the mixture can be treated as two overlapping continua, with $\rho = \tilde{\rho}_f + \tilde{\rho}_p$.

The physical configuration of the mixture is determined by the magnitude of Φ relative to the *packing fraction*,[22] Φ_{max}. If $\Phi < \Phi_{max}$, the mixture is a suspension of particles, while if $\Phi = \Phi_{max}$ it is a fluidized bed or porous medium. It is not possible to achieve $\Phi > \Phi_{max}$ without deforming the particles. If the particles are spheres randomly packed, $\Phi_{max} \approx 0.6$.

In general, the intrinsic densities are functions of p and T and vary with position and time. However, it is common to assume that the density of the particles is constant. We will limit our attention to binary mixtures having dense particles: $\rho_f < \rho_p$. Commonly, the particles are relatively very dense, so that ξ may be unit order or large even though $\Phi \ll 1$.

6.4.2 Sedimentation

If the particles of phase p are sufficiently dilute and small, sedimentation speed w_s (relative to the fluid phase) of an individual spherical particle of radius R is given by *Stokes' formula*:

$$w_s = \frac{2gR^2(\rho_p - \rho_f)}{9\eta},$$

where g is the acceleration of gravity and η is the viscosity of phase f. With $\rho_f < \rho_p$, the sedimentation velocity is downward. This formula is reasonably accurate provided the Reynolds number of the flow, $\text{Re} \equiv w_s R\rho_f/\eta$, is sufficiently small (roughly $\text{Re} < 500$). If the Reynolds number is greater than this, the sedimentation speed is found from a balance of the buoyancy force, $(4/3)\pi R^3 g(\rho_p - \rho_f)$ and the turbulent drag force[23] $4\pi R^2 \rho_f C_D w_s^2$, where C_D is a dimensionless coefficient[24] depending on the shape of the particle:

$$w_s = \sqrt{\frac{gR(\rho_p - \rho_f)}{3C_D\rho_f}}.$$

For a spherical particle $C_D \approx 1/2$.

As an array of particles sediments at low Reynolds number, each particle creates a flow field[25] that may impinge on its neighbors, affecting their rate of sedimentation. This is a tricky issue (e.g., see Guazzelli and Hinch, 2011). If the particles are modeled as a homogeneous random array of spheres (Batchelor, 1972), the mean sedimentation speed of the spheres is smaller[26] than that for an isolated sphere: $w_s(1 - 6.55\Phi)$. However, the sedimentation speed of a cluster of spheres is larger than that for an isolated sphere.[27] This may explain why heavy rainfall is not uniform, but instead falls in sheets.

[22] The value of Φ at which particles touch each other.
[23] See § 23.2.
[24] See § 11.5.1 and § 23.7.
[25] The flow field of a sphere of radius R sedimenting at speed w_s decays with radial distance r as $w_s R/r$.
[26] This is called *hindered settling*.
[27] Experiments show that the sedimentation speeds of arrays of particles tend correlate within 20 inter-particle separations, so that a particle affects thousands of its neighbors.

6.4.3 Mixture Mass Conservation

The continuity equation for the mixture as a whole is

$$\frac{\partial \rho}{\partial t} + \nabla \bullet (\rho \mathbf{v}) = 0 .$$

We may treat the mixture as two inter-penetrating continua with $\rho = \tilde{\rho}_f + \tilde{\rho}_p$ and $\rho \mathbf{v} = \tilde{\rho}_f \mathbf{v}_f + \tilde{\rho}_p \mathbf{v}_p$ and split the continuity equation in two, writing

$$\frac{\partial \tilde{\rho}_f}{\partial t} + \nabla \bullet (\tilde{\rho}_f \mathbf{v}_f) = -\rho_p s_\Phi \qquad \text{and} \qquad \frac{\partial \tilde{\rho}_p}{\partial t} + \nabla \bullet (\tilde{\rho}_p \mathbf{v}_p) = \rho_p s_\Phi ,$$

where $\rho_p s_\Phi$ is a source term quantifying the creation of particles from the fluid phase, due to change of phase. Note that these equations have the form of the general conservation law.[28]

Using $\tilde{\rho}_p = \Phi \rho_p$ and assuming that the particle density is constant, the particle conservation equation becomes

$$\frac{\partial \Phi}{\partial t} + \nabla \bullet (\Phi \mathbf{v}_p) = s_\Phi .$$

The particle velocity is related to the (macroscopic) fluid velocity by $\mathbf{v}_p = \mathbf{v}_f - w_s \mathbf{1}_z$, where $\mathbf{1}_z$ is the upward unit vector and w_s is the (downward) sedimentation speed. If the flow is turbulent, there is an additional small-scale flow,[29] the effect of which may be parameterized by an eddy diffusivity κ_Φ. Altogether,

$$\frac{\partial \Phi}{\partial t} + \nabla \bullet (\Phi \mathbf{v}_f) = s_\Phi + \frac{\partial}{\partial z}(\Phi w_s) + \nabla \bullet (\kappa_\Phi \nabla \Phi) .$$

Transition

Rheology provides the coupling between kinematics and dynamics, and leads to more specific forms of the momentum equations for elastic and fluid bodies. In this chapter we have developed constitutive equations governing the rheological behavior of linear elastic solids, linear viscous (Newtonian) fluids and nonlinear power-law-creep fluids. This provides us with the tools to quantify and analyze waves and flows. However, before diving into that, in the next chapter we will investigate the spatial structure of the atmosphere, oceans and mantle and the equations governing motions that are a perturbation on the static state.

[28] See Appendix D.5.
[29] See § 23.3.

7

Static State and Perturbations

Planetary atmospheres, oceans and interiors are dominantly in a state of hydrostatic equilibrium, with most dynamic processes being perturbations of that state. With this in mind, in this chapter, we will separate the elastic and fluid governing equations into static balance and perturbation equations. Static reference-state variables are indicated by a subscript r while perturbations have a prime. Note that reference-state variables often are functions of position – particularly spherical radius or elevation. We begin in § 7.1 with a presentation of the equations governing a static state. Next, in § 7.2 we briefly discuss the structures of the atmosphere, oceans and mantle. Then in § 7.3 we develop a set of equations quantifying small departures from the static state.

7.1 Static State

The static state of an elastic solid is simply the reference state with $\mathbf{u} = \mathbf{0}$ and $\mathbb{S}_r = -p_r\mathbb{I}$, while the static state of a fluid body is $\mathbf{v} = \mathbf{0}$ and again $\mathbb{S}_r = -p_r\mathbb{I}$. The equation of conservation of mass is identically satisfied in the static state for both elastic and fluid bodies, while the momentum equation for both is simply the *hydrostatic balance*

$$\nabla p_r = \rho_r \mathbf{g} \qquad \text{or equivalently} \qquad \frac{\mathrm{d}p_r}{\mathrm{d}z} = -\rho_r g,$$

where $g = \|\mathbf{g}\|$ and z is the upward coordinate.

Thermodynamics tells us that the density is a function of the pressure, temperature and composition.[1] We will ignore compositional effects for the time being and consider $\rho_r = f(p_r, T)$. A necessary condition for a fluid body to be in a static state is obtained by taking the curl of the momentum equation:

$$\nabla p_r \times \nabla \rho_r = \mathbf{0} \qquad \text{or equivalently} \qquad \nabla p_r \times \nabla T = \mathbf{0}.$$

Isothermal and isobaric surfaces must coincide in a static (barotropic) state. This is possible in a non-rotating, self-gravitating body, but if the body is rotating and density depends on temperature, a static state is not possible; gravity requires elliptic isobars, while thermal

[1] See § 5.2 and Appendix E.5.

conduction still produces spherical isotherms. This situation occurs, for example, in stellar interiors; they are in continual Eddington-Sweet motion.[2]

With the functional relation between ρ_r and p_r known, it is fairly straightforward to determine the static state of the atmosphere, oceans and Earth's interior. These structures are presented and discussed in the following section.

7.2 Static Structure

In the following three subsections, we briefly describe the static structure of the oceans, atmosphere and mantle.

7.2.1 Structure of Earth's Atmosphere

Earth's atmosphere is divided into distinct dynamic layers, with the *troposphere* extending from the ground to an elevation of roughly 10 km. Above the troposphere in succession are the *stratosphere*, *mesosphere* and *thermosphere*. We will focus primarily on the troposphere, but we need to begin by discussing the stability of the stratosphere.

The stratosphere is a stably stratified layer of the atmosphere, with temperature increasing with altitude. This thermal structure results from a balance between solar heating, primarily at short wavelengths, and radiative cooling, primarily at longer wavelengths. All other things being equal, a similar thermal structure would prevail in the troposphere due to this thermal radiative balance. That is, if it were left alone the troposphere would be a quiet stable layer. But it is not; convective motions driven by surface solar heating and evaporation of water churn the troposphere into a nearly adiabatic state, in a process called *penetrative convection*. The thickness of the troposphere is determined by the mean height to which convective clouds can rise through an atmosphere that wants to be stable; this height is called the *tropopause*. The stable stratification of the stratosphere causes the tops of strong convective clouds to splay horizontally upon reaching the tropopause, creating anvil clouds.

Earth's atmosphere consists primarily of the diatomic gases nitrogen and oxygen, with small amounts of other gases. The atmospheric gas that is dynamically important is water vapor. As a rising column of air cools adiabatically, water vapor condenses and releases latent heat that warms and expands the surrounding air, reinforcing the upward motion.

Earth's atmosphere behaves roughly as an ideal diatomic gas[3] having $\alpha T = 1$ and $c_p = 7R_s/2$. In this case the adiabat simplifies to[4]

$$\frac{\mathrm{d}T}{\mathrm{d}z} = -\frac{2g}{7R_s} \approx -9.8\frac{\mathrm{K}}{\mathrm{km}}.$$

[2] The von Zeipel paradox states that a star cannot be both rotating rigidly and in hydrostatic equilibrium. The resolution is that the fluid is baroclinic and has meridional motions.

[3] See Appendix E.8.5.

[4] Recall that $R_s = 287\ \mathrm{J{\cdot}kg^{-1}{\cdot}K^{-1}}$; see Appendix E.9.2 and Table B.5.

This is the *dry adiabat*. It quantifies the rate that a rising parcel cools and a sinking parcel warms. This is why it is cold on mountain tops and why the Santa Ana winds, which blow down from the high deserts to Los Angeles, are so warm.

Actually, this is only part of the story. The dry adiabat occurs only when there is no condensation of water vapor. As noted above, water vapor condenses as a parcel of moist air rises, releasing its latent heat of vaporization and reducing the rate that temperature decreases with elevation. This reduced rate of decrease is called the wet or *moist adiabat*.[5] Since the rate of cooling with elevation is less for the moist adiabat than for the dry, the moist adiabat leads to increased temperatures at high elevations.

The actual rate that temperature decreases with height in the troposphere, called the *mean lapse rate*, is a weighted average of the moist and dry adiabats:[6]

$$\left.\frac{\mathrm{d}T}{\mathrm{d}z}\right|_{mean} = -\gamma_a \approx -6.5\times 10^{-3}\,\frac{\mathrm{K}}{\mathrm{m}} = -6.5\,\frac{\mathrm{K}}{\mathrm{km}}\,.$$

This mean lapse rate describes an atmosphere that is thermally stably stratified (if vertical displacements do not involve condensation of water vapor), and capable of sustaining internal waves; see § 14.3. The lapse rate is fairly uniform throughout the troposphere, permitting integration of the lapse-rate equation to yield

$$T \approx T_a - \gamma_a z = T_a(1 - z/z_T)\,, \qquad \text{where} \qquad z_T = T_a/\gamma_a$$

is the thermal scale height. The ground-level temperature T_a may be a function of horizontal position. Horizontal variations of temperature are related to vertical variations in the geostrophic flow, known as the thermal wind.[7]

We can substitute this known thermal structure and the ideal gas law[8] $p = \rho R_s T$, where R_s is the specific gas constant for the atmosphere (including moisture), into the hydrostatic differential $\mathrm{d}p = -g\rho\,\mathrm{d}z$ to obtain an equation for the variation of pressure with elevation:

$$\mathrm{d}p = -\frac{gp}{R_s T_a(1 - z/z_T)}\mathrm{d}z\,.$$

This equation is readily integrated:

$$p = p_a\left(1 - \frac{z}{z_T}\right)^{\beta}, \qquad \text{where} \qquad \beta = \frac{g}{R_s\gamma_a}\,.$$

Typically $T_a \approx 288$ K and $p_a = 1$ atm $\approx 10^5$ Pa. Accounting for roughly 2% vapor content by mass[9] $R_s \approx 290\,\mathrm{J{\cdot}kg^{-1}{\cdot}K^{-1}}$; with $g = 9.8\,\mathrm{m^2{\cdot}s^{-2}}$, and $\gamma_a = 6.5\times 10^{-3}\,\mathrm{K{\cdot}m^{-1}}$, $z_T \approx 44$ km and $\beta \approx 5.2$. Finally, we may substitute the equations for p and T into the ideal gas law

[5] See Appendix E.9.4.
[6] A subscript a has been added to γ to distinguish it from the ratio of specific heats.
[7] See § 8.5 and § 26.2.
[8] See Appendix E.8.3.
[9] See Appendix E.9.3.

to obtain the density structure:

$$\rho = \rho_a (1 - z/z_T)^{\beta - 1} ,$$

where ρ_a is the density of air at sea level.

These formulas are valid within the troposphere. With the tropopause height varying between 9 km at the poles to 17 km at the equator (on average) and with $z_T \approx 44$ km, the factor z/z_T does not exceed 0.4. The pressure scale height z_p is defined as the elevation at which the pressure is equal to $1/e$ times that at sea level. Using the pressure formula given above,

$$z_p = z_T \left(1 - \beta^{-1/(\beta-1)}\right),$$

with $\beta \approx 5.2$, $z_p \approx z_T/3 \approx 14$ km. Note that this value is somewhat larger than that calculated assuming an isothermal atmosphere.

According to these formulas, atop Mount Everest, at an elevation of 8848 m (with $z/z_T \approx 0.2$), the temperature is 57 degrees Celsius lower than at sea level, the air pressure is 1/3 atmospheric and the density is 40 percent of that at sea level. And often the mountain top is in the midst of the jet stream[10]; it's not an ideal vacation spot.

7.2.2 Structure of the Oceans

If they had been left alone, the oceans would have settled into a stable state with density increasing with depth due both to a temperature gradient (warm on top and cold at the bottom) and a salinity gradient (fresh on top and salty at the bottom). Actually, most of the ocean is in such a state. This static state is disrupted primarily due to the action of wind stress at the surface. This stress raises waves, which often break. As a wave breaks, the plunging crest creates strong turbulence that acts to mix the waters immediately beneath the surface, creating a *well-mixed top layer* above a *thermocline*, where the state of the ocean rapidly changes from dynamic (above) to static (below).

It is interesting to note the similarity between the structure of the atmosphere and oceans. Both would be stably stratified if it were not for the disturbances introduced at Earth's surface and both have well-mixed layers close the surface.

Mixed Layer

The temperature and salinity of water within the mixed layer at the top of the oceans are nearly uniform due to vigorous turbulent motions induced within the layer, principally by the breaking of surface waves. This layer largely overlaps and coincides with the surface Ekman layer (see § 25.5.4). The motions within the mixed layer are a form of *penetrative convection* – convective motions driven by a source of energy at a boundary that invade and mix a fluid that wants to be stable and stratified.

[10] See § 26.3.

The thickness of the mixed layer is affected by the amount of mixing produced by surface wave action and by the strength of solar heating. The mixed layer tends to be deeper where wind stresses and wave heights are larger and shallower where there is strong surface heating.

Thermocline

The *thermocline*, also called the *pycnocline* or *halocline*, is a layer below the mixed layer, separating it from the stratified deep oceans beneath. The thermocline typically has a sharp top and rather indistinct bottom. It exists because the static state of the ocean is disrupted within the mixed layer and the well-mixed conditions (just above the thermocline) differ from the static conditions (just below). The thickness of the thermocline is determined by a balance between the action of thermal diffusion, which tends to thicken the layer, and convective motions, which tend to erode and thin the layer.

The strength of the thermocline is dictated in large part by the temperature of the mixed layer. Since the mixed layer is heated by solar insolation, the thermocline is stronger in the tropics and in summer months and is typically absent in polar regions. When there is no temperature contrast between surface and deep waters, vertical currents, from the surface to great depth, can be driven by salinity differences, as discussed in § 27.4.

Deep Ocean

The portion of the oceans lying beneath the thermocline, which constitutes the vast majority, is called the *deep ocean*. Water within the deep ocean has (in comparison to that within and above the thermocline) uniform temperature (about 3.6°C) and salinity (34.7 parts per thousand by weight or equivalently $\xi = 0.0347$). However, there are small but dynamically significant departures from uniformity, primarily in the vertical direction.

The deep ocean has a stabilizing (cold below warm) thermal gradient of roughly 5×10^{-3} K·m^{-1} and a stabilizing (salty below fresh) salinity gradient of roughly 6×10^{-7} m^{-1} by weight. Together these produce a density gradient of roughly 6×10^{-5} kg·m^{-4}. This stably stratified fluid can sustain internal gravity waves having periods of roughly several hours.[11]

7.2.3 Ocean Depths

The oceans sit directly atop the mantle, so that their depths are determined primarily by the hydrostatic balance of the uppermost mantle. Earth's surface is in *plate-tectonic* motion,[12] with oceanic lithospheric plates moving rigidly from *spreading centers* at *mid-ocean ridges* to oceanic trenches, where they dip downward into the mantle. These plates are composed of initially hot material that has welled up near the mid-ocean ridges. As the material moves away from the ridges, it cools, solidifies, thickens and subsides. These processes are modeled in this section.

[11] See § 14.3.2.
[12] See § 30.1.

To model the ocean depths, let's begin by considering a static layer of material having thermal diffusivity[13] κ occupying the half-space $z < 0$. Initially the material is at a constant temperature T_I. At time $t = 0$, a lower temperature $T_I - \Delta T$ is applied at the top surface.[14] The variation of temperature in the layer obeys the one-dimensional heat equation[15]

$$\frac{\partial T}{\partial t} = \kappa \frac{\partial^2 T}{\partial z^2} .$$

The solution of this equation satisfying the initial and boundary conditions is

$$T = T_I - \Delta T - \Delta T \operatorname{erf}(z/\sqrt{\kappa t}) ,$$

where the *error function*,[16]

$$\operatorname{erf}(\zeta) \equiv \frac{2}{\sqrt{\pi}} \int_0^{\zeta} e^{-\grave{\zeta}^2} \, \mathrm{d}\grave{\zeta} ,$$

is a monotonic increasing function with $\operatorname{erf}(-\infty) = -1$, $\operatorname{erf}(0) = 0$ and $\operatorname{erf}(\infty) = 1$. For Earth's mantle $T_I \approx 1400$ K and $\Delta T \approx 1100$ K.

The amount of heat $Q(t)$ removed from a vertical column of the layer having horizontal area A is given by

$$Q = \rho c_p A \Delta T \Delta z ,$$

where

$$\Delta z = \int_0^{\infty} \left(1 - \operatorname{erf}(z/\sqrt{\kappa t})\right) \mathrm{d}z = \sqrt{\kappa t/\pi}$$

is a measure of the thickness of the cold portion of the layer, now identified as the *oceanic lithospheric*. Lithosphere at temperature $T_I - \Delta T$ has density $\rho + \Delta\rho$, where

$$\Delta\rho = \alpha \rho \Delta T$$

and α is the coefficient of thermal expansion. As the lithosphere cools, it contracts, with its surface subsiding a distance $z_w(t)$. As the lithosphere subsides, the ocean above, having density ρ_w, deepens by the same amount, z_w, which may be determined by requiring a hydrostatic balance:[17]

$$\Delta\rho \Delta z = (\rho - \rho_w) z_w ;$$

[13] $\kappa = k/(\rho c_p)$ where k is the thermal conductivity, ρ is density and c_p is the specific heat at constant pressure; see Appendix E.7.

[14] This temperature, close to the freezing temperature of water, is determined by the nature of deep-ocean currents; see § 27.4.

[15] See Appendix D.5.

[16] Here and below, a dummy variable of integration is denoted by a "grave" accent.

[17] Hydrostatic balance is assured because the cold lithosphere is bounded above by a fluid ocean and below by a relatively fluid layer called the *asthenosphere*; see Figure 7.1.

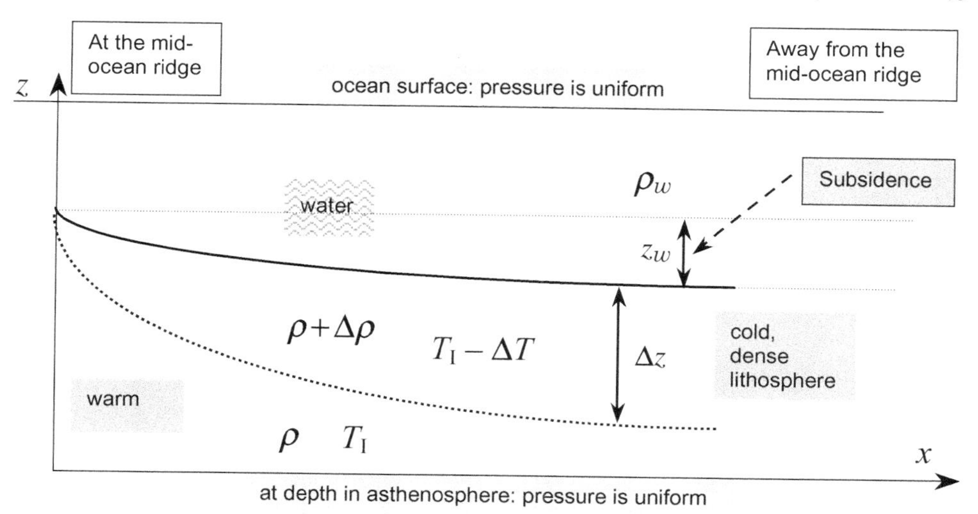

Figure 7.1 A schematic of the oceanic lithosphere cooling and subsiding as it moves away from a mid-ocean ridge.

that is,

$$z_w = \frac{\rho}{\rho - \rho_w} z_c, \qquad \text{where} \qquad z_c = \alpha \Delta T \sqrt{\frac{\kappa t}{\pi}}$$

is the vertical contraction of the cooled lithosphere. Note that increase in ocean depth is greater than the vertical contraction of the lithosphere because the weight of the added water magnifies lithospheric subsidence.

Although the oceanic lithosphere is stationary when viewed by an observer sitting on the ocean bottom, an observer sitting on the mid-ocean ridge sees this lithospheric plate moving away with speed v, cooling as it goes. Distance x from the mid-ocean ridge is related to elapsed time t by $x = vt$ and the mid-ocean observer sees a subsidence of the ocean floor given by

$$z_w = \alpha \Delta T \left(\frac{\rho}{\rho - \rho_w} \right) \sqrt{\frac{\kappa x}{\pi v}}$$

and measures a heat flux from the ocean floor decreasing with distance from the ridge proportional to $1/\sqrt{x}$. Lithospheric plates continue to thicken, cool and subside until they encounter a subduction zone, where their motion changes from predominantly horizontal to predominantly downward. After it turns the corner at a subduction zone and descends into the mantle, a lithospheric plate is called a *slab*.

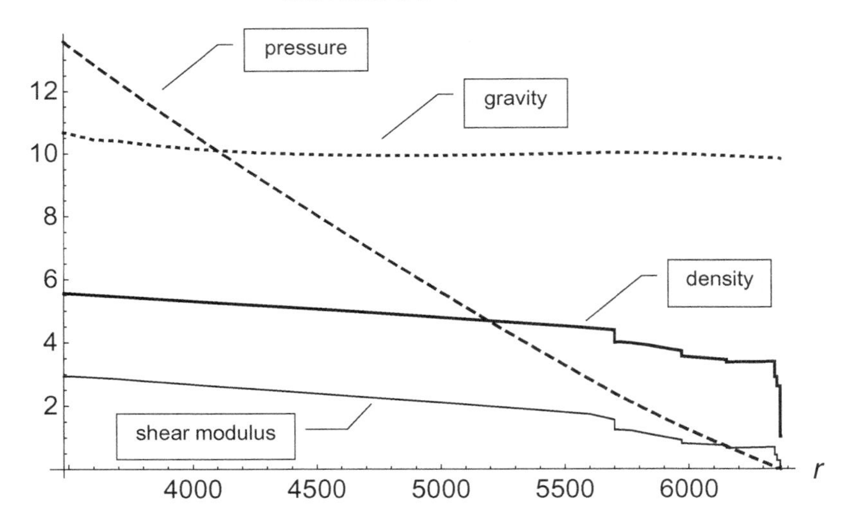

Figure 7.2 Plots of density (thick solid), pressure (dashed), gravity (dotted) and shear modulus (solid) versus radius within Earth's mantle, which extends from the core–mantle boundary at $r = 3480$ km to close to Earth's surface at $r = 6371$ km. Radius is in km, density in 10^3 kg·m^{-3}, pressure in Mbar (= 10^{11} Pa), gravity in m·s^{-2} and shear modulus in Mbar.

Mantle Cooling

Actually, we have been describing an important component of mantle flow and thermal evolution; the cold oceanic lithospheric plates are in fact the top boundary layer of a mantle that is convecting primarily because it is being cooled at the top. The oceanic lithosphere is colder and more dense than the underlying mantle. This increase in density is seen in Figure 7.2 as a leveling of the density profile just below the crust. This "blip" is actually a very important dynamic feature. The presence of denser material atop lighter is a dynamically unstable situation and the gravitational potential energy released as the cold slabs descend into the mantle at subduction zones drives mantle convection. The downward advection of cold slabs acts (on a geological time scale) to cool the bulk of the mantle; see Chapter 31.

The oceanic lithospheric plates are sufficiently cold that they act as elastic bodies, called *tectonic plates*.[18] This elasticity aids the plate-tectonic mode of mantle convection by transmitting the downward pull of the descending slabs to the horizontal oceanic lithosphere. Without this continuity of stress, the oceanic lithospheric plates would not move systematically toward the subduction zones. Perversely, this elasticity is also an impediment to plate-tectonic convection; there is a tendency for the tectonic plates to weld together and resist subduction.

[18] The total number of plates is somewhat uncertain, because the smaller ones are difficult to identify. By one count (Argus et al., 2011) there are 56 lithospheric plates.

In this subsection we have quantified the thermal and elastic structure of the top-most layer of the mantle (the lithosphere). In the following two sections, we will be considering the thermal and rheological state of the remainder of the mantle.

7.2.4 Structure of Earth's Mantle

Our knowledge of Earth's interior comes primarily from the analysis of seismic waves that travel through the mantle and core. This knowledge is summarized in the earth model known as PREM (Dziewonski and Anderson, 1981). Figure 7.2 shows the variation of density, pressure, gravity and shear modulus within the mantle as given by PREM. This graph reveals several interesting features of state of the mantle:

- the local acceleration of gravity within the mantle is nearly constant;
- pressure increases approximately linearly with depth;
- the density has discontinuities – due to phase transitions – at depths of 220, 400 and 660 km depth;
- apart from the discontinuities, density increases approximately linearly with depth in the mantle; and
- there is a small, but dynamically important, increase in density at the top of the mantle (this has been discussed in § 7.2.3).

A distinct layer (not evident in Figure 7.2), called the D″ (say "dee double prime") layer, occurs at the bottom of the mantle; this is likely a thermal boundary layer; see § 32.1.

The sharp decreases in density near Earth's surface represent an (average) ocean 3 km deep with density 1.02×10^3 kg·m^{-3} and a crust 21.4 km deep with density varying from 2.6 to 2.9 $\times 10^3$ kg·m^{-3}. (Note that this simple globally averaged model has the ocean sitting atop the continents, while in reality, oceans and continents sit side by side.) The continental crust is the buoyant residue of partial melting and fractionation that has accumulated over geological time.[19] This is in sharp dynamic contrast to the oceanic crust, which is the cold top region of a convecting mantle.

Mantle Thermal Structure

Our knowledge of the thermal structure of the mantle is summarized in Figure 7.3, which shows the temperature versus radius within the mantle, together with the variation of melting temperature, T_m. The mean temperature profile in the bulk of the mantle is roughly adiabatic;[20] this is a result of vigorous convection. The profiles in this figure are less definitive than the profiles of mechanical properties shown in Figure 7.2. Note that

- the mantle temperature shown in Figure 7.3 is an area average; we can expect significant departures from this average within descending slabs (discussed in § 31.4) and rising plumes (analyzed in § 32.2);

[19] See § 30.3.1.
[20] See § 29.2 and Figure 29.1.

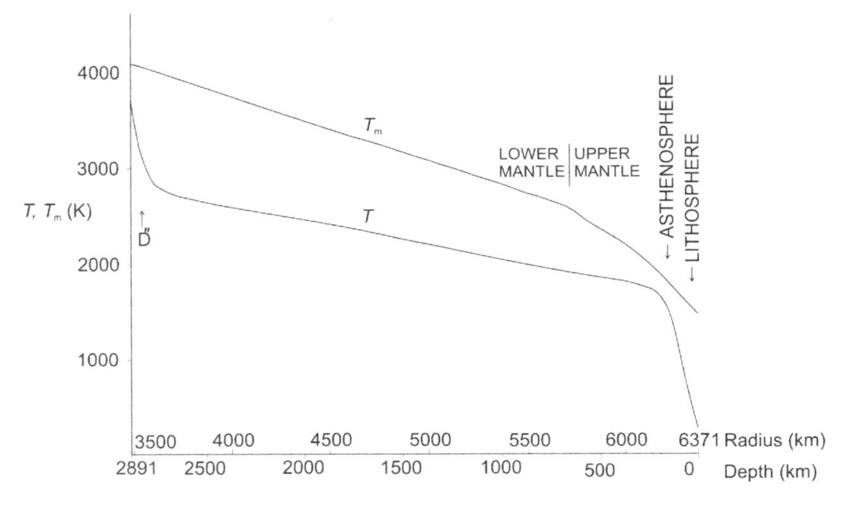

Figure 7.3 Variations in temperature T and melting temperature (eutectic temperature) T_m through the mantle. Figure reproduced with permission; to appear in Hodgkinson and Stacey (2017).

- the temperature deviates strongly from the adiabatic profile at the top and bottom of the mantle. These variations indicate boundary layers where heat is added at the bottom of the mantle in the D″ layer (see § 32.1) and extracted at the top (see § 7.2.3);
- the dimensionless ratio[21] T/T_m is largest – and the viscosity smallest – within the *asthenosphere* in the upper mantle and decreases with increasing depth within the mantle. Since this ratio controls the variation of viscosity, this means that mantle viscosity increases with depth (see § 29.2 and Appendix D.2.2);
- there is no large gradient of temperature at the boundary between the upper and lower mantles, indicating that heat is transferred via convection throughout the bulk of the mantle; and
- when extrapolated adiabatically to shallow depths, the adiabatic temperature intersects the melting curve. This means that mantle material partially melts if brought adiabatically to the surface, as happens beneath mid-ocean spreading centers.

The temperatures shown in this figure are somewhat uncertain. However, for purposes of making numerical estimates, we will assume that $T \approx 3700$ K and $T_m \approx 4100$ K at the base of the mantle. In § 30.7 we will investigate the mantle temperature profile in more detail.

7.2.5 Mantle Rheology

Earth's mantle is composed of silicates - complicated compounds containing oxygen, silicon and a number of other mostly non-metallic elements. Almost all of the mantle

[21] Called the *homologous temperature*, although it is a temperature ratio, not a temperature.

is below the solidus temperature[22] and in crystalline form; melting occurs only in isolated regions, such as beneath mid-ocean ridges and volcanoes, and silicates are rarely in a glassy state. Silicate crystals deform and flow when subject to deviatoric stresses in a process called *creep*. It is very likely that this is shear-thinning power-law creep, as described in § 6.2.3, with the rate of deformation increasing non-linearly as the stress increases, with the constant of proportionality being the *dynamic viscosity*, which depends on pressure, temperature and composition. These variations are discussed in appendices D.1 and D.2 from dynamic and thermodynamic points of view. In this section, we will map these behaviors onto the physical mantle.

An important feature of creep is the strong variation of viscosity with temperature. This variation is so strong that it is a dominant factor in mantle convection, with the issue of Newtonian or non-Newtonian flow being secondary.[23] By the same token, deformation and flow in the mantle is likely to be dominated by thin hot regions having relatively low viscosity. Heating due to viscous dissipation[24] is likely to be locally important in the mantle, confined to those thin regions where the rate of shear is large. This heating can help sustain these thin hot mobile regions. The combination of concentrated shear and nonlinear feedback between shear, elevated temperature and viscous heating makes numerical modeling of mantle flow a challenging task.

Mantle Rheological Structure

If we had to pick a single value for the viscosity of the mantle, it would be (roughly) $\eta \approx 3 \times 10^{22}$ Pa·s.[25] But there are significant and large departures from this mean, in both the radial and lateral directions, so that a single value for the viscosity has little meaning. Earth's mantle is divided into distinct layers and regions having differing elastic and rheological properties. The lithosphere at the top of the mantle is significantly colder than the mantle beneath, as we showed in the previous section; this region behaves as an elastic solid. Beneath the lithosphere is the asthenosphere, which has relatively low viscosity because the material is close to the melting point. Below that are the upper and lower mantles, which are separated by a discontinuity of density and other properties at 660 km depth (see Figure 7.2) but otherwise behave very similarly. There is good theoretical and observational evidence that the viscosity within the bulk of the mantle increases with depth, though there is no consensus regarding the amount of increase. A plausible value for the viscosity of the lower mantle is 3 to 4 $\times 10^{22}$ Pa·s (Čížková et al., 2012; King, 2016). At the very bottom of the mantle is the thin D$''$ layer, kept at an elevated temperature by the transfer of heat from the core. The viscosity at the base of the mantle can be several orders of magnitude less than that of the bulk of the lower mantle; see § 32.1. For example, using the formula for thermal activation of viscosity given in § 6.2.4, with $\beta \approx 27$, $T \approx 3700$ K and $T_m \approx 4100$ K,

22 Hence no liquid present; see Appendix E.10.2.
23 See Tozer (1972).
24 See § 4.5.1 and Appendix E.7.
25 See Figure 7 of King (2016).

a temperature decrease of 800 K (an oft-quoted value for the temperature contrast across the D″ layer) translates to a viscosity increase by a factor of roughly 3800. This means that a plausible value for the viscosity at the base of the mantle is 10^{19} Pa·s.

The viscosity of the mantle is controlled by the homologous temperature, $\upsilon_H = T/T_m$, where T_m is the melting temperature.[26] As we can see from Figure 7.3, the melting temperature varies roughly linearly with radius in the lower mantle. This may be parameterized by[27]

$$T_m \approx (6440 - 2320 r/r_c)\,\mathrm{K},$$

where $r_c = 3.48 \times 10^6$ m is the radius of Earth's core.

In addition to these averaged mantle features, there are two laterally localized features that play important roles in mantle convection. First, beneath subduction zones, slabs of cold lithosphere descend into the mantle. These act to cool the mantle. Second, beneath hotspot volcanoes,[28] warm material ascends from the D″ layer, conveying heat from the core toward Earth's surface.

Geochemical Reservoirs

We have two primary sources of evidence regarding the structure of the mantle; analysis of seismic rays that directly traverse the mantle (see Figure 7.2) and geochemical analysis of magmas. Geochemistry indicates that there are long-lived geochemical reservoirs within the mantle. This evidence seems to be at odds with the concept of whole mantle convection and has led to the suggestion that convection within the mantle is layered, with the lower mantle (beneath 660 km depth) being isolated from the upper mantle. However, this layering would require transfer of heat from lower to upper mantle via conduction and a conductive boundary layer would have a large temperature contrast (greater than 1000 K). Such a large temperature difference would be accompanied by strong vertical variations in seismic properties of the mantle. Such variations are not observed seismically.

An alternate model for the structure of long-lived reservoirs is based on the fact that mantle viscosity is strongly dependent on temperature (see § 6.2.4) and the realization that the mantle is not well mixed. That is, the mantle is very likely to have unorganized lateral heterogeneities in temperature, composition and viscosity as the result of past subduction of cold slabs and variations in the rate of radioactive heating.[29] Furthermore, since deformation and flow in the bulk of the mantle occurs preferentially in those regions having low viscosity, global estimates of mantle viscosity based on dynamics are likely to be biased toward lower values. Given the long thermal-relaxation times of subducted slabs, the location, structure and longevity of heterogeneities are strongly dependent on the history of plate tectonics, particularly on the geographic locations of past subduction.

[26] See Appendix E.10.2.
[27] See § 29.2.
[28] Various types of volcanoes are discussed in the introduction to Chapter 33.
[29] Subducted slabs are depleted in radioactive elements because these incompatible elements preferentially remain in liquid phases that rise to the surface at mid-ocean ridges and on volcanic arcs.

Those regions of the mantle having high viscosity tend not to participate in mantle convection and can act as long-lived "primitive" geochemical reservoirs. Support for this hypothesis of heterogeneous chemical reservoirs is provided by the relatively large turn-over times associated with mantle convection; see § 30.7.1.

This completes our survey of the static structure of Earth's atmosphere, oceans and mantle. In the following section, we investigate the equations governing departures from these static states.

7.3 Perturbation Equations

We will represent perturbations of the static state by a prime; that is, we shall write the density and pressure as

$$\rho = \rho_r + \rho' \qquad \text{and} \qquad p = p_r + p',$$

but we need not place primes on the displacement, $\mathbf{u}$, and velocity, $\mathbf{v}$, because their static parts are zero. Since the acceleration of gravity depends on the distribution of density within the body, we should allow for a perturbation of gravity. However, the perturbations of density we shall consider are either localized or vary harmonically on a short length scale. It follows that the perturbation of gravity in such cases is exceedingly small and may be ignored.

7.3.1 Elastic Perturbation Equations

The elastic perturbation equations are the equation of continuity

$$\rho' + \rho_r(\nabla_{\mathbf{X}} \bullet \mathbf{u}) = 0$$

and the perturbation Navier equation

$$\rho_r \frac{\partial^2 \mathbf{u}}{\partial t^2} = \rho' \mathbf{g} + \left(K + \tfrac{1}{3}\mu\right) \nabla_{\mathbf{X}}(\nabla_{\mathbf{X}} \bullet \mathbf{u}) + \mu \nabla^2_{\mathbf{X}} \mathbf{u}.$$

The Navier equation can be simplified a bit when applied to Earth's mantle. The gravitational force term is of magnitude $\rho_r g U/L$, where L is the length scale of the perturbation having displacement U. Let's compare this to the elastic term, of magnitude KU/L^2. The gravitational term is small provided $L \ll K/g\rho_r$. With $K \approx 5 \times 10^{10}$ kg·m·s^{-2}, $g \approx 10$ m·s^{-2} and $\rho_r \approx 2500$ kg·m^{-3},[30] we see that the gravitational term is small if $L \ll 2 \times 10^3$ km. For seismic body and surface waves, it is safe to ignore this term, giving

$$\rho_r \frac{\partial^2 \mathbf{u}}{\partial t^2} = \left(K + \tfrac{1}{3}\mu\right) \nabla_{\mathbf{X}}(\nabla_{\mathbf{X}} \bullet \mathbf{u}) + \mu \nabla^2_{\mathbf{X}} \mathbf{u}.$$

[30] These are near-surface values; K is much greater at depth.

Note that the Navier equation does not contain ρ' or p'. This equation can be solved for $\mathbf{u}$, then ρ' can be found from the continuity equation $[\rho' = -\rho_r(\nabla_{\mathbf{X}} \bullet \mathbf{u})]$ and p' follows from the perturbation of the equation of state given in § 5.5:

$$p' = K\rho'/\rho_r = -K(\nabla_{\mathbf{X}} \bullet \mathbf{u}).$$

7.3.2 Fluid Perturbation Equations

Now let's consider the fluid perturbation equations. Subtracting out the static state and assuming that $|\rho'| \ll \rho_r$, we have

$$\frac{\partial \rho'}{\partial t} + \nabla \bullet (\rho_r \mathbf{v}) = 0 \qquad \text{and}$$
$$\rho_r \frac{\mathrm{D}\mathbf{v}}{\mathrm{D}t} + 2\rho_r \mathbf{\Omega} \times \mathbf{v} = \rho' \mathbf{g} - \nabla p' + [\nabla \bullet (\nu \nabla)]\mathbf{v}$$

for a Newtonian fluid.

The *Boussinesq approximation* consists of omitting the time derivative from the continuity equation and treating the fluid as incompressible everywhere except in the buoyancy term $\rho'\mathbf{g}$; this is related to the anelastic approximation.[31]

These equations need to be supplemented by the equation of state. Barring shocks, wave motions occur isentropically and the perturbation equation of state is simply

$$p' = K\rho'/\rho_r.$$

On the other hand, flows often occur over long time scales and may be affected by variations of temperature and/or composition. In such cases the perturbation equation of state is

$$\rho' = -\rho_r \alpha T' + \rho_r \beta_T p' + \rho_\xi \xi'.$$

As noted in § 5.2, this equation introduces new variables, T' and ξ', representing perturbations of temperature and material (water vapor for the atmosphere and salt for the oceans), each of which requires its own governing equation; see § 5.3 and § 5.4. Consideration of this complication is deferred until § 24.3 in the case of the atmosphere and § 27.4 in the case of the oceans.

7.4 Boundary Layers

Flows on and within Earth are driven by fluxes of momentum and energy (that is, heat). In many cases these fluxes originate on - or are transferred across - boundaries that act to inhibit flows. Consequently, the dynamics of flows close to boundaries differs from that of flows in the interior of the fluid body (atmosphere, ocean or mantle).[32] In particular,

[31] See § 3.4.1.

[32] The *interior* is that region of the body sufficiently far from boundaries that the dynamically limiting influence of boundaries is negligible.

diffusion is typically important within a thin *boundary layer* close to the boundary and unimportant in the dynamics of the fluid interior. Boundary layers are typically thin because the associated diffusivity is small. We will encounter boundary-layer flows in our studies of katabatic winds (§ 22.2.5 and § 24.1), flows of rotating fluids (Chapter 25, § 27.3 and § 28.4.3), heat transfer to the mantle (§ 32.1) and lava flows (§ 35.1). In addition, boundary-layer-like structures arise in the cylindrical geometry of atmospheric vortices (§ 28.4.2) and mantle plumes (§ 32.2). A brief introduction to boundary-layer theory is found in Appendix G.2.

7.5 Dynamic Stability

In many circumstances, perturbations of the static state result from a dynamic instability of that state, leading to convective motions within the fluid. In this section we develop a necessary condition for dynamic instability.

Let's begin by noting that the static state requires a balance of gravity and the pressure gradient: $\nabla p_r = \rho_r \mathbf{g}$. If this state is to be stable, a parcel of fluid, when displaced vertically, should experience a gravitational force tending to return it to its initial position. That is, a parcel, initially in equilibrium with its surroundings at elevation z, when displaced vertically upward should be denser than its surroundings: $\rho_s(z + \Delta z) < \rho_p(z + \Delta z)$, where Δz is the magnitude of the upward displacement and subscripts p and s denote parcel and surroundings, respectively. Let's suppose that the parcel displacement occurs rapidly, so that exchange of heat with its surroundings is negligibly small; the entropy of the parcel remains constant. The displaced parcel experiences the same pressure as its new surroundings, so its density difference is due to the difference in entropy. From Appendix E.5.1, we have that $\partial\rho/\partial s = -\alpha T \rho / c_p$, where s is the specific entropy, T is the temperature, α is the coefficient of thermal expansion and c_p is the specific heat at constant pressure; see Appendix E.4.2. Since c_p, T, α and ρ are all positive, the stability condition becomes $s_p(z+\Delta z) < s_s(z+\Delta z)$. But since s_p is constant, $s_p(z+\Delta z) = s_s(z)$ and the stability condition becomes $s_s(z) < s_s(z + \Delta z)$ or equivalently $\mathrm{d}s/\mathrm{d}z > 0$; the specific entropy must increase with height if the fluid is to be dynamically stable.

We are halfway there; the next step is to turn this inequality into a condition on the temperature gradient. This is accomplished using the equation of state for entropy found in Appendix E.5: $\mathrm{d}s = (c_p/T)\,\mathrm{d}T - (\alpha/\rho)\,\mathrm{d}p$. Using this and the hydrostatic equation $\mathrm{d}p = -\rho g \mathrm{d}z$, the necessary condition for dynamic stability is

$$\frac{\mathrm{d}T}{\mathrm{d}z} > \frac{\mathrm{d}T_A}{\mathrm{d}z}, \qquad \text{where} \qquad \frac{\mathrm{d}T_A}{\mathrm{d}z} = -\frac{\alpha T g}{c_p}$$

is the adiabatic temperature gradient.[33] If the temperature decreases with elevation less rapidly than the adiabat, the fluid layer is dynamically stable. Note, in particular, that an

[33] See § 5.5.

isothermal state is dynamically stable. We expand on this point in § 30.5, showing that mantle convection is not driven by uniform radioactive heating.

The opposite statement is a necessary condition for dynamic instability. However, it is not sufficient if diffusion of heat (and material) alters the density of the displaced parcel and/or viscosity impedes the motions. A horizontal layer of fluid of depth D is convectively stable unless the *Rayleigh number*,

$$\mathrm{Ra} \equiv g\alpha(\Delta T)D^3/\nu\kappa\,,$$

where ν is the kinematic viscosity, κ is the thermal diffusivity and ΔT is the temperature excess (compared with the adiabat) of the bottom boundary compared with the top, exceeds a critical value that depends on the thermal and dynamic conditions at the fluid boundaries.

A horizontal layer of fluid with isothermal rigid boundaries is convectively unstable if $\mathrm{Ra} > 1100$. In Earth's atmosphere, with[34] $g \approx 9.8$ m·s^{-2}, $\alpha = 1/T \approx 0.0035$ K^{-1}, $D \approx 10^4$ m and $\nu \approx \kappa \approx 2 \times 10^{-5}$ m^2·s^{-1}, a temperature difference of 10^{-16}K is sufficient to initiate convective instability. In the oceans, with $\alpha \approx 2 \times 10^{-4}$ K^{-1}, $D \approx 3600$ m, $\nu \approx 10^{-6}$ m^2·s^{-1} and $\kappa \approx 1.4 \times 10^{-7}$ m^2·s^{-1}, a temperature difference of 10^{-17}K is sufficient to initiate convective instability. In Earth's mantle, with $\alpha \approx 5 \times 10^{-5}$ K^{-1}, $D \approx 2.9 \times 10^6$ m, $\nu \approx 10^{19}$ m^2·s^{-1} and $\kappa \approx 10^{-6}$ m^2·s^{-1}, a temperature difference of 1K is sufficient to initiate convective instability. It is readily seen that the atmosphere, oceans and mantle are in vigorous thermal convection.

The nature of the convective motions is determined in large part by the *Reynolds number*,

$$\mathrm{Re} \equiv UD/\nu\,,$$

where U is a typical flow speed. When Re exceeds a critical value, the flow is turbulent. Flows in the atmosphere and oceans are invariably turbulent whereas the large viscosity of silicates makes $\mathrm{Re} \ll 1$ in the mantle. Although mantle flow is laminar, it is chaotic, due to the strong dependence of viscosity on temperature; see § 7.2.5.

Transition

In this chapter we presented and discussed the equations governing hydrostatic balance for Earth's atmosphere, oceans and mantle, investigated the static structure of these regions, then developed the perturbation equations governing small departures from hydrostatic equilibrium. Due to the presence of the inertia term, these equations are nonlinear and remain very challenging to analyze and solve.

In the next chapter we discuss the nature of the perturbation equations for rotating fluids in more detail.

[34] See Table B.5.

8
Introduction to Rotating Fluids

Rotation of Earth plays a dominant role in the dynamics of its atmosphere and oceans on time scales of a day or more. In a rotating (non-inertial) frame, the dynamic effect of rotation is represented a virtual force, called the *Coriolis force*, in the Navier–Stokes equation.[1] A dominant effect of rotation is to produce waves and flows that are themselves rotating relative to a frame of reference fixed to Earth. This tendency is most readily observed in the atmosphere, which contains large-scale rotating air masses called "highs" and "lows" by meteorologists, and an array of smaller and often stronger rotating systems, including hurricanes (or cyclones), tornadoes, water spouts, and dust devils. As we shall see, rotation of Earth plays a direct role in large-scale systems, but only an indirect role in the smaller systems.

In this chapter we will modify the perturbation equations given in § 7.3.2 so that they apply more specifically to the atmosphere and oceans and in doing so will encounter a number of concepts and effects that are introduced by rotation. The Coriolis force complicates the analysis of rotating fluids in several ways; apart from introducing an extra term in the Navier–Stokes equation, it couples the horizontal components of this equation, so that motions are generally not unidirectional. It will take some care and effort to decipher the dynamic constraints that rotation imposes on the flow, so we shall proceed carefully, beginning in § 8.1 with the discussion of a number of rotational concepts. Then in § 8.2, we will present the equations governing flow in a rotating fluid. The rotation of the fluid itself, represented by the vorticity, is an important quantity of interest; the vorticity equation is presented and discussed in § 8.3. Often horizontal motions in a thin layer of fluid are independent of the vertical coordinate and can be quantified by vertically averaged equations; the vertically averaged continuity equation is developed in § 8.4. Large-scale motions in the atmosphere and oceans may be categorized as geostrophic or quasi-geostrophic. Geostrophic flows result from a balance between the Coriolis and pressure terms, as described in § 8.5, while quasi-geostrophic flows result from a balance between the inertial, Coriolis and pressure terms, as described in § 8.6.

[1] See § 6.2.2 and Appendix C.1.3. The dynamical effect of the rotation of Earth on the atmosphere and oceans is quantified by the Coriolis parameter, f, which is introduced and discussed in § 8.1.5.

8.1 Rotational Concepts

As the old saying goes, "you can't tell the players without a scorecard". So before delving into the mathematics of rotating fluids, let's discuss several concepts related to rotation, beginning with those that pertain to rigid bodies (rotation, revolution and angular velocity) in the following subsection, then discussing angular momentum in § 8.1.2, circulation, vorticity and pressure in § 8.1.3 and § 8.1.4 and finally the Coriolis parameter in § 8.1.5.

8.1.1 Rotation, Revolution and Angular Velocity

Rotation is circular rigid-body motion of a body about a fixed axis running through its center of mass, while *revolution* is orbital motion of a body about an axis not passing through its center of mass. To illustrate the distinction between rotation and revolution, note that Earth rotates about its axis of symmetry at a rate[2] 7.2921×10^{-5} rad·s^{-1} and it revolves about the center of mass of the Earth-Moon system (which lies a distance 4671 m from Earth's center of mass) at a rate $1/27.3$ revolutions per day or 2.664×10^{-6} rad·s^{-1}.

Rotation and revolution are very similar concepts that commonly are lumped together. Each is represented by a magnitude (an *angular speed*), usually denoted by Ω, and an axis orientation, denoted by a unit vector $\mathbf{1}_\Omega$ parallel to the axis. This magnitude and direction may be combined into an *angular velocity vector* $\boldsymbol{\Omega} = \Omega \mathbf{1}_\Omega$.

8.1.2 Angular Momentum

Now let's consider a small parcel of mass M revolving about an axis. (The parcel might be part of a larger body that is rotating or revolving about this axis.) The angular momentum of the parcel is $M\varpi^2\Omega = Mv\varpi$, where ϖ is the distance to the rotation axis and $v = \varpi\Omega$ is the speed of the parcel. Unless a torque is applied to the parcel its angular momentum is conserved. If the parcel is in orbit about a massive object, it experiences no torque and it must move such that $v\varpi$ is constant.[3]

The angular momentum of a parcel on – and moving with – Earth's surface is a maximum at the equator and zero at the poles (which delineate the axis of rotation). If the parcel moves away from the equator on Earth's surface its distance from the axis decreases, and – if no torques are applied – its rate of rotation must increase. This eastward motion relative to Earth's surface is seen as deflection to the {right, left} in the {northern, southern} hemisphere. Conversely, as the parcel moves toward the equator on Earth's surface the distance to the axis increases and the rotation rate decreases. This westward motion relative to Earth's surface is again seen as deflection to the {right, left} in the {northern, southern} hemisphere. However, since a pressure gradient can exert a torque on a fluid parcel, the angular momentum of a fluid parcel in the atmosphere or oceans is generally not conserved.

[2] This is once every 0.997 days or 366.24 times per year.
[3] This statement is equivalent to Kepler's second law of planetary motion.

8.1.3 Circulation and Vorticity

Now let's move on to two fluid-related concepts that are somewhat similar to rotation and revolution. *Circulation* is defined in relation to the motion of fluid around a closed circuit, denoted by C; specifically the circulation Γ is

$$\Gamma \equiv \oint_C \mathbf{v} \bullet \mathbf{dl},$$

where $\mathbf{v}$ is the velocity of the fluid and $\mathbf{dl}$ is a line segment directed along the circuit. The *vorticity vector* is the curl of the velocity field: $\boldsymbol{\omega} = \boldsymbol{\nabla} \times \mathbf{v}$. If a body is rotating rigidly with rate Ω, the circulation of a parcel at radial distance ϖ is $\Gamma = 2\pi\varpi^2\Omega$, while the magnitude of its vorticity is 2Ω. It is possible for a body of fluid to have a finite circulation and zero vorticity (almost everywhere); this curious state is analyzed in § 28.3.1. Note that the circulation has dimensions of length^2/time, which is the same as the kinematic viscosity.

The circulation is related to the vorticity vector by the area integral[4]

$$\Gamma = \iint_A \boldsymbol{\omega} \bullet \mathrm{d}\mathbf{A},$$

where $\mathrm{d}\mathbf{A}$ is an infinitesimal area vector pointing normal to surface A, which is bounded by the closed circuit, C.

In the study of the dynamics of the atmosphere and oceans, we need to keep track of two vorticities: that associated with the rotation of Earth and that associated with motions relative to a frame of reference fixed with respect to the solid Earth. The velocity associated with the Earth's rotation is $\Omega\varpi\,\mathbf{1}_\phi$, where $\mathbf{1}_\phi$ is a unit vector pointing eastward and ϖ is the distance to the rotation axis. Using the formula for the curl found in Appendix A.3.4, this motion has vorticity $2\boldsymbol{\Omega}$, where (by convention) $\boldsymbol{\Omega}$ points northward. Since Earth rotates eastward, its motion is counterclockwise when viewed by a non-rotating observer located above the surface and north of the equator. That is, {positive, negative} vorticity is associated with {counterclockwise, clockwise} motion in the northern hemisphere.

8.1.4 Circulation and Pressure

It is easy to demonstrate the relation between circulation and pressure by inducing rotation of a fluid within a container (perhaps by stirring a martini). The surface of the fluid is raised at the periphery of the container. This extra depth of fluid is supported by a pressure that increases in the radial direction. It is easy to show, using the momentum equation developed in § 4.6, that the pressure distribution is given by $p = p_0 + \rho\Omega^2\varpi^2/2$.

If we ignore the effect of Earth's rotation, this pressure distribution is the same whether we stir our martini clockwise or counterclockwise. However, if we stirred it very slowly so its rotation rate were comparable to that of Earth, we would be able to discern an

[4] Obtained using *Kelvin's circulation theorem*.

asymmetry.[5] The rotation of Earth, by itself, induces a positive radial pressure gradient within the martini. If we stir our martini in the same direction as Earth's rotation, we will see a decrease of pressure at the center. But if we were to stir it against Earth's rotation, the two rotations would tend to cancel and we would see an increased pressure at the center.

Now let's think of weather systems instead of martinis. We all know from watching the weather reports on TV that (in the northern hemisphere) low-pressure systems are accompanied by an counterclockwise rotation; the flow around a low-pressure center is the same sense as Earth's rotation. On the other hand, clockwise rotation in the northern hemisphere is opposite to Earth's rotation; the net effect is an increase of pressure at the center of a clockwise rotation.

8.1.5 Coriolis Parameter

The vorticity associated with Earth's rotation is a constant. However, this is not the full story. Since the atmosphere and oceans are thin fluid layers on Earth's surface, only the locally vertical component of vorticity is dynamically important.[6] This component is called the *Coriolis parameter*:

$$f(\theta) = 2\mathbf{\Omega} \bullet \mathbf{1}_z = 2\Omega \cos\theta \,,$$

where θ is the colatitude and $\mathbf{1}_z$ is a unit vector pointing locally upward. As shown in Figure 8.1, this parameter is odd about the equator and positive in the northern hemisphere. The magnitude of the Coriolis parameter is a maximum at the poles, $|f|_{max} = 1.414 \times 10^{-4}$ rad·s^{-1}, and decreases to zero at the equator – opposite to the latitudinal variation of angular momentum. At mid-latitudes $|f| \approx 10^{-4}$ rad·s^{-1}.

If we consider only the linear acceleration and Coriolis terms, the momentum equation simplifies to[7]

$$\frac{\partial \mathbf{v}}{\partial t} = -f\mathbf{1}_z \times \mathbf{v}$$

on Earth's surface. In the northern hemisphere, where $f > 0$, the Coriolis force causes fluid to deflect to the right, while in the southern hemisphere, where $f < 0$, the Coriolis force causes fluid to deflect to the left, as we have noted previously.

f-Plane and β-Plane Approximations

The Coriolis parameter is a function of θ, or equivalently northward distance from the equator $y = R(\pi/2 - \theta)$, where R is the radius of Earth. That is,

$$f(y) = 2\Omega \sin(y/R) \,.$$

[5] Of course it would have to be a really big martini to see the difference.
[6] See § 8.2.1.
[7] See § 6.2.1.

- In the *f-plane approximation*, the parameter f is assumed to be constant: $f = f_0$, where the subscript 0 denotes a constant value. This is a reasonable approximation when considering waves and flows that are restricted to a small range of latitude.
- The *β-plane approximation* consists of expanding $f(y)$ in powers of y about a mid-latitude value denoted by y_0, and keeping the first two (constant and linear) terms of the expansion:

$$f(y) = f_0 + \beta(y - y_0), \qquad \text{where}$$
$$f_0 = 2\Omega \sin(y_0/R) = 2\Omega \cos\theta_0 \qquad \text{and}$$
$$\beta = 2\Omega \cos(y_0/R)/R = 2\Omega \sin\theta_0/R$$

  are usually treated as specified constants. The variations of f and β with colatitude are illustrated in Figure 8.1.
- In the *equatorial β-plane* approximation, $f_0 = y_0 = 0$ and $f = \beta y$ with $\beta = 2\Omega/R$.

8.2 Equations Governing Rotating Flows

In the previous section we introduced the Coriolis parameter and explained its two main flavors: f-plane and β-plane. Now let's take a step backward and investigate how this parameter arises and how it fits within the basic equations governing flows in the atmosphere and oceans. In this process, we will simplify the governing equations using the *thin-layer approximation*.

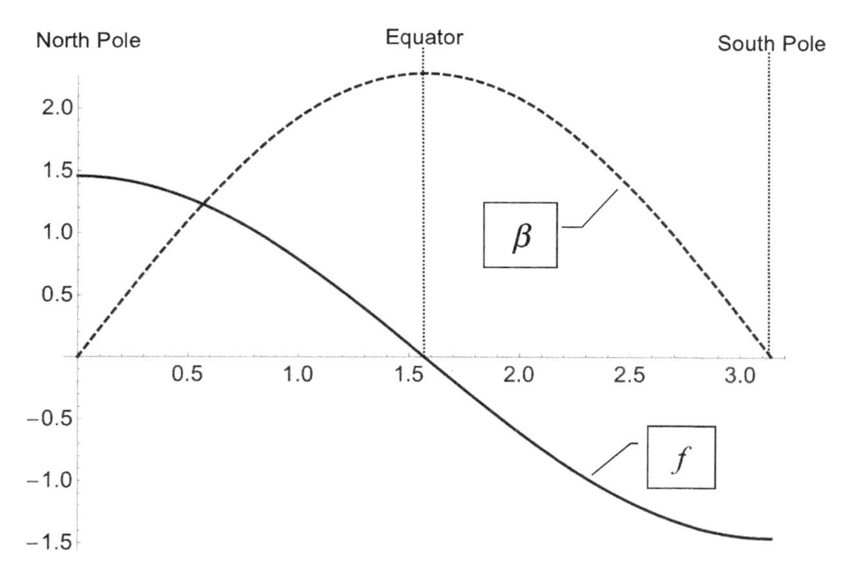

Figure 8.1 Variations of the Coriolis parameter f (solid line) and its derivative β (dashed line) versus colatitude θ. Note that f changes sign across the equator while β is positive everywhere (except at the poles). f is in units of 10^{-4} s^{-1} and β is in units of 10^{-11} m$^{-1}\cdot$s^{-1}.

Let's start with the fluid perturbation equations developed in § 7.3.2. Key steps in the development of the thin-layer equations are

- adoption of the anelastic approximation;[8]
- assuming the reference-state density is a function only of z; and
- separation of vectors into vertical and horizontal components.

Writing

$$\mathbf{g} = -g\mathbf{1}_z, \qquad \mathbf{v} = \mathbf{v}_H + w\mathbf{1}_z \qquad \text{and} \qquad 2\boldsymbol{\Omega} = f_H\mathbf{1}_y + f\mathbf{1}_z,$$

where $\mathbf{1}_y$ and $\mathbf{1}_z$ point locally northward and up, respectively, with

$$f = 2\Omega\cos\theta \qquad \text{and} \qquad f_H = 2\Omega\sin\theta,$$

these governing equations may be expressed as

$$\nabla_H \bullet \mathbf{v}_H + \frac{\partial(\rho_r w)}{\rho_r \partial z} = 0,$$

$$\frac{\mathrm{D}\mathbf{v}_H}{\mathrm{D}t} + f\mathbf{1}_z \times \mathbf{v}_H + f_H w \mathbf{1}_y \times \mathbf{1}_z = -\frac{1}{\rho_r}\nabla_H p' + \mathcal{V}(\mathbf{v}_H)$$

and $$\frac{\mathrm{D}w}{\mathrm{D}t} + f_H\mathbf{1}_y \times \mathbf{v}_H = -\frac{1}{\rho_r}\frac{\partial p'}{\partial z} - \frac{g}{\rho_r}\rho' + \mathcal{V}(w),$$

where $$\mathcal{V}(\cdot) \equiv \frac{\partial}{\partial z}\left(\nu\frac{\partial(\cdot)}{\partial z}\right) + \nu\nabla_H^2(\cdot)$$

is a linear differential operator quantifying the viscous force, $\nu = \eta/\rho$ is the kinematic viscosity[9] and the reference density ρ_r may be a function of height z. The vertical viscous term is important in the dynamics of Ekman layers (see Chapter 25) and in the study of shear-flow instability (see Appendix G.1) while the horizontal viscous term is important in determining the structure of upper-ocean currents (see § 27.3) and intense vortices (see § 28.4.2).

We have a set of four equations for five scalar unknowns: $\mathbf{v}_H$, w, p' and ρ'. The fifth equation is the equation of state for density presented in § 5.2. But we'll put that aside for the moment and focus in the next section on simplifying the governing equations in hand.

8.2.1 Thin-Layer Approximation

The atmosphere and oceans are thin, with vertical extent much smaller than horizontal. This ratio of scales provides us with a small dimensionless parameter that can be used as a basis of simplification. That is, supposing that the vertical and horizontal scales of motion

[8] See § 3.4.1.
[9] In writing this version of the momentum equation, we have ignored the variation of density across thin vertical boundary layers.

associated with waves and flows are D and L, respectively, the essence of the thin-layer approximation is to assume that $D/L \ll 1$.

Now let's see how this condition affords simplification of the governing equations. With the spatial scalings introduced above, plus using U to represent the magnitude of the horizontal velocity, we have $\mathbf{v}_H \sim U$, $\nabla_H \sim 1/L$ and $\partial/\partial z \sim 1/D$. It follows from the continuity equations that $w \sim DU/L$. The potential balances in the horizontal momentum equation (among inertial, vertical Coriolis, horizontal Coriolis, pressure and viscous forces) may be expressed symbolically as

$$\frac{U^2}{L} \sim fU \sim f_H \frac{D}{L} U \sim \frac{p'}{\rho_r L} \sim \nu U \frac{1}{D^2}.$$

In this balance, the p' term is passive, and we can focus on the size of the horizontal Coriolis force relative to the vertical Coriolis or inertial force. In mid-latitudes, where $f_H \sim f$, the horizontal Coriolis force is small compared with the vertical Coriolis force provided $D \ll L$. In equatorial regions with $f \sim \beta L$ and $f_H \sim \beta R$,[10] the horizontal Coriolis force is small provided $\sqrt{DR} \ll L$. The viscous term is important provided $D \sim \sqrt{\nu/f}$;[11] this is the Ekman-layer scale discussed in § 25.1. Note that the order of magnitude of the perturbation pressure in a rapidly rotating system (having $U \leq fL$) is $p'/\rho_r \sim fUL$.

A similar scale analysis of the vertical momentum equation, with the buoyancy term temporarily ignored, gives

$$\frac{U}{L}\frac{D}{L} U \sim f_H U \sim \frac{fUL}{D} \sim \nu \frac{U}{DL}.$$

It is readily seen that with the thin-layer approximation the horizontal Coriolis, inertia and viscous terms are negligibly small compared with the pressure term. Restoring the buoyancy term in the vertical component of the momentum equation, the thin-layer equations are

$$\nabla_H \bullet \mathbf{v}_H + \frac{\partial(\rho_r w)}{\rho_r \partial z} = 0, \qquad 0 = -\frac{1}{\rho_r}\frac{\partial p'}{\partial z} + \frac{g}{\rho_r}\rho'$$

and

$$\frac{\mathrm{D}\mathbf{v}_H}{\mathrm{D}t} + f\mathbf{1}_z \times \mathbf{v}_H = -\frac{1}{\rho_r}\nabla_H p' + \mathcal{V}(\mathbf{v}_H).$$

Isobaric Surface and Pressure Coordinate

The variable coefficient ρ_r can be troublesome when applying these equations to atmospheric flows. This difficulty is alleviated by restructuring these equations. As they stand, they are a set of three equations[12] for the four dependent variables $\mathbf{v}_H$, w, ρ' and p' as functions of the independent variables $\mathbf{x}_H$, z and t. They can be rewritten with p

[10] Recall that $R \approx 6 \times 10^6$m is the radius of Earth.

[11] The viscous term is also important if $L \sim \sqrt{\nu/f}$; this occurs in the radial boundary layer of a tornado; see § 28.4.2. In this extreme case the thin-layer approximation is no longer valid.

[12] The three equations need to be supplemented with the density equation of state.

replacing z as an independent variable and with the isobaric surface $Z(\mathbf{x}_H, p, t)$ replacing p' as a dependent variable. The modified equations, developed in appendices D.8.1 and D.8.2 are

$$\nabla_H \bullet \mathbf{v}_H + \frac{\partial w_p}{\partial p} = 0, \qquad \frac{\partial Z}{\partial p} = -\frac{1}{\rho_r g}$$

$$\text{and} \qquad \frac{\mathrm{D}\mathbf{v}_H}{\mathrm{D}t} + f\mathbf{1}_z \times \mathbf{v}_H = -g\nabla_H Z + \mathcal{V}(\mathbf{v}_H),$$

where $w_p = -g\rho_r w$ is a scaled vertical speed (positive downward) and

$$\frac{\mathrm{D}}{\mathrm{D}t} = \frac{\partial}{\partial t} + \mathbf{v}_H \bullet \nabla_H + w_p \frac{\partial}{\partial p}.$$

8.2.2 Local Cartesian Coordinates

It is common practice in the study of waves and flows in the atmosphere and oceans to introduce a local Cartesian frame of reference, affixed to Earth, with x being distance eastward from an arbitrary longitude, y distance northward from the equator and z distance upward from Earth's surface[13] and write

$$\mathbf{v} = u\mathbf{1}_x + v\mathbf{1}_y + w\mathbf{1}_z \qquad \text{and} \qquad \boldsymbol{\Omega} = \Omega\left(\sin\theta\,\mathbf{1}_y + \cos\theta\,\mathbf{1}_z\right).$$

Switching to local Cartesian coordinates, the thin-layer equations developed in § 8.2.1 become

$$\frac{\partial u}{\partial x} + \frac{\partial v}{\partial y} + \frac{\partial(\rho_r w)}{\rho_r \partial z} = 0, \qquad 0 = \frac{1}{\rho_r}\frac{\partial p'}{\partial z} + g\rho',$$

$$\frac{\mathrm{D}u}{\mathrm{D}t} - fv = -\frac{1}{\rho_r}\frac{\partial p'}{\partial x} + \mathcal{V}(u) \quad \text{and} \quad \frac{\mathrm{D}v}{\mathrm{D}t} + fu = -\frac{1}{\rho_r}\frac{\partial p'}{\partial y} + \mathcal{V}(v)$$

where

$$\frac{\mathrm{D}}{\mathrm{D}t} = \frac{\partial}{\partial t} + u\frac{\partial}{\partial x} + v\frac{\partial}{\partial y} + w\frac{\partial}{\partial z}$$

or equivalently

$$\frac{\partial u}{\partial x} + \frac{\partial v}{\partial y} + \frac{\partial w_p}{\partial p} = 0, \qquad \frac{\partial Z}{\partial p} = -\frac{1}{\rho_r g},$$

$$\frac{\mathrm{D}u}{\mathrm{D}t} - fv = -g\frac{\partial Z}{\partial x} + \mathcal{V}(u) \quad \text{and} \quad \frac{\mathrm{D}v}{\mathrm{D}t} + fu = -g\frac{\partial Z}{\partial y} + \mathcal{V}(v)$$

[13] These coordinates are related spherical coordinates by $\{\mathrm{d}x, \mathrm{d}y, \mathrm{d}z\} = R\{\sin\theta\,\mathrm{d}\phi, -\mathrm{d}\theta, \mathrm{d}r\}$, where R is the radius of Earth, θ is the colatitude and ϕ is the longitude. Comparing with Appendix A.3.5, we have $\{\mathbf{1}_x, \mathbf{1}_y, \mathbf{1}_z\} = \{\mathbf{1}_\phi, -\mathbf{1}_\theta, \mathbf{1}_r\}$ and $\{u, v, w\} = \{v_\phi, -v_\theta, v_r\}$.

where

$$\frac{\mathrm{D}}{\mathrm{D}t} = \frac{\partial}{\partial t} + u\frac{\partial}{\partial x} + v\frac{\partial}{\partial y} + w_p\frac{\partial}{\partial p}.$$

8.3 Vorticity and Pressure Equations

The thin-layer equations presented in § 8.2.2 are a set of three equations for three unknowns: u, v and Z. Two useful auxiliary equations may be formed from these equations by taking the curl of the horizontal components of the momentum equation, thereby eliminating Z to obtain the *vorticity equation* (in § 8.3.1) or by taking the divergence of these components to obtain the *pressure equation* (in § 8.3.2).

8.3.1 Vorticity Equation

The vorticity equation is constructed by eliminating Z from the y derivative of the x component of the momentum equation using the x derivative of the y component. The result may be expressed as

$$\frac{\mathrm{D}\zeta}{\mathrm{D}t} + \nabla_H \bullet (f\mathbf{v}_H) = \mathcal{V}(\zeta)$$

or equivalently

$$\frac{\mathrm{D}\zeta}{\mathrm{D}t} + \beta v = -f\left(\frac{\partial u}{\partial x} + \frac{\partial v}{\partial y}\right) + \mathcal{V}(\zeta),$$

where

$$\begin{aligned}\zeta &= \mathbf{1}_z \bullet \boldsymbol{\omega} = \mathbf{1}_z \bullet \nabla_H \times \mathbf{v}_H \\ &= -\nabla_H \bullet (\mathbf{1}_z \times \mathbf{v}_H) = \frac{\partial v}{\partial x} - \frac{\partial u}{\partial y}\end{aligned}$$

is the vertical component of the vorticity vector, commonly called the *vorticity*. In the bulk of the atmosphere and oceans, the viscous term is negligibly small. The term involving f represents *vortex stretching*; horizontal convergence of fluid acts as a source of vorticity in a rotating fluid.

8.3.2 Pressure Equation

The pressure equation is constructed by adding the x derivative of the x component of the momentum equation and the y derivative of the y component of the momentum equation. The result of this operation is

$$g\nabla_H^2 Z = f\zeta + \beta u - \frac{\mathrm{D}}{\mathrm{D}t}\left(\frac{\partial u}{\partial x} + \frac{\partial v}{\partial y}\right) + \mathcal{V}\left(\frac{\partial u}{\partial x} + \frac{\partial v}{\partial y}\right).$$

In mid-latitudes, the Coriolis term dominates the right-hand side and this equation simplifies to

$$g\nabla_H^2 Z = f\zeta \qquad \text{or equivalently} \qquad \nabla_H^2 p = \rho_r f\zeta\,.$$

This is a *Poisson equation*: a non-homogeneous Laplace equation. The solution has a sign opposite to that of the forcing. That is, a {positive, negative} vertical vorticity generates a {negative, positive} pressure and as we have seen in § 8.1.4 {counterclockwise, clockwise} circulation, as illustrated in Figure 8.2. Upward vertical motion, typically induced by convection, induces a horizontal convergence that concentrates vertical vorticity. The increase of spin accompanying this concentration is balanced by a pressure gradient pointing outward from a low-pressure center.

8.3.3 Discussion

The vorticity equation presented in § 8.3.1 shows that vorticity of a parcel of fluid is changed by

- meridional advection of the planetary vorticity (the term containing β);
- horizontal divergence or convergence (this effect is sometimes called vortex stretching); or
- the action of the viscous force.

Since $\beta = \partial f/\partial y$, the vorticity equation may be written as

$$\frac{\mathrm{D}\eta}{\mathrm{D}t} = -f\left(\frac{\partial u}{\partial x} + \frac{\partial v}{\partial y}\right) = f\frac{\partial w_p}{\partial p}\,,$$

where $\eta = \zeta + f$ is the *absolute vorticity* and the second version of the right-hand side is obtained using the continuity equation in pressure coordinates from § 8.2.1.

In the absence of viscous effects and horizontal convergence or divergence, the absolute vorticity of a parcel of fluid is conserved:

$$\frac{\mathrm{D}\eta}{\mathrm{D}t} = \frac{\mathrm{D}\zeta}{\mathrm{D}t} + \beta v = 0\,.$$

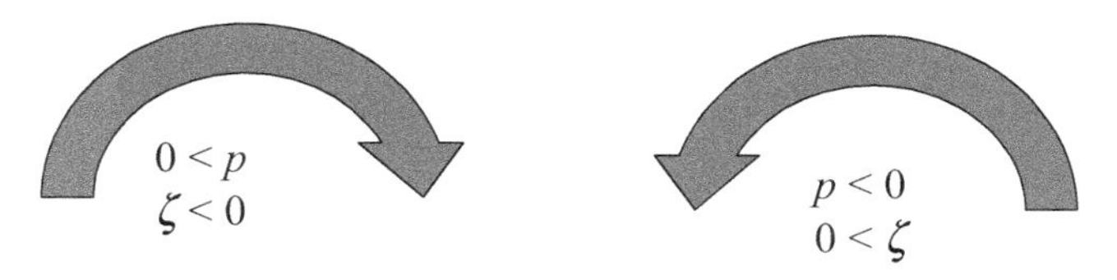

Figure 8.2 An illustration of the relation between the pressure, p, relative vertical vorticity, ζ and circulation (curved arrows) in the northern hemisphere. Circulation is clockwise around a high pressure system and counterclockwise around a low pressure system. The sense of circulation in the southern hemisphere is opposite to that shown.

According to this equation, northward motion ($v > 0$) induces a negative relative vorticity, which is associated with a positive geostrophic pressure and a clockwise circulation; see Figure 8.2. This change may be understood in terms of vortex stretching, as follows. Consider a column of fluid oriented parallel to the axis of rotation, having cross-sectional area A and height $D/\cos\theta$, where D is the (local vertical) depth of the fluid; see Figure 8.3. As this column is displaced northward its volume $V = AD/\cos\theta$ remains constant. Displacement northward is accompanied by a decrease in θ, an increase in $\cos\theta$ and a decrease in the height of the column of fluid. Since the volume remains constant, the cross-sectional area increases. This axial compression causes the vortex lines to diverge, leading to a decrease in the magnitude of ζ: $\zeta_2 < \zeta_1$.

8.4 Vertically Averaged Continuity Equation

If the horizontal velocity is independent of depth, the continuity equation from § 8.2.1 may be integrated in z from h_1 to h_2, giving

$$\nabla_H \bullet [(p_1 - p_2)\bar{\mathbf{v}}_H] = g\rho_1 \frac{\partial h_1}{\partial t} - g\rho_2 \frac{\partial h_2}{\partial t},$$

where subscripts 1 and 2 on p and ρ denote evaluation at $z = h_1$ and $z = h_2$, respectively, and an overbar has been added to $\mathbf{v}_H$ to remind ourselves that it is independent of z. This equation will be customized for application to the atmosphere and ocean in the following two subsections.

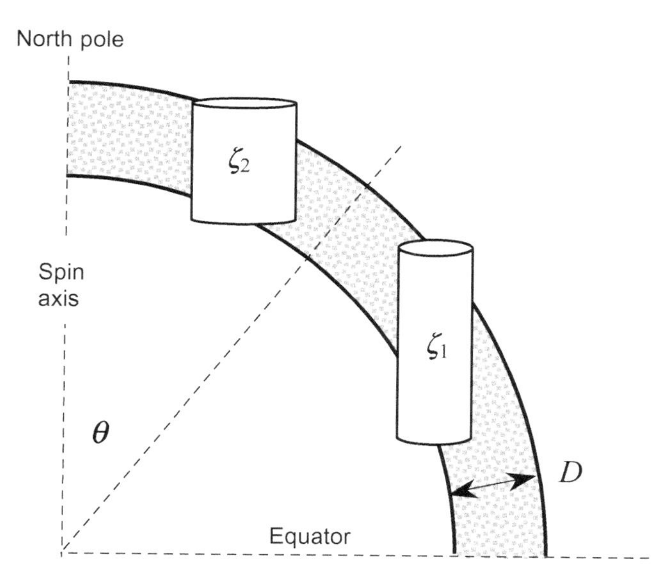

Figure 8.3 As a column of fluid aligned with the rotation axis is moved northward in the northern hemisphere, it compresses, causing a divergence of the vortex lines: $\zeta_2 < \zeta_1$. The local vertical depth of the fluid, denoted by D, and the volume of the column remain constant.

8.4.1 Application to the Atmosphere

The total atmosphere is averaged vertically provided the pressure and density at the top are zero, $p_2 = \rho_2 = 0$, and the pressure at the bottom is the local atmospheric pressure (not necessarily constant due to local topography.). Noting that the topography is independent of time ($\partial h_1/\partial t = 0$), the vertically averaged continuity equation may be expressed as

$$\nabla_H \bullet (D_s \bar{\mathbf{v}}_H) = 0, \qquad \text{where} \qquad D_s = \frac{p_1}{\rho_1 g} = \frac{R_s T_1}{g}$$

is the local *scale height* of the atmosphere: the increase in elevation over which the pressure decreases by a factor $1/e$. The alternate version for D_s has been obtained using the ideal gas law.[14] At sea level on Earth, $D_s \approx 8$ km, which is about the height of Mount Everest. This tells us that mountain ranges have a dominant effect on the flow of Earth's atmosphere.

The vertically averaged continuity equation for the atmosphere may be satisfied by the introduction of a scalar stream function:[15]

$$D_s \bar{\mathbf{v}}_H = \nabla \times (\Psi \mathbf{1}_z).$$

This satisfies the vertically averaged continuity equation and reduces the number of independent variables by one.

8.4.2 Application to the Oceans

Many types of waves and flows in the oceans may be modeled assuming the density is constant and the pressure balance is hydrostatic. In this case $p_1 - p_2 = \rho g(h_2 - h_1)$ and the vertically averaged continuity equation becomes

$$\nabla_H \bullet [(h_2 - h_1)\bar{\mathbf{v}}_H] = -\frac{\partial (h_2 - h_1)}{\partial t}.$$

There are two cases of interest to investigate: the layer is (a) the entire ocean depth and (b) the mixed layer above the thermocline.

For case (a), suppose the layer lies between $z = h_1 = -D(\mathbf{x}_H)$ and $z = h_2 = h(\mathbf{x}_H, t)$, where D is the equilibrium depth (when $h = 0$) of the ocean, as illustrated in Figure 8.4. In this case $h_2 - h_1 = D + h$ and the continuity equation becomes

$$(D + h)\nabla_H \bullet \bar{\mathbf{v}}_H = -\frac{\partial h}{\partial t} - \bar{\mathbf{v}}_H \bullet \nabla_H (D + h).$$

For case (b), suppose the layer lies between $z = h_1 = -D + h_b(\mathbf{x}_H, t)$ and $z = h_2 = h_t(\mathbf{x}_H, t)$, where now D is the equilibrium depth of the thermocline. In this case, the

[14] See Appendix E.8.
[15] See Appendix C.4.

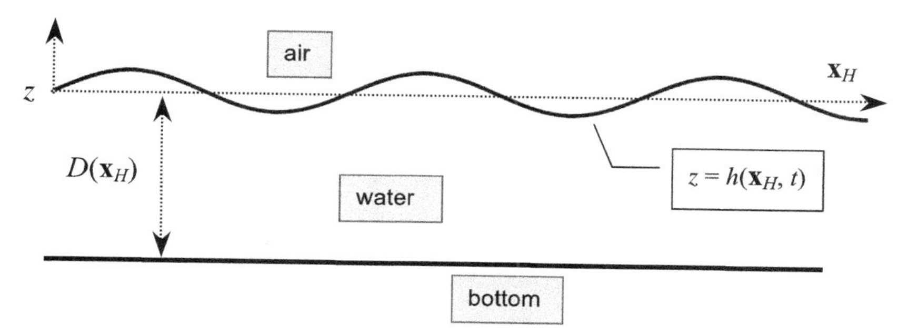

Figure 8.4 Schematic of a wave of height $h(\mathbf{x}_H, t)$ on the surface of an ocean of equilibrium depth $D(\mathbf{x}_H)$.

continuity equation becomes

$$(D + h_t - h_b)\nabla_H \bullet \bar{\mathbf{v}}_H = \frac{\partial h_b}{\partial t} + \bar{\mathbf{v}}_H \bullet \nabla_H h_b - \frac{\partial h_t}{\partial t} - \bar{\mathbf{v}}_H \bullet \nabla_H h_t.$$

Note that both of these continuity equations may be expressed in conservation form:[16]

$$\frac{\partial D_T}{\partial t} + \nabla \bullet (D_T \bar{\mathbf{v}}_H) = 0,$$

where D_T is the total depth of the fluid.

8.5 Geostrophic Balance

The geostrophic balance occurs when the flow is horizontal and steady and the inertial and viscous terms are negligibly small. In this case the horizontal momentum equation derived in § 8.2.1 simplifies to

$$f\mathbf{1}_z \times \mathbf{v}_H = -(1/\rho)\nabla_H p = g\nabla_p Z,$$

where Z is a constant pressure surface[17] and the subscript p means holding pressure constant.[18] In the atmosphere $\rho \approx 1$ kg·m^{-3} and a typical wind speed is $\|\mathbf{v}_H\| \approx 10$ m·s^{-1}; a geostrophic flow of this magnitude is balanced by a pressure gradient $\|\nabla_H p\| \approx 10^{-3}$ Pa·m^{-1} $\approx 10^{-5}$ atm·km^{-1}. The typical variation of atmospheric pressure associated with weather systems is several percent. A pressure difference of roughly 5% is maintained by a geostrophic flow of 10 m/s and width of 500 km. In the ocean, $\rho \approx 10^3$ kg·m^{-3} and a typical flow speed is $\|\mathbf{v}_H\| \approx 10^{-2}$ m·s^{-1}, so once again $\|\nabla_H p\| \approx 10^{-3}$ Pa·m^{-1} $\approx 10^{-5}$ atm·km^{-1}. A pressure difference that is 5% of atmospheric is compensated by a sea-surface elevation change of about 5 cm.

[16] See Appendix D.5.
[17] See § 8.2.1 and Appendix D.8.1. Often the geopotential surface $\Phi = gZ$ is used in place of Z.
[18] Typically the distinction between ∇_p and ∇_H is negligible. However, this distinction is important in developing the thermal wind equations; see § 26.2.

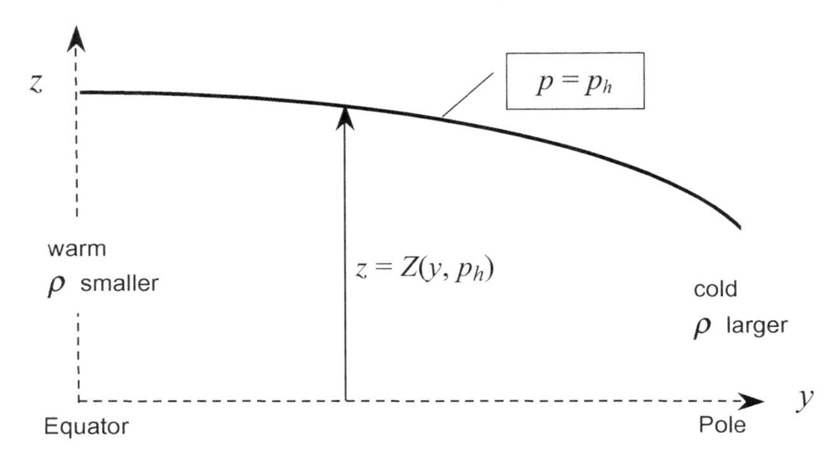

Figure 8.5 An illustration of the variation of the height Z of an isobaric surface with distance north from the equator due to the variation of density with latitude resulting from the latitudinal variation of temperature.

The horizontal momentum equation is readily solved for the *geostrophic flow* $\mathbf{v}_H$ by taking the cross product of the geostrophic balance with $\mathbf{1}_z$:

$$\mathbf{v}_H = \frac{1}{\rho f}\mathbf{1}_z \times \nabla_H p = -\frac{g}{f}\mathbf{1}_z \times \nabla_p Z\,.$$

In the geostrophic balance, the velocity vector is horizontal and parallel to isobars. According to this equation, it is not possible to have horizontal flow in the atmosphere or ocean without a pressure gradient. The pressure field associated with the vertical vorticity is governed by the divergence of the horizontal momentum equation:[19] $\nabla_H^2 p = \rho f \zeta$, where ζ is the vertical component of vorticity.

The isobaric (constant-pressure) surfaces tend to be lower at the poles than at the equator, as illustrated in Figure 8.5. Similarly, an isobaric surface is high over warm regions and low over cold regions. Various isobaric surfaces $Z(p_0)$, particularly that for 500 millibars, are used in the analysis of weather systems.

8.5.1 Geostrophic Flow in the Atmosphere

In Earth's mid-latitude atmosphere, planetary-scale geostrophic flow is predominantly from west to east (that is, the flow is *westerly*); such a flow is balanced by a pressure gradient pointing toward the equator. Using spherical coordinates $\{r, \theta, \phi\}$ with θ being the colatitude and ϕ being longitude (increasing eastward), a westerly flow may be expressed as

$$\mathbf{v}_H = u(r,\theta)\,\mathbf{1}_\phi\,.$$

[19] See § 8.3.2.

This flow is balanced by a pressure

$$p(r,\theta) = p_0 - \rho R \int^{\theta} f(\grave{\theta}) u(r,\grave{\theta})\, \mathrm{d}\grave{\theta},$$

where R is the radius of Earth. In the tropics the geostrophic flow is predominantly from east to west (the *easterlies* or *trade winds*); see § 26.1 and Figure 26.1.

The horizontal solution for $\mathbf{v}_H$ applies to each elevation in the atmosphere. If the density is a function only of pressure then the factor $\nabla_H p/\rho$ and the horizontal velocity are the same at all elevations; such flow is called *barotropic*. If the density depends on both pressure and temperature, then the flow is said to be *baroclinic* and the horizontal velocity varies with elevation. The nature of this flow is explored in § 26.2.

8.5.2 Geostrophic Flow in the Ocean

The isobaric surface Z has been introduced in the context of atmospheric dynamics. When we are studying ocean dynamics, the top of the ocean is a constant-pressure surface of particular interest, where $p = p_a$ (the atmospheric pressure), and we can replace Z with h (the sea-surface elevation). Now the steady geostrophic equation becomes

$$f\mathbf{1}_z \times \mathbf{v}_H = -g\nabla_H h \qquad \text{or equivalently} \qquad \mathbf{v}_H = g\mathbf{1}_z \times \nabla_H h/f.$$

Note that flow is normal to the gradient of the surface and is independent of depth.

This simple equation ignores a number of subtleties and complications that affect actual flows within the ocean, such as

- vertical variations of density that separate the ocean into a well-mixed upper layer and stably stratified bottom layer (see § 7.2.2);
- wind stresses applied to the surface (see § 11.5.1, §25.5.4 and Chapter 27);
- temporal variations of the acceleration of gravity that drive the tides (see § 12.2);
- viscous forces that are important in Ekman layers near horizontal boundaries (see § 25.5.3 and § 25.5.4); and
- variations of density that drive thermohaline currents (see § 27.4).

8.6 Quasi-Geostrophic Equations

We now turn our attention to the small deviations from the geostrophic balance that lead to *quasi-geostrophic* (QG) waves and flows. These motions are governed by the inviscid version of the thin-layer equations developed in § 8.2.2. Adopting the β-plane approximation the governing equations, expressed in local Cartesian coordinates, are

$$\frac{\partial u}{\partial x} + \frac{\partial v}{\partial y} + \frac{\partial w_p}{\partial p} = 0,$$

$$\frac{\mathrm{D}u}{\mathrm{D}t} - f_0 v - \beta(y - y_0)v = -g\frac{\partial Z}{\partial x}$$

$$\text{and} \qquad \frac{\mathrm{D}v}{\mathrm{D}t} + f_0 u + \beta(y - y_0)u = -g\frac{\partial Z}{\partial y}.$$

In the study of wave motions, the momentum equations are often linearized by neglecting the nonlinear portion of the inertia terms:

$$\frac{\partial u}{\partial t} - f_0 v - \beta(y - y_0)v = -g\frac{\partial Z}{\partial x}$$

$$\text{and} \qquad \frac{\partial v}{\partial t} + f_0 u + \beta(y - y_0)u = -g\frac{\partial Z}{\partial y}.$$

8.6.1 QG Scale Analysis

Let's take a moment and estimate the sizes of terms in the QG equations using scale analysis.[20] In order to do so, we must make some assumptions regarding the structure of the flow. Specifically we will assume that

- the flow is isotropic in the horizontal plane (this permits consideration of only one component of the momentum equation) with $u \sim v \sim U$ and $x \sim y \sim L$, where L and U are characteristic length scale and speed, respectively;
- the timescale is determined by the material derivative ($t \sim L/U$);
- the trigonometric functions are of unit order; and
- the pressure terms are in balance with the primary Coriolis terms ($Z \sim |f_0|LU/g$).

Now let's consider the structure of the horizontal momentum equations on a sphere. They each contain six terms, four of which are given in the QG equations: inertia, primary Coriolis, beta and pressure, plus two that we have already ignored: curvature and viscous. Since the pressure term is the same magnitude as the primary Coriolis, we need not consider that term in the scale analysis. Using the primary Coriolis term as the yardstick, the force ratios are as follows:

$$\frac{\text{inertia}}{\text{Coriolis}} = \frac{U}{L|f_0|} \equiv \mathrm{Ro}, \qquad \frac{\text{beta}}{\text{Coriolis}} = \frac{\beta L}{|f_0|},$$

$$\frac{\text{viscous}}{\text{Coriolis}} = \frac{\nu}{L^2|f_0|} \equiv \mathrm{Re} \quad \text{and} \quad \frac{\text{curvature}}{\text{Coriolis}} = \frac{U}{R|f_0|},$$

where R is the radius of Earth, Ro is the *Rossby number* and Re is the *Reynolds number*. At mid-latitudes the parameter values are $|f_0| \approx 10^{-4}\ \mathrm{s}^{-1}$, $R = 6.37 \times 10^6$ m, $\beta \approx 1.6 \times 10^{-11}\ \mathrm{m}^{-1}\cdot\mathrm{s}^{-1}$ and $\nu \approx 1.5 \times 10^{-5}\ \mathrm{m}^2\cdot\mathrm{s}^{-1}$. Estimating that $L \approx 10^6$ m and $U \approx 10\ \mathrm{m}\cdot\mathrm{s}^{-1}$,

[20] See § 2.4.5.

we have

$$\mathrm{Ro} \approx 0.1\,, \qquad \beta L/|f_0| \approx 0.16\,,$$

$$\mathrm{Re} \approx 1.5 \times 10^{-13} \quad \text{and} \quad U/R|f_0| \approx 0.016\,.$$

It is evident that the viscous force based on a laminar viscosity is exceedingly small; even if we had used the turbulent viscosity,[21] this term would be small. In addition, the magnitude of curvature term is only a few percent of the Coriolis term and may be neglected. However, both the inertia and β terms are at least 10 percent as large as the Coriolis term and are potentially important.

Transition

The present chapter investigated the important dynamic factors affecting rapid rotating fluids that have a small vertical extent.

This is the end of Part II, in which we have investigated the kinematics, dynamics, equation of state and rheology of a continuous body, divided the governing equations into static state and perturbation and introduced some basic concepts related to rapidly rotating fluids. In the following parts we will apply these equations to a variety of waves and flows, beginning in Part III with the study of waves that are not affected by Earth's rotation.

[21] See § 23.5.2.

Part III
Waves in Non-Rotating Fluids

Our investigation of waves is divided into two parts depending whether rotation of the Earth has a part to play. In the present part we limit our attention to those waves unaffected by rotation of Earth, with those waves that are influenced by this rotation investigated in Part IV. We begin our investigation in the following chapter with an introduction to – and brief survey of – the types of waves that occur on and within Earth. The simplest type of wave, the sound wave, is quantified and discussed in § 9.3. The most common type of elastic wave is the seismic wave; these waves occur continuously as the brittle portions of Earth's crust rub against one another. Seismic waves that are unaffected by Earth's surface, called body waves, are considered in § 10.1 and § 10.2, while those that affected the surface are considered in § 10.3 and § 10.4. While body waves in fluids (we are thinking of air and water) are essentially sound waves, edge waves in fluids (particularly water) are distinctly different from elastic edge waves. We investigate deep-water waves from a mathematical perspective in Chapter 11 and in § 11.5 discuss how these waves occur and interact on the oceans. The rich variety of waves occurring in shallow water is investigated in Chapters 12 and 13. Our survey of non-rotating waves concludes in Chapter 14 with an investigation of capillary, interfacial and internal waves.

9

Introduction to Waves

We will limit our investigation to *mechanical* waves, and will not consider hydromagnetic waves that occur in the upper regions of the atmosphere and in Earth's core. Mechanical waves – capable of transmitting an elastic deformation or fluid motion without appreciable loss – exist due to the presence of a force that acts to restore the body to its initial or reference state.[1] Perhaps the most familiar mechanical wave is the standing wave produced by plucking a stretched string or wire; this is the basis of all stringed musical instruments. We know from basic mechanics that the tone produced by a vibrating string of length L is characterized by the frequency ω, with $\omega \propto \sqrt{\sigma/\rho}/L$, where σ is the tension and ρ is the density per unit length of the string. The restoring force is the tension in the string; the frequency produced by a guitar string increases as the magnitude of the restoring force increases. (An instrument that is flat is brought into tune by tightening the string.)

If an unbounded (or very long) string is plucked, a traveling wave is generated, having speed proportional to $\sqrt{\sigma/\rho}$. As with frequency, the wave speed increases as the magnitude of the restoring force increases. We will see that this is a general property of waves; waves having large restoring forces travel rapidly, while those with weak restoring forces travel slowly. For example, waves in air, water or Earth's interior driven by compression of the medium travel relatively rapidly, while waves driven by gravity acting on a displaced surface travel relatively slowly.

In elastic bodies, compressibility and elasticity provide the necessary restoring forces to sustain both longitudinal and transverse waves. The restoring force for longitudinal fluid waves may be either fluid compressibility (as for sound waves) or gravity acting on the displacement of a free surface. In the latter case, the displacement of the surface provides the requisite "compressibility". Transverse waves can occur in a rotating fluid, with the Coriolis force being the restoring force, and evanescent transverse waves can occur in a viscous fluid; see Appendix F.3.

There is a wide variety of possible mechanical waves that can occur in the interior or near a boundary of a continuous body, so it is helpful to begin by categorizing them. The primary categorization is the location within the medium: throughout the body (*body waves*) or confined near an edge of the body (*edge waves*). The secondary categorization

[1] In addition, evanescent waves, not involving a restoring force, are possible: see Appendix F.3.

is the direction of displacement relative to the direction of propagation (longitudinal or transverse). Longitudinal waves have displacements primarily in the direction of wave propagation, while transverse waves have significant lateral (horizontal) displacements, normal to the direction of propagation. The tertiary categorization is the rheological nature of the medium: elastic or fluid. This categorization gives us eight possible types of non-rotating waves. In the following summary the nature of the restoring force for each type of wave and the section in which the wave is considered are indicated in parentheses.

- Body waves
  - Longitudinal body waves
    - Longitudinal elastic body waves. The most familiar example of these are seismic P waves (elastic compressibility; § 10.1).
    - Longitudinal fluid body waves. The most familiar example of these are sound waves (fluid compressibility; § 9.3).
  - Transverse body waves
    - Transverse elastic body waves. The most familiar example of these are seismic S waves (elastic rigidity; § 10.2).
    - Transverse fluid body waves. Since these waves rely on the elastic properties of medium, they do not exist in a non-rotating fluid. However, the Coriolis force provides the necessary "elasticity" in a rotating fluid, permitting the existence of Rossby waves (Coriolis force; § 15.3).
- Edge waves
  - Longitudinal edge waves
    - Longitudinal elastic edge waves. The most familiar example of these are seismic Rayleigh waves (elastic compressibility; § 10.3).
    - Longitudinal fluid edge waves. The most familiar example of these are deep-water waves that occur commonly on oceans and other open bodies of water (gravity; Chapter 11). Shallow-water waves (gravity; Chapters 12 and 13) are included in this category.
  - Transverse edge waves
    - Transverse elastic edge waves. The most familiar example of these are seismic Love waves (elastic rigidity; § 10.4).
    - Transverse fluid edge waves. As with transverse fluid body waves, these waves rely on the elastic properties of medium; proper transverse edge waves do not exist in a non-rotating fluid. However, the Coriolis force permits the existence of equatorial and topographic Rossby waves (Coriolis force; § 17.3 and § 18.2), and viscosity permits the existence of evanescent waves; see Appendix F.3.

Beyond this, there are a various other types of non-rotating waves to consider, depending on complications of dynamics or geometry. These include

- the tides, which are forced waves in the ocean driven by the gravitational attraction of Moon and Sun (§ 12.2)

- seiches, which are resonant waves in partially enclosed basins (§ 12.3)
- capillary waves, which involve surface tension (§ 14.1)
- interfacial waves, which involve two layers of fluid (§ 14.2); and
- internal gravity waves, which rely on a stable density stratification (§ 14.3).

Before delving into the analyses of these waves, it may be helpful to discuss the behavior of waves (in § 9.1) and introduce some approaches to analyzing waves (in § 9.2).

9.1 General Wave Behavior

In the present context, a wave is a transient displacement of a continuous body characterized by a frequency ω and a wavenumber k; these are related to the period P and wavelength Λ of the wave by

$$\omega P = 2\pi \qquad \text{and} \qquad k\Lambda = 2\pi\,.$$

The physical nature of a wave is encapsulated in its *dispersion relation*, $\omega = \omega(k)$, which is extracted from the governing equations. With the dispersion relation in hand, we can readily calculate two wave speeds: the *phase speed*, U_p, and the *group speed*, U_g, defined as[2]

$$U_p \equiv \omega/k \qquad \text{and} \qquad U_g \equiv \mathrm{d}\omega/\mathrm{d}k\,.$$

The phase speed, as the name implies, is the speed at which lines of constant phase travel. To clarify this, consider a wave varying with x and t as $\cos\theta$, where $\theta = kx - \omega t$ is the phase of the wave.[3] An observer moving with speed U_p sees the (local) phase remain constant. For example, a surfer successfully riding a wave moves with its phase speed.

The group speed is the speed at which the energy of the wave travels. Propagation of wave energy is not apparent when considering a uniform wavetrain, because there is no spatial variation of wave amplitude. The role of the group speed is made apparent if we consider a *wave packet* having an amplitude that varies slowly with position, such as

$$q = q_0 e^{-\chi^2}\cos\theta\,, \qquad \text{where} \qquad \chi = |x - U_g t|/L\,,$$

q is any dependent variable associated with the wave, q_0 is a constant and L is a spatial distance satisfying $1 \ll kL$, so that the envelope $e^{-\chi^2}$ varies on a spatial scale much greater than the wavelength. The wave energy, quantified by the envelope of the packet, moves with the group speed, while the magnitude of q oscillates within the envelope as $\theta = kx - \omega t$ varies.

If the dispersion relation is linear (that is, if ω is a linear function of k), there is no distinction between phase and group speeds: $U_p = U_g$ and the wave speed is independent of k. Such waves are said to be *non-dispersive*. That is, if a packet contains non-dispersive

[2] The scalar phase and group speeds are the magnitudes of the phase and group vectors; this relation is considered in Appendix F.1, which discusses wave theory in more detail.

[3] The symbol θ, which is used to represent spherical angle in Chapter 8, is here being used to represent phase.

waves having differing values of k, the waves all move with the same speed and the packet does not broaden (or disperse) with time. However, if the dispersion relation is nonlinear then $U_p \neq U_g$ and both the phase and group speeds depend on k; in this case the waves are said to be *dispersive*. Now waves within a packet having differing values of k move at differing speeds and the packet broadens (disperses) with time. This is the case, for example, for deep-water waves, which we investigate in Chapter 11; for these waves $U_p = 2U_g$; individual waves travel faster than an isolated packet of waves. A surfer riding a deep-water wave within a packet moves to the leading edge, where the waves peter out.

We will see that sound waves (studied in § 9.3) and seismic P and S body waves (studied in § 10.1 and § 10.2) have speeds independent of the wavenumber and so are non-dispersive. An initial disturbance is propagated at the wave speed without change of form. This property of P and S waves makes them particularly useful in deciphering the internal structure of Earth. On the other hand, edge waves (studied in § 10.3 and § 10.4) have speeds which depend on the wavelength and so are dispersive. This makes seismic edge waves more difficult to use in deciphering the structure of the material they have traversed.

With some preliminary understanding of the possible behaviors of waves in hand, it is time to move on in the next section to a discussion of the usual approaches to analyzing waves.

9.2 Orientation to Wave Analysis

As we have seen, the Navier–Stokes equation is nonlinear. However, typically wave motion within a fluid is a small deviation from a known (often static) state of the body and a set of linear equations governing this perturbation can be obtained by a standard procedure, as described in Chapter 7.[4] The approach to analyzing the linear perturbation equations depend on several factors, including whether

- the coefficients of the equations are constant or functions of position;[5]
- the waves are body waves or edge waves;
- the spatial domain is finite or infinite; and
- the waves are dispersive or non-dispersive.[6]

The analysis begins with an *ansatz*: an assumed form (educated guess) for the solution. Since the coefficients in the governing partial differential equations are independent of time, the time dependence is harmonic (sine and cosine) and/or exponential. If the coefficients are independent of position, the spatial dependence is also harmonic, though this may include real exponentials if edge waves are under consideration. If the coefficients depend on position, then the spatial dependence in that variable must be determined from the governing equations.[7]

[4] We will consider nonlinear shallow-water waves in Chapter 13.
[5] The coefficients are invariably independent of time.
[6] This may not be known until after the analysis.
[7] For example, as in the study of equatorial waves in Chapter 17.

It is common practice in wave analysis to represent the dependent variables as a Fourier sum or integral (depending whether the spatial domain is finite or infinite) of harmonic fundamental solutions. A major advantage of the Fourier representation is that it turns calculus (the governing equations) into algebra (the dispersion relation). The theory behind these representations is summarized in Appendix A.5.

Let's consider the simplest case with elastic or fluid wave motion governed by a linear partial differential equation with constant coefficients and look at waves traveling in one Cartesian direction x. Such a set of equations admits a simple harmonic *fundamental solution* of the form $\cos\theta = \mathrm{Re}[e^{i\theta}]$ with $\theta = kx - \omega t$,[8] representing a wave traveling in the direction of increasing x with speed $U = \omega/k$. An important property of this fundamental solution is that it is an element of a *complete set* of functions, permitting an arbitrary initial condition to be expressed as a linear combination of the elements of this set.[9]

It is readily apparent that, with the assumed waveform, each x derivative is replaced by a factor ik and each t derivative is replaced by a factor $-i\omega$. Consequently, the partial differential equation governing the wave behavior is transformed into a polynomial algebraic relation involving k and ω; this is the dispersion relation. Note that

- the dispersion relation will contain real coefficients if the each term in the governing equation contains an even number of derivatives;
- if a term in the equation contains an odd number of derivatives, ω will have a complex part, indicating that wave energy is being degraded with time or distance; and
- the order of the polynomial in ω represents the number of wave modes, except that some types of waves - such as sound waves - have two identical modes (representing waves traveling in two opposite directions).

The full solution to a wave problem is a combination of individual harmonics, with the form being dictated by the nature of the wave (dispersive or non-dispersive), as well as the spatial extent of the domain. The various forms of harmonic solution are summarized as follows.

9.2.1 Non-Dispersive

If the waves are non-dispersive, then the appropriate ansatz for a dependent variable q is

$$q(x,t) = \tilde{q}(x \pm Ut)$$

and $U = \omega/k$ is the wave speed. The function $\tilde{q}$ is determined by the initial condition:

$$\tilde{q}(x) = q(x,0)\,.$$

[8] Where $\mathrm{Re}[e^{i\theta}]$ refers to the real component of $e^{i\theta}$.
[9] See Appendix A.5.

9.2.2 Dispersive, Infinite Domain

If the coefficients in the governing equation are independent of position, the wave is dispersive and domain is infinite, the ansatz for a dependent variable q is

$$q(x,t) = \int_k \tilde{q}(k) e^{i\theta} dk + \text{c.c.},$$

where $\theta = kx - \omega t$ and "c. c." means the complex conjugate of the preceding term (This is necessary to make $q(x,t)$ real.). The function $\tilde{q}$ is determined by Fourier inversion of the initial condition:

$$\int_k \tilde{q}(k) e^{ikx} dk + \text{c.c.} = q(x,0).$$

For more detail, see Appendix A.5.

9.2.3 Dispersive, Finite Domain

If the coefficients are independent of position, the wave dispersive and domain finite, the ansatz for a dependent variable q is

$$q(x,t) = \sum_k \tilde{q}_k \cos(\theta + \theta_k) = \tfrac{1}{2} \sum_k \tilde{q}(k) e^{ikx - i\omega t + i\theta_k} + \text{c.c.}.$$

The discrete values of k are determined by conditions at the boundary of the domain. The coefficients $\tilde{q}_k$ and θ_k are determined by Fourier inversion of the initial condition:

$$\sum_k \tilde{q}_k \cos(kx + \theta_k) = \tfrac{1}{2} \sum_k \tilde{q}(k) e^{ikx + i\theta_k} + \text{c.c.} = q(x,0).$$

When considering edge waves the formulation should include the exponential factor e^{mz}, with z pointing from the edge into the medium. If the fluid is rotating, it may be necessary to include the harmonic term e^{ily} in the ansatz.

This completes our introductory survey of waves. This chapter concludes in the following section with an investigation of the simplest form of wave: the sound wave.

9.3 Sound Waves

The most familiar, ubiquitous and useful form of waves are sound waves in air. These are classified as longitudinal fluid body waves, and their propagation is governed by a simplified form of the perturbation equations of conservation of mass and momentum for fluids that were presented in § 7.3.2, together with the equation of state.

Let's consider a planar packet of sound waves traveling in the x direction through air of constant density, with the speed of the fluid denoted by u: $\mathbf{v} = u\mathbf{1}_x$. The Coriolis, gravitational and viscous forces are unimportant and will be neglected. Also, with the fluid speed small, the nonlinear inertia term in the momentum equation is negligibly small. Now $\nabla = \mathbf{1}_x \partial/\partial x$ and the perturbation governing equations simplify to

$$\frac{\partial \rho'}{\partial t} + \rho_0 \frac{\partial u}{\partial x} = 0 \qquad \text{and} \qquad \rho_0 \frac{\partial u}{\partial t} = -\frac{\partial p'}{\partial x},$$

plus the equation of state $\rho_0 p' = K\rho'$, where K is the incompressibility and ρ_0 is the constant density of unperturbed air. The most important variable is pressure; our ears sense variations of pressure and our brains turn them into recognizable signals. Eliminating the other two variables, we see that the pressure is governed by the *wave equation*:

$$\rho_0 \frac{\partial^2 p'}{\partial t^2} - K \frac{\partial^2 p'}{\partial x^2} = 0.$$

This equation has fundamental solution of the form $p' = \tilde{p}\cos(kx - \omega t)$ provided ω satisfies the *dispersion relation*

$$\omega = \sqrt{K/\rho_0}\,k.$$

The phase and group speeds of the wave are identical and equal to the *speed of sound*, denoted by U_s:

$$U_s = U_p = U_g = \sqrt{K/\rho_0}.$$

The fundamental solution describes a progressive wave moving in the direction of increasing x. The wave equation has two time derivatives, so we expect two solution modes. The second mode is a progressive wave traveling in the opposite direction: $p' = \tilde{p}\cos(kx + \omega t)$; altogether

$$p'(x,t) = \tilde{p}_+ \cos(kx + \omega t) + \tilde{p}_- \cos(kx - \omega t).$$

Using well-known trigonometric identities, these two progressive wave solutions may be converted to standing-wave solutions

$$p'(x,t) = \tilde{p}_C \cos(kx)\cos(\omega t) + \tilde{p}_S \sin(kx)\sin(\omega t).$$

Standing waves occur when progressive waves become trapped in an enclosure having a length equal to a certain multiple of the length of the wave, causing *resonance*. This is the principle behind woodwind and brass musical instruments and pipe organs. Resonant waves can have particularly large amplitude; anyone who likes to sing in the shower can attest to this.

The plane progressive wave solutions may be obtained by factoring the wave equation:

$$\frac{\partial^2 p'}{\partial t^2} - U^2 \frac{\partial^2 p'}{\partial x^2} = \left(\frac{\partial}{\partial t} - U\frac{\partial}{\partial x}\right)\left(\frac{\partial}{\partial t} + U\frac{\partial}{\partial x}\right)p' = 0;$$

this is satisfied if

$$\frac{\partial p'}{\partial t} \pm U \frac{\partial p'}{\partial x} = 0.$$

The wave equation given above governs the behavior of planar waves, dependent on one Cartesian coordinate and time. Typically sound waves originate from a point source in three dimensions and emanate away in the (spherical) radial direction, r. In this case the wave equation becomes[10]

$$\rho_0 \frac{\partial^2 p'}{\partial t^2} - K \frac{1}{r^2} \frac{\partial}{\partial r} \left(r^2 \frac{\partial p'}{\partial r} \right) = 0 .$$

The outward progressive wave has the form

$$p'(r,t) = \frac{r_0}{r} \tilde{p} \cos(kr - \omega t) .$$

Note that the amplitude decays as $1/r$; this decay is the result of geometric spreading of the wave. If the wave direction were reversed, the result would be an *implosion*, in which the amplitude of the wave increases without bound as the wavefront approaches $r = 0$.[11]

Sound waves are *non-dispersive*; their phase speed is independent of the frequency. This means that waves of different frequencies travel as a coherent package and do not disperse. This feature makes sound waves vitally important for communication. That is, a general wavetrain having the initial form[12]

$$p'(x,0) = p_I(x) = \int_k \tilde{p}(k) \cos(kx) \mathrm{d}k$$

is given by

$$p'(x,t) = p_I(x - ut) = \int_k \tilde{p}(k) \cos(k(x - Ut)) \mathrm{d}k$$

at a later time, where $\tilde{p}(k)$ codifies, among other sounds, all words in all languages. If U had depended on k the message would have become garbled.

The apparent speed, U_A, and frequency, ω_A, of sound, measured by an observer moving in the direction of the wave at speed U_O, are

$$U_A = U - U_O = U(1 - Ma) \qquad \text{and} \qquad \omega_A = \omega - kU_O ,$$

where $Ma = U_O/U$ is the *Mach number*. The moving observer sees the length $\Lambda = 2\pi/k$ of the wave unchanged, but measures a change in frequency ω_A. This is the well-known *Doppler effect*. As the observer's speed (in the direction of the wave) increases, the measured speed and frequency of the wave is correspondingly slower. An observer moving with (or faster than) the speed of sound (having $1 \leq Ma$), cannot hear anything shouted from behind.

[10] See Appendix A.3.5.

[11] Such waves are very rare in nature. The supernova explosion of a massive star is preceded by an implosive collapse of its iron-nickel core. Also, they are the basis of one means of developing fusion power, called *inertial confinement* and implosive waves initiated by the collapse of small bubbles play a central role in the curious phenomenon of *sonoluminescence*.

[12] This is a *Fourier integral*; see Appendix A.5.

By the same token, sounds emitted by an object moving at or faster than U cannot propagate ahead of the object. Instead, they pile up and form a *shock wave*; a wave of finite amplitude that travels faster than a sound wave. Shock waves are created naturally in Earth's atmosphere by lightning bolts, the impact of comets and asteroids and explosive eruptions of andesitic volcanoes.

9.3.1 Speed of Sound in Air

In the introduction to this chapter, we noted that the speed of a compressive wave depends on the rigidity of the material. In this section we quantified that relation, finding that $U = \sqrt{K/\rho}$. Let's briefly consider the implications of this relation for the speed of sound in the atmosphere.

To a good approximation, the atmosphere behaves as an ideal gas with[13]

$$U = \sqrt{\gamma R_s T},$$

where T is the absolute temperature, γ is the ratio of specific heats and R_s is the specific gas constant. Air is predominantly diatomic with[14] $\gamma = 7/5$ and $R_s \approx 287.1\ \text{m}^2\cdot\text{s}^{-2}\cdot\text{K}^{-1}$, giving $U \approx 343\ \text{m}\cdot\text{s}^{-1}$ or $1126\ \text{ft}\cdot\text{s}^{-1}$. Sound travels 1 km in roughly 3 seconds or equivalently 1 mile in 5 seconds. This provides a useful method for determining the distance from a bolt of lightning by counting the seconds that elapse between arrivals of the lightning flash and thunder. The value of R_s increases with water content,[15] so that the speed of sound is slightly greater in moist air. The presence of fog acts to dampen the propagation of sound waves, causing sounds to seem muffled.

In comparison to air, water is significantly less compressible (that is, more incompressible), with the result that, in spite of the increased density of water, the speed of sound in water is $U \approx 1.5\ \text{km}\cdot\text{s}^{-1}$,[16] which is roughly 1 mile per second or about five times the speed of sound in air. We will see in § 10.1 that the speed of sound in silicate rocks is roughly 30 times that in air: nearly $6\ \text{km}\cdot\text{s}^{-1}$.[17]

Transition

This chapter has provided us with an orientation to waves in non-rotating media, outlined several approaches to modeling waves and discussed some of the possible wave behaviors. Following this orientation, in § 9.3 we investigated sound waves, which are compressive waves involving longitudinal displacement of the fluid. A variety of waves that occur in an elastic body, such as Earth's mantle, are investigated in the next chapter.

[13] From Appendix E.8.6.
[14] See Appendix E.8.5 and Table B.5. The value quoted for R_s is for dry air.
[15] See Appendix E.9.3.
[16] See Table B.5.
[17] See Table 10.1.

10
Elastic Waves

As noted in the introduction to this part, there are four flavors of elastic waves: longitudinal and transverse body waves and longitudinal and transverse edge waves. These are investigated in the first four sections of this chapter: longitudinal body waves in § 10.1, transverse body waves in § 10.2, longitudinal edge waves in § 10.3 and transverse edge waves in § 10.4. This chapter includes a summary of elastic body waves (that is, seismic body waves) in Earth's interior in § 10.5.

In the previous chapter, we dealt with compressive waves in a fluid with the longitudinal speed of individual particles being a function of position and time: $u(x,t)$. In this chapter we will investigate waves in an elastic medium, with the dependent variable being displacement rather than speed. To facilitate comparison of the elastic-wave equations with the sound-wave equation studied in § 9.3, we will denote longitudinal displacement with the same symbol (u) as was used to represent speed.

10.1 Longitudinal Body Waves

Longitudinal body waves can exist in both fluid and elastic bodies. Longitudinal body waves in fluids are just sound waves, which have been considered in § 9.3. This section considers longitudinal body waves in elastic bodies. These waves are governed by the perturbation Navier equation presented in § 7.3.1. Again, let's consider a planar packet of waves traveling in the x direction, with the displacement denoted by u: $\mathbf{u} = u\mathbf{1}_x$. Now the Navier equation becomes

$$\rho_r \frac{\partial^2 u}{\partial t^2} = \left(K + \frac{4}{3}\mu\right) \frac{\partial^2 u}{\partial x^2}.$$

As with sound waves, this equation has fundamental solutions of the form $u = \tilde{u}\cos(kx \pm \omega t)$, provided ω satisfies the dispersion relation

$$\omega = \sqrt{\frac{K}{\rho_r}\left(1 + \frac{4\mu}{3K}\right)}\, k.$$

The phase and group speeds of these non-dispersive waves are identical; $U_g = U_p = U_l$, where

$$U_l \equiv \sqrt{\frac{K}{\rho_r}\left(1+\frac{4\mu}{3K}\right)}.$$

In seismology these are called *P waves* – P for primary; they are the first arrivals at a point distant from an earthquake. Note that P waves are non-dispersive. While this property is desirable in sound waves, it is less so for seismic waves, because it means that the destructive seismic energy does not disperse and these waves can cause significant damage at great distances.[1] Note that the elastic term causes the phase speed to be greater than in a fluid having the same bulk modulus.

An initial planar wavetrain of general shape, expressed as a Fourier integral

$$u(x,0) = u_I(x) = \int_k \tilde{u}(k)\cos(kx)\mathrm{d}k,$$

is at a later time given by

$$u(x,t) = u_I(x - U_l t) = \int_k \tilde{u}(k)\cos\big(k(x - U_l t)\big)\mathrm{d}k.$$

10.2 Transverse Body Waves

Now let's consider a planar packet of waves traveling in the x direction, but with the displacement, denoted by v, in the transverse (y) direction : $\mathbf{u} = v\mathbf{1}_y$. The Navier equation reduces to

$$\rho_r \frac{\partial^2 v}{\partial t^2} = \mu \frac{\partial^2 v}{\partial x^2}.$$

This equation has fundamental solutions of the form $v = \tilde{v}\cos(kx \pm \omega t)$, provided

$$\omega = \sqrt{\mu/\rho_r}\,k.$$

The phase and group speeds of these non-dispersive waves are identical; $U_g = U_p = U_t$, where

$$U_t \equiv \sqrt{\mu/\rho_r}.$$

In seismology these are called *S waves* – S for secondary;[2] they travel more slowly and arrive later than P waves. Again, a packet of general shape may be expressed in terms of a Fourier integral or sum.

[1] Actually seismic wave amplitudes diminish due to geometric dispersion.
[2] These are also called shear waves.

10.3 Longitudinal Edge Waves

Edge waves, as the name implies, exist near the edge of a spatial domain. They are also called surface waves. Let's consider the elastic body having a plane boundary (an edge) at $z = 0$ and occupying the region $z < 0$ (z points upward). As before, we will consider waves propagating in the x direction, but now allow the displacement to depend on z as well as x and t. We are interested in waves which behave sinusoidally in x and t, as do body waves, but also decay exponentially in the z direction, i.e., normal to the boundary.

Longitudinal edge waves, called *Rayleigh waves*, have displacements in the x and z directions. The displacement vector, $\mathbf{u}$, associated with these waves is curl-free and may be expressed in terms of a scalar potential.[3] Let's consider a fundamental harmonic mode of the form

$$\mathbf{u} = \nabla\phi \qquad \text{with} \qquad \phi = \tilde{\phi}\cos(kx \pm \omega t)e^{mz},$$

where $\tilde{\phi}$ is a constant (not dependent on x, z or t) and m is a vertical wavenumber. Note that if $m < 0$ the exponential factor grows with distance from the edge. Such a solution is unphysical and we must discard it. The Navier equation is satisfied provided

$$\omega = \sqrt{\frac{K}{\rho_r}\left(1 + \frac{4\mu}{3K}\right)}\sqrt{k^2 - m^2}.$$

The phase speed of a Rayleigh wave is

$$U_p = \sqrt{\frac{K}{\rho_r}\left(1 + \frac{4\mu}{3K}\right)}\sqrt{1 - \frac{m^2}{k^2}}$$

and the surface displacement is given by

$$\mathbf{u} = -\tilde{\phi}k\sin(kx \pm \omega t)\mathbf{1}_x - \tilde{\phi}m\cos(kx \pm \omega t)\mathbf{1}_z.$$

The Rayleigh wave exists only if $0 < m < k$ and travels more slowly than a P wave. In the limit $m \to 0$ the Rayleigh wave becomes a P wave. When $m = k$, the wave is stationary, and the solution is not of the form of a wave if $k < m$.

Note that Rayleigh waves are dispersive; waves having differing values of k (for the same value of m) travel at differing speeds and so disperse;

$$U_g = \sqrt{\frac{K}{\rho_r}\left(1 + \frac{4\mu}{3K}\right)}\left(1 - \frac{m^2}{k^2}\right)^{-1/2}.$$

However, being confined to the near-surface region, they are less prone to geometric dispersion than are P waves. Dispersive waves are less useful for seismic inversion than are non-dispersive waves.

[3] See Appendix A.8.

Table 10.1. *Typical parameter values relevant to seismic body waves in Earth's interior. Note that 1 Gpa* = 10^9 $N{\cdot}m^{-2}$ = 10^9 $kg{\cdot}m^{-1}{\cdot}s^{-2}$.

Level in Earth	ρ (kg·m^{-3})	K (Gpa)	μ (Gpa)	U_l (km·s^{-1})	U_t (km·s^{-1})
Crust	2600	52	26.6	5.8	4.5
Upper mantle	3500	170	80	8.9	6.8
Lower mantle	5500	600	280	13.3	10.4
Outer core	11000	900	0	9.0	0
Inner core	13000	1400	170	11.2	5.1

10.4 Transverse Edge Waves

Transverse edge waves, called *Love waves*, have displacement in the transverse (y) direction. One harmonic mode is of the form

$$\mathbf{u}(x,z,t) = \tilde{v}\cos(kx \pm \omega t)e^{mz}\mathbf{1}_y\,,$$

where $\tilde{v}$ is constant (i.e., not dependent on x, z or t). The Navier equation is satisfied provided

$$\omega = \sqrt{\mu/\rho_r}\sqrt{k^2 - m^2}\,.$$

The phase speed of a Love wave is

$$U_p = \sqrt{\frac{\mu}{\rho_r}}\sqrt{1 - \frac{m^2}{k^2}}\,.$$

If $m = 0$, we recover the transverse body wave. As m increases, the speed of the Love wave decreases. When $m = k$, the wave is stationary, and the solution is not of the form of a wave if $k < m$.

We have found that elastic edge waves are simply elastic body waves modified by the presence of the boundary. In Chapter 11 we will see that fluid edge waves are physically and mathematically different from fluid body waves.

10.5 Seismic Body Waves

In seismology, the speeds of P and S waves, which we denote by U_l and U_t, are represented by α and β, respectively. Table 10.1 gives typical values of K, μ, U_l and U_t in Earth's interior.

Observations of the travel time of seismic body waves as functions of distance between source (earthquake) and receiver (seismograph) are the principal source of information about Earth's interior. (The form or shape of the wave also provides useful information.)

Table 10.2. *Labeling of seismic body waves in Earth's interior. For more detail, see Storchak et al. (2003).*

Location	Longitudinal wave	Transverse wave
Mantle	P	S
Outer core	K	—
Inner core	I	J
Reflected from distant surface	Repeat P	Repeat S
Reflected from near surface	p	s
Reflected from core–mantle boundary	c	c
Reflected from inner-core boundary	i	—

The procedure of analyzing seismic data to obtain information about the structure of Earth's interior is called *seismic inversion*. For example, the existence of Earth's core, previously inferred from geomagnetic and geodetic studies, was confirmed in 1906 by Richard Oldham, who analyzed the reflection and diffraction of seismic waves caused by the core.[4] Subsequently the liquidity of the outer core was detected and confirmed by the observation that U_t and hence μ are zero there (creating an S-wave shadow zone) and in 1936 Inge Lehmann discovered the solid inner core by analyzing the behavior of P waves. The spherically symmetric structure of Earth inferred from seismic inversion is embodied in the PREM model (Dziewonski and Anderson, 1981) (e.g., see Figure 7.2).

Seismic wave nomenclature has been established to identify the path of a P or S wave as it traverses the mantle, outer core and inner core, as shown in Table 10.2. These symbols are used in sequence to identify paths of seismic rays[5] in Earth's interior, as illustrated in Figure 10.1. Note that seismic waves can change character upon reflection.

We have focused our attention on traveling seismic body waves. Seismic waves can also occur as standing waves, called *free oscillations*. Investigation of free oscillations is beyond the scope of this monograph.

The P and S waves involve horizontal motions, in the direction of propagation for P waves and transverse for S waves. These are felt as swaying of the ground. Up and down motions during earthquakes are generated by Rayleigh waves; see § 10.3.

10.5.1 Effects of Seismic Waves

As seismic waves expand away from the epicenter of an earthquake,[6] their amplitudes diminish due to geometric attenuation. In an ideal medium the amplitudes of body waves

[4] Wave *diffraction* is the bending of a wave around an obstacle.
[5] A seismic ray is a line normal to the front of a seismic wave. (The first arrival of a seismic wave follows the ray path through Earth's interior.)
[6] The *epicenter* is the point on Earth's surface immediately above the location of an earthquake.

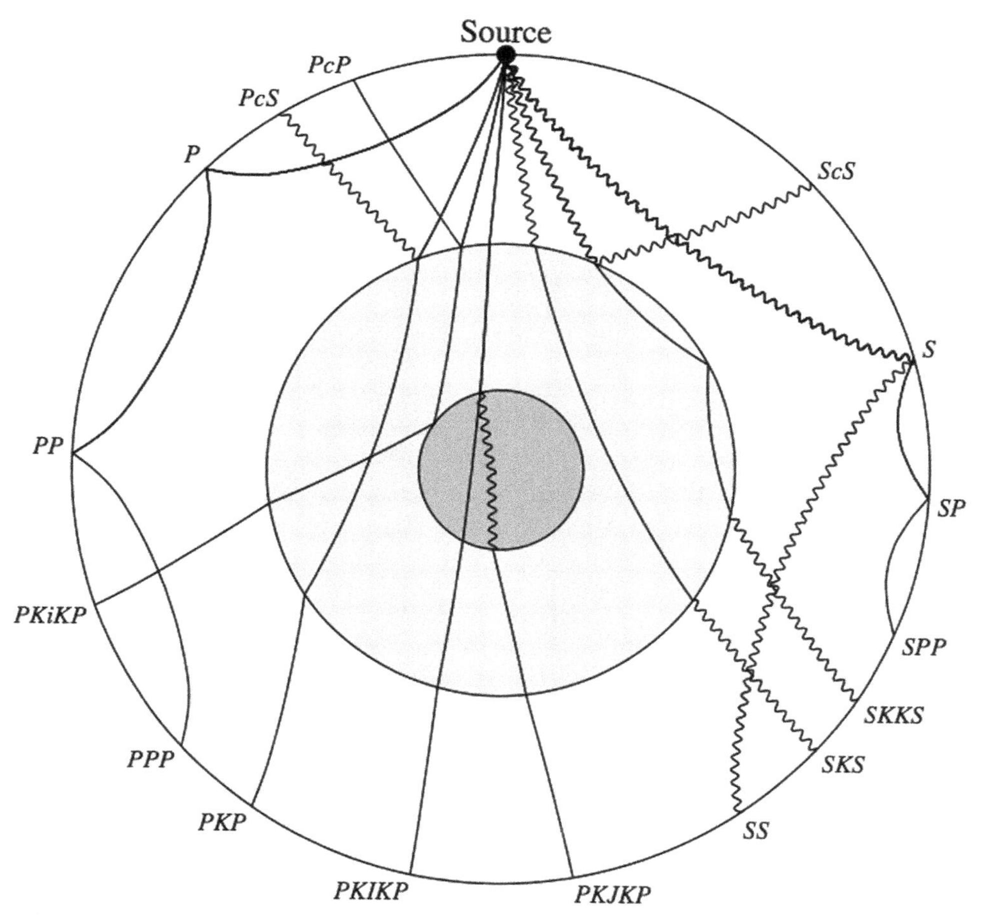

Figure 10.1 A sampling of seismic rays through Earth's interior showing nomenclature and typical path shapes. Wiggly lines indicate S waves. Reproduced, with permission, from Shearer (2009).

diminish as the inverse square of distance from the epicenter, while edge-wave amplitudes diminish more slowly, as the inverse of the distance. This means that the potentially damaging effects of edge waves increase relative to body waves as distance from the epicenter increases. The amplitudes of body and edge waves are also diminished as they traverse various structural inhomogeneities that dissipate and disperse the seismic wave energy. This loss of energy is quantified by the *seismic quality factor* Q, defined as $Q = 2\pi(E/\Delta E)$, where E is the wave energy and ΔE is the energy loss per cycle.

The magnitudes, and – by implication – the damaging effects, of seismic waves are measured by the *Richter scale*. This scale assigns a numerical value, ranging from 1 to 9, to an earthquake, with increasing values denoting increasingly strong earthquakes. Since there are several ways to measure the strength of earthquakes (e.g., ground motion,

Table 10.3. *Richter magnitude scale for rating the strength of earthquakes.*

Numerical Value	Description	Effects	Frequency
1.0–1.9	Micro	Not felt	Continual
2.0–2.9	Minor	Barely felt	10^6/year
3.0–3.9	Minor	Often felt	10^5/year
4.0–4.9	Light	Noticeable shaking	10^4/year
5.0–5.9	Moderate	Slight damage	10^3/year
6.0–6.9	Strong	Moderate damage	10^2/year
7.0–7.9	Major	Most buildings damaged	10/year
8.0–8.9	Great	Severe damage	1/year
9.0+	Great	Near total destruction	0.1/year

body-wave amplitude, energy released) there are in fact several Richter scales, but these scales have been calibrated so that they give similar magnitudes. Because the amplitudes and energies of earthquakes vary greatly, the Richter scale is logarithmic with base 10, so that an earthquake of magnitude $M + 1$ is 10 times stronger than one of magnitude M. Interestingly the frequencies decay similarly, with an earthquake of magnitude $M - 1$ occurring 10 times more frequently than one of magnitude M. The Richter scale tends to "saturate" about magnitude 8 because the rupture zone of a great earthquake necessarily must be large and hence not localized to a specific location. That is, the Richter scale is designed to measure the strength of an earthquake that occurs "over there" and is inaccurate if the measurement is made within the rupture zone.

Transition

In this chapter we have described four types of elastic body waves: longitudinal body waves having displacements in the direction of wave propagation, transverse body waves, having (horizontal) displacements at right angles to the direction of propagation, longitudinal edge waves and transverse edge waves, and have shown that body waves are a limiting case of edge waves. Body waves can travel as plane waves having constant amplitude everywhere on the wavefront, whereas the amplitudes of elastic edge waves decay exponentially with distance from an edge or boundary of the body. Longitudinal body waves (called P waves in seismology) travel faster than transverse body waves (seismic S waves). We also discussed the nature of and rating rating scale for seismic waves.

This completes our study of elastic waves; in the next chapter we turn our attention to water waves, which are a form of edge wave that are superficially similar to elastic edge waves, but are dynamically quite different.

11

Deep-Water Waves

Water waves are fluid edge waves; they also are called *surface waves* or *gravity waves*. These waves arise from a much different mechanism than do edge waves in elastic bodies. This is evident from the fact that water waves travel much more slowly than do sound waves. However, they do have the same basic structure as elastic edge waves: harmonic variation in the horizontal direction and exponential decay with depth. Water waves occur on and near the free surface; they cannot occur in a closed container. These waves occur because gravity acts to pull down the water that stands higher than average and push aside water lying beneath, as illustrated in Figure 11.1. This figure is a "snapshot"; the wave might be progressive (moving either to the left or right) or standing (with fixed nodes) or some combination of these two limiting cases.

The adiabatic incompressibility and kinematic viscosity of water are $K \approx 2 \times 10^9$ Pa and $\nu \approx 10^{-6}$ m^2·s^{-1}. The pressure at the bottom of a typical ocean, which is 4200 m deep, is about 4×10^7 Pa. This compresses water by about 2%. Since we are interested in surface waves, it is safe to treat water as completely incompressible. Viscous effects are important only near solid boundaries (and then only within thin boundary layers) or if the water is turbulent (as it is in breakers and whitecaps). So it is safe to treat water as inviscid in the study of non-breaking waves.

Waves on open water are almost always progressive (i.e., moving in a certain direction). Standing waves occur when two progressive waves travel past each other in opposite directions, such as commonly occurs near a vertical breakwater. We will consider progressive waves in this chapter, with standing waves considered in § 12.3.

We will find in § 11.1 that the equation governing motions within water is not the wave equation, but Laplace's equation.[1] The wave character of the solution comes from the boundary conditions at the surface. These conditions are developed in § 11.2.

11.1 Water Wave Equation

As with sound and seismic waves, let's focus our attention on plane waves traveling in the x direction, with motion and variation in both the horizontal (x) and vertical (z) directions.

[1] See Appendix A.6.

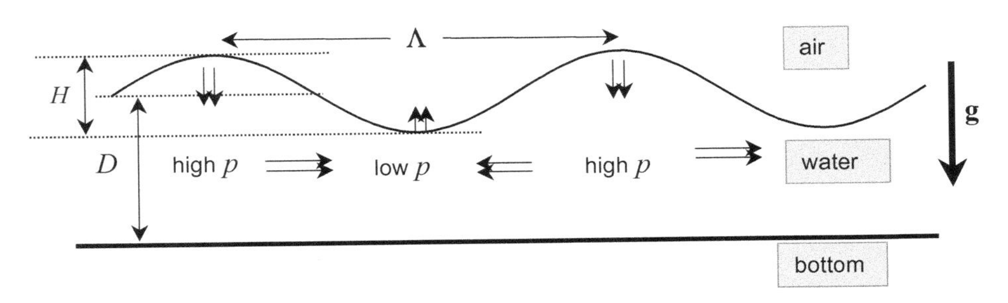

Figure 11.1 Schematic of a harmonic wave of height H and wavelength Λ on a layer of water of mean depth D. The double arrows show the direction of internal flow in response to the displacement of the surface.

An inviscid fluid which is initially irrotational remains so.[2] The constraint of irrotationality is ensured by writing the velocity as the gradient of a scalar, denoted by ϕ and called the *velocity potential*:[3]

$$\mathbf{v} = u\mathbf{1}_x + w\mathbf{1}_z = \frac{\partial \phi}{\partial x}\mathbf{1}_x + \frac{\partial \phi}{\partial z}\mathbf{1}_z.$$

Note that the vorticity is

$$\boldsymbol{\omega} = \left(\frac{\partial u}{\partial z} - \frac{\partial w}{\partial x}\right)\mathbf{1}_y = \left(\frac{\partial^2 \phi}{\partial x \partial z} - \frac{\partial^2 \phi}{\partial x \partial z}\right)\mathbf{1}_y = \mathbf{0}$$

and that ϕ has SI units $\mathrm{m}^2 \cdot \mathrm{s}^{-1}$.

The velocity potential "solves" the momentum equation,[4] and the remaining equation to solve is the continuity equation, $\nabla \bullet \mathrm{v} = 0$, which becomes *Laplace's equation*:

$$\frac{\partial u}{\partial x} + \frac{\partial w}{\partial z} = \frac{\partial^2 \phi}{\partial x^2} + \frac{\partial^2 \phi}{\partial z^2} = 0.$$

This equation is valid within the water, which occupies the half space $z < h(x,t)$, where h represents the deflection of the surface due to waves. In the absence of waves, $h = 0$ and the quiescent fluid occupies the half space $z < 0$.

The term $\partial u/\partial x$ represents the compression of fluid in the longitudinal direction. In sound waves this term is balanced by fluid compressibility,[5] represented by the time derivative of density. For surface water waves surface deflection provides the requisite "compressibility". Also, the time derivative of density in the compressible continuity equation previously provided one of the two time derivatives appearing in the wave

[2] See Appendix D.10.
[3] See Appendix A.8.2.
[4] See Appendix D.10.
[5] See Chapter 9.3.

equation. The assumption of incompressibility has removed this time derivative; we must find another to replace it.

11.2 Surface Boundary Conditions

Our governing equation is Laplace's equation, which generally yields smooth static solutions, not waves. How are we to get waves? Recall that we are studying surface waves; the answer must come from consideration of the boundary conditions at the surface of the water. We have two conditions at the boundary to consider: one kinematic and one dynamic. The kinematic condition is that the boundary is a material surface; by definition water does not cross its surface. The dynamic condition is that the pressure at the surface is equal to that in the fluid above. In the present case that fluid is air and the total pressure at the surface is equal to the atmospheric pressure.

The wave is characterized by a displacement of the surface from its rest position, which we shall define as the plane $z = 0$, with the un-displaced fluid occupying the half space $z < 0$. The surface of the fluid is a *material surface*; a particle initially on the surface remains there. With the position of the liquid surface given by $z - h(x,t) = 0$, the surface boundary condition[6] is obtained from the material derivative $\mathrm{D}(z-h)/\mathrm{D}t = 0$:

$$\text{at} \quad z = h(x,t): \quad w = \frac{\partial h}{\partial t} + u\frac{\partial h}{\partial x}$$

or in terms of the velocity potential,

$$\text{at} \quad z = h(x,t): \quad \frac{\partial \phi}{\partial z} = \frac{\partial h}{\partial t} + \frac{\partial \phi}{\partial x}\frac{\partial h}{\partial x}.$$

Notice a time derivative reappearing in the problem. Also, note that this condition is applied at the unknown location of the water surface. This makes the condition nonlinear and very difficult to analyze without making a simplifying assumption.

Now let's develop the dynamic boundary condition. In the absence of waves, the static pressure, p_r, within the water is given by the hydrostatic equation: $p_r = p_a - \rho g z$, where p_a is the atmospheric pressure and ρ is the (constant) density of water. When waves occur, the pressure within the water is given by the Bernoulli equation:[7]

$$p = p_a - \rho\left(gz + \frac{\partial \phi}{\partial t} + \frac{1}{2}v^2\right),$$

where $v = \|\mathbf{v}\| = \|\nabla\phi\|$ is the magnitude of the velocity. At the surface of the water (at $z = h$), the pressure is equal to the atmospheric pressure and

$$\text{at} \quad z = h(x,t): \quad gh + \frac{\partial \phi}{\partial t} + \frac{1}{2}\left(\frac{\partial^2 \phi}{\partial x^2} + \frac{\partial^2 \phi}{\partial z^2}\right) = 0.$$

[6] From § 3.5.

[7] See Appendix D.6.1.

11.2.1 Linearization

We have developed two surface boundary conditions: one kinematic and one dynamic. At the moment, they are rather difficult to handle; both are nonlinear and are applied at a moving, unknown boundary: $z = h(x,t)$. Moving-boundary problems are notoriously difficult to tackle. We need to transform these into linear conditions applied at a known position. This is accomplished by assuming that the deflection of the surface and the associated fluid motions are small. Let

$$h = \varepsilon h_1 + \varepsilon^2 h_2 + \ldots \qquad \text{and} \qquad \phi = \varepsilon\phi_1 + \varepsilon^2\phi_2 + \ldots,$$

where ε is a small parameter.[8] In addition, we may expand ϕ evaluated at $z = h$ in Taylor series about $z = 0$:

$$\phi_1(x,h,t) = \phi_1(x,0,t) + \varepsilon\frac{\partial\phi_1}{\partial z}(x,0,t)h_1 + \ldots.$$

Keeping terms linear in powers of ε, the kinematic and dynamic boundary conditions become[9]

$$\frac{\partial\phi_1}{\partial z}(x,0,t) = \frac{\partial h_1}{\partial t} \qquad \text{and} \qquad \frac{\partial\phi_1}{\partial t}(x,0,t) = -gh_1\,.$$

Taking the time derivative of the second (dynamic) condition and eliminating $\partial h_1/\partial t$ between the two boundary conditions, we have

$$\text{at} \qquad z = 0: \quad \frac{\partial^2\phi_1}{\partial t^2} = -g\frac{\partial\phi_1}{\partial z}\,.$$

It is comforting to see two time derivatives appearing in this condition; this should give us two wave modes.

Of course the dominant-order governing equation is

$$\frac{\partial^2\phi_1}{\partial x^2} + \frac{\partial^2\phi_1}{\partial z^2} = 0$$

and the pressure is given by

$$p_1 = p_a - \rho\left(gz + \frac{\partial\phi_1}{\partial t}\right).$$

From now on, we shall drop the subscript 1.

11.3 Surface Waves

Surface waves on an incompressible liquid such as water are governed by Laplace's equation in the liquid interior (i.e., $z < 0$) and by the boundary condition applied at the

[8] The symbol ε, called "curly epsilon," is a variant of ϵ; see Appendix A.7.
[9] In terms of the x and z component of velocity, these conditions are $w = \partial h/\partial t$ and $\partial u/\partial t = -g\partial h/\partial x$ at $z = 0$.

surface ($z = 0$). Laplace's equation admits solutions which are oscillatory (i.e., sines and cosines) in one spatial direction and monotonic (i.e., exponentials) in the other. A single Fourier component is

$$\phi(x,z,t) = \tilde{\phi}(k)\cos(kx \pm \omega t)e^{kz},$$

where again k is an arbitrary wavenumber and ω is the associated frequency. Laplace's equation dictates that the exponential z variation has the same wavenumber as the harmonic x variation, but does not constrain the frequency. The frequency is determined by the boundary condition to be

$$\omega = \sqrt{gk}.$$

This is the dispersion relation for liquid edge waves (i.e., surface water waves).[10] The phase speed of these waves is

$$U_p = \omega/k = \sqrt{g/k},$$

while the group speed is

$$U_g = \mathrm{d}\omega/\mathrm{d}k = \tfrac{1}{2}\sqrt{g/k}.$$

Also,

$$\begin{aligned}
u &= \frac{\partial\phi}{\partial x} = -\tilde{\phi}k\sin(kx \pm \omega t)e^{kz},\\
w &= \frac{\partial\phi}{\partial z} = \tilde{\phi}k\cos(kx \pm \omega t)e^{kz},\\
p &= p_a - \rho\left(gz + \frac{\partial\phi}{\partial t}\right)\\
&= p_a - \rho\left(gz + \tilde{\phi}\omega\sin(kx \pm \omega t)e^{kz}\right)
\end{aligned}$$

and

$$h = -\frac{1}{g}\frac{\partial\phi}{\partial t}(x,0,t) = -h_0\sin(kx \pm \omega t),$$

where

$$h_0 = \tilde{\phi}\omega/g = \tilde{\phi}\sqrt{k/g}$$

is the wave amplitude. It is readily seen that the magnitude of the velocity is

$$v(x,z,t) = \sqrt{u^2 + w^2} = \tilde{\phi}ke^{kz}$$

with $v(x,0,t) = \tilde{\phi}k$.

[10] See § 9.1.

The trajectory of a parcel of water, induced by a passing harmonic wave in an infinitely deep layer of water, is a circle having amplitude decaying exponentially with depth as illustrated in part (a) of Figure 11.2. The direction of motion around the trajectories is the same as the wave direction; if the wave is traveling to the {right, left}, the particle paths are {clockwise, counterclockwise}. Since water has the greatest potential energy at the wave crest, it follows that the transport of wave energy is in the direction of particle motion at the crest.

11.4 Discussion of Waves

We have formulated the boundary condition for surface water waves assuming that the deflections of the water surface are small, without any comparison, and this is a bit unsatisfying. If someone asks us "compared with what?", at the moment we have no answer. Actually we have made two simplifying assumptions: the kinetic energy of the wave is small compared with the potential energy and the angle of the surface deflection is small; that is,

$$v^2(x,0,t) \ll 2gh_0 \qquad \text{and} \qquad h_0 \ll \Lambda = 2\pi/k.$$

Using the solution for a single harmonic, we can see that the second condition implies the first; both are satisfied provided

$$\tilde{\phi} \ll 2\pi\sqrt{g/k^3}.$$

We can readily observe that this condition is generally well satisfied; the amplitudes of surface waves on water are generally much less than their wavelengths.

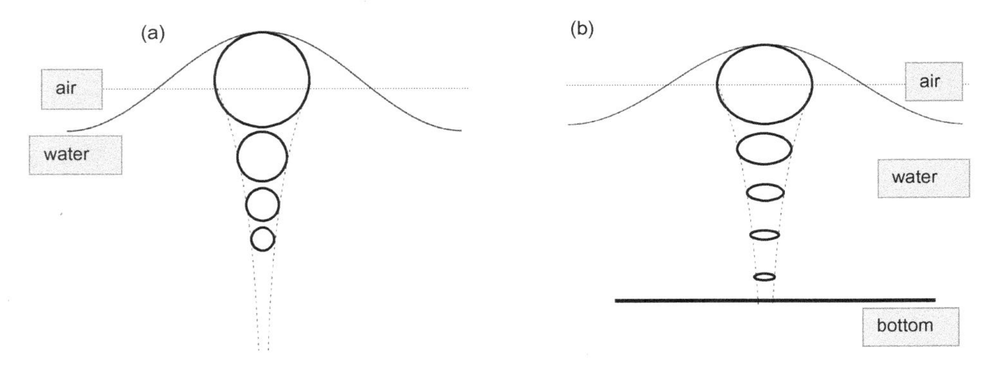

Figure 11.2 Part (a): Water of infinite depth. Trajectories (solid circles) of water parcels due to the passage of a harmonic wave. The amplitudes of the trajectories decay exponentially with depth (as illustrated by the dotted lines), with the rate of decay equal to the length of the wave. Part (b): Water of finite depth. Trajectories (solid ellipses) of water parcels due to the passage of a harmonic wave. The amplitudes of the trajectories decay exponentially with depth (as illustrated by the dotted lines), with the rate of decay equal to the length of the wave. In addition, amount of flattening of the ellipses increases with depth, with the trajectories being horizontal at the bottom.

We have developed a mathematical solution for a harmonic wave having an amplitude h_0. Much of the literature dealing with ocean waves uses the wave height H instead of the amplitude. For a harmonic wave, $H = 2h_0$. Also, since the wavelength $\Lambda = 2\pi/k$ is easier to envisage than the wavenumber k, it is preferable express the phase and group speeds as

$$U_p = \sqrt{\frac{g\Lambda}{2\pi}} \quad \text{and} \quad U_g = \frac{1}{2}\sqrt{\frac{g\Lambda}{2\pi}}.$$

Note that the phase speed is equal to the speed of an object that has been dropped a vertical distance $\Lambda/4\pi$. Also, the period P (i.e., the time interval between successive crests or troughs) of a wave is given by

$$P = U_p/\Lambda = \sqrt{2\pi\,\Lambda/g}.$$

Now let's consider the nature of deep water waves. We have seen that the phase speed is twice the group speed for surface waves. Wave energy propagates with the group speed. An observer moving with a packet of waves sees individual waves moving from the back of the packet to the front. This phenomenon can be observed by tossing a rock into still water and observing the circular packet of surface waves emanating from the entry point. Individual waves in the packet advance to the leading edge and disappear while new waves continually appear, as if by magic, at the trailing edge. The packet also widens as it advances outward, due to dispersion.

It is interesting that the speeds and periods of waves are independent of the density of the liquid. The speed of waves on water or mercury are the same, provided the wavelength is the same. This independence occurs because we have ignored the density of air (for water $\rho_a/\rho \approx 1.3 \times 10^{-3}$). Of course this result fails for waves of small length such that surface tension becomes important[11] or if the overlying liquid has a density comparable to that of the underlying liquid.[12]

11.5 Ocean Waves

In this section, we try to make some sense of the behavior of waves that occur on the oceans and other large bodies of water, such as seas and lakes.

Waves are generated and sustained by wind stresses. The precise process of wave generation is complicated,[13] but in essence small-scale turbulence of the atmosphere ruffles the water surface, producing waves having small wavelength and amplitude. Subsequently, nonlinear interactions between these waves and the turbulent air flow cause a shift in the phase of the surface pressure relative to the surface displacement as shown in Figure 11.3, with the result that a net horizontal force is exerted on the water by the wind and energy is fed from the wind into the surface waves. This force is similar to form drag

[11] See Appendix D.4.
[12] See § 14.2.
[13] Known as the *Miles-Phillips mechanism*.

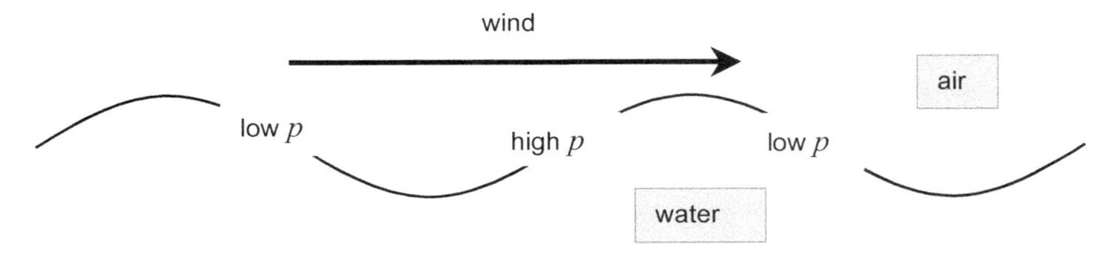

Figure 11.3 Schematic of a harmonic wave driven by the wind. Air turbulence and separation cause a shift in the phase of the dynamic air pressure relative to the shape of the wave.

in aeronautics.[14] As the wind continues to blow, the dominant waves have progressively longer length and larger height.

It takes a certain time and distance to build up a set of waves. As energy is fed into the waves, their length, height and speed depend on three factors:

- the down-wind distance over which the wind can act, called the *fetch*;
- the duration of time the wind has been blowing; and
- the strength of the wind, with stronger wind being capable of developing and sustaining bigger waves.

Waves of very small amplitude are approximately sinusoidal. As the wave height increases (with wavelength Λ held constant), the wave form becomes distorted from sinusoidal due to the action of nonlinear terms in the surface boundary condition that we have ignored. In particular, the wave crest is sharpened and the wave trough is elongated. When $H \approx \Lambda/7$ the crest forms a cusp, with the angle of the surface from the horizontal at the cusp being about $\pi/6$ radians or 30 degrees. Such a wave is said to be *fully developed*. Fully developed waves are discussed in § 11.5.2, following consideration of wind stresses in the next section. Then ocean waves are discussed in § 11.5.3 and in § 11.5.4 we consider the behavior of waves onshore.

11.5.1 Wind Stress

Waves grow as they acquire energy from the wind. Wind can supply energy in two ways: by exerting a tangential shear stress and a form drag. The tangential stress requires action of viscosity and leads to rotational motion in the water. Since viscosity is very small, the tangential stress is small and will be ignored. Form drag arises because the air pressure on the windward face of the wave is larger than that on the lee face; see Figure 11.3.

It follows from dimensional analysis that the stress, τ_W, exerted by the wind is proportional to the density of air, ρ_a, and the square of a speed. Commonly this speed

[14] See § 23.7.

is taken to be the wind speed at a height of 10 m above the mean sea surface, U_{10},[15] giving

$$\tau_W = C_D \rho_a U_{10}^2,$$

where the constant C_D is the (dimensionless) drag coefficient. Published values vary from $C_D \approx 0.001$ for smaller values of U_{10} to ≈ 0.002 for larger speeds, with considerable scatter. Typical values consistent with this equation are[16]

- $\tau_W \approx 0.1$ Pa;
- $C_D \approx 0.0013$; and
- $U_{10} \approx 8\ \mathrm{m{\cdot}s^{-1}} \approx 16$ knots.

There is a systematic increase of drag with wind speed; a reasonable parameterization is

$$C_D = \left(0.61 + 0.063\, U_{10}\ \mathrm{s \cdot m^{-1}}\right) \times 10^{-3},$$

for $6\ \mathrm{m \cdot s^{-1}} < U_{10} < 22\ \mathrm{m \cdot s^{-1}}$ (Smith, 1980). We will discuss the drag coefficient in § 23.7.

Beaufort Wind and Douglas Sea Scales

The magnitudes of winds over and waves upon the oceans, and the relative risks they can pose, have long been a concern to mariners. These magnitudes and associated risks associated with winds are codified in the *Beaufort scale*, as given in Table 11.1. Winds of hurricane force are further categorized using the Saffir–Simpson scale; see Table 28.1.

Waves that are actively driven by the wind are called *sea* or *wind sea*, while waves that are coasting along without being driven are called *swell*. Seas and swells are classified by the *Douglas Sea Scale*, containing ten degrees ranging from 0 to 9; see Table 11.2. Wind speeds are given in a variety of units; note that

$$1\ \text{kph} = 0.911344\ \text{fps} = 0.621371\ \text{mph} = 0.539957\ \text{kt} = 0.2777778\ \text{mps},$$

where kph is kilometers per hour, fps is feet per second, mph is miles per hour, kt is knots and mps is meters per second.

11.5.2 Fully Developed Waves

After the wind has been blowing sufficiently long or hard over a sufficiently large fetch, the waves become *fully developed* and don't get any bigger. As energy is fed into a wave of maximum amplitude, it begins to break – forming a turbulent whitecap. The times and distances necessary for full wave development depend on the wind strength as given in Table 11.3. Real waves on open water are a jumble of simple ideal waves and roughly 10

[15] More precisely, the wind speed is defined as the average speed of the wind over a 10-minute period at a height of 10 meters above the surface.
[16] Recalling that $\rho_a = 1.27\ \mathrm{kg{\cdot}m^{-3}}$; see Table B.5.

Table 11.1. *Beaufort wind force scale for rating the speed and danger of winds over the open ocean. Speeds in miles per hour are numerically 15% larger than in knots. Wind speeds of force 12 or greater are categorized by the enhanced Fujita and Saffir–Simpson scales; see Table 28.1.*

Numerical Value	Description	Speed ($m{\cdot}s^{-1}$)	Speed (knots)
0	Calm	< 0.3	<1
1	Light air	0.3–1.5	1– 3
2	Light breeze	1.6–3.3	4– 6
3	Gentle breeze	3.4–5.5	7–10
4	Moderate breeze	5.5–7.9	11–16
5	Fresh breeze	8.0–10.7	17–21
6	Strong breeze	10.8–13.8	22–27
7	Near gale	13.9–17.1	28–33
8	Fresh gale	17.2–20.7	34–40
9	Strong gale	20.8–24.4	41–47
10	Storm	24.5–28.4	48–55
11	Violent storm	28.5–32.6	56–63
12	Hurricane force	>32.7	>64

Table 11.2. *Douglas sea scale for rating the danger of waves. Wave heights quantify seas, but not swells.*

Degree	Height (m)	Wind Sea Description	Swell Description
0	0	Glassy	No swell
1	0–0.1	Calm	Very low
2	0.1–0.5	Smooth	Low
3	0.5–1.25	Slight	Light
4	1.25–2.5	Moderate	Moderate
5	2.5–4.0	Rough	Moderate rough
6	4.0–6.0	Very rough	Rough
7	6.0–9.0	High	High
8	9.0–14.0	Very high	Very high
9	14.0–	Phenomenal	Confused

percent of real waves have a (momentary) height at least twice the average height. The highest wave reliably recorded had a height of 112 ft $\approx$ 34 m. This required a wind greater than 75 km/hr ($\approx$ 21 $m{\cdot}s^{-1}$ $\approx$ 40 knots) sustained for many days.

Table 11.3. *Wind-driven waves. U is the wind speed measured at a height of 10 m. 1 knot = 0.51444 m·s^{-1} and 1 nautical mile = 1.852 km. Adapted from Bascom (1964).*

Wind (Knots)	Fetch (naut. miles)	Time (hours)	Height (m)	Phase Speed (knots)	Largest 10% (m)
10	10	2.4	0.3	3.34	0.55
15	34	6.0	0.8	5.45	1.5
20	75	10	1.5	7.46	3.0
25	160	16	2.7	10.0	5.5
30	280	23	4.3	12.6	8.5
40	710	42	8.5	17.8	17.4
50	1420	69	14.6	23.3	30.2
U	$U^{3.08}$	$U^{2.08}$	$U^{2.414}$	$U^{1.207}$	$U^{2.48}$

The last line in Table 11.3 gives the observed power-law relation between that variable and the wind speed, U_{10}. Note that the fetch is the product of the development time and the wind speed, which is reassuring. It is shown in Appendix F.5 that traveling waves have kinetic and potential energy in equal amounts and the energy per unit area in a single harmonic wave is proportional to the square of the wave height. With height nearly proportional to $U^{5/2}$, the wave energy varies roughly as U^5.[17] This explains, in part, the destructive power of strong storms.

The phase speed of fully developed waves is given by

$$U_p = \sqrt{7gH/(2\pi)} \approx \sqrt{gH}\,.$$

This is the speed of an object that has been dropped from a height equal to the wave amplitude $h_0 = H/2$. This expression may be inverted to obtain an expression for the wave height in terms of the phase speed:

$$H \approx U_p^2 g\,.$$

It is easier to measure the wave speed than height; this equation may be used to estimate wave height from observation of wave speed.[18]

Putting numbers for water on Earth into the formulas for phase speed and period, we have that

$$U_p = 4.5\frac{\text{km}}{\text{hr}}\sqrt{\frac{\Lambda}{\text{m}}} \qquad \text{and} \qquad P = 0.8\,\text{s}\sqrt{\frac{\Lambda}{\text{m}}}\,.$$

[17] The energy of a single harmonic varies as U^4; the extra U arises due to nonlinear effects.

[18] The calculation is eased by setting $g = 10$ m·s^{-2}.

A wave of length {1, 10, 100, 1000} m travels approximately {4.5, 14, 45, 140} km/hr and has a period of {0.8, 2.5, 8, 25} s. These speeds are very slow compared with the speed of sound in water ($\approx$ 5260 km/hr). The speed of a surface water wave would be comparable to that of a sound wave in water if the water wave had a wavelength on the order of 10^6 m. But this length is much greater than the depth of the ocean, in which case the analysis of this section is invalid. This deficiency is remedied in the following chapter.

There is a limit to the size of waves driven by wind of a given strength, U; the wind cannot drive waves traveling faster than the wind itself, that is, $U > U_p$. It is natural (based on dimensional analysis) to assume that the waves are fully developed when their phase speed is a fraction, χ, of the wind speed:

$$U_p = \chi U,$$

where χ must lie between 0 and 1. With this parameterization, the average height of waves driven by a wind of speed U is given by

$$H = \frac{2\pi \chi^2 U^2}{7g} \approx \frac{\chi^2 U^2}{g}.$$

The data in Table 11.3 indicate that $\chi \approx 1/2$ for the average wave height. However, there is a systematic variation of χ with U: $\chi \propto U^{0.2} - U^{0.25}$, which is likely due to nonlinear effects.

11.5.3 Discussion of Ocean Waves

We have been focusing primarily on the behavior of a single train of waves, whereas the ocean surface is a jumble of waves that can interact in various ways. These interactions can result in *rogue waves* that have extraordinary height and can be very transient or surprisingly persistent. There are occasional reports of rogue waves. For example, Ernest Shackleton and five colleagues encountered one on their epic voyage in April, 1916, from Elephant Island to South Georgia Island in a small (6.9 m long) lifeboat, the *James Caird*. To quote from Lansing (1959):

At midnight ... Shackleton himself assumed the helm ... His eyes were just growing accustomed to the dark when he turned and saw a rift of brightness in the sky astern. He called to the others to tell them the good news that the weather was clearing to the southwest.

A moment later he heard a hiss, accompanied by a low, muddled roar, and he turned and looked again. The rift in the clouds, actually the crest of an enormous wave, was advancing rapidly toward them. He spun around and instinctively pulled his head down.

"For god's sake, hold on!" he shouted. "It's got us!"

For an instant nothing happened. The *Caird* simply rose higher and higher, and the dull thunder of this enormous breaking wave filled the air.

And then it hit ...

Another report of a rouge wave is found in the log entry for January 12, 1953, of the *Felicity Ann*, a 7 m sloop (Davison, 1956):

In one of the lulls ... I heard a roaring sound in the distance like surf breaking on a shore, and looked out of the cabin to see astern and coming towards us at a tremendous rate a line of white stretching across the sea from horizon to horizon. There was no escaping it, and I watched, transfixed, from the cabin, looking aft over the stern. As it drew nearer it grew higher and higher, and as it caught up with us, towering, tremendous, roaring, and breaking, I ducked below and pulled the hatch down. The ship was picked up and thrown on her beam ends, and it seemed for a moment as if we had had it...

Due to dispersion, a jumbled packet of waves, created locally by a storm, spreads as it radiate away from the storm, with speed of individual waves increasing with the wavelength; longer waves travel faster and also are subject to the least dissipation. Common values for deep-water waves on the open oceans are $T = 10 - 20$ sec and $\Lambda = 150 - 600$ m. The leading waves, called *swell*, are usually quite regular, arriving at a distant shore with relatively uniform amplitude and period. Their long wavelength makes them susceptible to *wave focusing* by shoals near the shore or by variations in oceanic currents, causing their amplitudes to be anomalously high locally along the shore. The combined effect of long period and focusing makes such waves rather treacherous. Unwary people are occasionally swept from the shore by such waves.

11.5.4 Storm Surge

In open seas the momentum and energy produced by wind stress is radiated away from the storm. Near shore, it is a different story, as this wind stress is responsible for the storm-surge experienced on shore as a storm approaches. As the wind blows shoreward, water piles up near shore; the height of the storm surge is determined by a balance between the on-shore wind stress, τ_W, acting over a fetch of magnitude L, and the hydrostatic force, $\rho g H_s^2/2$, produced by an excess of water depth (that is, a storm surge) of depth H_s:

$$H_s = \sqrt{\frac{2L\tau_W}{\rho g}} \approx \sqrt{\frac{2LC_D\rho_a}{\rho g}}\,U\,.$$

Note that the height varies linearly with the wind speed and as the square root of the fetch. We know that $\rho_a/\rho = 1.2 \times 10^{-3}$, $g = 9.8$ m·s^{-2} and $C_D \approx 0.0013$. Now,

$$H_s \approx 0.02\sqrt{\frac{L}{\text{km}}}\left(\frac{U}{\text{m/s}}\right)\text{m}\,.$$

If $L = 100$ km and $U = 30$ m·s^{-1} ($= 58.3$ knots) then $H_s \approx 6$ m. Often the actual storm surge is somewhat less than this because of along-shore flow away from the center of the storm and because of rip currents. The actual height of storm surges depends on the variation of water depth off shore and on the curvature of the coastline. Shallow water inhibits the flows that act to diminish the height of the surge, as does a coastline shaped like an embayment.

For example, both conditions hold in Apalachee Bay south of Tallahassee, Florida, where predicted storm surges are particularly high.

11.6 Water of Finite Depth

So far we have considered waves on the surface of a body of water having infinite depth. In this section, we consider waves in the surface of a body of water having an impervious bottom at a finite depth, $z = -D$. The previous equations and conditions still hold, but we must also apply the condition that the vertical velocity is zero at the bottom:[19]

$$w(x, -D, t) = \frac{\partial \phi}{\partial z}(x, -D, t) = 0.$$

Previously we considered the solution to Laplace's equation having exponential decay with depth. Now we must include the exponential having growth with depth and the fundamental solution becomes

$$\phi(x, z, t) = \tilde{\phi} \cos(kx \pm \omega t) \frac{\cosh(kz + kD)}{\cosh(kD)},$$

with

$$u(x, z, t) = -\tilde{\phi} k \sin(kx \pm \omega t) \frac{\cosh(kz + kD)}{\cosh(kD)},$$
$$w(x, z, t) = \tilde{\phi} k \cos(kx \pm \omega t) \frac{\sinh(kz + kD)}{\cosh(kD)}$$

and again

$$h(x, t) = -\tfrac{1}{2} H_0 \sin(kx \pm \omega t),$$

with $H_0 = 2\tilde{\phi}\omega/g$. Note that ϕ is an even function of $z + D$; this behavior is necessary to satisfy the bottom boundary condition.

The surface boundary condition gives the dispersion relation

$$\omega = \sqrt{gk \tanh(kD)}$$

and the phase and group speeds become

$$U_p = \sqrt{\frac{g}{k} \tanh(kD)} \qquad \text{and} \qquad U_g = \sqrt{g} \frac{\tanh(kD) + kD \operatorname{sech}^2(kD)}{2\sqrt{k \tanh(kD)}}.$$

In the limit $D \to \infty$, this reduces to the dispersion relation for deep-water waves. As the water depth decreases, the phase speed slows monotonically, while the group speed initially increases, then decreases in step with the phase speed, as shown in Figure 11.4. The trajectories of fluid parcels are now ellipses, flattened in the z direction, with the amount of

[19] We had implicitly applied the condition $w \to 0$ as $z \to -\infty$; now we must make this condition explicit.

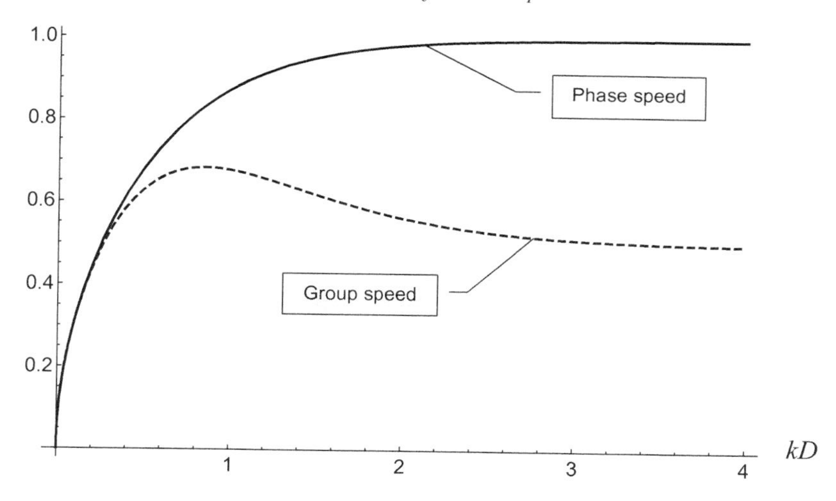

Figure 11.4 Plots of phase speed (solid line) and group speed (dashed line) versus water depth, D, normalized to the deep-water phase speed.

flattening increasing with depth, as illustrated in part (b) of Figure 11.2. The nature of this solution in the limit $kD \ll 1$ is investigated in Chapter 12.

Transition

In this chapter we have investigated an important class of waves: surface water waves on deep water. We have seen that the wave equation arises from a combination of the kinematic and boundary conditions at the top of the water; motions within the water body are governed by Laplace's equation. Then in § 11.5 we investigated the nature of wave growth on the open ocean and the behavior of waves near shore and in § 11.6 waves in water of finite depth. In the following chapter, we investigate the behavior of surface water waves when the water depth is much less than the length of the wave.

12

Linear Shallow-Water Waves

An interesting special case of waves of finite depth presented in § 11.6 is the shallow-water limit $kD \ll 1$ or equivalently $D \ll \Lambda/2\pi$; see Figure 12.1. In this limit, the fundamental mode developed in § 11.6 becomes

$$u(x,z,t) = -\tilde{\phi} k \sin(kx \pm \omega t) \qquad \text{and}$$

$$w(x,z,t) = \tilde{\phi} k \cos(kx \pm \omega t)\,(kz + kD)\,,$$

with the dispersion relation now given by

$$\omega = \sqrt{gD}k\,.$$

Shallow water waves are non-dispersive, with speed

$$U = U_p = U_g = \sqrt{gD}$$

and period

$$P = \Lambda/\sqrt{gD}\,.$$

The speed of the linear shallow-water wave is equal to the free-fall speed of an object that has dropped a distance equal to 1/2 the depth of the water.

Note that in the shallow-water limit the velocity potential ϕ becomes independent of z to dominant order; the single harmonic mode is

$$\phi(x,z,t) = \tilde{\phi}\cos(kx \pm \omega t)\,.$$

This satisfies the wave equation

$$\frac{\partial^2 \phi}{\partial t^2} = U^2 \frac{\partial^2 \phi}{\partial x^2}\,.$$

The longitudinal speed $u = \partial\phi/\partial x$ obeys the same equation: $\partial^2 u/\partial t^2 = U^2 \partial^2 u/\partial x^2$.

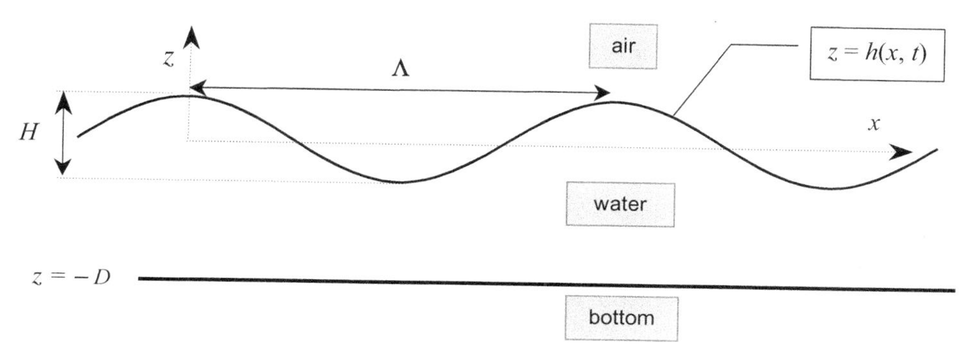

Figure 12.1 Schematic of a harmonic shallow wave of height H and amplitude $H/2$ on a layer of water of depth D. This figure is similar to Figure 11.1.

We have considered a single harmonic mode. Since these waves are non-dispersive, a linear shallow-water wavetrain of general form may be expressed as a Fourier integral.[1]

Several types of shallow-water waves of particular interest occur in the oceans: tsunamis (discussed in the following section) and luni-solar tides (discussed in § 12.2). In addition, harbors and other nearly-enclosed basins experience sloshing motions called seiches (see §12.3). An interesting class of shallow-water waves, coastal waves, is investigated in §12.4.

12.1 Tsunamis

When an event such as an earthquake or landslide perturbs a large surface area of ocean (or other large body of water), a *tsunami* is created. Tsunamis have very long wavelengths - sufficiently long to behave as shallow-water waves on the deep ocean. The average depth D of the oceans is 5 km, giving a tsunami wave speed of roughly 220 m/s = 800 km/hr. The period of a tsunami is given by $P = \Lambda/U$. If for example Λ = 80 km, the period is 6 minutes. This is enough time for curious people to walk out from shore during the ebb phase of a tsunami and be drowned as the next wave crest comes ashore.

The wave crest seen in vertical cross-section (parallel to the direction of propagation) has an excess area of water proportional to the product of wave height and length: $A \approx \Lambda H/\pi$. Typical values might be H = 0.3 m, Λ = 80 km and $A \approx$ 8,000 m^2. As the wave passes, this amount of water moves back and forth each period. This is accomplished by a "sloshing" speed V given by $VD \approx A/P = \Lambda H/(\pi P)$. With the above estimates, $V \approx$ 0.0044 m/s = 16 m/hr. A tsunami typically has an imperceptible amplitude and produces an imperceptible water speed in deep water.

12.1.1 Tsunami Near Shore

Now suppose a tsunami wave moves into shallowing water near shore, as shown in Figure 12.2. As the leading portion of the wave crest moves into shallow water, its speed

[1] See § 9.2.1 and Appendix A.5.

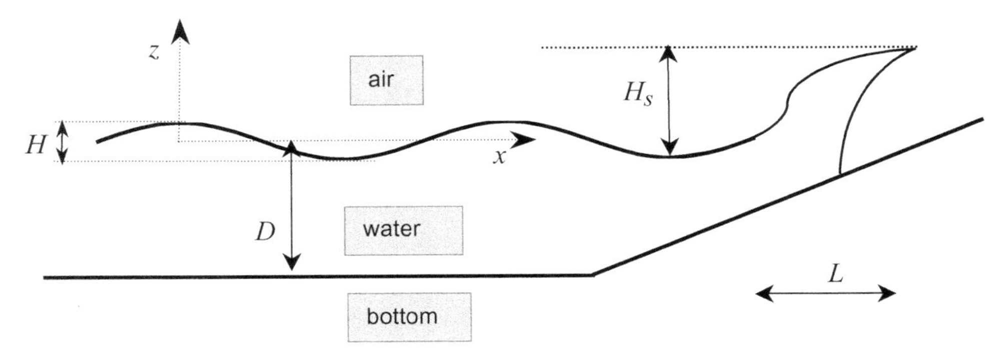

Figure 12.2 Schematic of a tsunami approaching shore in shallowing water.

decreases, and the remaining portion "catches up," causing the wave to deepen and steepen. During this process, the excess area is preserved, creating a wave of height H_s and length $L \ll \Lambda$ such that $H_s L \approx A$ or $H_s \approx \Lambda H/(\pi L)$. With $L \ll \Lambda$, the result is $H \ll H_s$; the wave height increases dramatically. For example, the Tohoku tsunami of March 11, 2011, which devastated the region around Fukushima, Japan, had $H \approx 2$ m and $\Lambda \approx 200$ km, giving $A \approx 1.3 \times 10^5$ m^2 (See Løvholt et al. 2012). This was sufficient to produce a wave roughly 40 m high having an on-shore length of 3 km.

A tsunami wave comes onshore as a *flash flood*, the dynamics of which are investigated in § 21.4.

12.2 Luni-Solar Tides

So far we have studied free waves that propagate through a continuous body without forcing. The luni-solar tides considered in this section are an example of forced waves that are excited and maintained by gravitational forces that vary with time.[2] Free waves can have arbitrary frequencies, while the tides have frequencies dictated by the orbital motions of the Moon and Sun. "The tide" is the variation of sea level, h, typically measured at a point along a coastline; it is the superposition of a number of tidal modes characterized by frequency and amplitude:

$$h(t) = \sum_j h_j \cos(\omega_j t)\,,$$

where ω_j and h_j are the frequency and amplitude of the jth tidal mode. The tidal frequencies are dictated by astronomy, while the amplitudes are dictated by the relative locations of the Moon and Sun and by the geometry of the coastline and topography of the sea floor.

[2] The tides are affected by rotation and so should rightfully be placed in Part IV. However the qualitative discussion presented in this section does not include mention of the Coriolis parameter.

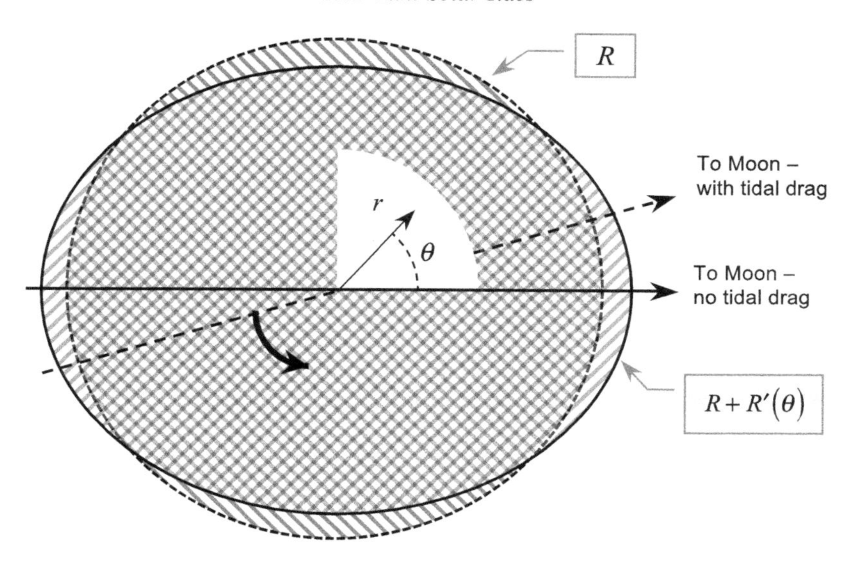

Figure 12.3 Deformation of an initially spherical "water-world" Earth (dashed circle), having radius R, due to the presence of the Moon. Earth rotates counterclockwise about an axis normal to the plane of the paper, as indicated by the curved arrow. The solid horizontal line points toward the Moon in the case there is no tidal drag; the dashed line points to the moon in the case of finite tidal drag. The deformed planet is a prolate spheroid having radius $R + R'(\theta)$, rotationally symmetric about the horizontal line. The amount of deformation is exaggerated for clarity.

12.2.1 Simple Tides

Tides are fairly simple in theory,[3] but rather complicated in practice. We shall begin with a very simple case, then add complications as we develop our understanding. Let's begin by considering "water world": a planet completely covered by an ocean. The gravitational attraction of the Moon perturbs this world's state of spherical equilibrium, producing a tidal displacement of the water surface given by

$$h_M(\theta) = \Pi_M R(3\cos^2\theta_M - 1)\,, \qquad \text{where} \qquad \Pi_M = \frac{M_M R^3}{2M_E L_M^3}$$

is a parameter quantifying the relative lunar tidal range,[4] θ_M is the lunar colatitude (relative to a radial line extending from Earth to the Moon – see Figure 12.3), M_E is the mass of Earth, M_M is the mass of the Moon, R is the radius of Earth and L_M is the distance to the Moon. The displacement of Earth's surface is a prolate spheroid with maximum elongation on the ray extending from Earth to the Moon.

Writing $M_E = 4\pi\rho_E R^3/3$ and $M_M = 4\pi\rho_M R_M^3/3$, where ρ_E is the mean density of Earth, ρ_M is the mean density of the Moon and R_M is the radius of the Moon, the lunar-tidal-range

[3] The equations governing tides on a rotating planet are presented in Appendix F.6.
[4] Maximum displacement minus minimum displacement.

parameter may be expressed as

$$\Pi_M = \frac{\rho_M}{16\rho_E}\gamma_M^3 ,$$

where $\gamma_M = 2R_M/L_M$ is the angular width of the full Moon as seen in the night sky.

The Sun also induces tides

$$h_S(\theta) = \Pi_S R(3\cos^2\theta_S - 1) , \qquad \text{where} \qquad \Pi_S = \frac{\rho_S}{16\rho_E}\gamma_S^3$$

quantifies the relative solar tidal range, ρ_S is the mean density of the Sun, $\gamma_S = 2R_S/L$ is the angular width of the Sun and θ_S is the solar colatitude (relative to a radial line extending from Earth to the Sun).

For the Earth-Moon system, $\gamma_M = 0.009033$ radians, $\rho_M/\rho_E = 3.34/5.51 = 0.6062$, $\Pi_M = 5.58 \times 10^{-8}$ and the maximum lunar tidal range is 0.355 m. It is an astronomical coincidence that $\gamma_M \approx \gamma_S$. It follows that the solar tides are smaller than the lunar tides by the ratio of their densities: $\rho_S/\rho_M = 1.41/3.34 = 0.422$; the maximum solar tidal range is 0.150 m. When Moon and Sun are aligned the tidal range is 0.505 m,[5] and when they are in opposition, it is only 0.205 m.[6] Actual tidal ranges are affected by the presence of continents and the variations in ocean depth.

If Earth were not rotating with respect to the Moon (and Sun), the tides would be static. However, Earth rotates on its axis approximately once per day[7] so that the tides appear as traveling waves. To simplify the analysis of Earth rotation, let's consider only the lunar tides, assume that Earth's rotation axis is perpendicular to the ray joining the centers of Earth and Moon and consider tides in the equatorial plane. Now the lunar colatitude varies linearly with time: $\theta_M = \Omega t + \theta$, where Ω is Earth's rate of rotation and θ is the initial colatitude, with θ increasing to the East (like longitude), and

$$h_M(\theta, t) = \Pi_M R\left(3\cos^2(\Omega t + \theta) - 1\right) .$$

This is a traveling *semi-diurnal* shallow-water wave, having two maxima and two minima per day. The wave moves westward, in the opposite direction to Earth's rotational motion, in order to maintain a constant position relative to the Moon. On Earth the continents, particularly the Americas, act to almost completely block this progressive wave. As a result, the tides behave more like standing waves rather than traveling waves; see § 12.3.

12.2.2 Tidal Modes

As noted above, "the tide" is the superposition of a number of modes having various frequencies that derive from the orbital motions of Earth, Moon and Sun. The dominant frequencies are due to Earth's spin and orbital rotation and the Moon's orbital rotation. Less

[5] This is called the *spring tide*.
[6] This is called the *neap tide*.
[7] Actually 366.24 times per year.

Table 12.1. *Astronomical frequencies and periods that affect the tides. "d" means day, "y" means year.*

Description	Symbol	Frequency (s^{-1})	Period
Earth spin[8]	ω_d	7.2722×10^{-5}	one day
Moon orbital rotation	ω_m	2.6617×10^{-6}	27.32158 d
Earth orbital rotation	ω_y	1.9911×10^{-7}	365.242 d
variation of Moon's perigee	ω_p	2.2505×10^{-8}	8.861 y
Moon's nodal precession	ω_n	-6.95×10^{-9}	18.613 y

Table 12.2. *Tidal modes, arranged by frequency. "hr" means hour; "d" means day.*

Symbol	Frequency (s^{-1})	Period	Amplitude (m)	Description
K2	1.4584×10^{-4}	11.97 hr	0.0307	Luni-solar
S2	1.4544×10^{-4}	12.00 hr	0.11284	Main solar
M2	1.4052×10^{-4}	12.421 hr	0.24233	Main lunar
N2	1.3785×10^{-4}	12.66 hr	0.04640	Elliptical lunar
K1	7.292×10^{-5}	23.93 hr	0.14156	Luni-solar
O1	6.759×10^{-5}	25.82 hr	0.10051	Main lunar
P1	6.496×10^{-5}	26.87 hr	0.04684	Main solar
Mf	5.32×10^{-6}	13.66 d	0.04174	Lunar fortnightly
Mm	2.64×10^{-6}	27.55 d	0.02203	Lunar monthly
Ssa	3.98×10^{-7}	182.62 d	0.01945	Solar semiannual
Sa	1.99×10^{-7}	365.24 d	0.0032	Solar annual

dominant frequencies include the Moon's nodal precession and variation of the Moon's perigee.[9] Further complications include the 23.5 degree tilt of Earth's rotation axis relative to the ecliptic plane and the 5.1 degree tilt of the Moon's orbit relative to the ecliptic plane.[10] The most prominent forcing frequencies are summarized in Table 12.1. The tidal modes have frequencies that are sums and differences of these astronomical frequencies. Historically the existence of the tidal modes was established by time-series analysis of tidal records. The most important of these are listed in Table 12.2.[11] The most energetic tidal mode is the main lunar mode: M2.

[8] Relative to the Sun.
[9] *Perigee* is the point on the Moon's elliptic orbit that is closest to Earth.
[10] The *ecliptic plane* is the apparent path of the Sun relative to the stars.
[11] A more complete listing of tidal modes is found in the Wikipedia entry *Theory of tides*.

The variation of tidal height near the coast is due to several factors. The effect that causes tsunamis to have large amplitude near coasts[12] is not very important because the tidal period is sufficiently long that the ocean has time to adjust to the presence of shallow water near shore and the buildup in wave amplitude is much less pronounced. The tidal amplitudes depend strongly on the shape of the coastline and the variation of ocean depth away from the coast and locally may be strongly affected by resonant interaction between the natural frequency of a bay or inlet (such as the Bay of Fundy) and the forcing frequency of the tide. This resonance leads to the formation of seiches in bays and inlets, as explained in § 12.3. Before investigating seiches, let's complete our discussion of tides by considering tidal damping in the following section.

12.2.3 Tidal Damping

As we have seen, the gravitational pulls of the Moon and Sun produce a number of oceanic tidal modes. The surface disturbance associated with modes produced by the Sun remain in phase with the Sun and travel around Earth once per day. Similarly the surface disturbance associated with lunar modes remain in phase with the Moon and travel around Earth a little less (due to the orbital motion of the Moon) than once per day. If there were no dissipation of tidal energy, the surface deflections associated with each mode would be centered on its generating planetary body. However, dissipative tidal motions in shallow seas (particularly the European shelf, Patagonian shelf and Bering Sea) affect the phase of the tidal modes, causing the modes to be dragged eastward by the rapidly rotating Earth. As a consequence, an observer on Earth sees the tides lagging behind the generating body.

Tidal friction dissipates roughly 3.75×10^{12} W. Since the tides are in steady state, they must have a counterbalancing source of energy. This source is the kinetic energy (roughly 2×10^{29} J) stored in the rotation of Earth. As rotational energy is dissipated by the tides, the rate of Earth's rotation is gradually decreasing, so that the day is gradually lengthening with the current rate being about 0.0017 s/century.[13] The rate of increase in the length of day has been highly variable in the geological past because the magnitude of tidal friction is sensitive to the mean ocean level, which has fluctuated as the ice ages have waxed and waned.

As tidal drag slows the rotation of Earth, the tidal bulge exerts a forward pull on the Moon, transferring angular momentum – and orbital energy – from Earth to Moon, causing the Moon to recede. The current rate of lunar recession is about 4 cm/year, curiously similar to the mean rate of continental drift (that is, tectonic plate motion). There is observational evidence that a day was 21.9 hours long some 620 million years ago, and computer simulations suggest that the day was only 6 hours long immediately after the Moon formed 4.53×10^9 years ago. Tidal drag will cease if the rotation rate of Earth slows such that Earth and Moon are in synchronous rotation or if the oceans evaporate due to warming of the Earth.

[12] See § 12.1.1.

[13] This rate is a dimensionless number $\approx 5.4 \times 10^{-13}$.

12.3 Seiches

A *seiche* is a standing wave that occurs on a body of water that is entirely or partially enclosed. Up to this point, we have been concerned primarily with traveling waves, but as noted in § 9.3, two waves of equal amplitude traveling in opposite directions form a standing wave. The purpose of this section is to investigate the fundamentals of a standing wave, by considering free motions of water having mean depth D in a bay having mean length L in the x direction; its size in the transverse direction is irrelevant. Suppose further that the body of water is enclosed at $x = 0$ and open to the ocean at $x = L$.

Superposing two traveling waves of the form presented in § 11.6, the velocity potential of a standing wave is of the form

$$\phi(x,z,t) = \tilde{\phi}\cos(kx)\cos(\omega t)\frac{\cosh(kz+kD)}{\cosh(kD)},$$

with $\omega = \sqrt{gD}k = 2\pi\sqrt{gD}/\Lambda$. The corresponding x and z velocity components are

$$u(x,z,t) = \frac{\partial\phi}{\partial x} = -\tilde{\phi}k\sin(kx)\cos(\omega t)\frac{\cosh(kz+kD)}{\cosh(kD)}$$

and

$$w(x,z,t) = \frac{\partial\phi}{\partial z} = \tilde{\phi}k\cos(kx)\cos(\omega t)\frac{\sinh(kz+kD)}{\cosh(kD)}.$$

The horizontal velocity is zero at the closed end of the bay (at $x = 0$) and is a maximum at $x = \pi/2k = \Lambda/4$, where Λ is the wavelength of the wave. If that maximum is located at the open end (at $x = L$) and if the frequency of the tidal forcing is equal (or nearly equal) to the frequency of the standing wave, a resonant wave of exceptionally large amplitude occurs within the bay. The presence of a seiche is most evident by viewing the vertical displacement of the water surface. This displacement, and the vertical component of velocity are greatest at the closed end of the bay.

In theory a resonant interaction occurs provided

$$\omega = \frac{\pi\sqrt{gD}}{2L}(1+4j),$$

for $j = 0, 1, 2, \ldots$, but in practice only the fundamental mode, having $j = 0$, has appreciable amplitude; the higher modes are dampened by dissipative effects.

12.4 Coastal Waves

So far, we have been investigating linear shallow-water waves in a layer of constant depth; these waves can move non-dispersively in any horizontal direction. This isotropy is broken in the near-shore environment with the water depth varying with distance from shore. These near-shore waves, of necessity, move parallel to the shore.[14]

[14] Linear waves moving toward the shore become nonlinear and break in the surf zone.

12.4.1 Coastal Wave Equations

Coastal waves are governed by the vertically averaged continuity and horizontal momentum equations. The former is given in § 8.4.2:

$$\frac{\partial h}{\partial t} + \nabla_H \bullet \big((D+h)\bar{\mathbf{v}}_H\big) = 0,$$

where $D(\mathbf{x}_H)$ is the undisturbed depth of the fluid and $h(\mathbf{x}_H, t)$ is the surface deflection. Assuming that the period of the wave is much less than one day, the Coriolis term is negligibly small and horizontal momentum equation is simply[15]

$$\frac{\partial \bar{\mathbf{v}}_H}{\partial t} = -g\nabla_H h,$$

where we have set the constant-pressure surface Z equal to h. Eliminating $\bar{\mathbf{v}}_H$ from the continuity equation using the momentum equation, we have a single equation for the surface deflection:

$$\frac{\partial^2 h}{\partial t^2} - g\nabla_H \bullet \big((D+h)\nabla_H h\big) = 0.$$

This nonlinear equation is difficult to solve; we need to simplify it. We may linearize it by assuming that the deflection of the interface is much smaller than the mean water depth: $|h| \ll D$ and consider the linear equation

$$\frac{\partial^2 h}{\partial t^2} - g\nabla_H \bullet (D\nabla_H h) = 0,$$

but it still has a variable coefficient. This difficulty is manageable, provided $D(\mathbf{x}_H)$ is simple enough.

Let's introduce local Cartesian coordinates with x pointing along shore and y pointing from shore toward deeper water and assume that the water depth increases linearly with distance from shore:

$$D = \varepsilon y,$$

where ε is the slope of the ocean bottom. A typical value of ε is 0.01 rad (0.5 degree), but it can be as small as 0.001 rad, for example off the west coast of Florida. Now the governing equation becomes

$$\frac{\partial^2 h}{\partial t^2} - \varepsilon g y\left(\frac{\partial^2 h}{\partial x^2} + \frac{\partial^2 h}{\partial y^2}\right) - \varepsilon g\frac{\partial h}{\partial y} = 0.$$

[15] This is the non-rotating version of the quasi-geostrophic equations given in § 8.6.

The coefficients in this equation are independent of x and t, which permits us to express the dependent variable using the Fourier transform; let

$$h(x,y,t) = \int_k \tilde{h}(y;k) e^{\mathrm{i}kx - \mathrm{i}\omega t} \mathrm{d}k + \text{c.c.}.$$

Now the equation becomes an ordinary differential equation, with k acting as a parameter:

$$y\frac{\mathrm{d}^2\tilde{h}}{\mathrm{d}y^2} + \frac{\mathrm{d}\tilde{h}}{\mathrm{d}y} + \left(\frac{\omega^2}{\varepsilon g} - yk^2\right)\tilde{h} = 0.$$

Note that in the limit $y \to \infty$, this equation simplifies to $\mathrm{d}^2\tilde{h}/\mathrm{d}y^2 = k^2\tilde{h}$, with solution $\tilde{h} = h_0 e^{-ky}$. The minus sign has been chosen to satisfy the decay condition.

12.4.2 Coastal Waves Solution

Equations with variable coefficients are more difficult to solve than those with constant coefficients and – with few notable exceptions – do not have analytic solutions. Among the notable exceptions are cases, such as the present, in which the coefficients are linear functions of the independent variable. When we encounter an equation such as this, it is often necessary to transform the dependent and/or independent variable to get the equation in canonical, or standard, form.[16] If we let

$$\tilde{h}(y;k) = h_0(k) e^{-y^*/2} h^*(y^*), \qquad \text{where} \qquad y^* = 2ky$$

is a dimensionless independent variable, the equation becomes

$$y^* \frac{\mathrm{d}^2 h^*}{\mathrm{d}y^{*2}} + (1 - y^*)\frac{\mathrm{d}h^*}{\mathrm{d}y^*} + jh^* = 0,$$

where

$$j \equiv \frac{\omega^2}{2\varepsilon g k} - \frac{1}{2} \qquad \text{or equivalently} \qquad \omega = \pm\sqrt{(2j+1)\varepsilon g k}$$

is the dispersion relation.

If we restrict the constant j to be a positive integer,[17] the equation is known as *Laguerre's equation* and has an infinite set of polynomial solutions known as *Laguerre polynomials*, denoted by L_j: $h^* = L_j(y^*)$. The polynomials may be generated using the *Rodrigues formula*

$$L_j(y^*) = \frac{1}{j!}\left(\frac{\mathrm{d}}{\mathrm{d}y^*} - 1\right)^j y^{*j}.$$

[16] An useful guide in this process is Murphy (1960).

[17] If j is not an integer, the equation is known as *Kummer's equation*, a particular form of a class of equations known as *hypergeometric equations*; see §13.2(i), p. 322 of Olver et al. (2010). Note that Olver et al. is available on the world wide web at http://dlmf.nist.gov.

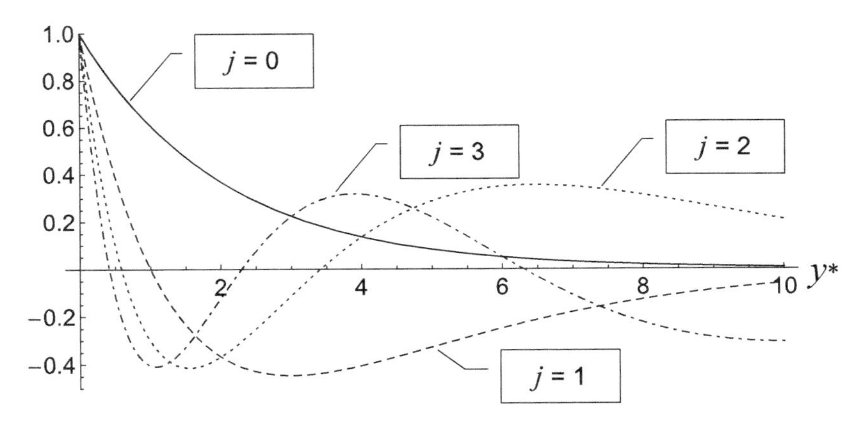

Figure 12.4 The first four coastal-wave solutions (solid, dashed, dotted and dash-dot lines, respectively) as a function of $y^* = 2ky$.

The first four of these polynomials, normalized to satisfy $L_j(0) = 1$, are

$$L_0(y^*) = 1, \quad L_1(y^*) = -y^* + 1, \quad L_2(y^*) = \tfrac{1}{2}\big(y^{*2} - 4y^* + 2\big)$$

$$\text{and} \qquad L_3(y^*) = \tfrac{1}{6}\big(-y^{*3} + 9y^{*2} - 18y^* + 6\big).$$

The structure of the first four of the modes $e^{-y^*/2}h^*(y^*)$ is illustrated in Figure 12.4. The Laguerre polynomials are a *complete set of orthogonal polynomials*;[18] any arbitrary function of y^* can be represented as a sum of Laguerre polynomials. Consequently, an arbitrary initial condition can be expressed as a sum of these modes. Putting it the other way around, an arbitrary initial disturbance sorts itself out into a set of Laguerre polynomials or modes, each of which travels at a distinct speed.

The dispersion relation $\omega = \pm\sqrt{(2j+1)\varepsilon gk}$ describes two identical modes moving to the left or right along the shore. These waves are dispersive; like deep-water waves, their group speed is half that of the phase speed:

$$U_p = \pm\sqrt{(2j+1)\varepsilon g/k} \qquad \text{and} \qquad U_g = U_p/2.$$

Note that the wave speed increases with index j. As with seiches, typically the fundamental mode (having $j = 0$) has the largest amplitude (higher-order modes decay more rapidly with time due to dissipative effects).

The wave period is given by

$$P = \sqrt{2\pi\Lambda/(\varepsilon g)},$$

where Λ is the wavelength. If, for example, $\varepsilon \approx 0.06$, $P \approx \sqrt{10\Lambda/\mathrm{m}}$ s. With $\Lambda = \{100, 10^3, 10^4, 10^5\}$ m, $P \approx \{30, 100, 300, 10^3\}$ s. Since $\varepsilon \ll 1$, coastal waves travel more

[18] For an explanation of orthogonal polynomials, see Dettman (1969).

slowly, and persist much longer, than deep-water waves. This makes them vulnerable to the effects of rotation. This complication is addressed in § 18.2. As $y^* \to 0$, $D \to 0$ but h^* does not. Consequently the assumption that $|h| \ll D$ is violated close to the shore, where the wave becomes nonlinear. Quantification of this behavior is beyond the scope of this elementary survey.

Transition

In this chapter we have studied linear shallow-water waves, which include tsunamis, the tides, seiches and coastal waves. In the next chapter, we investigate nonlinear shallow-water waves.

13

Nonlinear Shallow-Water Waves

So far we have studied linear deep-water and shallow-water waves. Now we shall turn our attention to nonlinear shallow-water waves, particularly solitary waves. We begin in the following section by re-formulating the shallow-water wave equations to include effects of nonlinearity and dispersion that were ignored in the previous chapter. Then in § 13.2 these equations are scaled and non-dimensionlized. This procedure introduces two dimensionless parameters representing nonlinearity and dispersion. Four limiting cases are identified and studied in the follow four sections, the most interesting of which is the nonlinear dispersive case investigated in § 13.6, involving solitary waves.

13.1 Re-Formulation

In order to develop the equations governing nonlinear shallow-water waves, let's begin with the full nonlinear formulation for surface-water waves in a body of water having depth D. The governing equation, valid for $-D < z < h$ remains Laplace's equation:[1]

$$\frac{\partial^2 \phi}{\partial x^2} + \frac{\partial^2 \phi}{\partial z^2} = 0,$$

where ϕ is the velocity potential. The nonlinear kinematic and dynamic conditions at the free surface are[2]

$$\text{at} \qquad z = h(x,t): \quad \frac{\partial h}{\partial t} - \frac{\partial \phi}{\partial z} + \frac{\partial \phi}{\partial x}\frac{\partial h}{\partial x} = 0$$

$$\text{and} \quad \frac{\partial \phi}{\partial t} + gh + \frac{1}{2}\left(\frac{\partial \phi}{\partial x}\right)^2 + \frac{1}{2}\left(\frac{\partial \phi}{\partial z}\right)^2 = 0.$$

The bottom boundary condition is simply

$$\text{at} \qquad z = -D: \quad \frac{\partial \phi}{\partial z} = 0.$$

The surface boundary conditions are a pair of coupled nonlinear partial differential equations, applied at an unknown position $z = h(x,t)$. Nonlinear equations are even more

[1] See § 11.1.
[2] From § 11.2.

difficult to solve than linear equations with variable coefficients, so we have our work cut out for us in attempting to analyze and solve these equations. This provides us with a chance to use some of the tools of the trade, including

- scaling the variables;[3]
- using dimensional analysis to determine the number of dimensionless parameters;
- non-dimensionalizing the variables and equations;
- introducing a small parameter;
- linearizing the problem based on the small parameter; and
- streamlining the notation.

13.2 Scaling and Non-Dimensionalization

Let's begin to use our tools by scaling the dependent and independent variables h, ϕ, x, z and t. To begin with, it is obvious that we should scale x with the length of the wave Λ, z with the water depth D and h with the maximum wave height, h_m. We can recover the balance of the linear equations by scaling t with $\Lambda/\sqrt{gD}$ and ϕ with $h_m\Lambda\sqrt{g/D}$. That is, we can write

$$x = \Lambda x^*, \qquad z = Dz^*, \qquad h = h_m h^*$$
$$t = \Lambda/\sqrt{gD}\,t^* \qquad \text{and} \qquad \phi = h_m\Lambda\sqrt{g/D}\,\phi^*,$$

where an asterisk denotes a dimensionless variable.

The scaling process involves four dimensional parameters: three lengths (D, h_m, Λ) and an acceleration (g). These parameters involve two fundamental dimensions: length and time. According to the Buckingham Pi theorem[4] we can construct two dimensionless parameters from the three lengths. We shall choose[5]

$$\alpha = h_m/D \qquad \text{and} \qquad \beta = D^2/\Lambda^2.$$

We shall see that α and β are measures of nonlinearity and dispersion, respectively. The linear surface boundary conditions developed in § 11.2.1 are recovered if we set $\alpha = \beta = 0$. In what follows, we will assume that $0 \le \alpha \ll 1$ and $0 \le \beta \ll 1$.

The next step is to introduce the scaled dimensionless variables into the governing equation and boundary conditions. The result is a set of equations littered with asterisks, whose only purpose is to remind us that the variables are dimensionless. But we can recognize this by the presence of the parameters α and β. So the asterisks are rather superfluous and in the interest of neatness, we will drop them and write the dimensionless

[3] See § 2.4.5.
[4] See § 2.4.4.
[5] If we had followed the instructions in § 2.4.4, these parameters would be named Π_1 and Π_2, but this would make for difficult reading.

equations as

$$\text{at} \quad z = \alpha h(x,t): \quad \beta\frac{\partial h}{\partial t} - \frac{\partial \phi}{\partial z} + \alpha\beta\frac{\partial \phi}{\partial x}\frac{\partial h}{\partial x} = 0$$

$$\text{and} \quad \beta h + \beta\frac{\partial \phi}{\partial t} + \frac{\alpha\beta}{2}\left(\frac{\partial \phi}{\partial x}\right)^2 + \frac{\alpha}{2}\left(\frac{\partial \phi}{\partial z}\right)^2 = 0;$$

$$\text{for} \quad -1 < z < \alpha h(x,t): \quad \beta\frac{\partial^2 \phi}{\partial x^2} + \frac{\partial^2 \phi}{\partial z^2} = 0;$$

$$\text{at} \quad z = -1: \quad \frac{\partial \phi}{\partial z} = 0.$$

This problem is to be solved for the velocity potential $\phi(x,z,t;\alpha,\beta)$ and surface deflection $h(x,t;\alpha,\beta)$. Note that the parameter α multiplies the nonlinear terms, while β multiplies every term not containing a z derivative.

If α were zero, the problem would reduce to that for linear shallow-water waves studied in Chapter 12, with ϕ independent of z to dominant order in β. This suggests an expansion of ϕ in Taylor series[6] in powers of β. At the same time, let's assume a form of ϕ that satisfies the bottom boundary condition, writing

$$\phi = \sum_{j=0} \frac{(1+z)^{2j}}{(2j)!}\phi_j\beta^j,$$

where the variables $\phi_j(x,t)$ are independent of z. The governing equation is satisfied provided

$$\phi_j = -\frac{\partial^2 \phi_{j-1}}{\partial x^2} = (-1)^j\frac{\partial^{2j}\phi_0}{\partial x^{2j}}.$$

That was fairly easy; we have solve the governing equation in terms of a single function of integration, $\phi_0(x,t;\alpha)$, with

$$\phi = \sum_{j=0}(-1)^j\frac{(1+z)^{2j}}{(2j)!}\frac{\partial^{2j}\phi_0}{\partial x^{2j}}\beta^j.$$

All the action is in the surface boundary conditions which serve to determine ϕ_0 and h. Substituting the β series into the boundary conditions and assuming that $\beta \ll 1$ (keeping only terms that contain β to the zeroth or first power), these equations become

$$\frac{\partial h}{\partial t} + \frac{\partial}{\partial x}\left((1+\alpha h)\frac{\partial \phi_0}{\partial x} - \frac{\beta}{6}(1+\alpha h)^3\frac{\partial^3 \phi_0}{\partial x^3}\right) = 0$$

[6] See Appendix A.4. Note that this expansion does not rest on the assumption that $\beta \ll 1$.

and

$$\frac{\partial \phi_0}{\partial t} + h + \frac{\alpha}{2}\left(\frac{\partial \phi_0}{\partial x}\right)^2$$
$$= \frac{\beta}{2}(1+\alpha h)^2 \left(\frac{\partial^3 \phi_0}{\partial t \partial x^2} + \alpha \frac{\partial \phi_0}{\partial x}\frac{\partial^3 \phi_0}{\partial x^3} - \alpha \left(\frac{\partial^2 \phi_0}{\partial x^2}\right)^2\right).$$

Note that we have not assumed that $\alpha \ll 1$ – yet. The surface boundary conditions have morphed into two governing equations for two unknown functions: $h(x,t)$ and $\phi_0(x,t)$. These equations can be cast in somewhat more familiar form by differentiating the second with respect to x and replacing $\partial \phi_0 / \partial x$ by u:

$$\frac{\partial h}{\partial t} + \frac{\partial}{\partial x}\left((1+\alpha h)u - \frac{\beta}{6}(1+\alpha h)^3 \frac{\partial^2 u}{\partial x^2}\right) = 0$$

and

$$\frac{\partial u}{\partial t} + \frac{\partial}{\partial x}\left(h + \frac{\alpha}{2}u^2\right) = \frac{\beta}{2}\frac{\partial}{\partial x}\left((1+\alpha h)^2 \left(\frac{\partial^2 u}{\partial t \partial x} + \alpha u \frac{\partial^2 u}{\partial x^2} - \alpha \left(\frac{\partial u}{\partial x}\right)^2\right)\right).$$

Now let's look at some special cases by limiting the values of α and β, starting with $\alpha = \beta = 0$.

13.3 Linear, Non-Dispersive

Setting $\alpha = \beta = 0$, the problem simplifies to the case of linear, non-dispersive shallow water waves, with the two conditions becoming

$$\frac{\partial h}{\partial t} + \frac{\partial u}{\partial x} = 0 \qquad \text{and} \qquad \frac{\partial u}{\partial t} + \frac{\partial h}{\partial x} = 0.$$

These may be combined into a single wave equation

$$\frac{\partial^2 u}{\partial t^2} = \frac{\partial^2 u}{\partial x^2}.$$

This is just the shallow-water case developed in the previous chapter. The wave has a dimensionless phase speed of unity and is non-dispersive.

13.4 Nonlinear and Non-Dispersive

If $\beta = 0$, the conditions become

$$\frac{\partial h}{\partial t} + \frac{\partial}{\partial x}\left((1+\alpha h)u\right) = 0 \qquad \text{and} \qquad \frac{\partial u}{\partial t} + \frac{\partial}{\partial x}\left(h + \frac{\alpha}{2}u^2\right) = 0.$$

This set of nonlinear equations govern flows in rivers and channels. We will consider them later in Chapter 19. The factor $1+\alpha h$ is the local depth of fluid. According to the dispersion relation for shallow-water waves, waves travel faster in deeper fluid, causing the leading edges of waves to steepen.

13.5 Linear and Dispersive

If $\alpha = 0$, the conditions simplify to

$$\frac{\partial h}{\partial t} + \frac{\partial u}{\partial x} = \frac{\beta}{6}\frac{\partial^3 u}{\partial x^3} \qquad \text{and} \qquad \frac{\partial u}{\partial t} + \frac{\partial h}{\partial x} = \frac{\beta}{2}\frac{\partial^3 u}{\partial x^2 \partial t}.$$

These linear equations are readily combined into a single equation for u:

$$\begin{aligned}\frac{\partial^2 u}{\partial t^2} &= \frac{\partial^2 u}{\partial x^2} + \frac{\beta}{6}\frac{\partial^2}{\partial x^2}\left(3\frac{\partial^2 u}{\partial t^2} - \frac{\partial^2 u}{\partial x^2}\right) \\ &\approx \frac{\partial^2 u}{\partial x^2} + \frac{\beta}{3}\frac{\partial^4 u}{\partial x^4}.\end{aligned}$$

The dimensionless dispersion relation for waves satisfying this equation is (reinstating the asterisks)

$$\omega^* = \sqrt{1 + \tfrac{1}{3}\beta k^{*2}}\, k^*.$$

We can recover the dimensional form from this by using the scaling for x and t, $\omega^* = \Lambda\omega/\sqrt{gD}$ and $k^* = \Lambda k$:

$$\begin{aligned}\omega &= \sqrt{gD}\sqrt{1 + \tfrac{1}{3}\beta\Lambda^2 k^2}\, k \\ &= \sqrt{gD}\sqrt{1 + \tfrac{1}{3}D^2 k^2}\, k.\end{aligned}$$

The parameter β causes the dispersion relation to become nonlinear waves now are dispersive, with long-length waves traveling faster.

13.6 Nonlinear and Dispersive

Finally consider the case that both α and β are non-zero, but now limit α to be small compared with unity: $\alpha \ll 1$. Discarding terms containing both α and β, the equations are

$$\frac{\partial h}{\partial t} + \frac{\partial u}{\partial x} = \frac{\beta}{6}\frac{\partial^3 u}{\partial x^3} - \alpha\frac{\partial}{\partial x}(hu)$$

and

$$\frac{\partial u}{\partial t} + \frac{\partial h}{\partial x} = \frac{\beta}{2}\frac{\partial^3 u}{\partial t \partial x^2} - \frac{\alpha}{2}\frac{\partial u^2}{\partial x}.$$

These are called the *Boussinesq equations*.[7] They have been written with the dominant terms on the left hand side and the perturbation terms on the right. To dominant order,

[7] Not to be confused with the Boussinesq approximation, which is something quite different; see § 7.3.2.

these equations describe a pair of non-dispersive wave modes traveling a speed equal to unity: one wave traveling in the positive x direction having $h = u = f_+(x-t)$ and the second traveling in the negative x direction having $h = -u = f_-(x+t)$, where $f_\pm$ describe arbitrary waveforms.

We can differentiate the first equation with respect to x and the second with respect to t and subtract the results, obtaining

$$\left(\frac{\partial}{\partial t} - \frac{\partial}{\partial x}\right)\left(\frac{\partial u}{\partial t} + \frac{\partial u}{\partial x}\right) = \frac{\beta}{2}\frac{\partial^4 u}{\partial t^2 \partial x^2} - \frac{\beta}{6}\frac{\partial^4 u}{\partial x^4} + \alpha\frac{\partial^2}{\partial x^2}(hu) - \frac{\alpha}{2}\frac{\partial^2 u^2}{\partial x \partial t}$$

To make further progress, let's restrict our attention to the wave traveling in the positive x direction; at dominant order for this wave $h = u$ and $\partial/\partial t = -\partial/\partial x$. Now we can replace h by u in the perturbation terms on the right; this introduces an error that is second order[8] in α and β – negligibly small. Similarly, we can replace $\partial/\partial t$ by $-\partial/\partial x$ everywhere except in the factor $\partial u/\partial t + \partial u/\partial x$ on the left-hand side; again this introduces a negligibly small error. Now the equation becomes

$$\frac{\partial}{\partial x}\left(\frac{\partial u}{\partial t} + \frac{\partial u}{\partial x}\right) = -\frac{\beta}{6}\frac{\partial^4 u}{\partial x^4} - \frac{3\alpha}{4}\frac{\partial^2 u^2}{\partial x^2}.$$

This is readily integrated once to give

$$\frac{\partial u}{\partial t} + \frac{\partial u}{\partial x} + \frac{3}{2}\alpha u\frac{\partial u}{\partial x} + \frac{\beta}{6}\frac{\partial^3 u}{\partial x^3} = 0$$

This is known as the *Korteweg–de Vries equation*,[9] or more simply the KdV equation.

The KdV equation is a nonlinear partial differential equation with the remarkable property that it has an exact analytic solution. We know that if $\alpha = \beta = 0$ the KdV equation has a solution in the form of a progressive wave of arbitrary shape: $u(x,t) = f(x-t)$. The nonlinear term (containing α) acts to steepen the wave form while the dispersive term (containing β) acts to widen it. As it happens, there a particular form of f such that these two effects precisely cancel! Allowing for the possibility that the wave travels at a different speed than linear shallow-water waves, let

$$u(x,t) = f(\zeta), \qquad \text{where} \qquad \zeta = x - (1+\varepsilon)t$$

and ε is a small perturbation of the linear wave speed. Note that $f(0) = 1$ is the maximum value of f.[10] Now the KdV equation becomes a third-order ordinary differential equation in f:

$$\beta f''' = 6\varepsilon f' - 9\alpha f f',$$

[8] A second-order error is of order α^2, $\alpha\beta$ and/or β^2.

[9] Often this equation is presented without the term $\frac{\partial u}{\partial x}$. This is obtained by the transformation $u(x,t) = \sigma(x,t) - 2/(3\alpha)$.

[10] Since $u = h$ to dominant order and h has been scaled such that its maximum is unity.

where a prime[11] now denotes differentiation with respect to ζ. This is to be solved subject to suitable symmetry and decay conditions. We shall seek a solution that is symmetric about $\zeta = 0$ and tends to zero as $\zeta \to \pm\infty$; specifically

$$f(0) = 1\,, \qquad f'(0) = 0 \qquad \text{and} \qquad f(\pm\infty) = 0\,.$$

It is readily seen that a first integral of this equation is[12]

$$2\beta f'' = 12\varepsilon f - 9\alpha f^2\,.$$

This equation has an integrating factor: f'. Multiplying the equation with this factor, we can obtain a second integral:[13]

$$\beta(f')^2 = 3(2\varepsilon - \alpha f)f^2\,.$$

The conditions at $\zeta = 0$ require

$$\varepsilon = \alpha/2\,.$$

Now the governing equation becomes

$$\beta(f')^2 = 3\alpha(1 - f)f^2\,.$$

The solution of this is simply

$$f(\zeta) = \operatorname{sech}^2\left(\sqrt{\frac{3\alpha}{4\beta}}\,\zeta\right).$$

In terms of (dimensionless) h and u, the solution is

$$h(x,t) = u(x,t) = \operatorname{sech}^2\left(\sqrt{\frac{3\alpha}{4\beta}}\left(x - t - \frac{\alpha}{2}t\right)\right).$$

This describes a *soliton*: a non-dispersive solitary wave that propagates without change of shape or amplitude. This function is plotted in Figure 13.1 for the case that $3\alpha = 4\beta$. Note that the wave is very compact, with the amplitude decaying exponentially as $|\zeta|$ becomes large. This wave travels coherently with a speed somewhat larger than a linear wave, with the increased speed being due to nonlinearity. The speed of the soliton is given by

$$U = \sqrt{gD}\left(1 + \frac{\alpha}{2}\right) = \sqrt{gD}\left(1 + \frac{h_m}{2D}\right) \approx \sqrt{g(D + h_m)}\,.$$

[11] This is a clash of commonly employed notations; see § 7.3. Rather than introduce a non-standard symbol for either a perturbation or a derivative, we use a prime to denote both and remind ourselves to be alert to what a prime means in a given situation.

[12] The condition at infinity requires the constant of integration to be zero.

[13] Again, the condition at infinity requires the constant of integration to be zero.

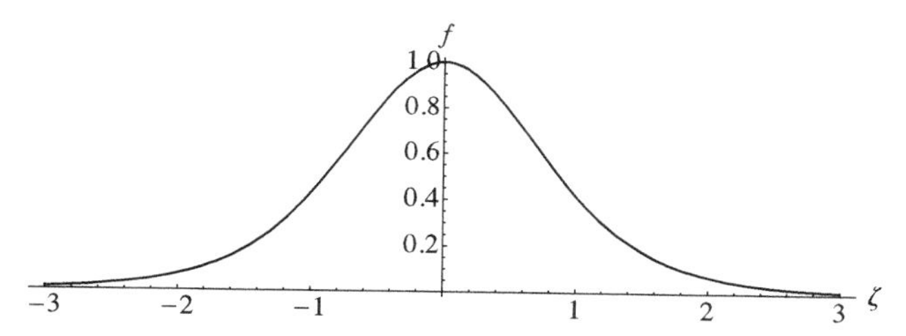

Figure 13.1 A plot of the solitary wave.

The speed of the soliton is the same as that of a linear wave in water of depth $D+h_m$; the solitary wave thinks the water has a depth equal to that at its crest.

Switching back to dimensional form, the surface displacement is

$$h = h_m \operatorname{sech}^2\left(\frac{x-Ut}{\Lambda}\right) \quad \text{with} \quad \Lambda = D\sqrt{\frac{4D}{3h_m}}.$$

and U as previously given. As the wave height h_m increases, the wave travels faster and becomes narrower. Note that the conditions $D \ll \Lambda$ and $h_m \ll D$ are consistent.

Let's put some numbers in to see how this theory fits with observation. In his description of his original sighting of a solitary wave on Union Canal in Scotland in 1834,[14] John Scott Russell gives wave height h_m = "a foot to a foot and a half" (0.3-0.45 m), wave speed $\sqrt{gD}$ = "some eight or nine miles an hour" (3.5-4.0 m/s) and wave length Λ = "some thirty feet long" (9 m), but reported no measurement of D. The Wikipedia entry for Union Canal lists the depth D of 5 feet ($\approx$ 1.5 m). A linear shallow-water wave would travel at a speed of 3.88 m/s on a canal of this depth, consistent with Russell's observation. Also, with these numbers, $\alpha \approx 0.1$ and $\beta \approx 0.03$; both are small, so the theory is applicable.

A bit of history is in order. In the early nineteenth century goods were commonly conveyed by canal barges, pulled by horses plodding along the canal embankment. Typically the bow of the barge pushed a bow wave ahead and the stern of the barge sat deep in the water, so that the horse had to pull the barge "uphill". However, it was found that a sudden acceleration of the barge (produced by a startled horse, say) could cause the barge to rise up on – and ride upon – its own bow wave, which took the form of a solitary wave. This had two distinct advantages: drag was reduced (in spite of the increased speed) and the barge rode higher relative to the canal bottom, permitting it to progress in shallower water without bottoming out.

A distinguishing feature of the soliton that sets it apart from other waves we have studied is the associated displacement. A soliton involves a deformation of the water surface having a non-zero average. Consequently as the wave passes, there is a net displacement of water in the direction in which the wave propagates. However, the water is stationary far

[14] e.g., see the Wikipedia entry for John Scott Russell. For more detail, see https://archive.org/stream/transactionsofro14royal#page/47/mode/1up and subsequent pages.

from the wave. In Part V, we will consider a variety of shallow-water motions that entail flow of water at large distances. Among other things, this will permit us to investigate shallow-water hydraulic shock waves.

Transition

The main focus of this chapter has been the soliton – a solitary wave that propagates without change of shape or amplitude. It is an interesting example of a nonlinear partial differential equation that has an exact analytic solution.

In the next chapter, which is the last chapter of Part III, we investigate a number of other types of waves that occur in non-rotating fluids, including capillary, interfacial and internal gravity waves.

14

Other Non-Rotating Waves

This chapter contains analyses and discussions of waves that occur in non-rotating fluids as a result of physical effects that we have ignored up till now, including (in the following section) capillary waves that involve surface tension, interfacial waves (in § 14.2) that occur on a sharp interface between two fluids and internal gravity waves (in § 14.3) that occur in a fluid whose density varies smoothly with depth.

14.1 Capillary Waves

Capillary waves are small-scale surface waves that are affected by surface tension.[1] The presence of surface tension modifies the dynamic boundary condition at the surface of a fluid. In the absence of surface tension this condition is simply equality of pressure across the surface. With surface tension, the pressure condition becomes

$$p = p_a + \gamma_s \left(\frac{1}{r_1} + \frac{1}{r_2} \right),$$

where r_1 and r_2 are the principal radii of curvature of the surface, defined as positive if the fluid is on the concave side, and γ_s is the surface tension.[2] If the wave is planar and harmonic (with surface deflection $h \propto \cos(kx)$, say) and has small amplitude, then $1/r_1 \approx k^2 h$ and $1/r_2 = 0$. The surface tension of water (below air) varies with temperature and is roughly 7.28×10^{-2} N·m^{-1} = 7.28×10^{-2} kg·s^{-2} at room temperature. The linearized kinematic and dynamic boundary conditions become[3]

$$\text{at} \quad z = h: \quad \frac{\partial \phi}{\partial z} = \frac{\partial h}{\partial t} \quad \text{and} \quad \left(g + \frac{\gamma_s k^2}{\rho} \right) h + \frac{\partial \phi}{\partial t} = 0.$$

It is readily seen that the capillary wave solution is just the linear solution presented in § 11.3, with g replaced by $g + \gamma_s k^2/\rho$. Specifically, the dispersion relation now is

$$\omega = \sqrt{gk + \gamma_s k^3/\rho}$$

[1] See Appendix D.4.

[2] Usually the surface tension coefficient is denoted by γ, but here we have added the subscript s to distinguish it from the ratio of specific heats.

[3] See § 11.2. The kinetic energy term has been neglected.

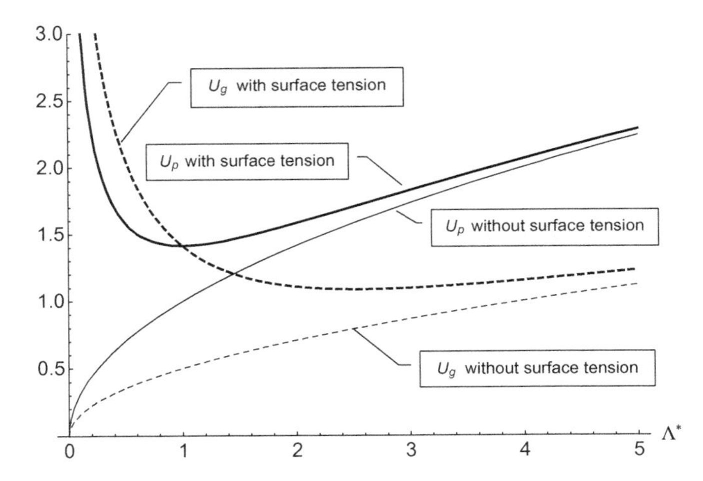

Figure 14.1 Plots of the dimensionless phase (solid line) and group (dashed line) speeds versus the dimensionless wavelength. The thick lines are the solutions for capillary waves and the thin lines depict the solution without surface tension effects.

and the phase and group speeds are

$$U_p = \sqrt{\frac{g}{k} + \frac{\gamma_s k}{\rho}} \qquad \text{and} \qquad U_g = \frac{g + 3\gamma_s k^2/\rho}{2\sqrt{gk + \gamma_s k^3/\rho}}.$$

This solution is better understood by writing the phase and group speeds in dimensionless form using the wavelength, Λ, as the independent parameter rather than the wavenumber.[4] Let

$$U_p = U_0 U^*, \qquad U_g = U_0 U_g^* \qquad \text{and} \qquad k = \frac{2\pi}{\Lambda_0 \Lambda^*},$$

where

$$U_0 = \sqrt{\frac{g\Lambda_0}{2\pi}}, \qquad \Lambda_0 = 2\pi\sqrt{\frac{\gamma_s}{g\rho}}$$

and an asterisk denotes a dimensionless variable. With $\gamma_s = 0.0728$ kg$\cdot$s^{-2}, $g = 9.8$ m$\cdot$s^{-2} and $\rho \approx 10^3$ kg$\cdot$m^{-3}, $\Lambda_0 \approx 0.017$ m and $U_0 \approx 0.163$ m$\cdot$s^{-1}. Now the dimensionless phase and group speeds are

$$U_p^* = \sqrt{\frac{1}{\Lambda^*} + \Lambda^*}, \qquad \text{and} \qquad U_g^* = \frac{3 + (\Lambda^*)^2}{2\sqrt{\Lambda^* + (\Lambda^*)^3}}.$$

These speeds are plotted versus the dimensionless wavelength in Figure 14.1. Note that

[4] $k = 2\pi/\Lambda$.

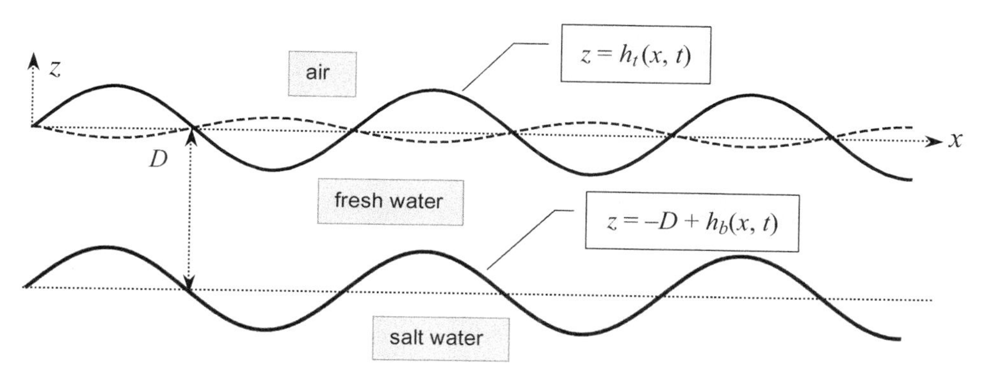

Figure 14.2 A schematic showing the geometry of the fresh-water layer of mean thickness D lying atop a layer of salty water of infinite depth, with the interfacial waves illustrated by the harmonic curves. There are two possible modes of waves on the top surface, a fast mode denoted by the solid curve and a slow mode denoted by the dashed curve.

- the effects of surface tension tend to zero as the length of the wave becomes large (as $\Lambda^* \to \infty$);
- U_p^* has a minimum value of $\sqrt{2}$ when $\Lambda^* = 1$;
- U_g^* has a minimum value of $(6\sqrt{3}-9)^{1/4} \approx 1.08626$ at $\Lambda^* = \sqrt{3+2\sqrt{3}} \approx 2.542$; and
- the phase speed of capillary waves is less than the group speed if $\Lambda^* < 1$, opposite to the case of deep-water waves.

Surface tension provides a lower bound on the possible phase and group speeds of water waves. The slowest waves have phase speed about 23 cm/s and group speed about 17.6 cm/s.

14.2 Interfacial Waves

Interfacial waves are very similar to deep-water waves, except that the overlying fluid is less-dense water, rather than air. These waves often occur within fjords where fresh water lies atop sea water, or on the *thermocline* within the ocean,[5] where warmer water lies atop cooler water. In this section we will consider a physical problem, depicted in Figure 14.2, consisting of three layers of fluid, with air (denoted by subscript a) on top, a layer of fresh water (denoted by subscript t) in the middle and a layer of salt water (denoted by b) on the bottom. The fluids have constant densities with the atmospheric density ρ_a set equal to zero. The bottom layer has infinite thickness and, in the absence of disturbances, the middle (fresh) layer has thickness D. Upward distance is measured from the undisturbed top of the upper (fresh-water) layer. When the interfaces are disturbed, their deviations are given by $h_t(t)$ and $h_b(t)$; that is the disturbed interfaces are located at $z = -D + h_b(t)$ and $z = h_t(t)$, as illustrated in Figure 14.2. The goal of this section is to quantify the harmonic deflections

[5] See § 7.2.2.

of the interfaces. We will find that these deflections can occur in two modes: a fast mode in which the deflections of the two interfaces are in phase and of similar magnitude and a slow mode in which the deflections are out of phase, with the disturbance of the lower interface having an amplitude greater than that of the upper interface.

The mathematical formulation and solution of this problem builds on the analysis presented in Chapter 11, modified to account for the density of the fresh-water layer and motions within that layer. This problem is rather complicated, so we shall proceed to formulate and solve it slowly and carefully.

14.2.1 Formulation

The formulation for interfacial waves employs the following simplifying assumptions:

- viscosity is negligibly small;
- the layer having depth D is thin (compared with Earth's radius);
- the densities of the two layers are constant;
- deviations from static equilibrium are small;
- flow is confined to the $x-z$ plane and independent of y;
- the fluid flow is irrotational;
- the deflections of the interfaces are small, so that the boundary conditions may be applied at the undeformed interfaces; and
- the horizontal velocity is independent of depth within the upper layer.

The condition that the flow remain irrotational is satisfied by the introduction of a velocity potential[6]

$$\mathbf{v} = \nabla\phi = \frac{\partial\phi}{\partial x}\mathbf{1}_x + \frac{\partial\phi}{\partial z}\mathbf{1}_z .$$

The fluid motions in the two layers above and below their common interface are governed by Euler's equation.[7] Ignoring the nonlinear inertial and Coriolis terms, setting $\mathbf{g} = -g\mathbf{1}_z$ and introducing the velocity potential, this becomes

$$\rho\frac{\partial\nabla\phi}{\partial t} = -\rho g\mathbf{1}_z - \nabla p .$$

With the density constant within each layer, this equation can be integrated, giving the linearized Bernoulli equation,[8]

$$\rho\frac{\partial\phi}{\partial t} + p = \rho g(H - z) ,$$

where the head H is constant, having differing values in the two layers. This equation serves to determine the pressure in the two layers, with suitable subscripts (t for the top layer and b for the bottom) on ρ, p, ϕ and H.

[6] See § 11.1.
[7] See § 6.2.1.
[8] See Appendix D.6.

Following the procedure described in § 7.3, we may express the pressure as a sum of a static (reference) part and a perturbation, $p = p_r + p'$, and the Bernoulli equation splits into two parts:

$$p_r = \rho g(H - z) \qquad \text{and} \qquad \rho\frac{\partial \phi}{\partial t} + p' = 0,$$

with the static reference state given by

$$p_r = p_a + \begin{cases} -\rho_t g z & \text{for} \quad -D < z < 0 \\ \rho_t g D - \rho_b g z & \text{for} \quad z < -D \end{cases}$$

and

$$H = \begin{cases} p_a/\rho_t g & \text{for} \quad -D < z < 0 \\ (p_a + \rho_t g D)/\rho_b g & \text{for} \quad z < -D, \end{cases}$$

where p_a is the atmospheric pressure. Note that p_r is continuous at $z = -D$, while H is not. The static pressure profile is depicted in Figure 14.3 as a dashed line. The total pressure is a continuous function of depth, as is the static pressure. It follows that the perturbation pressure is also continuous. However, since the density is not continuous, neither is $\partial\phi/\partial t$.

Vertical variation of the perturbation pressure, p', is induced by vertical acceleration, and the cumulative effect of this variation is represented by the term $\rho\partial\phi/\partial t$. With wave amplitudes assumed to be small, the vertical acceleration is also small, but its effect is cumulative. For example, the effect of vertical acceleration acts to dampen the magnitude

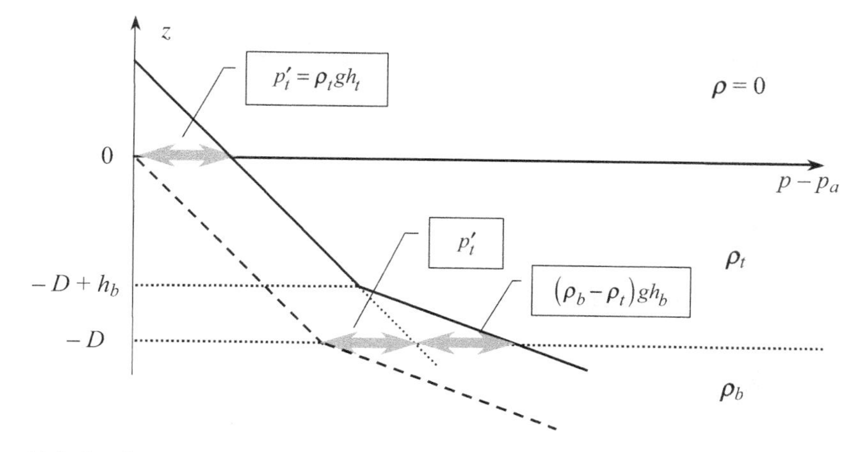

Figure 14.3 A schematic of pressure versus depth. The dashed line is the static pressure profile, the solid line is the actual profile and the horizontal distance between these two lines depicts the magnitude of the pressure perturbation. Specific pressure perturbations are shown as gray arrows, with p_b' equaling the sum of the two arrows shown at $z = -D$. Note that in general the magnitude of p_t' at $z = -D$ is not equal to its magnitude at $z = 0$.

of the pressure disturbance with depth, permitting a surface wave to sit atop a quiescent ocean of infinite depth. However, locally, on a vertical scale of the size of the wave itself, the vertical acceleration can be ignored and the pressure perturbation can be treated as constant. That is, the perturbation pressure p'_t at the location of the undeformed top surface (at $z = 0$) is very close in magnitude to the weight of the overburden of water of depth h_t, having density ρ_t:

$$\text{at} \quad z = 0: \quad p'_t = \rho_t g h_t \quad \text{and} \quad \frac{\partial \phi_t}{\partial t} + g h_t = 0.$$

This pressure balance is illustrated in Figure 14.3. Now consider the bottom interface, located at $z = -D + h_b$. The total pressure is continuous across this interface. Since the static pressure is continuous everywhere, this condition reduces to continuity of the perturbation pressure:

$$\text{at} \quad z = -D + h_b: \quad p'_b = p'_t.$$

The magnitude of p'_b at $z = -D$ is equal to the perturbation pressure at $z = -D + h_b$ (that is, $p'_t = -\rho_t \partial\phi_t/\partial t$) plus the weight of the overburden, $(\rho_b - \rho_t) g h_b$; altogether,

$$\text{at} \quad z = -D: \quad p'_b = p'_t + (\rho_b - \rho_t) g h_b$$

$$\text{and} \quad \rho_b \left(\frac{\partial \phi_b}{\partial t} + g h_b \right) = \rho_t \left(\frac{\partial \phi_t}{\partial t} + g h_b \right).$$

These pressures are depicted schematically in Figure 14.3.

In sum, the problem consists of the following equations and conditions, arranged in order of increasing depth:

$$\text{at} \quad z = 0: \quad \frac{\partial \phi_t}{\partial t} + g h_t = 0 \quad \text{and} \quad \frac{\partial h_t}{\partial t} = \frac{\partial \phi_t}{\partial z};$$

$$\text{for} \quad -D < z < 0: \quad \frac{\partial^2 \phi_t}{\partial x^2} + \frac{\partial^2 \phi_t}{\partial z^2} = 0;$$

$$\text{at} \quad z = -D: \quad \rho_b \left(\frac{\partial \phi_b}{\partial t} + g h_b \right) = \rho_t \left(\frac{\partial \phi_t}{\partial t} + g h_b \right),$$

$$\frac{\partial \phi_t}{\partial z} = \frac{\partial \phi_b}{\partial z} \quad \text{and} \quad \frac{\partial h_b}{\partial t} = \frac{\partial \phi_b}{\partial z};$$

$$\text{for} \quad z < -D: \quad \frac{\partial^2 \phi_b}{\partial x^2} + \frac{\partial^2 \phi_b}{\partial z^2} = 0;$$

$$\text{as} \quad z \to -\infty: \quad \phi_b \to 0.$$

Note that the problem has four z derivatives and two interfaces of unknown location in z. These require six boundary conditions in z; we have two at the upper interface, three at the lower and one as $z \to -\infty$. Note also that the problem has four time derivatives, indicating that we are modeling a flow having four wave modes. However, these – like sound waves – occur in pairs, so that there are only two distinct wave modes.

Limitation

The assumption that the dominant force balance in the vertical is hydrostatic limits the validity of the analysis and solution to waves for which $\partial^2 h/\partial t^2 \ll g$ or for waves having frequency $\omega \ll \sqrt{g/h}$. We are seeking waves that are modifications of the deep-water waves, for which $\omega = \sqrt{gk}$ or smaller. This implies that the analysis below is valid provided $kh \ll 1$. This restriction is in fact the same as holds for the deep-water wave equations; see § 11.4.

14.2.2 Ansatz and Solution

This problem is to be solved for ϕ_t and ϕ_b as functions of x, z and t and h_t and h_b as functions of x and t. Since the problem is linear with constant coefficients, the solution can be expressed as a linear combination of fundamental harmonic modes. Let's consider one such mode, with ϕ_t and ϕ_b varying as $\cos\theta$, where $\theta = kx - \omega t$. The velocity potentials satisfy Laplace's equation, with exponential variation in z. Further, the bottom boundary condition requires ϕ_b to be a positive exponential function of z. So let's try the ansatz

$$\phi_t(x,z,t) = \left(\tilde{\phi}_t \cosh(kz + kD) + \tilde{\phi}_b e^{kz+kD}\right) \cos\theta$$

and

$$\phi_b(x,z,t) = \tilde{\phi}_b e^{kz+kD} \cos\theta\,.$$

These forms satisfy the governing equations within the layers, the bottom boundary condition and continuity of vertical velocity at the fluid interface. The interface displacements are out of phase with the velocity potentials; selecting their amplitudes to satisfy the dynamic conditions (these are the conditions involving g), these displacements may be written as

$$h_t(x,t) = -\frac{\omega}{g}\left(\cosh(kD)\tilde{\phi}_t + e^{kD}\tilde{\phi}_b\right) \sin\theta$$

and

$$h_b(x,t) = \frac{\omega}{g}\left(\frac{\rho_t}{\rho_b - \rho_t}\tilde{\phi}_t - \tilde{\phi}_b\right) \sin\theta\,.$$

The kinematic boundary conditions at the top of each layer are satisfied provided

$$\left(1 - \frac{\omega^2}{kg}\right) e^{kD}\tilde{\phi}_b + \left(\sinh(kD) - \frac{\omega^2}{kg}\cosh(kD)\right)\tilde{\phi}_t = 0$$

and

$$\left(1 - \frac{\omega^2}{kg}\right)\tilde{\phi}_b + \frac{\omega^2}{kg}\frac{\rho_t}{\rho_b - \rho_t}\tilde{\phi}_t = 0\,.$$

This set of two linear homogeneous algebraic equations for $\tilde{\phi}_t$ and $\tilde{\phi}_b$ has a non-trivial solution only if the determinant of the coefficient matrix is zero. After some manipulation[9] this gives us the dispersion relation

$$\left(1 - \frac{\omega^2}{gk}\right)\left(\frac{\omega^2}{gk}(\rho_b \coth(kD) + \rho_t) - \rho_b + \rho_t\right) = 0\,.$$

[9] Using the identity $e^{kD} = \cosh(kD) + \sinh(kD)$.

This equation is quadratic in ω^2/gk and describes two pairs of wave modes.

The fast pair of modes has the same dispersion relation as deep-water waves: $\omega = \sqrt{gk}$. The interfaces deflect synchronously with

$$h_t/h_b = e^{kD} \qquad \text{and} \qquad \tilde{\phi}_t = 0.$$

This mode is unaffected by the stratification.

The slow pair of modes has dispersion relation

$$\omega = \sqrt{gk}\sqrt{\frac{\rho_b - \rho_t}{\rho_t + \rho_b \coth(kD)}}.$$

The interface deflections are out of phase with

$$\begin{aligned} \frac{h_t}{h_b} &= (\rho_b - \rho_t)\frac{\cosh(kD)\tilde{\phi}_t + e^{kD}\tilde{\phi}_b}{(\rho_b - \rho_t)\tilde{\phi}_b - \rho_t\tilde{\phi}_t} \\ &= -(\rho_b - \rho_t)\frac{\cosh(kD)\big(\coth(kD) - 1\big)\rho_b + e^{-kD}\rho_t}{\rho_t\big(\rho_t + \coth(kD)\rho_b\big)} \\ \text{and} \qquad & \rho_t\tilde{\phi}_t + \big(2\rho_t + \coth(kD)\rho_b - \rho_b\big)\tilde{\phi}_b = 0. \end{aligned}$$

The deflection of the upper surface is small relative to that of the lower when $\rho_b - \rho_t \ll \rho_b$. If the bottom layer has finite thickness D_b, then the dispersion relation for the slow wave becomes

$$\omega = \sqrt{gk}\sqrt{\frac{\rho_b - \rho_t}{\coth(kD_b)\rho_t + \coth(kD)\rho_b}}.$$

Let's see what this result can tell us about the upper ocean. The slow mode has a speed significantly slower than that of the usual surface wave because the density contrast across the thermocline is much smaller than that between water and air at the top of the ocean. Because of this slower wave speed, a ship underway can dissipate much of its power in generating internal gravity waves, leaving little to drive the ship forward. This phenomenon, called *dead water*, was first described by Fridtjof Nansen: "When caught in dead water *Fram* appeared to be held back, as if by some mysterious force, and she did not always answer the helm. In calm weather, with a light cargo, *Fram* was capable of 6 to 7 knots. When in dead water she was unable to make 1.5 knots. We made loops in our course, turned sometimes right around, tried all sorts of antics to get clear of it, but to very little purpose."

14.2.3 Thin-Layer Limit

The solution for the slow mode simplifies a bit in the thin-layer limit, which is characterized by $kD \ll 1$. In this limit $\coth(kD) \approx 1/kD$ and the dispersion relation for the slow mode simplifies to

$$\omega = k\sqrt{gD(\rho_b - \rho_t)/\rho_b}.$$

In this limit, the wave is non-dispersive[10] with phase and group speed given by

$$U_s = \sqrt{gD(\rho_b - \rho_t)/\rho_b}\,.$$

Also, the ratio of surface deflections becomes

$$\frac{h_t}{h_b} = -\frac{\rho_b - \rho_t}{\rho_t}\,.$$

This ratio of deflections describes a purely hydrostatic balance, with the acceleration term in the vertical component of the momentum equation being negligibly small.

14.3 Internal Gravity Waves

So far we have investigated waves on a single interface (e.g., between water and air)[11] and on two interfaces (e.g., between air, fresh water and salt water).[12] These investigations could be expanded to three or more discrete interfaces, but the algebra would quickly get out of hand. Instead, in this section we consider internal waves in a fluid that is stably stratified, due to either an increase in potential temperature[13] or a decrease in salt content with increasing elevation. We want to investigate waves that are perturbations of this static state.

Ignoring

- the Coriolis term;
- nonlinear terms;
- viscous terms; and
- compressibility of the fluid

and assuming the wave motion is in the $x-z$ plane with $\mathbf{v} = u\mathbf{1}_x + w\mathbf{1}_z$, the governing equations are[14]

$$\frac{\partial \rho'}{\partial t} + \frac{\mathrm{d}\rho_r}{\mathrm{d}z}w + \rho_r\left(\frac{\partial u}{\partial x} + \frac{\partial w}{\partial z}\right) = 0,$$

$$\rho_r\frac{\partial w}{\partial t} = -\frac{\partial p'}{\partial z} - g\rho' \qquad \text{and} \qquad \rho_r\frac{\partial u}{\partial t} = -\frac{\partial p'}{\partial x},$$

where once again a subscript r denotes a static reference-state variable that depends only on z and a prime denotes a perturbation associated with the wave. This is a set of three equations for four unknowns: u, w, p' and ρ'. Normally the fourth equation is provided by the equation of state relating density and pressure, but with the fluid being incompressible,

[10] Just as shallow-water waves in a homogeneous fluid; see Chapter 12.
[11] In Chapter 11.
[12] In § 14.2.
[13] See § 5.6.
[14] From § 4.7.

the velocity is non-divergent and the fourth equation comes from splitting the continuity equation in two:

$$\frac{\partial u}{\partial x} + \frac{\partial w}{\partial z} = 0 \qquad \text{and} \qquad \frac{\partial \rho'}{\partial t} + \frac{\mathrm{d}\rho_r}{\mathrm{d}z} w = 0 .$$

Stable stratification of an incompressible fluid implies that $\mathrm{d}\rho_r/\mathrm{d}z < 0$. As the fluid rises ($w > 0$), an observer sitting at a fixed elevation sees the density increase at a rate $-w\mathrm{d}\rho_r/\mathrm{d}z$. After a time interval Δt, the fluid has risen a distance $h = w\Delta t$ and the density at a fixed elevation has increased by an amount $\rho' = -h\mathrm{d}\rho_r/\mathrm{d}z$; that is,[15]

$$\rho = \rho_r - \left(\frac{\mathrm{d}\rho_r}{\mathrm{d}z}\right) h .$$

Substituting this density equation of state into the continuity equation and the time derivative of the vertical component of the momentum equation and noting that $\partial h/\partial t = w$, we have

$$\rho_r \frac{\partial^2 w}{\partial t^2} = -\frac{\partial^2 p'}{\partial t \partial z} + g\frac{\mathrm{d}\rho_r}{\mathrm{d}z} w .$$

This equation, together with the continuity equation and x component of the momentum equation, form a set of three equations for three unknowns: u, w and p'.

14.3.1 Brunt–Väisälä Frequency

The strength of stratification is normally measured by the *Brunt–Väisälä frequency*, defined by

$$N^2 \equiv -\frac{g}{\rho_r}\frac{\mathrm{d}\rho_r}{\mathrm{d}z} .$$

This gives a real value of N provided $\mathrm{d}\rho_r/\mathrm{d}z < 0$: the density decreases with increasing elevation. If the density increases with height, then N is imaginary and the mode of motion is exponential growth rather than sinusoidal; this describes an unstable state.

The Brunt–Väisälä frequency is the maximum frequency of internal oscillations within a stratified fluid. We have seen[16] that the density gradient of the deep ocean is $\mathrm{d}\rho_r/\mathrm{d}z \approx -6 \times 10^{-5}$ kg·m^{-4}. With $g \approx 10$ m·s^{-2} and $\rho_r \approx 10^3$ kg·m^{-3}, we have $N \approx 8 \times 10^{-4}$ rad·s^{-1}; the period of vertical oscillations in the ocean is $2\pi/N \approx 3$ hours.

14.3.2 Internal Waves

Now let's investigate the nature of internal gravity waves that propagate in an arbitrary direction in the x–z plane, treating the coefficients g, $\rho_r = \rho_0$ and N as constants. Let's look

[15] If the fluid were compressible, the pressure perturbation would induce an additional change of density given by the adiabatic equation of state (see § 7.3.2): $\rho_r p'/K$, where K is the adiabatic incompressibility. Recall that K has the same dimensions as pressure and $K = \infty$ for an incompressible fluid.

[16] In § 7.2.2.

for time-dependent harmonic solutions having wavenumber k and frequency ω, so that $\partial^2/\partial x^2 \to -k^2$ and $\partial^2/\partial t^2 \to -\omega^2$. The vertical component of the momentum equation may be expressed as

$$\left(\omega^2 - N^2\right) w = \frac{1}{\rho_0}\frac{\partial^2 p'}{\partial t \partial z}.$$

At the moment, we have three equations (continuity and x and z components of momentum) for three unknowns, u, w and p'. Let's combine these into a single equation involving a single variable. First eliminating u between the continuity and x momentum equations, we have

$$p' = -\frac{\rho_0}{k^2}\frac{\partial^2 w}{\partial z \partial t}.$$

Next, using this to eliminate p' from the vertical component of the momentum equation, we have a single linear ordinary differential equation for a single unknown, w:

$$\frac{\partial^2 w}{\partial z^2} - \left(1 - \frac{N^2}{\omega^2}\right) k^2 w = 0.$$

This equation admits a solution harmonic in z, with $w = \tilde{w}\cos(kx + mz - \omega t)$ provided the dispersion relation is

$$\omega = N\frac{k}{\sqrt{k^2 + m^2}}.$$

Internal waves are transverse waves, with motion parallel to the lines of constant phase. The phase and group speeds are

$$U_p = N\frac{1}{\sqrt{k^2 + m^2}} \quad \text{and} \quad U_g = N\frac{m^2}{\left(k^2 + m^2\right)^{3/2}}.$$

If $m = 0$, we recover the vertical solution $\omega = N$; motion and lines of constant phase are vertical and the phase and group speeds are $U_p = N/k$ and $U_g = 0$. If $m > 0$, $0 < U_g < U_p$; lines of constant phase are tilted from the vertical and the frequency of oscillation is reduced, because the effect of gravity is smaller. As the lines of constant phase approach horizontal, the wave frequency approaches zero.

14.3.3 Quantification of Coefficients

Internal gravity waves in the atmosphere, oceans and other large bodies of water having small amplitude are governed by linearized continuity and momentum equations that contain coefficients g, K and ρ_r. In this subsection we investigate the variation of these coefficients with z. Since the atmosphere and oceans are thin, we can safely treat g as constant.

Water is essentially incompressible and we can set $K = \infty$ when considering internal waves in water. On the other hand, air is quite compressible; we have seen in § 5.6 that $K = \gamma p$, where γ is the ratio of specific heats, with $\gamma = 7/5$.[17] We can quantify waves in both water and air by writing $K = \gamma p_r$ with γ equal to 7/5 for air and ∞ for water. This introduces a new variable into the formulation: $p_r(z)$, but it is related to ρ_r by the hydrostatic equation $\mathrm{d}p_r = -g\rho_r$, so we can focus on the variation of ρ_r.

Now let's consider ρ_r (and p_r). We have seen in § 7.2.1 that within the atmosphere

$$\rho_r = \rho_a\,(1 - z/z_T)^{\beta-1} \qquad \text{and} \qquad p_r = p_a\,(1 - z/z_T)^{\beta}\,,$$

where z is elevation above sea level, $z_T \approx 44$ km is the thermal scale height and $\beta \approx 5.2$. In the oceans, the variation of ρ_r with depth depends on the temperature and salinity constant and the pressure structure is not dynamically relevant. Within the oceans, density varies most rapidly within the pycnocline, which may be thought of as a thermocline of finite thickness. Current theory applies if the density profile is linear in z – or approximately so on the vertical scale associated with the wave.

An important factor introduced in the previous section is the natural frequency $N = \sqrt{-g\mathrm{d}\rho_r/\rho_r\mathrm{d}z}$. In that section we implicitly assumed N to be constant. In the troposphere, $N = \sqrt{(\beta-1)g/(z_T - z)}$. The scale of variation of N in the troposphere is roughly the scale height, $z_T \approx 44$ km. Invariably internal waves have vertical scale much less than this magnitude, so that we can treat N as constant within the troposphere, having a magnitude $N \approx 0.02\ \mathrm{s}^{-1}$.

Transition

This chapter contains brief summaries of capillary, interfacial and internal waves, which are in effect generalizations of surface water waves introduced in Chapter 11. All of the waves considered in Chapters 12 through 14 act on time scales sufficiently short (much less than one day) that the rotation of Earth does not play an important role in their dynamics. In the next part, we will consider waves that act on timescales of a day or more, so that the rotation of Earth is important in their dynamics.

[17] See Appendix E.8.5.

Part IV

Waves in Rotating Fluids

Waves exist due to the presence of a restoring force. In Part III, we investigated a variety of waves that occur in the atmosphere, oceans and Earth's mantle due to the action of:

- compressibility (sound waves in § 9.3 and elastic compressive waves in § 10.1 and § 10.3);
- elastic rigidity (elastic transverse waves in § 10.2 and § 10.4);
- gravitation:
  - at a single sharp interface (deep-water waves in Chapter 11 and shallow-water waves in Chapters 12 and 13);
  - at two sharp interfaces (interfacial waves in § 14.2); and
  - in a continuously stratified fluid (internal gravity waves in § 14.3); and
- surface tension (capillary waves in § 14.1).

If the fluid is rotating relative to an inertial frame of reference, the Coriolis force also acts as a restoring force and, as we shall see in this part, plays a crucial role in a number of types of waves. In fact, there is a bewildering array of waves because the Coriolis force comes in two flavors: f-plane and β-plane (see § 8.1.5) and can act alone or in conjunction with other effects, such as coastlines or bottom topography. In addition, the Coriolis force can modify the forms of some non-rotating waves, in effect creating new types of waves. So how can we make sense of all these? Let's begin our survey in Chapter 15 with an analysis of three "pure" rotating waves: geostrophic (f-plane), inertial and quasi-geostrophic (β-plane) waves.

Following this introduction, we will investigate the effect of the Coriolis force on several of the waves we previously studied. Rotation affects those waves that persist for a day or more, while shorter-period (higher frequency) waves are unaffected. Those waves having compressibility, elasticity or surface tension as the restoring force have high frequency and are unaffected by rotation. However, interfacial, stratified and (occasionally) deep-water waves have sufficiently low frequencies to be affected by rotation; these modified waves are considered in Chapter 16. Rotation permits the existence of novel types of waves, including equatorial Kelvin and Rossby waves (see Chapter 17) and topographic waves

Table IV.1. *Mathematical structure of equations governing various types of waves. C means constant Coriolis parameter (f plane); V means variable Coriolis parameter (β plane); v is the transverse speed; w is the vertical speed. Waves with no interface deflection are transverse. The thin-layer approximation implies a hydrostatic pressure balance in the vertical.*

Wave type	Chapter or section	f	v	$\partial/\partial y$	w	Thin-layer approx.
Deep water	11	$=0$	$=0$	$=0$	$\neq 0$	no
Shallow water	12 & 13	$=0$	$=0$	$=0$	$\neq 0$	yes
Interfacial	14.2	$=0$	$=0$	$=0$	$\neq 0$	yes
Internal	14.3	$=0$	$=0$	$=0$	$\neq 0$	no
Geostrophic	15	C	$\neq 0$	$\neq 0$	$=0$	yes
Inertial	15.2	C	$\neq 0$	$=0$	$=0$	no
Rossby	15.3	V	$\neq 0$	$\neq 0$	$=0$	yes
Poincaré	16.1	C	$\neq 0$	$=0$	$\neq 0$	no
Rotating interfacial	16.2	C	$\neq 0$	$=0$	$\neq 0$	no
Rotating internal	16.3	C	$\neq 0$	$=0$	$\neq 0$	no
Equatorial Kelvin	17.2	V	$=0$	$\neq 0$	$\neq 0$	yes
Equatorial Rossby	17.3	V	$=0$	$\neq 0$	$\neq 0$	yes
Coastal Kelvin	18.1	C	$=0$	$\neq 0$	$\neq 0$	yes
Topographic Rossby	18.2	C	$\neq 0$	$\neq 0$	$\neq 0$	yes

(see Chapter 18). For reference and orientation, the various types of waves are summarized in Table IV.1.

We will continue our practice of previous chapters and look for waves traveling in the x direction. In the absence of rotation, fluid waves have no transverse component and we were able to set $v = 0$ in in our previous investigations. The presence of rotation induces transverse motion in most – but not all – rotating waves.

15

Geostrophic, Inertial and Rossby Waves

In this chapter, we investigate "pure" rotating waves that occur in the absence of boundaries and without complicating effects such as interfaces or stratification. Assuming the fluid is incompressible, homogeneous and inviscid and neglecting nonlinear inertial terms, the governing equations, written in local Cartesian coordinates, are:[1]

$$\frac{\partial u}{\partial x}+\frac{\partial v}{\partial y}+\frac{\partial w}{\partial z}=0, \qquad \frac{\partial w}{\partial t}=-\frac{1}{\rho_r}\frac{\partial p'}{\partial z},$$

$$\frac{\partial u}{\partial t}-fv=-\frac{1}{\rho_r}\frac{\partial p'}{\partial x} \quad \text{and} \quad \frac{\partial v}{\partial t}+fu=-\frac{1}{\rho_r}\frac{\partial p'}{\partial y}$$

with $f(y)$ plotted in Figure 8.1. Commonly, f is approximated by $f \approx f_0+\beta(y-y_0)$.

These equations govern the motions of three types of "pure" waves: geostrophic (investigated in § 15.1), inertial (investigated in § 15.2) and quasi-geostrophic (investigated in § 15.3) waves.

15.1 Geostrophic Waves

The simplest type of rotating wave, the *geostrophic wave*, is so simple that if we blink we might miss seeing it. In § 8.5 we blinked – by omitting the time-derivative term from the horizontal momentum equation. This term has been included in the equations above. But the geostrophic-wave equations are simpler than this; the wave generates no pressure perturbation. With $p'=0$, the vertical momentum equation requires $w=0$ and the horizontal momentum equations simplify to

$$\frac{\partial u}{\partial t}-fv=0 \quad \text{and} \quad \frac{\partial v}{\partial t}+fu=0.$$

These equations have general solution

$$u=u_0\cos(ft) \quad \text{and} \quad v=-u_0\sin(ft),$$

[1] From § 8.6. Recall that the local Cartesian coordinates $\{x,y,z\}$ are {east, north, up} with respective velocity components $\{u,v,w\}$.

with the continuity equation requiring u_0 to be a constant. Note that the Coriolis term f is a function of latitude (or equivalently distance from the equator, y; see Figure 8.1).

What have we found? An inertial wave,[2] constrained to move on a thin shell, with a frequency equal to the Coriolis parameter, f. The frequency of the wave is maximum at the poles and decreases to zero at the equator. A geostrophic wave is in effect a rigid rotation of the fluid about an axis tilted with respect to Earth's axis of rotation, viewed by an observer fixed to Earth, except that the fluid axis is a function of latitude. This tendency for circular motion is an intrinsic feature of waves in rotating fluids. Note that the period of this wave is the same as the interval of time it takes a *Foucault pendulum* to rotate by π radians.

The fluid trajectories are circles of radius u_0/f, with particles moving clockwise in the northern hemisphere (and counterclockwise in the southern hemisphere), as seen by an observer above the wave. The fluid particles conserve their angular momentum, so that their eastward speed, relative to Earth, increases as they move northward, deflecting the trajectory to the right. When a particle is on the northern half of its circular trajectory, it is going eastward faster than Earth. As a particle moves southward, its eastward speed decreases, again deflecting the trajectory to the right. When it is on the southern half, it is going slower; an Earth-bound observer sees this as westward motion.

The geostrophic wave is a transverse wave with the Coriolis force being the restoring mechanism. This wave is excited in the oceans by storms, but it is not evident in atmospheric motions. Pressure gradients and variations of f with latitude come into play within the atmosphere, transforming this wave into a quasi-geostrophic Rossby wave; see § 15.3. Before investigating Rossby waves, in the next section let's consider inertial waves, which are a generalization geostrophic waves involving a pressure perturbation.

15.2 Inertial Waves

Inertial waves are a generalization of the geostrophic wave, having non-zero pressure perturbations. By adopting the f-plane approximation (treating f as constant) and assuming that u, v, w and p' are independent of y we may consider waves propagating in the east-west (x) direction. Now the governing equations presented at the beginning of this chapter simplify to

$$\frac{\partial u}{\partial x}+\frac{\partial w}{\partial z}=0, \qquad \frac{\partial w}{\partial t}=-\frac{1}{\rho_r}\frac{\partial p'}{\partial z},$$

$$\frac{\partial u}{\partial t}-f_0 v=-\frac{1}{\rho_r}\frac{\partial p'}{\partial x} \quad \text{and} \quad \frac{\partial v}{\partial t}+f_0 u=0.$$

Further assuming ρ_r is constant, it is a simple matter to eliminate v, w and p' to obtain a single linear partial differential equation for u:

$$\frac{\partial^4 u}{\partial t^2 \partial x^2}+\frac{\partial^4 u}{\partial t^2 \partial z^2}+f_0^2\frac{\partial^2 u}{\partial z^2}=0.$$

[2] See § 15.2 immediately below and Appendix F.7.

This equation has harmonic solutions of the form $u = \tilde{u}\cos(kx + mz - \omega t)$ provided ω satisfies the dispersion relation

$$\omega = \frac{f_0 m}{\sqrt{k^2 + m^2}}.$$

The horizontal phase and group speeds (in the x direction) of the inertial wave are

$$U_p = \frac{f_0 m}{k\sqrt{k^2 + m^2}} \quad \text{and} \quad U_g = -\frac{f_0 k m}{\left(k^2 + m^2\right)^{3/2}}.$$

If $k = 0$, these simplify to the geostrophic wave investigated in the previous section. These waves are dispersive unless $k = m$. A more general derivation and analysis of inertial waves is found in Appendix F.7.

15.3 Rossby Waves

Rossby waves are geostrophic waves that are affected by pressure gradients and the β effect; such waves are termed quasi-geostrophic (QG). Rossby waves in the atmosphere are of planetary scale and play a dominant role in the dynamics of the atmosphere; these waves cause the latitudinal variation of the jet stream and underly the changes in weather that we all observe.

Consider QG horizontal flow (in the $x - y$ plane). Ignoring topography, the continuity equation,

$$\frac{\partial u}{\partial x} + \frac{\partial v}{\partial y} = 0,$$

is automatically satisfied by the introduction of a stream function;[3] let

$$u = \frac{\partial \Psi}{\partial y} \quad \text{and} \quad v = -\frac{\partial \Psi}{\partial x}.$$

The scalar vorticity equation[4] simplifies to

$$\frac{\mathrm{D}_H \eta}{\mathrm{D}t} = 0,$$

with

$$\eta = \zeta + f \approx -\nabla_H^2 \Psi + f_0 + \beta(y - y_0)$$

where

$$\nabla_H^2 = \frac{\partial^2}{\partial x^2} + \frac{\partial^2}{\partial y^2}$$

[3] See appendices C.4 and C.5.1.

[4] See § 8.3.3.

is the horizontal Laplacian operator and

$$\frac{\mathrm{D}_H}{\mathrm{D}t} \approx \frac{\partial}{\partial t} - \frac{\partial \Psi}{\partial x}\frac{\partial}{\partial y}.$$

Ignoring the nonlinear term, the vorticity equation becomes

$$\frac{\partial}{\partial t}\nabla_H^2 \Psi + \beta \frac{\partial \Psi}{\partial x} = 0.$$

Let's look for harmonic solutions of this equation by writing

$$\Psi(x,y,t) = \sum_{k,l} \tilde{\Psi}(k,l)\cos(kx + ly - \omega t).$$

The governing equation yields the dispersion relation

$$\omega(k,l) = -\beta \frac{k}{k^2 + l^2}.$$

The phase and group speeds of Rossby waves (in the zonal direction) are

$$U_p = \frac{\omega}{k} = -\beta \frac{1}{k^2 + l^2} \qquad \text{and} \qquad U_g = \frac{\partial \omega}{\partial k} = \beta \frac{k^2 - l^2}{(k^2 + l^2)^2}.$$

Note that

- Rossby waves exist due to the variation of the Coriolis parameter with latitude, quantified by the parameter β;
- the phase speed of Rossby waves is *westward* relative to still air. Rossby waves move eastward due to the general eastward motion of the atmosphere in mid-latitudes; and
- Rossby waves are dispersive, with wave components having crests aligned principally north-south moving eastward and wave components having crests aligned principally east-west moving westward.

The Rossby wave is a transverse wave; its restoring mechanism is explained in the following subsection.

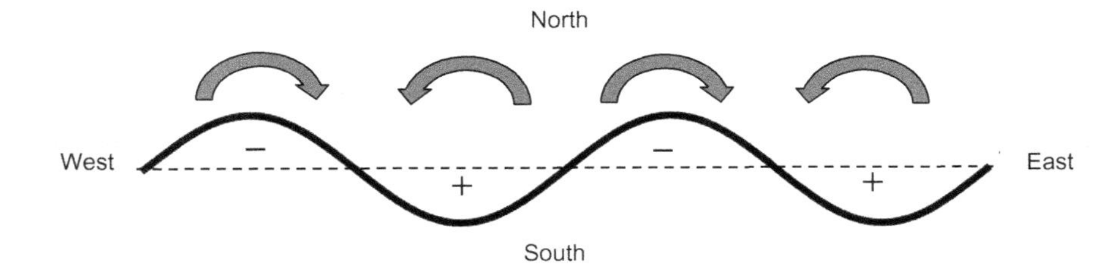

Figure 15.1 A schematic of the flow (curved arrows) induced by northward and southward displacement of fluid having constant absolute vorticity. The sign of the relative vorticity, ζ, is denoted by the $\pm$ signs.

15.3.1 Rossby-Wave Mechanism

The potential for wave motion exists when there is a mechanism acting to restore a parcel to its original position. The mechanism by which a Rossby wave exists and propagates can be understood in terms of conservation of absolute vorticity:[5]

$$\eta = \zeta + f(\theta) = \text{constant},$$

where θ is the colatitude. It is helpful to recall[6] that a negative vorticity is associated with a clockwise motion (when viewed from above) and that f increases northward. A region of the atmosphere or ocean that is displaced northward has a relative vorticity ζ that is negative. The clockwise rotation of fluid associated with this negative vorticity strengthens the northward displacement on the western side of the region and weakens it on the eastward side. By the same token, a southward displacement induces an counterclockwise rotation that reinforces the displacement on the western side of the region. To expand on this idea, suppose that the displacement of the fluid to the north and south of an equilibrium position is in the form of a sinusoid, as illustrated in Figure 15.1. The regions displaced {northward, southward} have {negative, positive} relative vorticity and the associated circulation is {clockwise, counterclockwise}, as denoted by the curved arrows. The further displacement induced by the this flow causes the wave to propagate westward.

Transition

In this chapter, we have investigated three "pure" rotating waves: geostrophic (in § 15.1), inertial (in § 15.2) and Rossby (in § 15.3) waves and found that these waves are characteristically transverse, having circular or oscillatory horizontal motions. In the next chapter, we investigate how this tendency modifies the shallow-water, interfacial and internal gravity waves that we investigated in Chapter 12, § 14.2 and § 14.3.

[5] That is, vorticity as measured in an inertial frame of reference.
[6] See § 8.1.3 and Figure 8.1.

16

Rotationally Modified Waves

In this chapter we investigate the effect of rotation on three types of non-rotating waves that have relatively low frequencies and so are susceptible to the action of the Coriolis force. Linear shallow-water waves, when modified by the Coriolis force, are called *Poincaré waves*; these are investigated in the following section. Rotationally modified interfacial and internal-gravity waves are studied in § 16.2 and § 16.3, respectively.

16.1 Poincaré Waves

As we have seen in Chapter 12, shallow-water waves in the open ocean move with speed $\sqrt{gD} \approx 200$ km·hr^{-1} and are capable of traversing an ocean in a matter of hours. However, if these waves persist for long enough (perhaps by moving onto a continental shelf), they can be affected by Earth's rotation; such waves are called Poincaré waves. We will investigate the influence of rotation on shallow-water waves in this section.

Poincaré waves are governed by case (a) of the averaged continuity equation presented in § 8.4.2 and the linearized quasi-geostrophic equations presented in § 8.6, with the constant-pressure surface Z set equal to the deflected water surface located at elevation h. Assuming that

- the equilibrium water depth is constant;
- the surface deflection is small ($|h| \ll D$);
- horizontal flow is independent of depth z; and
- the Coriolis term is constant ($\beta = 0$),

the governing equations are

$$\frac{\partial u}{\partial x} + \frac{\partial v}{\partial y} = -\frac{1}{D}\frac{\partial h}{\partial t},$$

$$\frac{\partial u}{\partial t} - f_0 v = -g\frac{\partial h}{\partial x} \qquad \text{and} \qquad \frac{\partial v}{\partial t} + f_0 u = -g\frac{\partial h}{\partial y}.$$

These equations are linear with constant coefficients, so we anticipate harmonic behavior in x, y and t. Since some dependent variables are differentiated and others not, their phases are not readily apparent. With this in mind, we will apply the ansatz as suggested in § 9.2.2; let's assume the wave travels in the x direction with

$$q(x,t) = \int_k \tilde{q}(k) e^{ikx - i\omega t} dk + \text{c.c.},$$

for $q = h, u$ and v. Note that we are assuming that the wave does not vary in the y direction, but it does have motion (v) in that direction. Now the equations become, in matrix form

$$\begin{bmatrix} k & 0 & -\omega/D \\ -i\omega & -f_0 & gik \\ f_0 & -i\omega & 0 \end{bmatrix} \begin{bmatrix} \tilde{u} \\ \tilde{v} \\ \tilde{h} \end{bmatrix} = \begin{bmatrix} 0 \\ 0 \\ 0 \end{bmatrix}.$$

This set of three linear homogeneous algebraic equations has a non-trivial solution only if the determinant of the coefficient matrix is zero; this requires that the dispersion relation be given by

$$\omega^2 = gDk^2 + f_0^2.$$

If $f_0 = 0$, the dispersion relation is that for shallow-water waves moving non-dispersively in the positive or negative x direction at speed $U = \sqrt{gD}$; see Chapter 12. On the other hand, if D is sufficiently small, $\omega \approx f_0$ and the Poincaré wave behaves like a geostrophic wave.

The addition of rotation makes the shallow-water waves dispersive with the phase speed faster and the group speed slower:

$$U_p = \frac{\omega}{k} = \sqrt{U^2 + \frac{f_0^2}{k^2}} \qquad \text{and} \qquad U_g = \frac{d\omega}{dk} = \frac{U^2}{U_p}.$$

In addition, rotation induces motion in the y direction that is lacking when rotation is unimportant: $\tilde{v} = -if_0\tilde{u}/\omega$.

In the absence of rotation, fluid motion associated with shallow-water waves is just a back-and-forth sloshing. With the addition of rotation, the fluid motion includes a lateral circular movement. This extra movement may be thought of as a geostrophic wave (see § 15.1) distorted by the presence of a pressure gradient, induced by the surface deflection.

16.2 Rotating Interfacial Waves

Interfacial waves (studied in § 14.2) travel much more slowly – and take a longer time to travel a specified distance – than shallow-water waves because the density contrast across the thermocline is much smaller than at the top of the ocean. If the lifetime of an interfacial wave is a day or longer, its structure and dynamics are influenced by Earth's rotation. In this section, we investigate the effect of the Coriolis force on an interfacial wave traveling in a general horizontal direction x.[1] This is a rather complicated problem, so the tasks of

[1] In the f-plane approximation, dynamics and motions in the horizontal plane are isotropic.

formulation and solution of the equations governing rotating interfacial waves are separated into two subsections below.

16.2.1 Formulation

With the addition of the Coriolis term to the momentum equation, the fluid is no longer irrotational and we cannot use the velocity-potential formulation of § 14.2.1. Also, we have seen that the Coriolis force engenders a lateral fluid motion in the case of shallow-water waves. We must anticipate this to occur for interfacial waves as well; we cannot assume that $v = 0$. As in our study of interfacial waves, subscripts t and b will denote top-layer and bottom-layer variables, respectively. We will employ the following simplifications:

- viscous forces are negligibly small;
- the densities in the two layers are constant;
- deflections of the interfaces are small; and
- variation in the y direction is zero (although the speed in the y direction is non-zero).

but will not assume that the upper layer is thin compared with the wavelength of the waves (this assumption would gain us little simplification). The pressure conditions developed in § 14.2.1 still apply

$$\text{at} \quad z = 0: \qquad p_t' = \rho_t g h_t \qquad \text{and}$$

$$\text{at} \quad z = -D: \qquad p_b{}' = p_t' + g(\rho_b - \rho_t) h_b \,.$$

In each layer the continuity equation is simply

$$\frac{\partial u}{\partial x} + \frac{\partial w}{\partial z} = 0 \,.$$

This may be satisfied by introducing a stream function; let

$$u = \frac{\partial \Psi}{\partial z} \qquad \text{and} \qquad w = -\frac{\partial \Psi}{\partial x} \,.$$

The condition of continuity of the vertical component of velocity at the bottom interface is assured by specifying that $\Psi_t = \Psi_b$ at $z = -D$. The horizontal momentum equation in each layer is

$$\frac{\partial^2 \Psi}{\partial t \partial z} - f_0 v = -\frac{1}{\rho}\frac{\partial p'}{\partial x} \qquad \text{and} \qquad \frac{\partial v}{\partial t} + f_0 \frac{\partial \Psi}{\partial z} = 0 \,.$$

Eliminating v between these two equations, we have

$$\frac{\partial^3 \Psi}{\partial t^2 \partial z} + f_0^2 \frac{\partial \Psi}{\partial z} = -\frac{1}{\rho}\frac{\partial^2 p'}{\partial t \partial x} \,.$$

This, plus the vertical component of the momentum equation, govern flow in each layer. Altogether the problem consists of

$$
\begin{array}{lll}
\text{at} & z=0: & -\dfrac{\partial \Psi_t}{\partial x} = \dfrac{\partial h_t}{\partial t} \quad \text{and} \quad p_t' = g\rho_t h_t\,; \\
\text{for} & -D<z<0: & \dfrac{\partial^2 \Psi_t}{\partial t \partial x} = \dfrac{1}{\rho_t}\dfrac{\partial p_t'}{\partial z} \quad \text{and} \quad \dfrac{\partial^3 \Psi_t}{\partial t^2 \partial z} + f_0^2 \dfrac{\partial \Psi_t}{\partial z} = -\dfrac{1}{\rho_t}\dfrac{\partial^2 p_t'}{\partial t \partial x}\,; \\
\text{at} & z=-D: & \Psi_t = \Psi_b, \qquad -\dfrac{\partial \Psi_b}{\partial x} = \dfrac{\partial h_b}{\partial t} \\
\text{and} & & p_b' = p_t' + g(\rho_b - \rho_t)h_b\,; \\
\text{for} & z<-D: & \dfrac{\partial^2 \Psi_b}{\partial t \partial x} = \dfrac{1}{\rho_b}\dfrac{\partial p_b'}{\partial z} \quad \text{and} \quad \dfrac{\partial^3 \Psi_b}{\partial t^2 \partial z} + f_0^2 \dfrac{\partial \Psi_b}{\partial z} = -\dfrac{1}{\rho_b}\dfrac{\partial^2 p_b'}{\partial t \partial x}\,; \\
\text{as} & z\to -\infty: & \{\Psi_b, p_b'\} \to 0.
\end{array}
$$

16.2.2 Solution

These equations and boundary conditions are to be solved for the six dependent variables h_t, h_b, Ψ_t, p_t', Ψ_b and p_b'. The first two of these are functions of x and t, while the remainder depend on x, z and t. Note that we are again assuming that the variables do not depend on y. Since these equations are linear with constant coefficients, the fundamental solutions are harmonic: sinusoidal in x and t and exponential in z. The x and t structures will be identical in the two layers, but the z structures may differ. Let's look for fundamental solutions of the form

$$\{\Psi(x,z,t), h(x,t)\} = \{\tilde{\Psi}(z), \tilde{h}\}\sin(kx - \omega t)$$

and[2]

$$p'(x,z,t) = \frac{k\omega}{m}\rho\tilde{p}(z)\cos(kx - \omega t)$$

in each layer, where[3]

$$m = k\omega / \sqrt{|\omega^2 - f_0^2|}\,.$$

is – as we shall soon see – the vertical wavenumber. In the non-rotating case ($f_0 = 0$), this expression simplifies to $m = k$, in agreement with the analysis of interfacial waves in § 14.2. Note that $k \le m$; rotation causes the wave to decay more rapidly with depth. The geostrophic wave of § 15.1 is recovered in the limit $\omega^2 \to f_0^2$ and $k \to 0$.

[2] The scaling factor has been included in the formulation for p' to make the following mathematical development a bit neater.

[3] The absolute value ensures that we have the decaying mode, with $m > 0$.

The fundamental solutions must satisfy

$$
\begin{array}{llll}
\text{at} & z=0: & k\tilde{\Psi}_t = \omega \tilde{h}_t & \text{and} \quad k\omega \tilde{p}_t = mg\tilde{h}_t; \\
\text{for} & -D<z<0: & \dfrac{\mathrm{d}\tilde{p}_t}{\mathrm{d}z} = m\tilde{\Psi}_t & \text{and} \quad \dfrac{\mathrm{d}\tilde{\Psi}_t}{\mathrm{d}z} = m\tilde{p}_t; \\
\text{at} & z=-D: & \tilde{\Psi}_t = \tilde{\Psi}_b, & \quad k\tilde{\Psi}_b = \omega \tilde{h}_b \\
\text{and} & & k\omega\rho_b\tilde{p}_b = k\omega\rho_t\tilde{p}_t + mg(\rho_b-\rho_t)\tilde{h}_b; & \\
\text{for} & z<-D: & \dfrac{\mathrm{d}\tilde{p}_b}{\mathrm{d}z} = m\tilde{\Psi}_b & \text{and} \quad \dfrac{\mathrm{d}\tilde{\Psi}_b}{\mathrm{d}z} = m\tilde{p}_b; \\
\text{as} & z\to-\infty: & \{\tilde{\Psi}_b, \tilde{p}_b\} \to 0. &
\end{array}
$$

Following the solution procedure employed in § 14.2.2, let's assume that

$$
\begin{aligned}
\tilde{\Psi}_t &= -\tilde{\phi}_b e^{(mz+mD)} - \tilde{\phi}_t \sinh(mz+mD), \\
\tilde{p}_t &= -\tilde{\phi}_b e^{(mz+mD)} - \tilde{\phi}_t \cosh(mz+mD) \\
\text{and} \qquad \tilde{\Psi}_b &= \tilde{p}_b = -\tilde{\phi}_b e^{(mz+mD)},
\end{aligned}
$$

where the constants $\tilde{\phi}_b$ and $\tilde{\phi}_t$ have been selected so that the present notation is synchronous with that introduced in in § 14.2.2.

These forms again satisfy the governing equations within the layers, the asymptotic bottom condition and continuity of vertical velocity at the bottom interface. It remains to satisfy the kinematic and dynamic conditions at the two interfaces. Selecting the phases and amplitudes of the interface displacements to satisfy the dynamic conditions (those containing g), they become

$$
\begin{aligned}
\tilde{h}_t &= \frac{k\omega}{mg}\left(-\tilde{\phi}_t\cosh(kD) - \tilde{\phi}_b e^{kD}\right) \\
\text{and} \qquad \tilde{h}_b &= \frac{k\omega}{mg}\left(-\tilde{\phi}_b + \frac{\rho_t}{\rho_b-\rho_t}\tilde{\phi}_t\right).
\end{aligned}
$$

The kinematic conditions at the top of each layer are satisfied provided

$$
\begin{aligned}
\left(1-\frac{\omega^2}{mg}\right)\tilde{\phi}_b &= e^{-kD}\left(\frac{\omega^2}{mg}\cosh(kD) - \sinh(kD)\right)\tilde{\phi}_t \\
\text{and} \qquad \left(1-\frac{\omega^2}{mg}\right)\tilde{\phi}_b &= -\frac{\omega^2}{mg}\frac{\rho_t}{\rho_b-\rho_t}\tilde{\phi}_t.
\end{aligned}
$$

These imply either $m=\omega^2/g$ and $\tilde{\phi}_t=0$ or

$$
\frac{\omega^2}{mg}(\rho_b\coth(kD)+\rho_t) - \rho_b + \rho_t = 0.
$$

This equation, together with $m = k\omega/\sqrt{|\omega^2-f_0^2|}$, is the dispersion relation. Note that if $f_0=0$, then $m=k$ and the dispersion relation is identical to that obtained in § 14.2.2.

This dispersion relation has two types of modes, a fast mode, which is essentially a rotational modification of the deep-water wave studied in § 16.1, and a slow mode. In general, the dispersion relation for the slow mode is transcendental, due to the factor $\coth(mD)$. To avoid this complication, we will limit our attention to the case $mD \ll 1$. Now the dispersion relation for the slow modes becomes

$$\frac{\omega^2}{m^2 D}\rho_b + \frac{\omega^2}{m}\rho_t - g(\rho_b - \rho_t) = 0$$

$$\text{or} \quad \left(|\omega^2 - f_0^2|\right)\alpha + \omega\sqrt{|\omega^2 - f_0^2|} - \alpha k^2 U_s^2 = 0,$$

where

$$U_s = \sqrt{gD(\rho_b - \rho_t)/\rho_b}$$

is the phase and group speed of the non-rotating, non-dispersive slow mode obtained in § 14.2.3 and

$$\alpha \equiv \frac{\rho_b}{kD\rho_t}$$

is a dimensionless parameter. Note that in the limit $kD \ll 1$, $1 \ll \alpha$ and the dispersion relation simplifies to

$$|\omega^2 - f_0^2| = k^2 U_s^2 .$$

In the slow-rotation limit, defined by $f_0 < kU_s$, the wave has frequency $\omega = \sqrt{f_0^2 + k^2 U_s^2}$; this is the slow interfacial wave of § 14.2 modified by rotation. The effect of rotation is to increase the wave frequency. This wave is dispersive, with rotation increasing the phase speed and decreasing the group speed. In the fast-rotation limit, defined by $kU_s < f_0$, the wave has frequency $\omega = \sqrt{f_0^2 - k^2 U_s^2}$; this may be thought of as a geostrophic wave (see § 15.1) affected by the presence of the interface, which acts to slow its frequency. Note that the wave becomes dispersive.

16.3 Rotating Internal Gravity Waves

Now let's investigate how rotation affects the internal gravity waves that we investigated in § 14.3. We will again look for waves traveling in the x direction, but must allow for the possibility that the transverse speed is non-zero. Assuming the Coriolis term is constant (that is, adopting the f-plane approximation[4]) and writing $\mathbf{v} = \mathbf{v}_H + w\,\mathbf{1}_z$, the governing

[4] See § 8.1.5.

equations are[5]

$$\rho_r \frac{\partial^2 w}{\partial t^2} = -\frac{\partial^2 p'}{\partial t \partial z} + g \frac{\mathrm{d}\rho_r}{\mathrm{d}z} w, \qquad \nabla_H \bullet \mathbf{v}_H + \frac{\partial w}{\partial z} = 0$$

and $$\rho_r \left(\frac{\partial \mathbf{v}_H}{\partial t} + f_0 \mathbf{1}_z \times \mathbf{v}_H \right) = -\nabla_H p'.$$

Let's introduce the Brunt–Väisälä frequency, $N = \sqrt{-g\mathrm{d}\rho_r/\rho_r\mathrm{d}z}$ and look for time-dependent harmonic solutions having wavenumber k and frequency ω, so that $\nabla_H^2 \to -k^2$ and $\partial^2/\partial t^2 \to -\omega^2$.[6] Now the vertical component of the momentum equation may be expressed as

$$\rho_r \left(\omega^2 - N^2 \right) w = \frac{\partial^2 p'}{\partial t \partial z}.$$

Eliminating $\mathbf{1}_z \times \mathbf{v}_H$ from the horizontal momentum equation using the cross product of that equation with $\mathbf{1}_z$, we have

$$\mathbf{v}_H = \frac{1}{\rho_r (f_0^2 - \omega^2)} \left(f_0 \mathbf{1}_z \times \nabla_H p' - \nabla_H \frac{\partial p'}{\partial t} \right).$$

The horizontal divergence of this is[7]

$$\nabla_H \bullet \mathbf{v}_H = \frac{k^2}{\rho_r (f_0^2 - \omega^2)} \frac{\partial p'}{\partial t}.$$

Now the continuity equation becomes

$$\frac{k^2}{\rho_r (f_0^2 - \omega^2)} \frac{\partial p'}{\partial t} + \frac{\partial w}{\partial z} = 0.$$

The time derivative of this may be expressed as

$$p' = \frac{\rho_r (f_0^2 - \omega^2)}{k^2 \omega^2} \frac{\partial^2 w}{\partial t \partial z}$$

and used to eliminate p' from the vertical component of the momentum equation, giving us a single equation,

$$\frac{\partial^2 w}{\partial z^2} - k^2 \frac{\omega^2 - N^2}{\omega^2 - f_0^2} w = 0,$$

for a single unknown, w.

[5] See § 8.2.1 and § 14.3.
[6] See Appendix A.5.3.
[7] Note that $\nabla_H \bullet (\mathbf{1}_z \times \nabla_H p') = \nabla_H p' \bullet (\nabla_H \times \mathbf{1}_z) - \mathbf{1}_z \bullet (\nabla_H \times \nabla_H p') = \mathbf{0}$; see Appendix A.2.6.

As in § 14.3, this equation admits a solution harmonic in z, with $w = \tilde{w}\cos(kx + mz - \omega t)$, but now the dispersion relation is

$$\omega = \sqrt{\frac{N^2k^2 + f_0^2 m^2}{k^2 + m^2}}.$$

If $f_0 = 0$, this dispersion relation simplifies to that developed in § 14.3 for non-rotating internal gravity waves, while if $N = 0$, it simplifies to the inertial wave investigated in § 15.2.

Transition

In this chapter we have investigated the effect of rotation on three types of waves: linear shallow-water waves, interfacial waves and internal-gravity waves. These waves, which are non-dispersive in the absence of rotation, become dispersive when the Coriolis force is important. In each case, rotation increases the phase speed and decreases the group speed.

In the following chapter, we will investigate rotating waves that occur close to the equator and rely on the β-plane effect for their existence; these waves have no non-rotating analog.

17

Equatorial Waves

Equatorial waves occur most prominently in the upper level (the mixed layer above the thermocline) of the oceans. In § 14.2 we identified a "slow" interfacial wave having a relatively large-amplitude deflection of the interface. The frequency of this mode is proportional to the square root of the difference in density between the fluid below and above the interface. As this difference becomes small, the wave period becomes large. If the period is of the order of, or larger than, a day these waves will be affected by the rotation of Earth. We will investigate this influence in the present section.

In particular, we will investigate *equatorial waves* that are deformations of the oceanic thermocline that are "trapped" near the equator due to the latitudinal variation of the Coriolis parameter.

17.1 Development of Equatorial-Wave Equations

We shall consider a thin layer (having depth D) of less dense fluid ($\rho = \rho_t$) lying atop a deep layer of denser fluid ($\rho = \rho_b$) with the top of the layer at elevation $z = h_t(x,y,t)$ and the bottom at $z = -D + h_b(x,y,t)$; see Figure 14.2. The analysis is based on the following assumptions:

- the fluids are incompressible;
- the fluids are in hydrostatic balance;
- the density in each layer is constant;
- the horizontal velocity components in the thin layer are independent of depth;
- terms containing the advective factor $\mathbf{v}\cdot\nabla$ are negligibly small; and
- the equatorial β-plane approximation is valid:

$$f \approx \beta y, \qquad \text{where} \qquad \beta = 2\Omega/R \approx 2.2 \times 10^{-11}\,\mathrm{m}^{-1}\,\mathrm{s}^{-1}.$$

Employing the local Cartesian reference frame introduced in § 8.2.2, the continuity and momentum equations for the thin layer are

$$\frac{\partial \bar{u}}{\partial x} + \frac{\partial \bar{v}}{\partial y} + \frac{\partial w}{\partial z} = 0, \qquad \frac{\partial p}{\partial z} = -\rho_t g,$$
$$\frac{\partial \bar{u}}{\partial t} - \beta y \bar{v} = -\frac{1}{\rho_t}\frac{\partial p}{\partial x} \quad \text{and} \quad \frac{\partial \bar{v}}{\partial t} + \beta y \bar{u} = -\frac{1}{\rho_t}\frac{\partial p}{\partial y},$$

where variables with a bar atop are independent of depth z. Integrating the vertical momentum equation and applying the condition that the pressure is equal to the atmospheric pressure p_a at $z = h_t$, we have

$$p = p_a + \rho_t g(h_t - z).$$

Hydrostatic equilibrium is maintained provided

$$h_t = -\frac{\rho_b - \rho_t}{\rho_t} h_b.$$

As we saw in § 14.2, the deflections of the top and bottom surfaces are out of phase and the deflection of the top surface is smaller than that of the bottom surface. In the limit $\rho_b - \rho_t \ll \rho_t$, the deflection of the top surface is negligibly small. Using this simplification, the pressure in the layer may be expressed in terms of h_b:

$$p = p_a - g(\rho_b - \rho_t)h_b - g\rho_t z$$

and the horizontal momentum equations become

$$\frac{\partial \bar{u}}{\partial t} - \beta y \bar{v} = g' \frac{\partial h_b}{\partial x} \quad \text{and} \quad \frac{\partial \bar{v}}{\partial t} + \beta y \bar{u} = g' \frac{\partial h_b}{\partial y},$$

where $g' = (\rho_b - \rho_t)g/\rho_b$ is the reduced gravity.

The curl of the horizontal momentum equations gives us the linearized vertical vorticity equation:

$$\frac{\partial \zeta}{\partial t} + \beta \bar{v} = -\beta y \left(\frac{\partial \bar{u}}{\partial x} + \frac{\partial \bar{v}}{\partial y} \right),$$

where $\zeta = \partial \bar{v}/\partial x - \partial \bar{u}/\partial y$.

Integration of the continuity equation from the bottom of the layer ($z = -D + h_b$) to the top ($z = h_t$) and noting that $w = \partial h/\partial t$ at the top and bottom yields

$$\frac{\partial \bar{u}}{\partial x} + \frac{\partial \bar{v}}{\partial y} = \frac{1}{D + h_t - h_b} \left(\frac{\partial h_b}{\partial t} - \frac{\partial h_t}{\partial t} \right)$$

or, assuming that $|\rho_b - \rho_t| \ll \rho_b$, $|h_t - h_b| \ll D$ and $|h_t| \ll |h_b|$,

$$\frac{\partial \bar{u}}{\partial x} + \frac{\partial \bar{v}}{\partial y} = \frac{1}{D} \frac{\partial h_b}{\partial t}.$$

This continuity equation and horizontal components of the momentum equation form a set of three linear partial differential equations with coefficients dependent on y, involving three unknowns: $\bar{u}$, $\bar{v}$ and h_b, each of which is a function of x, y and t. In the following two sections we will simplify these equations and investigate two types of equatorial waves.

17.2 Equatorial Kelvin Waves

Equatorial Kelvin waves[1] have no north-south motions. Setting $\bar{v} = 0$, the vertical vorticity equation developed in the previous section simplifies to

$$\frac{\partial^2 \bar{u}}{\partial y \partial t} = \beta y \frac{\partial \bar{u}}{\partial x}.$$

This is a single linear partial differential equation for u as a function of x, y and t, with a coefficient dependent on y. The most general solution decaying with distance from the equator ($|y|$) can be expressed as

$$\bar{u}(x,y,t) = \tilde{u}(x - Ut)e^{-\beta y^2/2U},$$

where U is the phase (and group) speed of the (non-dispersive) wave and $\tilde{u}$ is an arbitrary function. We readily see that the wave amplitude decreases exponentially with increasing distance from the equator provided $U > 0$, which means that Kelvin waves move eastward. The amplitude of an equatorial Kelvin wave is symmetric about the equator and the wave seems to be trapped close to the equator.

The phase speed U may be determined from the continuity and x component of the momentum equation:

$$\frac{\partial \bar{u}}{\partial x} = \frac{1}{D}\frac{\partial h_b}{\partial t} \qquad \text{and} \qquad \frac{\partial \bar{u}}{\partial t} = g' \frac{\partial h_b}{\partial x}.$$

Eliminating h_b we have the simple wave equation

$$\frac{\partial^2 \bar{u}}{\partial t^2} - U^2 \frac{\partial^2 \bar{u}}{\partial x^2} = 0$$

with

$$U = \sqrt{g'D}.$$

This phase speed is similar in form to that for shallow-water waves, but with a much reduced gravity.[2]

Typically the mixed layer thickness is in the range 200 to 1000 m. The coefficient of thermal expansion of water is about 2×10^{-4} K^{-1}. A thermal contrast of 10 K, say, produces a density contrast of roughly 0.2%, so that $g' \approx 0.02$ m·s^{-2}. With $D \approx 200$ m, the formula above gives $U \approx 2$ m·s$^{-1} \approx 7$ km·hr^{-1}. The width of the Kelvin wave is roughly of magnitude $\sqrt{U/\beta}$. Recalling that $\beta \approx 2 \times 10^{-11}$ m^{-1} s^{-1}, the wave has a typical width of about 300 km (extending 150 km north and south of the equator). Such a wave can traverse the Pacific Ocean in about two months. An equatorial Kelvin wave in the form of

[1] These waves are distinct and different from the Kelvin waves produced by a moving object on the surface of a body of water; see Appendix F.4.

[2] See Chapter 12.

a thickened layer containing exceptionally warm water will cause an El Niño event when it reaches South America and splays northward and southward.

17.3 Equatorial Rossby Waves

Equatorial Rossby waves are a bit more complicated than Kelvin waves, because they have north-south, as well as east-west, motions. In quantifying these waves, we can use the Kelvin-wave solution for guidance. The vertically-averaged governing equations are a set of three linear partial differential equations for $\bar{u}$, $\bar{v}$ and h_b as functions of x, y and t.[3] Since the coefficients are functions of y, this variable needs special attention. Rossby waves are dispersive, so we need represent the x and t dependence of the solution using a Fourier integral, rather than the simple function $\tilde{u}(x-Ut)$ employed in the previous section; let

$$\{\bar{u}, h_b\} = \{\tilde{u}, \tilde{h}\}\cos\theta \qquad \text{and} \qquad \bar{v} = \tilde{v}\sin\theta\,,$$

where $\theta = kx - \omega t$ and the variables with tildes are functions of y and the parameter k. The full solution is an integral over k. Substituting this assumed form into the horizontal momentum and continuity equations, we have

$$\omega\tilde{u} - \beta y\tilde{v} = -kg'\tilde{h}\,, \qquad -\omega\tilde{v} + \beta y\tilde{u} = g'\frac{\mathrm{d}\tilde{h}}{\mathrm{d}y}$$

$$\text{and} \qquad -k\tilde{u} + \frac{\mathrm{d}\tilde{v}}{\mathrm{d}y} = \frac{\omega}{D}\tilde{h}\,.$$

It is readily verified that if we set $\tilde{v} = 0$, we recover the equations governing Kelvin waves with dispersion relation $\omega = \sqrt{g'D}k$.

We may eliminate $\tilde{u}$ and $\tilde{h}$ from these three equations and obtain a single equation for $\tilde{v}$:

$$\frac{\mathrm{d}^2\tilde{v}}{\mathrm{d}y^2} + \left(\frac{\omega^2}{g'D} - k^2 - \frac{\beta k}{\omega} - \frac{\beta^2}{g'D}y^2\right)\tilde{v} = 0\,.$$

This equation may be simplified by non-dimensionalizing y, k and ω; writing $y = (g'D/\beta^2)^{1/4}y^*$, $k = (\beta^2/g'D)^{1/4}k^*$ and $\omega = (\beta^2 g'D)^{1/4}\omega^*$, it becomes

$$\frac{\mathrm{d}^2\tilde{v}}{\mathrm{d}y^{*2}} + \left(\omega^{*2} - k^{*2} - \frac{k^*}{\omega^*} - y^{*2}\right)\tilde{v} = 0\,,$$

and the equation is no longer cluttered with the parameters g', D and β.

It is well known[4] that this equation has bounded (as $|y^*| \to \infty$) analytic solutions provided the constant coefficient is an odd positive integer; the condition

$$\omega^{*2} - k^{*2} - \frac{k^*}{\omega^*} = 2J + 1$$

[3] The horizontal momentum equations and continuity equation from § 17.1.

[4] This is *Weber's equation*; see Murphy (1960) or Zwillinger (1998).

is the dispersion relation for the waves. This equation has a sequence of analytic solutions of the form

$$\tilde{v}_J(y^*) = e^{-y^{*2}/2} \sum_{j=0}^{J} h_j y^{*2j}$$

provided J is a non-negative integer. The series is known as the *Hermite polynomial.* The first of these solutions (with $J = 0$) is

$$\tilde{v}_0 = v_0 e^{-y^{*2}/2} = v_0 e^{-\beta y^2/2U},$$

where v_0 is an arbitrary amplitude. Solutions for J greater than zero are readily obtained using the recurrence relation

$$\tilde{v}_J = \left(y^* - \frac{\mathrm{d}}{\mathrm{d}y^*} \right) \tilde{v}_{J-1}.$$

The first few of these solutions are

$$\tilde{v}_1 = v_1 y^* e^{-y^{*2}/2}, \quad \tilde{v}_2 = v_2 \left(y^{*2} - \tfrac{1}{2} \right) e^{-y^{*2}/2},$$

$$\tilde{v}_3 = v_3 \left(y^{*3} - \tfrac{3}{2} y^* \right) e^{-y^{*2}/2} \quad \text{and}$$

$$\tilde{v}_4 = v_4 \left(y^{*4} - 3y^{*2} + \tfrac{3}{4} \right) e^{-y^{*2}/2},$$

where the factors v_j are arbitrary constants.

The dispersion relation is cubic in ω^*, which tells us that this solution represents three modes of motion. In general, the three modes are coupled,[5] but if k^* is either small or large, the modes decouple. There are two *inertio-gravity modes* having

$$\omega^* = \pm\sqrt{k^{*2} + 2J + 1}$$

and one *equatorial Rossby wave* having

$$\omega^* = -\frac{k^*}{k^{*2} + 2J + 1}.$$

The inertio-gravity waves may travel either eastward or westward, while the equatorial Rossby wave travels only westward. These waves are dispersive, with the phase and group speeds of the inertio-gravity wave similar in magnitude to that of the Kelvin wave and the speeds of the equatorial Rossby wave being considerably smaller.

[5] And can be expressed analytically as explained in §1.11(iii) on p. 23 of Olver et al. (2010).

Transition

This chapter has been a brief introduction to two types of equatorial waves; waves that exist close to the equator due to the latitudinal variation of the Coriolis parameter. These waves have no non-rotating analogs. Equatorial Kelvin waves have particle trajectories within vertical planes of constant latitude, while equatorial Rossby waves are a bit more complicated, having particle trajectories with finite latitudinal extent. In the following chapter we consider oceanic waves that are affected by rotation and also by the presence of a coastline or the variation of water depth.

18

Coastal and Topographic Waves

In this chapter we investigate rotating waves that involve the structure of the basin that contains and confines the body of water: either a coastline or varying basin depth. Coastal Kelvin waves (see § 18.1) occur on and near the thermocline and are affected by the presence of a coastline, while topographic Rossby waves (§ 18.2) occur throughout the depth of the ocean and are "trapped" by variations in bottom topography. Both these waves are shallow-water waves with horizontal velocity independent of depth.

18.1 Coastal Kelvin Waves

The equatorial Kelvin waves studied in § 17.2 have fluid motions parallel to the equator and appear to be trapped close to that boundary. Coastal Kelvin waves are similar in that fluid motions are parallel to – and appear to be trapped near – the coastline, but the mechanism of trapping is somewhat different. Coastal Kelvin waves are governed by the same equations as equatorial Kelvin waves,[1] except that the Coriolis parameter is treated as constant, $f = f_0$, rather than varying linearly with y:

$$\frac{\partial \bar{u}}{\partial t} = g' \frac{\partial h_b}{\partial x}, \qquad f_0\, \bar{u} = g' \frac{\partial h_b}{\partial y} \qquad \text{and} \qquad \frac{\partial \bar{u}}{\partial x} = \frac{1}{D} \frac{\partial h_b}{\partial t},$$

where $g' = (\rho_b - \rho_t)g/\rho_b$ is the reduced gravity. Now x and y are local coordinates, oriented relative to the coastline rather than to latitude and longitude. Specifically, x is local distance along the coastline and y is distance into the water from the coast.[2]

These waves are non-dispersive, permitting us to assume[3]

$$\{\bar{u}, h_b\} = \{\tilde{u}, \tilde{h}\} e^{-ly},$$

where $\tilde{u}$ and $\tilde{h}$ are functions of $x - Ut$ and $U = \sqrt{g'D}$. The governing equations are satisfied provided

$$l = f_0/U \qquad \text{and} \qquad \tilde{u} = -(g'/D)\tilde{h}.$$

[1] See § 17.1.
[2] As we stand on the shore and gaze at the water, x increases to our right.
[3] See § 9.2.1.

An observer standing on shore sees these waves traveling to the right with the same speed as equatorial Kelvin waves. For example, when an equatorial Kelvin wave reaches South America, the warm water associated with an El Niño event subsequently moves northward as a coastal Kelvin wave.

18.2 Topographic Rossby Waves

Topographic Rossby waves occur in the ocean with the variation of depth replacing the β effect in stretching and compressing vortex lines. They may occur, for example, over the continental shelf or continental slope. These waves are the coastal waves considered in § 12.4, but with the addition of the Coriolis parameter. As before, we will use local Cartesian coordinates with x along the shore and y increasing offshore and the fluid layer extending from $z=-D$ to $z=h$, with D being a function of y. The governing equations are the linearized (assuming $|h| \ll D$) vertically integrated continuity equation[4] and the linearized quasi-geostrophic equations for the f plane.[5] Writing $\mathbf{v}_H = \bar{u}\mathbf{1}_x + \bar{v}\mathbf{1}_y$, where the overbar denotes variables independent of depth, these are

$$\frac{\partial h}{\partial t} + D\left(\frac{\partial \bar{u}}{\partial x} + \frac{\partial \bar{v}}{\partial y}\right) + \frac{\mathrm{d}D}{\mathrm{d}y}\bar{v} = 0,$$

$$\frac{\partial \bar{u}}{\partial t} - f_0\bar{v} = -g\frac{\partial h}{\partial x} \quad \text{and} \quad \frac{\partial \bar{v}}{\partial t} + f_0\bar{u} = -g\frac{\partial h}{\partial y}.$$

As an aside, the vorticity equation,[5]

$$\frac{\partial \zeta}{\partial t} = -\frac{f_0}{D}\left(\frac{\partial h}{\partial t} + \bar{v}\frac{\mathrm{d}D}{\mathrm{d}y}\right),$$

shows that motions toward deeper or shallower water, quantified by the term $-\bar{v}\mathrm{d}D/\mathrm{d}y$, act to stretch or compress vortex lines, in a manner similar to the β effect, with $f_0\mathrm{d}(\log D)/\mathrm{d}y$ replacing β; see § 15.3.1 and Figure 8.3.

We may readily obtain a single equation for the surface deflection, h, by solving the horizontal components of the momentum equation for $\bar{u}$ and $\bar{v}$ in terms of h:

$$\left(\frac{\partial^2}{\partial t^2} + f_0^2\right)\bar{u} = -\left(gf_0\frac{\partial}{\partial y} + g\frac{\partial^2}{\partial t\partial x}\right)h$$

and

$$\left(\frac{\partial^2}{\partial t^2} + f_0^2\right)\bar{v} = \left(gf_0\frac{\partial}{\partial x} - g\frac{\partial^2}{\partial t\partial y}\right)h,$$

then using these to eliminate $\bar{u}$ and $\bar{v}$ from the continuity equation:

$$\left(\frac{\partial^2}{\partial t^2} + f_0^2\right)\frac{\partial h}{\partial t} - gD\nabla_H^2\frac{\partial h}{\partial t} + g\frac{\mathrm{d}D}{\mathrm{d}y}\left(f_0\frac{\partial h}{\partial x} - \frac{\partial^2 h}{\partial t\partial y}\right) = 0.$$

[4] From § 8.4.2.

[5] From § 8.6, with the constant-pressure surface Z replaced by the surface elevation of the ocean, h.

The coefficients in this equation are independent of x and t, which permits us to express the dependent variable using the Fourier transform; let

$$h = \int_k \tilde{h} e^{ikx - i\omega t} dk + \text{c.c.},$$

where $\tilde{h}$ is a function of y and the wavenumber k. Now the equation becomes an ordinary differential equation:

$$Dg \frac{d^2 \tilde{h}}{dy^2} + g \frac{dD}{dy} \frac{d\tilde{h}}{dy} + \left(\omega^2 - f_0^2 + \frac{f_0 g}{\omega} k \frac{dD}{dy} - Dgk^2 \right) \tilde{h} = 0.$$

As in § 12.4, let's consider $D = \varepsilon y$. Now the governing equation becomes

$$y \frac{d^2 \tilde{h}}{dy^2} + \frac{d\tilde{h}}{dy} + \left(\frac{\omega^2 - f_0^2}{\varepsilon g} + \frac{f_0}{\omega} k - yk^2 \right) \tilde{h} = 0.$$

This equation is nearly identical to that considered in § 12.4. We can again transform it to Laguerre's equation by writing

$$\tilde{h} = e^{-ky} h^*(y^*) \qquad \text{and} \qquad y^* = 2ky.$$

The equation becomes

$$y^* \frac{d^2 h^*}{dy^{*2}} + (1 - y^*) \frac{dh^*}{dy^*} + Jh^* = 0,$$

where

$$J \equiv \frac{f_0}{2\omega} + \frac{\omega^2 - f_0^2}{2\varepsilon gk} - \frac{1}{2}.$$

The solutions are again the Laguerre polynomials provided J is a non-negative integer.

The dispersion relation, obtained from the definition of J, may be expressed as:

$$\underbrace{\frac{k_f}{k} \left(\frac{\omega}{f_0} \right)^3}_{\mathbf{1}} - \underbrace{\frac{k_f}{k} \left(\frac{\omega}{f_0} \right)}_{\mathbf{2}} - \underbrace{(2J+1) \frac{\omega}{f_0}}_{\mathbf{3}} + \underbrace{1}_{\mathbf{4}} = 0,$$

where $k_f = f_0^2/(\varepsilon g)$ is a characteristic wavenumber. This dispersion relation is cubic in ω, representing three wave modes. The general solution of this cubic may be expressed in closed form using formulas 1.11.12–1.11.15 on page 23 of Olver et al. (2010), but this solution is not very illuminating. It is helpful to simplify these modes in two limits, summarized as follows:

- $k_f \ll k$. In the short wavelength (large wavenumber) limit with J of unit order, term 2 is unimportant and there is

Figure 18.1 The topographic Rossby wave induced by hurricane Dennis led to a larger-than-predicted storm surge that caught many coastal residents in North Florida by surprise. Photo by the author.

  - a pair of fast modes having $\omega \approx \pm f_0\sqrt{(2J+1)k/k_f} = \pm\sqrt{(2J+1)\varepsilon g k}$ (balancing terms 1 and 3); and
  - a mode having $\omega \approx f_0/(2J+1)$ (balancing terms 3 and 4).
- $k \ll k_f$. In the long wavelength (small wavenumber) limit with J of unit order, term 3 is unimportant and there is
  - a pair of geostrophic modes[6] having $\omega \approx \pm f_0$ (balancing terms 1 and 2); and
  - a slow mode $\omega \approx f_0 k/k_f = \varepsilon g k/f_0$ (balancing terms 2 and 4).

This last mode is the topographic Rossby wave. It is non-dispersive and travels to the right, according to an observer standing on the shore, with a speed

$$U = \varepsilon g/f_0 .$$

The speed of this wave depends linearly on the slope ε. On a typical slope with $\varepsilon \approx 0.005$, $g \approx 10$ m·s^{-2} and $f_0 \approx 10^{-4}$ s^{-1}, the speed of a topographic wave is quite large: $U \approx 500$ m·s^{-1}. However, if the slope is very shallow, as occurs off the west coast of the Floridan peninsula, the speed is much smaller.

The M2 component of the tide excites a topographic Rossby wave that travels counterclockwise around the rim of the northern Pacific Ocean.[7] This tidal mode has a

[6] See § 15.1.
[7] See § 12.2 and Table 12.2.

period close to 12 hours. A wave moving at a speed of 500 m/s can travel a distance equal to half Earth's circumference in that time.

To demonstrate that the study of topographic Rossby waves is more than a dry mathematical exercise, let's consider Hurricane Dennis, which in July, 2005, moved northward in the Gulf of Mexico roughly parallel to the west coast of Florida. The predicted storm surge in Apalachee Bay of North Florida, some 275 km from the landfall near Pensacola, was 1 to 2 m, but the actual surge was considerably (more than 1 m) greater, surprising many coastal residents; see Figure 18.1. The excess surge has been attributed (Morey et al., 2006; see also Dukhovskoy and Morey, 2010) to the resonant interaction between the movement of the storm and a topographic Rossby wave. The storm traveled approximately 8 m/s. This speed is achieved by a topographic Rossby wave if $\varepsilon \approx 0.0001$.

Transition

In this chapter we have investigated two types of oceanic waves that are affected by the shape of the basin: coastal Kelvin waves and topographic Rossby waves. This completes our survey of waves in rotating fluids. We now turn our attention to one-dimensional flows in channels and down slopes. These flows are unaffected by rotation; the new ingredient is gravitational forcing.

Part V

Non-Rotating Flows

We now shift our attention from rotating waves to simple, non-rotating one-dimensional flows. We begin the next phase of our journey in the following chapter with an orientation to one-dimensional flows. Following this orientation, the appropriate forms of the continuity and momentum equations that governing these flows are developed in § 19.1. These equations are applied to steady and un-steady flows in a horizontal channel in Chapters 20 and 21, respectively. In these preliminary studies the channel is horizontal, so gravity has no effect (other than to keep the fluid in the channel). In Chapter 22 we allow the channel to slope downward and investigate how this affects the flow of simple materials. Virtually all natural flows of water and air are turbulent; we develop a simple model of turbulent flows in Chapter 23 and then conclude this part by applying this model to some simple flows in Chapter 24.

19

Orientation to One-Dimensional Flow

In Chapter 12 we studied motions of a shallow layer of fluid that do not involve net fluid flow. We return to the study of a shallow layer of fluid, now allowing for such flows. We will consider the motion of an inviscid incompressible fluid (e.g., water) in a channel oriented in the x direction of fixed width W, having a bottom of variable height, $z = -D(x)$, with the top of the water being given by $h(x,t)$, so that the depth of the water is $h(x,t) + D(x)$ see Figure 19.1. The mean flow is predominantly in the x direction, with secondary vertical (z) motions as needed to accommodate changes in water depth and elevation of channel bottom. The total flow may include turbulent eddies that do not contribute to the flow, but which do affect the momentum balance and degrade mechanical energy.

The channel may be slightly tilted with respect to the vertical, so that

$$\mathbf{g} = g\,(\sin\varepsilon\,\mathbf{1}_x - \cos\varepsilon\,\mathbf{1}_z) \approx g\,(\varepsilon\mathbf{1}_x - \mathbf{1}_z)\,,$$

where g is the magnitude of the local acceleration of gravity and $\sin\varepsilon \approx \varepsilon$ is the slope; see Figure 19.1. In Chapters 20 and 21, we will investigate flow in a horizontal channel: $\varepsilon = 0$, while in Chapters 22 and 23, we will look at flow in a channel that is tilted slightly downward: $0 < \varepsilon \ll 1$.

We shall see that one-dimensional flows can be characterized by two dependent variables that are governed by two equations best written in conservation form[1]

$$\frac{\partial a}{\partial t} + \frac{\partial F}{\partial x} = s\,,$$

where a is the distribution of a quantity of interest (such as mass or momentum), F is the flux of a in the x direction and s is a volumetric source (or sink, if $s < 0$) of a. These equations will be obtained by integrating the continuity and x component of the momentum equation over the depth of the layer of water.

Most of the wave motions we considered in previous chapters are governed by linear equations, the one exception being the nonlinear shallow-water waves investigated in Chapter 13. The one-dimensional motions studied in this part have more general time dependence than the wave motions considered previously and often are nonlinear.

[1] See Appendix D.5.

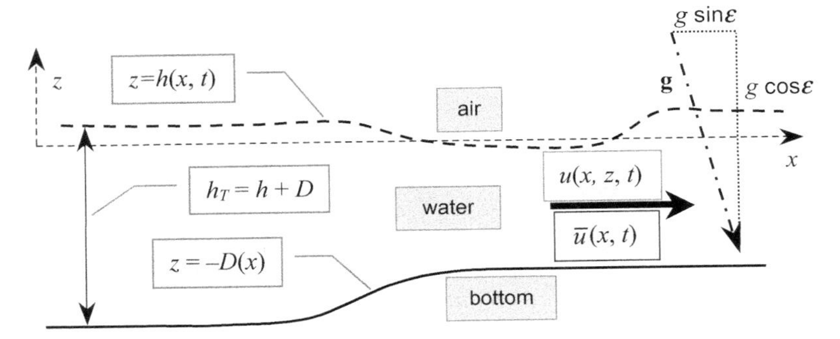

Figure 19.1 Vertical cross-section of one-dimensional flow in a channel. The coordinates are fixed relative to the channel bottom, but the zero-level for z is arbitrary. The channel bottom is at $z = -D(x)$, the top of the water is at $z = h(x,t)$ and the water depth is $h_T = h + D$. The channel may be tilted relative to the vertical by a small angle ε. The downstream speed profile is given by $u(x,z,t)$ and the mean downstream speed by $\bar{u}(x,t)$.

In Part III we studied waves that were a perturbation on a static state and we chose to measure vertical distances from the equilibrium water level. In the study of flows we lose this point of reference; there no longer is an equilibrium water level to use as our reference elevation. The channel bottom may be used as the vertical reference level, if it is flat, as is done in Chapter 20. In other cases, the vertical reference level is arbitrary and largely irrelevant.

The materials in this part are organized as follows. First, § 19.1 contains a review and summary of the relevant governing equations: conservation of mass and momentum. We shall develop three versions of the momentum equation, each of which is useful depending on circumstances. In the next two chapters (20 and 21) we consider flows that are not forced; i.e., there is no source of mechanical energy and there are no losses. Smooth one-dimensional flow of an irrotational fluid is investigated in Chapter 20, and it is found that smooth channel flow can be in one of two dynamic states (sub-critical or super-critical) depending on the speed of flow relative to the speed of shallow-water waves. The transition from sub- to super-critical flow is normally smooth; this is investigated in Chapter 20. The reverse transition from super- to sub-critical flow is abrupt and occurs across a *shock wave*; these transitions are investigated in Chapter 21. Most of these simplified investigations consider lossless flow in a horizontal channel, driven by conditions far upstream and far downstream, the exception being the study of flash floods in § 21.4, in which flow resistance plays a pivotal part. The remainder of this part focuses on the behavior of a layer of material as it moves down a slope. Gravitationally forced flows are studied in Chapter 22. Examples of such flows include rigid sliding, Newtonian viscous flow and non-Newtonian viscous flow. This part concludes with investigations of the nature of turbulent channel flow in Chapter 23 and some simple turbulent flows in Chapter 24.

19.1 One-Dimensional Flow Equations

In this section we will develop the vertically averaged equations governing one-dimensional flow of water, beginning with the continuity equation[2] and the x and z components of the momentum equation.[3] To a very good approximation, water may be treated as incompressible, so that the continuity equation simplifies to

$$\frac{\partial u}{\partial x}+\frac{\partial w}{\partial z}=0,$$

while the x and z components of the momentum equation for a gently sloping channel are

$$\rho\frac{\mathrm{D}u}{\mathrm{D}t}=-\frac{\partial p}{\partial x}+\varepsilon\rho g+\frac{\partial \tau}{\partial z} \qquad \text{and} \qquad \rho\frac{\mathrm{D}w}{\mathrm{D}t}=-\frac{\partial p}{\partial z}-\rho g,$$

where the material derivative[4] is

$$\frac{\mathrm{D}}{\mathrm{D}t}=\frac{\partial}{\partial t}+u\frac{\partial}{\partial x}+w\frac{\partial}{\partial z},$$

ε is the channel slope and τ is the $x-z$ shear stress.[5] The drag force will be ignored in Chapters 20 and 21, then reinstated in Chapters 22 and 23. In the following two sections, we will develop equations governing one-dimensional flow: the continuity equation in § 19.1.1 and the momentum equation in § 19.1.2.

19.1.1 Continuity Equation

The continuity equation is subject to the kinematic conditions that the bottom (at $z=-D$) and top (at $z=h$) of the fluid are material surfaces. Using a coordinate system fixed relative to the channel bottom[6] these conditions are

$$G_b(x,z,t)\equiv z+D(x)=0 \qquad \text{and} \qquad G_t(x,z,t)\equiv z-h(x,t)=0.$$

The material derivatives of these equations yield

$$\left.\frac{\mathrm{D}G_b}{\mathrm{D}t}\right|_{z=-D}=u_b\frac{\mathrm{d}D}{\mathrm{d}x}+w_b=0$$

and

$$\left.\frac{\mathrm{D}G_t}{\mathrm{D}t}\right|_{z=h}=-\frac{\partial h}{\partial t}-u_t\frac{\partial h}{\partial x}+w_t=0,$$

[2] From § 3.4.1.
[3] From § 4.7.
[4] From Appendix C.2.
[5] See § 4.5. If the drag force were due to molecular viscosity, τ would equal $\eta\partial u/\partial z$.
[6] So that D is independent of t.

where a subscript t denotes the top of the layer (at $z=h$) and subscript b denotes the bottom (at $z=-D$).

As in the study of shallow-water waves, we will concentrate on motions and variations predominantly in the x direction, but with the distinguishing feature being a non-zero spatially-averaged flow in that direction:

$$q(x,t) = \int_{-D}^{h} u(x,z,t)\,\mathrm{d}z = (D+h)\bar{u},$$

where q is the volume flow of water (in the x direction) per unit width (having units of length^2/time) and $\bar{u}(x,t)$ is the mean downstream speed.

Note that

$$w = w_b - \int_{-D}^{z} \frac{\partial u}{\partial x}\mathrm{d}\grave{z}, \qquad \text{where} \qquad w_b = -u_b\frac{\mathrm{d}D}{\mathrm{d}x}$$

is the vertical speed of fluid at the base of the channel. In particular integration of the continuity equation from $z=-D$ to $z=h$ yields

$$w_t - w_b = -\int_{-D}^{h} \frac{\partial u}{\partial x}\mathrm{d}z\,.$$

Using *Leibniz's integral rule*[7] and the boundary conditions on w, this may be expressed as

$$\frac{\partial h}{\partial t} + \frac{\partial q}{\partial x} = 0\,.$$

This equation relating h and q is a *conservation equation*[8] that lacks a source term; barring nuclear reactions, there are no sources or sinks of mass.

19.1.2 Momentum Equation

Now let's look at the momentum equations. We start with two, the x and z components of momentum, but are readily able to solve the z component, permitting us to focus on the x momentum equation. We will develop three versions of this equation, a lossless version (version (b)) which is useful in the investigation of steady, unforced channel flows in Chapters 20 and 21, and two more general versions that are used in the study of hydraulic shock waves.

We will limit our attention to situations in which the vertical acceleration is small compared with the acceleration of gravity: $|\mathrm{D}w/\mathrm{D}t| \ll g$, so that the vertical momentum

[7] $\partial/\partial x \int_{-D}^{h} u\mathrm{d}z = \int_{-D}^{h}(\partial u/\partial x)\mathrm{d}z + u_t\partial h/\partial x + u_b\mathrm{d}D/\mathrm{d}x$.

[8] See Appendix D.5.

balance is hydrostatic:

$$\frac{\partial p}{\partial z} = -\rho g.$$

The water pressure is equal to the atmospheric pressure p_a at the surface. This permits integration of the vertical momentum equation:

$$p(x,z,t) = p_a + \rho g\,(h(x,t) - z).$$

If we express the pressure as the sum of a static profile plus a perturbation, $p = p_r + p'$, it is readily seen that

$$p_r(z) = p_a - \rho g(z - h_0) \qquad \text{and} \qquad p'(x,t) = \rho g\,(h(x,t) - h_0).$$

where h_0 is the equilibrium water depth.

Using this solution for pressure, (the x component of) the momentum equation may be expressed as

$$\rho \frac{\mathrm{D}u}{\mathrm{D}t} = -\rho g \frac{\partial h}{\partial x} + \varepsilon \rho g + \frac{\partial \tau}{\partial z}.$$

This states that the fluid is accelerated by the action of three forces represented by the terms on the right-hand side: pressure, gravity and drag, respectively. If the channel is sloping, the downslope gravitational force acts to accelerate the channel flow, providing additional energy to the flow. The pressure term acts to redistribute kinetic energy, while the drag term always degrades kinetic energy to heat.[9]

In a manner similar to representing the pressure as the sum of a static part and perturbation, we can express the downstream speed u as the sum of a vertical mean $\bar{u}$ and a perturbation u'; let

$$u(x,z,t) = \bar{u}(x,t) + u'(x,z,t).$$

We will show in Chapter 23 that the effect of nonlinear interactions of velocity perturbations on the mean flow may be parameterized by the drag term, $\partial\tau/\partial z$. This provides justification for the neglect of the perturbation u' in this and the next two chapters. The vertically averaged momentum equation now becomes

$$\rho\left(\frac{\partial \bar{u}}{\partial t} + \bar{u}\frac{\partial \bar{u}}{\partial x}\right) + \rho g \frac{\partial h}{\partial x} = \varepsilon \rho g + \frac{\partial \tau}{\partial z}.$$

This averaged momentum equation is dynamically consistent provided $\partial\tau/\partial z$ is independent of z. Recall (see Figure 4.5) that a positive shear stress τ acts in the positive x direction on a surface of the fluid having an external normal in the positive y direction. That is, a positive stress exerted by the channel bottom acts on the fluid in the negative x

[9] Degradation of kinetic energy is an unavoidable consequence of the second law of thermodynamics; see Appendix E.

direction (counteracting the gravitational force exerted in the positive x direction due to the slope of the channel). For the time being, we will neglect the surface force exerted at $z = h$ due to the wind.[10] Now τ may be expressed as

$$\tau = \tau_b \frac{h - z}{D + h},$$

where τ_b is the (positive) drag stress at the bottom of the fluid, representing frictional resistance. Commonly this resistance is parameterized by

$$\tau_b = \lambda \rho \bar{u}^2 / 8,$$

where λ is a dimensionless coefficient called the *Darcy friction factor.*[11] The origin of this parameterization is investigated in Chapter 23. With this, the momentum equation becomes

$$\text{Version (a)}: \qquad \frac{\partial \bar{u}}{\partial t} + \bar{u}\frac{\partial \bar{u}}{\partial x} + g\frac{\partial h}{\partial x} = \varepsilon g - \frac{\lambda \bar{u}^2}{8(D+h)}$$

or equivalently

$$\frac{\partial \bar{u}}{\partial t} + g\frac{\partial H}{\partial x} = \varepsilon g - \frac{\lambda \bar{u}^2}{8(D+h)},$$

where

$$H(x,t) = h + \frac{\bar{u}^2}{2g}$$

is the head. The head is the elevation to which moving water would rise if it were brought to rest at atmospheric pressure. Speeds can be measured directly in terms of elevation using a *pitot tube*. The factor gH is the amount of mechanical energy per unit mass (that is, the *specific energy*); the factor $\bar{u}^2/2$ is the specific kinetic energy (the kinetic energy per unit mass)[12] and gh is the specific potential energy (the potential energy per unit mass). This formula provides a direct conversion of kinetic energy to height. A speed of 1, 3, 10, 30, 100 m/s (i.e., roughly 3.6, 10, 36, 100, 360 km/hr) produces a head of approximately 0.05, 0.5, 5, 50, 500 m. (This also tells us that the kinetic energy of an automobile traveling at 100 km/hr is equal to one dropped off a building 15 stories high.)

If the channel were horizontal and we were to neglect the frictional resistance term, the momentum equation would simplify to

$$\text{Version (b)}: \qquad \frac{\partial \bar{u}}{\partial t} + g\frac{\partial H}{\partial x} = 0.$$

This version of the momentum equation will be used in the following two chapters when we investigate the basic features of channel flow, including sub- and super-critical flow and hydraulic shock waves.

10 This term will be reinstated when we consider wind-driven flows is Chapter 27.

11 In the literature, the symbol f is often used in place of λ. An alternate formulation uses the *drag coefficient*: $C_D = \lambda/8$; see § 11.5.1.

12 See Appendix E.8.4. We are ignoring the contribution due to vertical speed.

Table 19.1. *Conservation forms for the continuity equation and two versions of the x component of the momentum equation; see Appendix D.5.*

Equation	Quantity	Flux	Source
Continuity	h	q	None
Version (a)	$\bar{u}$	gH	$\varepsilon g - \lambda \bar{u}^2/8(D+h)$
Version (c)	q	gM	$gh\mathrm{d}D/\mathrm{d}x + \varepsilon g(D+h) - \lambda \bar{u}^2/8$

The momentum equation can be written in conservation form by multiplying version (a) of the momentum equation by $D+h$ and using $\partial h/\partial t + \partial q/\partial x = 0$ from § 19.1.1; with $q = (D+h)\bar{u}$, we arrive at

$$\text{Version (c)}: \qquad \frac{\partial q}{\partial t} + \frac{\partial}{\partial x}(gM) = gh\frac{\mathrm{d}D}{\mathrm{d}x} + \varepsilon g(D+h) - \frac{\lambda}{8}\bar{u}^2,$$

where

$$M(x,t) = \tfrac{1}{2}h^2 + hD + (D+h)\bar{u}^2/g$$

is the *momentum function*. The term $gh\mathrm{d}D/\mathrm{d}x$ represents the hydrostatic force exerted on the fluid by a sloping bottom. This form is useful in investigating flow through a shock wave; see Chapter 21.

Summary and Discussion

To summarize, version (b) of the momentum equation is useful if the flow is lossless, while versions (a) and (c) are useful for flows with dissipation. The continuity equation and these two versions of the x component of the momentum equation are written in conservation form.[13] The conserved quantity, flux and source for each of these are summarized in Table 19.1.

If the flow is steady and uniform, it must also be independent of x. This requires both D and h to be constant. Without loss of generality, we can set $D = 0$. In this limit, the continuity and version (b) momentum equations are identically satisfied, while version (c) of the momentum equation requires that

$$\tau_b = \varepsilon \rho g h_0 \qquad \text{and} \qquad \bar{u} = \sqrt{8\varepsilon g h_0/\lambda}\,.$$

This simple equation gives the balance between the downslope component of gravity acting to accelerate the flow and the retarding force produced by the stress that the bottom exerts on the flow.

We recover the linear shallow-water wave equations studied in Chapter 12 by

[13] See Appendix D.5.

- neglecting kinetic energy ($\bar{u}^2 \ll 2gh$), channel slope ($\varepsilon = 0$) and drag ($\tau_b = 0$);
- setting the elevation $z = 0$ at the undisturbed level of the water surface;
- assuming the channel bottom is level (at $z = -D_0$); and
- assuming $|h| \ll D_0$.

With these assumptions the continuity and version-(b) momentum equations become

$$\frac{\partial h}{\partial t} + D_0 \frac{\partial \bar{u}}{\partial x} = 0 \qquad \text{and} \qquad \frac{\partial \bar{u}}{\partial t} + g \frac{\partial h}{\partial x} = 0.$$

The known solutions[14] of this set of equations are non-dispersive shallow-water waves having phase and group speed $U = \sqrt{gD_0}$. Since linear shallow-water waves are non-dispersive, any wave form will propagate without change of shape, at dominant order. However, since the speed of shallow-water waves increases with the depth of the water, nonlinear effects can cause the wave form to evolve with time. In particular, the front of a wave of deeper water moving into shallower water will tend to steepen with time, forming a hydraulic shock wave, which we will investigate in Chapter 21.

Transition

In this chapter we have oriented ourselves to the topic of one-dimensional flows and developed two equations governing these flows: the continuity equation and a scalar momentum equation. Three versions of the momentum equation are useful, depending on the circumstances. We will begin to develop an understanding of one-dimensional flows by investigating steady channel flow in the following chapter.

[14] See Chapter 12.

20
Steady Channel Flow

Although most rivers are turbulent and their flows are forced by a sloping river bottom, there are situations in which channeled flow is laminar, or nearly so, and in which the flow is not forced by a local slope. Examples include flow at the exit of a lake or reservoir, through a sluice gate at the bottom of a dam or in an estuary. Further, it is best to present the fundamental features of channel flow in a simplified manner, before adding obfuscating complications. In this chapter we will consider these simpler flows and postpone introduction of slope and drag until later chapters; we assume that the flow along the channel is independent of z and that the drag force is negligibly small: $\tau_b = 0$ and employ version (b) of the momentum equation from § 19.1.2.

We begin this chapter by analyzing smooth steady channel flow in § 20.1, wherein we classify flows as sub- or super-critical. In the second section, § 20.2, we introduce variation of the channel depth (specifically, a bump) and investigate its effect on the flow, including the smooth transition from the sub- to super-critical state.

20.1 Steady Flow in a Uniform Channel

In this section we consider the simplest possible channel flow: uniform steady flow in a channel of constant depth with no drag. With the channel bottom being flat, we can use this as the reference level and – without loss of generality – set $D = 0$, so that the (uniform) fluid depth is given by h. Now the continuity equation from § 19.1.1 and version-(b) momentum equation from § 19.1.2 are readily integrated to yield

$$\bar{u}h = q \qquad \text{and} \qquad \bar{u}^2 = 2g(H - h)\,.$$

The channel-flow problem has been reduced to bare bones; the flow speed $\bar{u}$ and water depth h are constants, independent of position x and z and time t. It would seem that nothing interesting can be discovered in this limiting case, but that is far from the truth, as we shall soon see.

We have a set of two nonlinear algebraic equations to be solved for $\bar{u}$ and h in terms of three parameters: gravity g, head H and specific volume flow q. This set may be simplified by doing a bit of dimensional analysis.[1] There are five quantities and two dimensions

[1] See Appendix B.1.

(length and time), which means that we can construct three dimensionless parameters. Two of these should be dimensionless versions of $\bar{u}$ and h and one should be a combination of g, H and q. Introducing

$$\bar{u} = \frac{q}{H} u^*, \qquad h = Hh^* \qquad \text{and} \qquad \Pi = \frac{q^2}{2gH^3},$$

the governing equations become

$$u^* h^* = 1 \qquad \text{and} \qquad h^* + \Pi u^{*2} = 1 .$$

Note that Π is a measure of the strength of the flow. If $\Pi = 0$, then $u^* = h^* = 1$ or equivalently $\bar{u} = 0$ and $h = H$.

We may eliminate u^* from our pair of equations and obtain a *constitutive equation* relating water depth (quantified by h^*) and strength of flow (quantified by Π):

$$\Pi = h^{*2}(1 - h^*) .$$

This is a cubic equation to be solved for positive values of h^*, with the stipulation that $0 \leq \Pi$. It is readily seen that the factor $h^{*2}(1-h^*)$ is positive (permitting Π to be positive) only in the interval $0 < h^* < 1$; see panel (a) of Figure 20.1. Further, this factor has a maximum value of $4/27$ at $h^* = 2/3$. This provides an upper limit, called the *critical value* and denoted by a subscript c, on the magnitude of Π:

$$\Pi_c = 4/27 \qquad \text{or equivalently} \qquad q_c = \sqrt{g(2H/3)^3} .$$

It is impossible to find positive real values of $\bar{u}$ and h, such that $q > q_c$. This is the first non-trivial result of our investigation of this very simple channel flow.

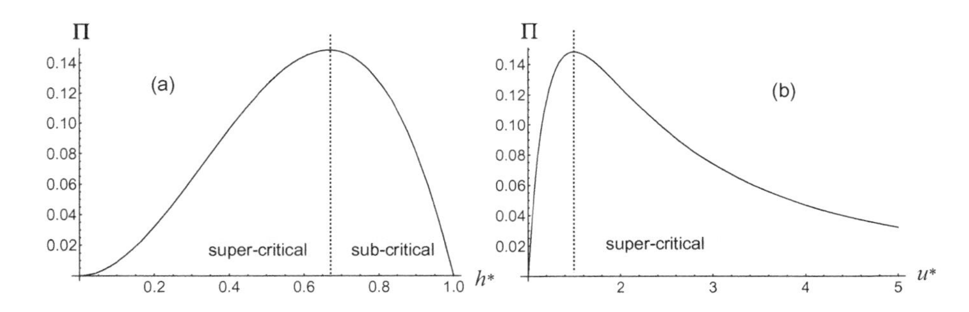

Figure 20.1 (a) A plot of dimensionless parameter Π versus dimensionless depth h^*. This is a portion of a cubic. The maximum value $\Pi = 4/27$ occurs for $h^* = 2/3$. Flow is sub-critical if $h^* > 2/3$ and super-critical if $h^* < 2/3$. (b) Plot of dimensionless parameter Π versus dimensionless speed u^*. The maximum value $\Pi = 4/27$ occurs for $u^* = 3/2$. Flow is sub-critical if $u^* < 3/2$ and super-critical if $u^* > 3/2$.

Table 20.1. *Shallow-flow characteristics*

Sub critical	Critical	Super critical
$\mathrm{Fr} < 1$	$\mathrm{Fr} = 1$	$\mathrm{Fr} > 1$
$2H/3 < h < H$	$h = 2H/3$	$0 < h < 2H/3$
$\bar{u} < \sqrt{gh}$	$\bar{u} = \sqrt{gh}$	$\bar{u} > \sqrt{gh}$

In the critical state,

$$h_c = 2H/3 \qquad \text{and} \qquad \bar{u}_c = \sqrt{2gH/3} = \sqrt{gh_c}\,.$$

This relation between speed and depth is just that obtained in Chapter 12 for the speed of shallow-water waves. Should this be a surprise to us?

The constitutive relation may be written in terms of the dimensionless speed u^* as

$$\Pi = (u^* - 1)/u^{*3}\,.$$

This relation is plotted in panel (b) of Figure 20.1. It is readily seen that, while the magnitude of Π is constrained, the possible values of $\bar{u}$ and h are not. That is, channeled flow can happily proceed at any speed and with any depth, but the combination $\Pi = (\bar{u}h)^2/2g(h + \bar{u}^2/2g)^3$ cannot exceed 4/27. The nature of this flow is characterized by the flow speed as measured by a stationary observer: someone standing beside the channel.

A second interesting property of this simple flow, seen in Figure 20.1, is that for values of Π less than the maximum, the equation admits two values of depth and speed. That is, for given values of H and q, the flow can be either deep and slow or shallow and fast. These two solutions are characterized by the magnitude of the *Froude number*

$$\mathrm{Fr} = \bar{u}/\sqrt{gh}\,.$$

The critical solution is characterized by $\mathrm{Fr} = 1$. On the slow or *sub-critical branch*, $\mathrm{Fr} < 1$ and on the fast or *super-critical branch*, $\mathrm{Fr} > 1$.

The physical importance of the Froude number can be understood by considering the behavior of a linear shallow-water wave superposed on the steady flow. These waves serve to convey information regarding changes in flow conditions. If $\mathrm{Fr} < 1$, linear waves can travel upstream against the flow, bringing information (to our observer standing beside the channel) about conditions downstream. If $\mathrm{Fr} = 1$, upstream-traveling waves are stationary relative to a standing observer, and if $\mathrm{Fr} > 1$ waves traveling against the flow are swept downstream (relative to our observer); consequently information about downstream conditions cannot be conveyed upstream. The Froude number is analogous to the Mach number introduced in § 9.3; in that case the medium was stationary and the observer moving, while here the medium is moving and the observer stationary.

20.2 Steady Flow in a Channel Having Variable Elevation

Now let's generalize the situation studied in the previous section by allowing the elevation of the channel bottom to vary with downstream distance: $z = -D(x)$, but keeping the horizontal flow steady and uniform and still neglecting drag. Now both $\bar{u}$ and h can vary with x, but this dependence is parametric, as the flow equations[2] again can be integrated to give us two algebraic equations:

$$\bar{u}h_T = q \qquad \text{and} \qquad \bar{u}^2 = 2g(H + D - h_T),$$

where $h_T = h + D$ is the depth of the water; see Figure 19.1.

We can non-dimensionalize the variables with

$$\bar{u} = (q/H)u^*, \quad h_T = Hh^*, \quad D = HD^* \quad \text{and} \quad \Pi = q^2/(2gH^3).$$

Now the non-dimensional governing equations become

$$u^*h^* = 1 \qquad \text{and} \qquad h^* + \Pi u^{*2} = 1 + D^*.$$

Eliminating u^*, we have the constitutive relation

$$\Pi = h^{*2}(1 + D^* - h^*).$$

With the flow being steady, q and H – and consequently Π – are constant, independent of x and t.

If D is constant (with $D = D^* = 0$), the problem reduces to that investigated in the previous section. If the bottom of the channel is not flat, but has a small bump, say, how does the fluid depth and speed respond? This question can be answered by taking the differential of the constitutive relation, with Π fixed and with D^* nearly equal to 0. Recalling that $h_c^* = 2/3$ this yields

$$\mathrm{d}h^* = \frac{h^*}{3(h^* - h_c^*)}\mathrm{d}D^* \qquad \text{or equivalently} \qquad \mathrm{d}h = \frac{h}{3(h - h_c)}\mathrm{d}D.$$

Recall that D is the channel depth, so that a bump on the channel bottom (i.e., a decrease in channel depth) is characterized by $\mathrm{d}D < 0$. The sign of the change in water depth ($\mathrm{d}h$) depends on which branch the solution is on. If it is on the super-critical branch (having relatively shallow water), the denominator is negative ($h < h_c$) and $\mathrm{d}h > 0$; the water gets deeper over a bump, while if it is on the slow branch, the denominator is positive ($h > h_c$) and the water gets shallower over a bump. In either case, *the flow is driven toward the critical state by a bump*. Of course, the opposite holds true: as the channel bottom deepens, the flow is driven away from critical; sub-critical flow becomes even slower, while super-critical flow becomes even faster.

[2] Continuity equation from § 19.1.1 and version (b) of the momentum equation from § 19.1.2.

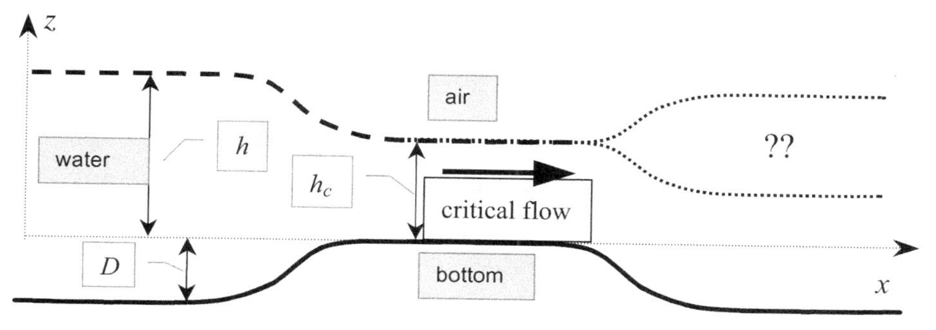

Figure 20.2 Flow encountering a smooth increase in bottom elevation of magnitude D. The flow is critical over the bump. As the channel deepens beyond the bump, the flow can return to sub-critical (upper dotted line), or transition smoothly to super-critical (lower dotted line).

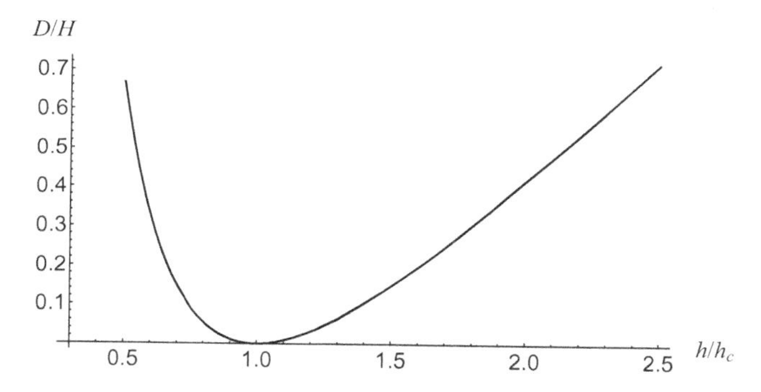

Figure 20.3 Relative height D/H of the bump necessary to achieve critical flow versus relative depth h/h_c of the incoming flow.

If the bump is sufficiently high, the flow reaches critical over the bump and can change character as it flows across the bump.[3] This is called a *transition*. Commonly an obstacle on the bottom induces sub-critical (i.e., slow and deep) flow to change to super-critical (i.e., fast and shallow) flow. The transition from sub to super-critical flow invariably occurs smoothly. The opposite transition, from super to sub-critical flow almost always occurs abruptly, by flowing through a hydraulic shock wave.

To determine the height of the bump necessary to induce critical flow, let's consider the situation sketched in Figure 20.2, with sub-critical upstream flow, having depth h, encountering a smooth shallowing of the channel of magnitude D, where the flow is critical. Without loss of generality we may set $z = 0$ where the flow is critical so that $h_c^* = 2/3$, $h_c = 2H/3$ and $\Pi_c = 4/27$. Solving the constitutive relation for D^* and noting

[3] This behavior is analogous to the dynamics of flow in a super-sonic nozzle.

that $\Pi = \Pi_c = h_c^{*2}(1 - h_c^*)$, the bump height is given by

$$D^* = \left(h_c^* - h^*\right)^2 \frac{h_c^* + 2h^*}{3h_c^* h^{*2}} \quad \text{or equivalently} \quad \frac{D}{H} = (h_c - h)^2 \frac{h_c + 2h}{3h_c h^2}\,.$$

The relative step size D/H is plotted versus the relative change of depth h/h_c in Figure 20.3. Note that the bump height D is positive whether h is greater or less than h_c. Why is that?

Transition

In this chapter we studied steady channel flow and found that a flow of depth h and mean speed $\bar{u}$, can be characterized by the magnitude of its Froude number, $\text{Fr} = \bar{u}/\sqrt{gh}$, as being in either a sub-critical (when $\text{Fr} < 1$) or super-critical (when $\text{Fr} > 1$) state and we saw how a bump in the channel bottom drives flow toward the critical state. If the bump is high enough it can trigger a smooth, steady transition from the sub-critical to the super-critical state. In the following chapter, we investigate the reverse transition, from super-critical to sub-critical flow.

21

Unsteady Channel Flow: Hydraulic Shock Waves

The principal aim of this chapter is to determine the transition from super-critical to sub-critical channel flow. While this transition could occur smoothly, as depicted in Figure 20.2 (with the flow arrow reversed), typically the transition is in the form of a *hydraulic shock wave* that is abrupt (that is, occurring over a small downstream interval) and spontaneous (that is, occurring without any prompting). Hydraulic shock waves are distinct from compressive shock waves which rely on the compressibility of water for the restoring force and which travel far faster than hydraulic shocks. Hydraulic shocks are impelled into motion by the greater water depth on the lee side of the shock. Shallow-water hydraulic shocks have been given differing names depending on the circumstance of their occurrence. Typically the name *hydraulic jump* is reserved for a shock that is stationary with respect to an observer standing on the ground. Hydraulic jumps commonly occur below a dam when water is released from the base of the reservoir by lifting a sluice gate slightly. Close to the gate the water flows rapidly, but then the depth increases abruptly at a fixed distance downstream from the gate as the flow abruptly transitions from super-critical to sub-critical. A hydraulic shock moving up a river due to the change in tide is called a *bore* or surge wave. Bores form in river channels during rising tide, most notably in the Amazon and Orinoco in South America that are reported to have bores as high as 4 m. A hydraulic shock having no water in front is called a *flash flood*. A flash flood wave moves at the mean speed of the water behind the wave.

Hydraulic shocks often are unsteady when viewed by an observer standing beside the channel, so we must now consider unsteady (time dependent) flows. The details of this transition are complicated, involving turbulent flow and degradation of mechanical energy within a localized downstream interval. In this chapter we will gloss over these details regarding the structure of the shock and investigate the kinematic and dynamic constraints on the uniform flow on either side of a shock. The specific volume flow, q, is conserved across the hydraulic shock. However, turbulent motions within the hydraulic shock scramble the flow, so that the head along a given streamline is not conserved across the shock. It follows that we cannot apply version (b), but instead must use version (c) of the momentum equation developed in § 19.1.2. However, since hydraulic shocks typically occur abruptly, variation of the channel bottom and slope of the channel are not important and can be ignored (that is, $D = \varepsilon = 0$). Now the governing equations are

$$\frac{\partial h}{\partial t} + \frac{\partial}{\partial x}(h\bar{u}) = 0 \quad \text{and} \quad \frac{\partial}{\partial t}(h\bar{u}) + \frac{\partial}{\partial x}\left(\frac{gh^2}{2} + h\bar{u}^2\right) = -\frac{\lambda}{8}\bar{u}^2,$$

where λ is the Darcy friction factor introduced in § 19.1.2.

21.1 Jump Conditions

A hydraulic shock is a singular region where the flow is turbulent and dissipative, and our simple equations are not valid within the shock. However, we do not need to consider this region directly, but can equate conditions on either side of the shock, such that mass and momentum are conserved. These conditions are commonly referred to as *jump conditions*. In this section, we will frame the general form of the jump conditions, then apply them to mass and momentum in the following two sections.

Consider the generic one-dimensional conservation equation

$$\frac{\partial a}{\partial t} + \frac{\partial F}{\partial x} = s,$$

where a is the amount of a measurable quantity, F is the flux of that quantity in the x direction and s is the volumetric source of a; see Table 19.1. The jump condition is obtained by integrating this equation across the shock from x to $x + \Delta x$:

$$\int_x^{x+\Delta x} \frac{\partial a}{\partial t}\, d\grave{x} + \Delta F = \int_x^{x+\Delta x} s\, d\grave{x}.$$

where the symbol Δ denotes the change in a quantity across the wave; that is, $\Delta F = F(x + \Delta x) - F(x)$.

We will assume the shock to be thin in the x direction. Since s is a volumetric source, the amount of a added in this thin region is negligible and we can ignore the term on the right-hand side. If we use a frame of reference fixed to the shock, and assume that strength of the shock does not vary with time, then the first term is zero and the jump condition simplifies to

$$\Delta F = 0.$$

This states that *the flux across a hydraulic shock is continuous when viewed in a frame of reference fixed to the shock.*

Let's reconsider the above calculation, relaxing the requirement that we move with the hydraulic shock. We can interchange the time derivative and x integration (still ignoring the source term) and write

$$\frac{\partial}{\partial t}\int_x^{x+\Delta x} a\, d\grave{x} + \Delta F = 0.$$

If the shock is moving at speed U_S relative to our observer, the change in the amount of a in our tiny space interval is given by

$$\frac{\partial}{\partial t}\int_x^{x+\Delta x} a\,\mathrm{d}\grave{x} = -U_S \Delta a$$

and the jump condition becomes

$$-U_S \Delta a + \Delta F = 0,$$

which readily gives the shock speed as measured by moving our observer:

$$U_S = \Delta F / \Delta a.$$

Hydraulic shocks are governed by conservation equations for mass and momentum, with two associated fluxes.[1] The constraints on the structure and speed of a hydraulic shock dictated by these equations are considered in the following two sections.

21.2 Kinematics of Hydraulic Shocks

In the continuity equation the quantity of interest is h and the associated flux is $q = \bar{u}h$. It follows from the jump condition developed in the previous section that

$$\Delta(\bar{u}h) = 0,$$

where $\bar{u}$ is the speed of flow measured by an observer moving with the shock. A stationary observer sees the hydraulic shock moving at speed U_S and writes this condition as

$$\Delta((\bar{u} - U_S)h) = 0 \qquad \text{or equivalently} \qquad \Delta(q - U_S h) = 0.$$

This condition is readily solved for the shock speed:

$$U_S = \Delta q / \Delta h,$$

which is in agreement with the formula for shock speed derived in the previous section.

To illustrate this, consider a situation with deep water, of depth h_1, moving with uniform speed from the left (i.e., toward positive x) into still water of depth h_2 in a channel having a bottom of constant elevation, as shown in Figure 21.1. The areas of the two stippled rectangles must be equal. The tall rectangle on the left has an area proportional to $h_1\bar{u}_1$ while that on the right has an area proportional to $(h_1 - h_2)U_S$, with the same constant of proportionality: Δt, the time interval. It is clear in this case that the speed of the shock cannot be less than the speed of the deep water on the left: $\bar{u}_1 \leq U_S$. Now consider the same situation in a frame of reference moving with the hydraulic, as shown in Figure 21.2. Again the two rectangles must have equal areas, with the tall one on the left being proportional to $h_1(\bar{u}_1 - U_S)$ and the short one on the right being proportional to $h_2 U_S$.

[1] If thermal effects were important we would have need of a third conservation equation, with associated flow and jump condition. In aeronautics the set of three jump conditions are known as the *Rankine Hugoniot conditions*.

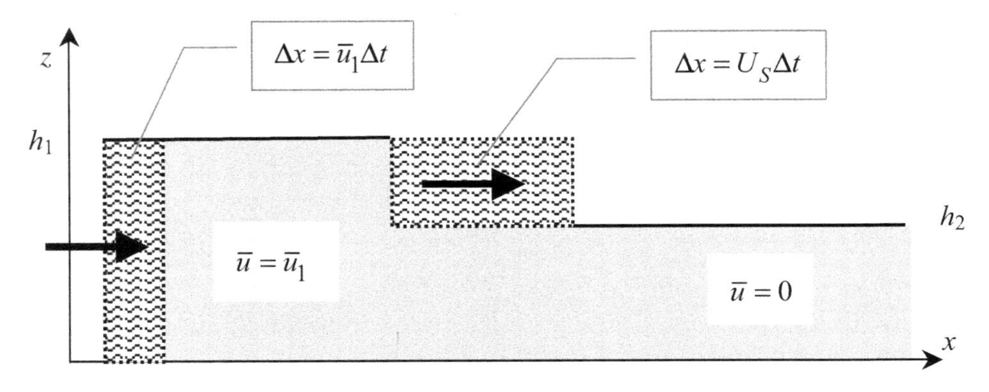

Figure 21.1 Schematic of a shallow-water hydraulic shock moving toward increasing x, as seen by an observer sitting in the water ahead of the shock. Δt is an arbitrary time interval.

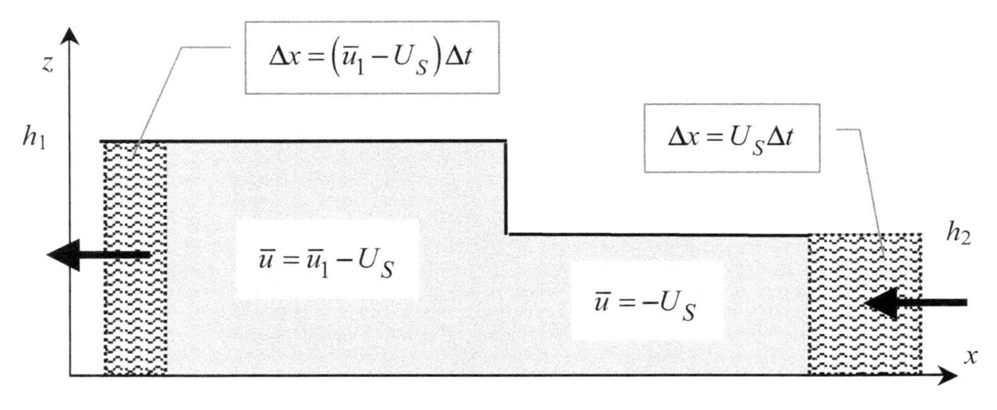

Figure 21.2 Schematic of a shallow-water hydraulic shock as seen by an observer moving with the shock.

The strength of a hydraulic shock is characterized by the change in water level as it passes. In order that the shock exist as a dynamic entity, it must be traveling faster than a shallow-water wave on a fluid of depth h_2 and slower than a wave on fluid of depth h_1:

$$\sqrt{gh_2} < U_S < \sqrt{gh_1}\,.$$

If this condition is not satisfied, the wave will broaden and dissipate. At this point, we cannot verify that this condition is satisfied, because we have not considered the dynamics of hydraulic shocks. In subsequent sections we will investigate the behavior of shocks and other features associated with river flow. In order to do so, we need a second equation relating the mean speed and depth, i.e., a constitutive relation $u(h)$. This come from conservation of momentum, which is considered in the following section.

21.3 Dynamics of Hydraulic Shocks

In § 20.1 we used version (b) of the momentum equation to explain the transition of flow from sub- to super-critical as it passes over a bump and § 21.2 we investigated the kinematics of hydraulic shocks. In this section, we will investigate their dynamics and investigate the transition of flow from super- to sub-critical using version (c) of the momentum equation.[2]

In version (c) of the momentum equation the amount being quantified is the specific volume flow of $q = \bar{u}h$, and its associated flux is $gM = \bar{u}^2 h + gh^2/2$. An observer fixed to the shock finds that this flux is continuous, with

$$\Delta(gM) = \Delta(\bar{u}^2 h + gh^2/2) = 0 .$$

On the other hand, an observer in motion relative to the shock sees changes in both $\bar{u}h$ and gM. Using the formula for shock speed from § 21.1, we see that these changes are related by the formula

$$U_S = \Delta(gM)/\Delta(\bar{u}h) .$$

This may be expressed as

$$M_S \equiv \frac{1}{g} h_1 \bar{u}_1 (\bar{u}_1 - U_S) + \frac{1}{2} h_1^2 = \frac{1}{g} h_2 \bar{u}_2 (\bar{u}_2 - U_S) + \frac{1}{2} h_2^2 ,$$

where M_S is the momentum flow across the shock and subscripts 1 and 2 denote conditions on either side of the shock, as shown in Figure 21.2.

21.3.1 Hydraulic Jumps

A hydraulic jump is a hydraulic shock wave that is stationary with respect to an observer standing beside the channel. In this case, $U_S = 0$ and the conservation laws yield

$$q_S = h_1 \bar{u}_1 = h_2 \bar{u}_2 \qquad \text{and} \qquad M_S = \frac{q_S^2}{h_1} + \frac{g}{2} h_1^2 = \frac{q_S^2}{h_2} + \frac{g}{2} h_2^2 .$$

Factoring out the trivial solution ($h_1 = h_2$) from the latter, we have

$$g h_1 h_2 (h_1 + h_2) = 2 q_S^2 \qquad \text{or} \qquad g (h_1 + h_2) = 2 \bar{u}_1 \bar{u}_2 .$$

Combining this with the continuity condition, we have that

$$h_1 = h_2 \left(\frac{\sqrt{1 + 8\mathrm{Fr}^2} - 1}{2} \right) \quad \text{and} \quad \bar{u}_1 = \bar{u}_2 \left(\frac{\sqrt{1 + 8Fr^2} + 1}{4\mathrm{Fr}^2} \right) ,$$

where

$$\mathrm{Fr} = \bar{u}_2 / \sqrt{g h_2}$$

[2] See § 19.1.2.

is the Froude number of the incoming flow; see Figure 21.2. If $\mathrm{Fr} > 1$, as we shall see it must, then $h_2 < h_1$ and $\bar{u}_2 > \bar{u}_1$; the incoming flow is shallow and fast.

The change in total head (i.e. energy) experienced as fluid moves from state 2 to state 1 is

$$\Delta H = \frac{1}{2g}\left(\bar{u}_2^2 - \bar{u}_1^2\right) + h_2 - h_1 .$$

Using the above solutions we have that

$$\Delta H = 28h_2\left(\frac{1-\mathrm{Fr}^2}{4\mathrm{Fr}^2 + 23 + 9\sqrt{1+8Fr^2}}\right).$$

This quantity is clearly negative when $\mathrm{Fr} > 1$, which is a physically reasonable result; mechanical energy is being converted to heat. The opposite case is unphysical, as it calls for the spontaneous conversion of heat into mechanical energy. That is, a shock cannot occur in a sub-critical flow (having $\mathrm{Fr} < 1$).

The form of the hydraulic shock depends on the Froude number of the incoming flow; according to Sellin (1969), if

- $1.0 < \mathrm{Fr} < 1.7$, the shock is undular (i.e., wavy);
- $1.7 < \mathrm{Fr} < 2.5$, the shock is transitional, with some of the undular waves breaking; and if
- $2.5 < \mathrm{Fr}$, the shock is abrupt, with a recirculating flow region, called a *reverse flow roller*, at the top. A floating object (such as a capsized canoe) may be held indefinitely in a roller.

21.3.2 Bores

A hydraulic shock moving into still water is called a bore. In this case $\bar{u}_2 = 0$, but $h_2 \neq 0$, and the conservation laws yield

$$U_S = \sqrt{\frac{gh_1}{2h_2}(h_1+h_2)} \quad \text{and} \quad \bar{u}_1 = (h_1 - h_2)\sqrt{\frac{g}{2h_1h_2}(h_1+h_2)} .$$

If $h_2 = h_1$, this reduces to the shallow-water wave formula with $U_S = \sqrt{gh_1}$. If $h_2 = 0$, the solution become singular. This singularity is resolved in the following section.

21.4 Flash Floods

A flash flood is a hydraulic shock that advances along dry ground. In this case $h_2 = 0$ and $\bar{u}_2$ is ill defined. Conservation of mass yields

$$U_S = \bar{u}_1 .$$

The structure of water depth in a flash flood may be determined using the continuity equation and version (a) of the momentum equation.[3] Setting $D = 0$, these are:

$$\frac{\partial h}{\partial t} + \frac{\partial (h\bar{u})}{\partial x} = 0 \qquad \text{and} \qquad \frac{\partial \bar{u}}{\partial t} + \bar{u}\frac{\partial \bar{u}}{\partial x} + g\frac{\partial h}{\partial x} = \varepsilon g - \frac{\lambda \bar{u}^2}{8h} .$$

[3] From § 19.1.1 and 19.1.2.

The principal force impelling a flash flood is the pressure gradient, represented by the term $g\partial h/\partial x$, while the friction term $-\lambda\bar{u}^2/8h$ provides the retarding force. The slope force εg also plays an important role in the force balance and requires us to consider three cases of flash floods:

- case (a): $0 < \varepsilon$ – in this case the flood flows down a dry riverbed, typically driven by up-gradient rainfall;
- case (b): $\varepsilon = 0$ – in this case flow advances over a horizontal bed, perhaps driven by a tsunami or dam break; and
- case (c): $\varepsilon < 0$ – in this case flow advances along an up-sloping bed, again possibly driven by a tsunami or dam break.

These three cases will be investigated by seeking similarity solutions in which the dependent variables $h(x,t)$ and $\bar{u}(x,t)$ are functions of a single independent variable $x^* = -k(x - U_S t)$, where k is a suitable wavenumber. This ansatz transforms our formidable set of partial differential equations into ordinary differential equations, which are far easier to analyze and solve. The similarity approach is based on two key features of the flow as seen by an observer standing on the ground watching the flash flood pass by:

- the wave front – and the water behind it – moves steadily at speed U_S; and
- the water elevation varies with x and t in the combination $x - U_S t$.

Based on these observations, we may set $\bar{u}(x,t) = U_S$. Further, realizing that $\partial/\partial t = -U_S\partial/\partial x$, the continuity equation is identically satisfied and the inertial term of the momentum equation is identically zero; the momentum equation reduces to

$$\frac{\partial h}{\partial x} = \varepsilon - \frac{h_0}{h} \qquad \text{where} \qquad h_0 = \frac{\lambda U_S^2}{8g}.$$

This equation is to be solved in the interval $x < U_S t$ subject to the condition that $h = 0$ at $x = U_S t$. The solution is readily found by separation of variables and may be expressed implicitly as

$$\frac{\varepsilon^2}{h_0}(x - U_S t) = \varepsilon\frac{h}{h_0} + \log\left(1 - \varepsilon\frac{h}{h_0}\right)$$

or in dimensionless form as

$$\begin{aligned} x^* &= -sh^* - \log(1 - sh^*), \quad \text{where} \quad h^* = |\varepsilon|h/h_0, \\ s &= \text{sign}(\varepsilon) \qquad\qquad\qquad \text{and} \qquad x^* = -\varepsilon^2(x - U_S t)/h_0. \end{aligned}$$

If $\varepsilon = 0$, the water surface is parabolic:

$$h = \sqrt{h_0(U_S t - x)}$$

or in dimensionless form as

$$h^* = \sqrt{2x^*}, \qquad \text{where} \qquad h^* = h/h_0 \qquad \text{and} \qquad x^* = -(x - U_S t)/h_0.$$

The structure of the advancing flow front (in dimensionless form) for each of the three cases is illustrated in 21.3. In all three cases, the depth of the fluid initially rises parabolically behind the shock. In case (a), the profile flattens out, asymptotically approaching a constant depth, while in cases (b) and (c), the depth increases indefinitely behind the front. In cases (b) and (c), other factors – not included in this simple model – enter to limit the growth of h.

In each case momentum and energy are provided to the front by the weight of the water and are dissipated by resistance to flow. The effect of channel slope adds momentum and energy in case (a) and detracts these in case (c). Close to the front, the effect of slope is small and all three profiles have the same parabolic structure.

Transition

In this chapter we have developed the conditions (called jump conditions) that must be satisfied by a flow undergoing an abrupt transition in speed and depth (called a hydraulic shock), and applied these to investigate several simple types of hydraulic shock, including hydraulic jump, bore and flash flood.

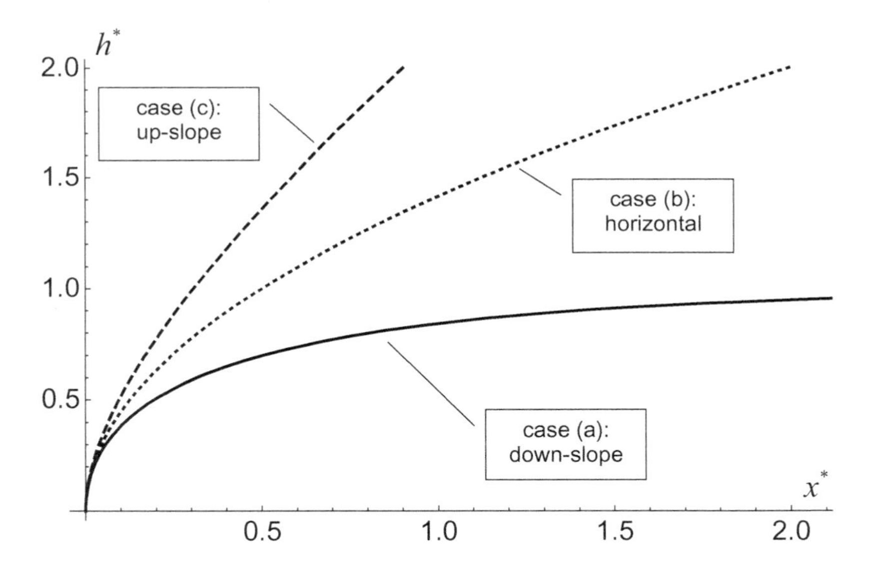

Figure 21.3 A plot of the dimensionless water height, h^*, versus the dimensionless distance, x^*, behind the flash-flood front for case (a): downslope (solid line), case (b): horizontal (dotted line) and case (c):up-slope (dashed line).

So far we have investigated a number of idealized (lossless) flows within a horizontal channel. In the following chapter we investigate more realistic flows with both a source and sink of energy. The source of energy is the gravitational potential energy released as a continuous body moves down a slope. The sink of energy is drag resistance provided by bottom boundary.

22

Gravitationally Forced Flows

In the previous two chapters, we investigated the nature of flow in a horizontal channel assuming (except for the study of flash floods in § 21.4) resistance to flow is negligibly small, and we were able to classify flows as either sub- or super-critical and to investigate the smooth transition from sub- to super-critical and the abrupt transition from super- to sub-critical. We focused on the flow of water (as opposed to other fluids) because it has a very low value of viscosity and in many cases may be treated as inviscid. However, all real flows involve resistance which tends to degrade the mechanical energy. Typically channeled flows are maintained by the release of gravitational potential energy as the material moves downslope.

In this chapter we consider the one-dimensional flow of a continuous material down a slope. The material might be water in a river, ice in a glacier, snow in an avalanche, debris in a rock slide, molten rock in a lava stream, mud in a landslide, muddy water in a turbidity current, etc. We will formulate the problem fairly generally in § 22.1, then specialize it to specific types of material in § 22.2.

22.1 Formulation

Most natural downslope flows, exemplified by rivers and glaciers, are much wider than they are deep and move along beds that are relatively flat. Typically a river is wider than deep by a factor of 100 (e.g., see Yalin, 1992). Also, rivers, glaciers and like flows typically are contained by banks that are raised and relatively steep. It follows that a first approximation of the flow bed is a shallow rectangle and that the velocity is independent of the cross-stream direction to dominant order.

Let's consider flow of material in a straight, flat-bottomed channel (at $z = 0$) of constant width; as previously, the downstream direction is x, the (unimportant) cross-stream direction is y and the (nearly) vertical direction is z. The height of the surface is represented by $z = h(x,t)$.[1] In this chapter we will focus on flows that are steady and invariant in the x direction, having h constant and $\mathbf{v} = u(z)\mathbf{1}_x$. Our goal is to determine the vertical structure of the flow, $u(z)$, driven by a down-gradient component of gravity.

[1] Here h is the total fluid depth, equal to $D + h$ as used in Chapter 21.

With no down-gradient variation, the continuity equation is identically satisfied and the fluid is not accelerating ($\mathrm{D}\mathbf{v}/\mathrm{D}t = \mathbf{0}$). Neglecting rotation ($\boldsymbol{\Omega} = \mathbf{0}$), the momentum equation (from § 4.7) reduces to a static force balance:

$$\mathbf{0} = \rho\mathbf{g} - \nabla p + \nabla \cdot \mathbb{S}' ,$$

where $\mathbb{S}'$ is the deviatoric stress tensor. Let's restrict our attention to channels having small slope ε (that is, $\varepsilon \ll 1$), so that $\mathbf{g} \approx g'\mathbf{1}_x - g\mathbf{1}_z$, where $g' = \varepsilon g$ is the reduced gravity, parallel to the ground. We want to investigate simple shear flows, parallel to – and close to – the ground. Outside the shear layer, the momentum equation simplifies to the static balance: $\mathbf{0} = \rho_r\mathbf{g} - \nabla p_r$, where a subscript r denotes the static reference state (which may depend on position). With the flow being a simple shear, only the $x-z$ elements (denoted by τ) of $\mathbb{S}'$ are non-zero and the shear force acts only in the x direction: $\nabla \cdot \mathbb{S}' = (\mathrm{d}\tau/\mathrm{d}z)\mathbf{1}_x$. Since the shear force does not upset the static balance normal to the ground, the reference-state pressure is imposed on the layer and the x component of the momentum equation simplifies to

$$\frac{\mathrm{d}\tau}{\mathrm{d}z} = (\rho_r - \rho)g' .$$

When this equation is applied to a discrete layer of material (as in §§ 22.2.1, 22.2.2 and 22.2.3) $\rho_r = 0$ and when applied to a layer of air (as in § 22.2.5), ρ_r is the density of air outside the shear layer.

As noted above, our goal is to determine $u(z)$, but at the moment our governing equation doesn't contain that variable. This deficiency will be remedied by the introduction of an appropriate stress–strain relation. But before doing so, let's investigate, in the following subsection, the vertical structure of the stress balance within a layer of material of constant density. This integrated stress balance will be applied to sliding in § 22.2.1, laminar viscous flow of a layer of Newtonian fluid in § 22.2.2 and laminar flow of a non-Newtonian fluid in § 22.2.3. Following a brief discussion of lava flows in § 22.2.4, we will revert to the form of momentum equation given above when studying katabatic winds in § 22.2.5.

22.1.1 Stress Balance

Let's consider the stress balance within a uniform layer of fluid, having constant density with $\rho_r = 0$. Assuming that there is no shear stress at the top of the layer ($\tau(h) = 0$), the momentum equation is readily integrated from bottom (at $z = 0$) to top (at $z = h$) to yield the bottom stress

$$\tau_b = \rho g' h .$$

It is common practice to express[2] this relation in terms of the *friction velocity*[3]

$$\hat{u} \equiv \sqrt{\tau_b/\rho} = \sqrt{g'h} = \sqrt{\varepsilon g h} .$$

[2] A caret ^ (also called a hat) will be used to denote parameters associated with the rough bottom boundary: $\hat{u}$, $\hat{z}$ and $\hat{\varepsilon}$.

[3] Also called the *shear velocity*. A misnomer; this is a speed, not a velocity.

Using this, the indefinite integral of the shear equation is

$$\tau/\rho = g'(h-z) = \hat{u}^2(1-z/h).$$

This states that the shear stress must vary linearly with depth in the flowing material, in order to compensate for the acceleration of gravity. This stress is provided by viscous forces for viscous materials such as ice (in glaciers) or magma flows. However, for relatively inviscid materials, such as liquid water, this stress is provided by turbulent flow. (We will investigate turbulence in Chapter 23.)

The nature of the flow depends on the stress–rate-of-strain relation of the flowing material. In the following section we will consider the flow of several types of material.

22.2 Some Laminar Flows

We begin in § 22.2.1 by considering steady rigid sliding and show that it is an unstable situation, not likely to persist. The simplest continuous flow, that of a Newtonian viscous liquid, is investigated in § 22.2.2. An example of a non-Newtonian fluid is provided by the flow of glaciers, considered in § 22.2.3. Lava flows are briefly considered in § 22.2.4, and we close out this set of examples with a study of laminar katabatic winds in § 22.2.5.

22.2.1 Sliding

Let's consider the case in which the material moves without internal deformation, so that u is uniform, independent of z. If the stress is the result of dry friction, it is given by

$$\tau_b = \mu_f F_N, \qquad \text{where} \qquad F_N = \rho g h$$

is the normal force per unit area (in the case that the slope is small) and μ_f is the (dimensionless) coefficient of sliding friction. The steady momentum equation is satisfied provided $\mu_f = \varepsilon$, where ε is the slope. In general, this condition is quite difficult to satisfy. If it is not satisfied, then the body experiences an acceleration (or deceleration) and u is a function of time. In this case, the momentum equation becomes

$$\frac{\partial u}{\partial t} = g(\varepsilon - \mu_f).$$

According to this, if $\mu_f \neq \varepsilon$, the body experiences a uniform acceleration or deceleration and the speed increases or decreases linearly with time. This situation cannot persist indefinitely. If the body is decelerating, motion and deceleration cease when u reaches zero. If the body is accelerating, it will either encounter some obstruction or will disintegrate due to vibrations as it moves over a rough bottom. In either case, accelerated sliding (of snow or rocks) often leads to catastrophe.

The flow of dry material out onto a flat surface is called *runout*. The runout distance can be estimated from energy considerations. A mass, M, of material starting at elevation Z has potential energy gMZ. This energy is used in moving the material a distance $(Z/\varepsilon)+L$ against the force of sliding friction; see Figure 22.1. A frictional force of $\mu_f gM$, exerted

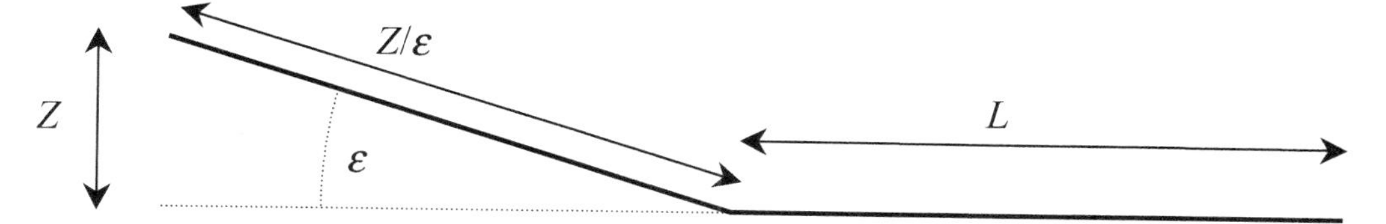

Figure 22.1 Schematic illustrating runout distance.

through that distance absorbs an energy $\mu_f gM\big((Z/\varepsilon)+L\big)$. Equating these two energies, we find that the runout distance is

$$L = Z\left(\frac{1}{\mu_f} - \frac{1}{\varepsilon}\right).$$

Of course, this is a very rough estimate. Other important factors are the cohesiveness of the body and variation in the effective friction force. Also, if the material is either boulders (that can roll) or a material that becomes liquefied as it moves down the slope, the runout distance can be much greater than this estimate. Liquefaction and mobilization of material that does not normally exhibit fluid behavior is an important factor in landslides and strong floods, contributing to the destructive power of these flows.

22.2.2 Laminar Viscous Flow

The simplest laminar-flow example is the flow of a Newtonian viscous fluid, which has the constitutive relation

$$\tau = \rho\nu\frac{\partial u}{\partial z},$$

where $\nu = \eta/\rho$ is the kinematic viscosity.[4] This coefficient measures the diffusion of momentum. If we again assume the flow speed and material depth to be independent of x and t (or vary slowly in these variables), u is a function only of z and the shear balance becomes

$$\nu\frac{\mathrm{d}u}{\mathrm{d}z} = g'(h-z).$$

This equation is to be solved subject to the condition that the material speed is zero at the bottom of the fluid, that is, $u=0$ and $z=0$. The solution to this problem is a simple quadratic in z:

$$\frac{u(z)}{\bar{u}} = 3\frac{z}{h} - \frac{3}{2}\left(\frac{z}{h}\right)^2, \qquad \text{where} \qquad \bar{u} = \frac{g'h^2}{3\nu}$$

is the mean speed. Note that:

[4] See § 6.2.2.

- the maximum speed occurs at the top of the flowing material. In the case of laminar flow, the maximum is 3/2 times the mean;
- the mean speed varies quadratically with depth (this is in contrast to the inviscid, un-driven flow investigated in § 20.1, wherein it was found that the critical speed in a horizontal channel varies as the square root of water depth);
- the volume flow is cubic in depth:[5]

$$q = \frac{g' h^3}{3\nu}; \quad \text{and}$$

- the average speed varies inversely with the viscosity.

Water has a particularly small viscosity and so would experience quite large velocities if the flow were to remain laminar. For example, with $\nu = 10^{-6}$ m^2·s^{-1}, $g \approx 10$ m·s^{-2}, ε = 0.0001 (so that $g' \approx 0.001$ m·s^{-2}) and $h = 10$ m, say, the average flow speed is predicted to be 30 km·s^{-1}. This is not very realistic; in reality the flow speed is held in check by turbulent stresses, as explained in Chapter 23.

22.2.3 *Flow of Glaciers*

The flow of glaciers down slopes provides an example of flow of a non-Newtonian fluid. Let's use Glen's law[6] with $n = 3$ ($\dot{\epsilon} = \tau^3/(2\eta\mu^2)$) as the stress–strain relation, where, with the present geometry, $\dot{\epsilon} = \mathrm{d}u/\mathrm{d}z$. Using the stress balance $\tau = \rho\hat{u}^2(1 - z/h)$, Glen's law gives us

$$\frac{\mathrm{d}u}{\mathrm{d}z} = \frac{5\bar{u}}{h^4}(h - z)^3, \qquad \text{where} \qquad \bar{u} = \frac{h\rho^3\hat{u}^6}{10\eta\mu^2}$$

is – as we shall see – the mean speed. The integral of this differential equation that satisfies the no-slip condition ($u = 0$ at $z = 0$) is

$$\frac{u}{\bar{u}} = \frac{5}{4} - \frac{5}{4}\left(1 - \frac{z}{h}\right)^4.$$

The maximum flow speed is 5/4 times the mean. The flow profile of non-Newtonian fluid obeying Glen's law is more blunt than that for a Newtonian fluid, as illustrated in Figure 22.2, which shows graphs of $u/\bar{u}$ versus z/h for non-Newtonian and Newtonian rheologies.

In practice, the flow of glaciers and ice sheets is quite complex, due to a number of factors, including the formation of crevasses and the presence of water, particularly at the base of the glacier where it is in contact with the bedrock. The issue of crevasses is beyond the scope of this book, because the glacier is no longer a continuous body; see § 1.2. Water at the base of a glacier can facilitate both *basal sliding* and deformation of the underlying bed. Again, these complications are beyond the scope of this book.

[5] In volcanology, this is known as Jeffreys's equation or the Jeffreys equation (but not Jeffrey's equation – Sir Harold would be displeased).

[6] See § 6.2.3.

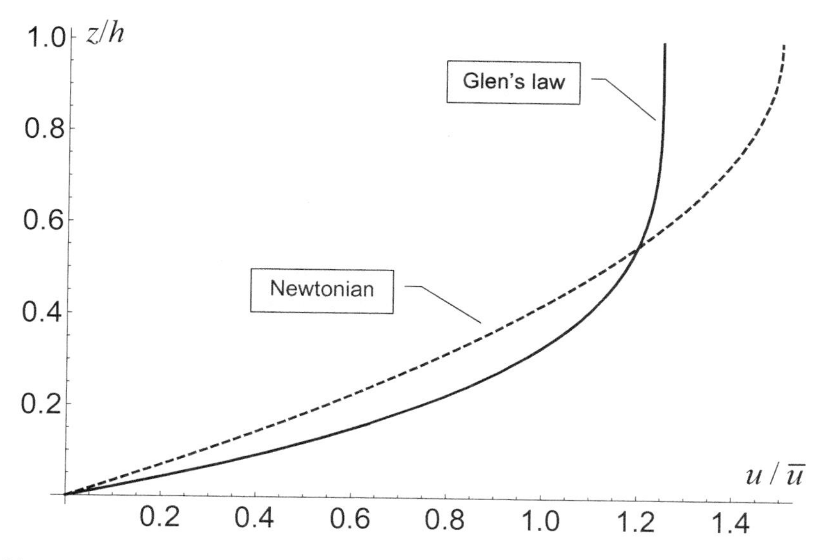

Figure 22.2 Plots of flow speed versus depth in a glacier (solid curve) in comparison with the viscous profile (dashed curve). Both profiles have the same mean speed.

22.2.4 Lava Flows

Molten lava[7] behaves as a Newtonian fluid, but with a viscosity that is highly dependent on composition (particularly the silica and water content) and temperature. Viscosities ranging from about 10 Pa·s (10 $kg{\cdot}m^{-1}s^{-1}$) for lava having low silica content (such as olivine basalts at Kilauea) to 10^{11} Pa·s for high silica (dry granitic) lava. With a density of 2.8×10^3 $kg{\cdot}m^{-3}$, the kinematic viscosity is about 4×10^{-3} $m^2{\cdot}s^{-1}$ for olivine and as high as 4×10^7 $m^2{\cdot}s^{-1}$ for granitic lava. This produces a speed of basaltic lava in a flow 1 m deep of 25 $m{\cdot}s^{-1}$ down a slope of 1%. This is quite fast. Under the same conditions, granitic lava will flow 10^{10} time slower, less than 1 cm/year! The viscosity of lava is very sensitive to temperature and to water content. The addition of 5% H_2O to a dry granitic lava reduces its viscosity by a factor of 10^6.[8]

The flow of lava with a temperature-dependent viscosity is investigated in § 35.1.

22.2.5 Laminar Katabatic Winds

In the previous subsections we have considered the downslope flow of a layer of material having a specified thickness. In this subsection we consider a more subtle flow: the downslope flow of a cold layer of air close to the ground, called a *katabatic wind*. In this case the thickness of the cold layer is not specified, but must be found as part of the solution.

[7] Lava is the non-volatile part of magma; it is the stuff that is left over after magma erupts onto the surface and its gases escape into the atmosphere.

[8] See Table 2.1, p. 23 of Williams and McBirney (1979).

In this section we will develop and solve a simple steady-state model of laminar katabatic wind down a slope. Air is quite transparent to thermal radiation, so that the ground temperature is maintained by radiation of heat to outer space, and the radiative exchange of heat between the ground and adjacent air is negligibly small. The air adjacent to the ground is cooled by conduction of heat and this cooled air flows down the slope. A steady state is possible in this situation provided the air is thermally stably stratified (warm air above cool), so that the flow advects warmer air from higher up the slope at a rate that matches the rate of conductive cooling. This is a form of *natural convection.*

Katabatic-Wind Equations

Let's consider a flat slope with x being the downslope coordinate and z being normal to x, pointing nearly vertical, with flow being entirely in the downslope direction: $\mathbf{v} = u(z)\mathbf{1}_x$. In this simple model, we will ignore compression of air and the associated variation of density with pressure. However, a key ingredient in the model is the variation of density with temperature. Since the air density is variable, we cannot directly find a first integral of the x momentum equation as we did in § 22.1.1. Instead we must revert to $\mathrm{d}\tau/\mathrm{d}z = (\rho_r - \rho)g'$, with the shear stress τ given by the laminar viscous formula $\tau = \eta \mathrm{d}u/\mathrm{d}z$, where η is the dynamic viscosity of air;[9] that is,

$$\frac{\mathrm{d}^2 u}{\mathrm{d}z^2} = \frac{\rho_r - \rho}{\eta} g' .$$

Now let's determine the variation of density that drives the downslope flow. Since the air is in hydrostatic balance, has uniform composition and we are ignoring compressibility, flow is driven entirely by thermally induced density differences. Recalling that $\mathrm{d}\rho = -\rho\alpha \mathrm{d}T$,[10] where α is the coefficient of thermal expansion, and noting that $-\alpha = 1/T$ for air,[11] a reasonable equation of state is

$$\rho = \rho_0(2T_0 - T)/T_0 ,$$

where ρ_0 is a constant density and T_0 is a constant temperature.

Steady katabatic winds occur only if the air far from the ground is stably stratified with warmer air lying above cooler air. Since the ground is sloping, this temperature variation is seen as a decrease of the reference-state temperature with increasing x, represented by $T_r = T_0(1 - \gamma_x x)$, where γ_x is a specified constant, quantifying the rate of decrease of temperature in the x direction. The corresponding reference density is $\rho_r = \rho_0(1 + \gamma_x x)$. Flow is driven by deviations from this reference state, induced by a cooler ground temperature: $T = T_0(1 - \gamma_x x) - \Delta T$, where ΔT is a specified (constant)

[9] See § 6.2.2.
[10] See § 5.2.
[11] See Appendix E.8.5.

temperature difference. The air temperature is given by

$$T = T_0(1 - \gamma_x x) - \Delta T T^*,$$

with $T^*(z)$ being the dimensionless temperature. We anticipate that $T^*(0) = 1$ and T^* decreases monotonically with increasing z, asymptotically approaching zero as $z \to \infty$. The variation of density due to this thermal structure may be expressed as

$$\rho \approx \rho_0(1 + \gamma_x x + (\Delta T/T_0)T^*)$$

and the momentum equation becomes

$$\nu \frac{d^2 u}{dz^2} = -\frac{\Delta T}{T_0} g' T^*,$$

where $\nu = \eta/\rho_0$ is the kinematic viscosity. The heat equation is a balance of downslope advection of warm air and thermal diffusive cooling provided by vertical diffusion of heat:

$$\kappa \frac{d^2 T^*}{dz^2} = \frac{T_0 \gamma_x}{\Delta T} u,$$

where κ is the thermal diffusivity.[12] These equations are to be solved on the domain $0 < z < \infty$, subject to the conditions

$$u(0) = 0, \qquad T^*(0) = 1, \qquad \text{and} \qquad u(\infty) = T^*(\infty) = 0.$$

Analysis and Solution

This pair of linear second-order ordinary differential equations may be combined into a single complex equation by multiplying the first equation by $-\mathrm{i}/U$, where $\mathrm{i} = \sqrt{-1}$ is the imaginary unit and

$$U = \frac{\Delta T}{T_0} \sqrt{\frac{\kappa g'}{\nu \gamma_x}}$$

is the velocity scale for the katabatic wind, and adding the result to the second equation:

$$\frac{d^2 \mathrm{W}}{dz^{*2}} = 2\mathrm{i}\mathrm{W},$$

where[13]

$$\mathrm{W} = T^* - \mathrm{i}\frac{u}{U} \qquad \text{and} \qquad z^* = \left(\frac{\gamma_x g'}{4\nu\kappa}\right)^{1/4} z = \frac{z}{h}.$$

[12] From Appendix E.7.
[13] Complex quantities are denoted by Roman (un-italicized) symbols.

This equation is to be solved on the interval $0 < z^* < \infty$ subject to the conditions

$$\mathrm{W}(0) = 1 \qquad \text{and} \qquad \mathrm{W}(\infty) = 0.$$

The solution is simply

$$\mathrm{W} = e^{-\sqrt{2\mathrm{i}}z^*}$$

or equivalently,

$$T^* = e^{-z/h}\cos(z/h) \qquad \text{and} \qquad u = Ue^{-z/h}\sin(z/h).$$

Profiles of T^* and u/U are plotted versus z/h in panel (a) of Figure 24.1. The values of temperature and downslope speed oscillate with z because downslope advection of warmer air causes an over-shoot of the temperature profile.

Discussion of Laminar Katabatic Winds

Let's estimate the magnitudes of the speed U and layer thickness h for laminar katabatic flow. The molecular values of kinematic viscosity and thermal diffusivity for air are nearly the same: $\nu \approx \kappa \approx 2 \times 10^{-5}$ m^2·s^{-1}, while plausible values of the other parameters are $\Delta T/T_0 \approx 0.02$, $g' \approx 0.05$ m·s^{-2} and $\gamma_x \approx 5 \times 10^{-6}$ m^{-1}. Altogether we roughly estimate that $U \approx 2$ m·s^{-1} and $h \approx 1/\sqrt{10}$ m.

Similar to water flowing in a channel,[14] katabatic wind is a flow having a characteristic speed and depth. This similarity raises the question whether a katabatic flow might become super-critical and subject to the hydraulic shocks we investigated in Chapter 21. Recall that a flow is super-critical – and prone to instability – if the Froude number exceeds unity: $\mathrm{Fr} = U/\sqrt{gh} > 1$. With the estimates in the previous paragraph and $g = 9.8$ m·s^{-2}, Fr is slightly larger than unity. As we saw in § 21.3.1, with this value of Fr the flow is prone to an undular instability, with some of the undulations possibly breaking. This instability is potential a source of irregularity and time dependence in katabatic flows.

With these speed and length scales, the Reynolds number of the flow, $\mathrm{Re} = Uh/\nu$, is greater than 10^4, indicating that the flow is in fact turbulent. We will investigate turbulent katabatic winds in § 24.1 and revisit the issue of flow instability using that more realistic model.

Transition

In this chapter we have investigated several examples of laminar flow down a slope including rigid sliding, Newtonian viscous flow, glaciers and laminar katabatic winds.

Most naturally occurring flows are turbulent. In the following chapter we develop a simple model of turbulence in simple shearing flow and in the subsequent chapter, use this theory to model several simple turbulent flows.

[14] Which we studied in Chapters 20 and 21.

23

A Simple Model of Turbulent Flow

In this chapter we address one of the most difficult problems in physical science: *turbulence*. Nearly all geophysical flows are turbulent. Much of our knowledge of turbulent flows is based on observation; a first-principle theory of turbulence does not exist, though there are a number of heuristic theories. We will consider turbulence in the context of channel flow of water (think of rivers and streams), but the concepts and models considered apply to many other geophysical flows, including motions of the atmosphere and oceans, as well as engineered flows, such as flows over airplanes and flows over and within turbines and rockets.

As water flows down a sloping channel, the downslope gravitational force imparts momentum and kinetic energy to the fluid. The force is balance by a drag force exerted by the bottom and the kinetic energy is dissipated as heat by viscosity. This requires the action of molecular viscosity. The trouble is, the viscosity of water is very small: $\nu \approx 10^{-6}$ $m^2 \cdot s^{-1}$. We have seen in § 22.2.2 that the speed of flow predicted by laminar-flow theory is very large; such speeds do not occur in natural settings. Turbulence invariably develops if a given flow is large or rapid enough and acts to retard the flow. In § 23.1 we discuss when and how turbulence arises and investigate the transition from laminar to turbulent flow. Then in § 23.2 we present and discuss the engineering approach to the quantification of turbulence.

Models of turbulent flow invariably characterize it as the sum of a "macroscopic" mean and a set of "microscopic" perturbations, though in practice this division is not very clean. The equations governing the mean flow, developed in § 23.3, contain perturbation terms which must be determined. While man-made channels can have relatively smooth bottoms, we will see in § 23.4 that natural channels are invariably hydraulically rough. The approach which is the best combination of simplicity and accuracy is mixing-length theory, coupled with a rough bottom; this is introduced in § 23.5. The resulting turbulent velocity profile for flow of water in a channel is investigated in § 23.6. This chapter concludes in § 23.7 and § 23.8 with some comments on the drag coefficient and the turbulent diffusion of heat.

23.1 Transition to Turbulence

Turbulence arises in fluids bodies that are too large or are flowing too fast for viscous forces to be effective in maintaining an orderly flow. The tendency for flow to become disorderly, that is, *unstable* and *turbulent*, is represented by a dimensionless parameter called the *Reynolds number* and denoted by Re; a flow becomes unstable when the Reynolds number exceeds a critical value. There are several forms for Re; for channel flow we may choose[1]

$$\mathrm{Re} = q/\nu\,.$$

For smooth channels, the critical value of Re is about 2000, but for rough channels it can be as small as 100. Using the laminar relation between q and h ($q = \varepsilon g h^3/3\nu$ from § 22.2.2), we find that the critical depth of water in a rough channel is given by

$$h_c = \left(\frac{300\nu^2}{\varepsilon g}\right)^{1/3} \approx \frac{3\times 10^{-4}}{\varepsilon^{1/3}}\ \mathrm{m}\,.$$

where ε is the slope of the channel. The approximation uses $\nu \approx 10^{-6}$ m^2·s^{-1} and $g \approx 10$ m·s^{-2}. Unless the slope is extremely small, virtually all natural flows of water are turbulent. On the other hand, the critical depth for olivine-basaltic lava flowing down a 1% slope is $\approx 1/3$ m and for dry granitic lava it is $\approx 1.6\times 10^6$ m.

Now let's try to understand why flows become turbulent, by thinking about the local conditions within a layer of fluid having total depth h that is flowing laminarly in the x direction and with speed u varying with depth z: $u(z)$. Recall from kinematics that the speed of motion is measured relative to an observer that may be in motion.[2] An observer moving with a parcel of fluid doesn't know about the total flow q, but does see nearby motion due to local shearing, which is quantified by

$$\left|\frac{\mathrm{d}u}{\mathrm{d}z}\right| = \frac{1}{t_s}\,,$$

where t_s is a shearing timescale. In the absence of viscous forces, the flow obeys Bernoulli's equation,[3] which states that the pressure and kinetic energy sum to a constant. This leads to the curious result that the parcel of fluid in a shearing flow sees the pressure "over there" as being lower. By the same token the parcel over there sees the pressure here as being lower. Since fluid flows from high pressure to low, the parcel here wants to go over there and the parcel over there wants to come here. And they do. But of course, every parcel of every size is trying to do the same, and they get in each other's way, resulting in a chaotic jumbled motion called turbulence. Moreover, as turbulence sets in, the motions continually create new patterns of shearing that engender further transverse motions. Actually, there is a method to this madness. Momentum and energy are fed into the fluid motion at the large

[1] Recall that q is the volume flow per unit width, having dimensions L^2·T^{-1}.
[2] See § 3.4.
[3] See Appendix D.6.

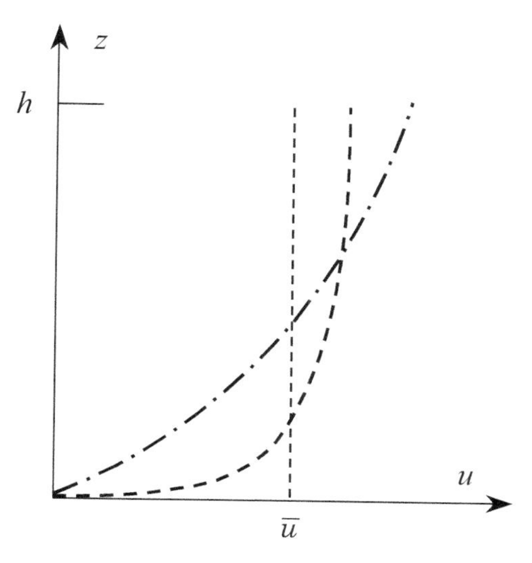

Figure 23.1 Schematic plots of a typical laminar (dash-dot) and turbulent (dashed) profiles with distance from the bottom in a layer of fluid of depth h. $\bar{u}$ is the mean speed.

scale, quantified here by the water depth, h, but viscosity is capable of acting only on small scales, of order $\sqrt{\nu t_s}$. Turbulence acts to chop the large-scale motion into successively smaller bits, until they are small enough for viscosity to do its business.[4] There is a large body of literature dealing with this cascade of energy down the size spectrum, but we will not be investigating turbulence at this level of detail.

When the channel flow becomes turbulent, the mean velocity profile becomes more blunt as shown in Figure 23.1. For the same mean flow, the turbulent profile has a much larger velocity gradient at $z = 0$ and consequently a much larger drag than the laminar profile. Putting it the other way around, for the same slope forcing and boundary drag, the turbulent flow has a much smaller mean flow.

The Navier–Stokes equation is a nonlinear vector partial differential equation. Analytic solutions of this equation are almost as rare as hen's teeth, and nearly all describe laminar flows. Many modern scientists say never mind searching for analytic solutions, let's just solve the system numerically. There are several problems with that approach. First, the range of flow scales in the atmosphere or oceans is enormous: from thousands of kilometers down to the smallest viscous scale. Second, except in rare circumstances, we do not need to know the detailed structure of the flow. We are interested in gross characteristics such as mean speed and the transport properties of the flow, which are best described by simple mathematical models. And a simple approximate analytic theory is capable of providing

[4] This cascade is eloquently stated in the well-known ditty by Lewis Richardson: "Big whorls have little whorls that feed on their velocity; and little whorls have lesser whorls and so on to viscosity."

more insight into the physical nature of the flow than is the output of a complicated numerical simulation. With this justification and fortification, let's turn our attention to development of a simple model of turbulent flow. This endeavor will be rewarded with analytic solutions of turbulent katabatic flow (in §24.1) and turbulent Ekman-layer flow (in §25.1.2).

23.2 Engineering Approach to Turbulent Channel Flow

Over the years, there have been a number of engineering approaches to the study of turbulent channel flow.[5] The best of these is the Darcy-Weisbach equation, which is a linear relation between the mean channel flow, denoted by $\bar{u}$, and the friction velocity $\hat{u} = \sqrt{\tau_b/\rho}$:[6]

$$\bar{u} = \sqrt{8/\lambda}\hat{u},$$

where λ is the Darcy friction factor.[7] An alternate way to write this relation is to solve explicitly for τ_b:

$$\tau_b = C_D \rho \bar{u}^2, \qquad \text{where} \qquad C_D = \lambda/8$$

is the *drag coefficient*.[8]

If the flow of depth h is driven by gravity in a channel having slope ε, then $\hat{u} = \sqrt{\varepsilon g h}$,[9]

$$\tau_b = \varepsilon \rho g h \qquad \text{and} \qquad \bar{u} = \sqrt{8\varepsilon g h/\lambda}.$$

In this case, the friction velocity is equal to the speed of an object that has been dropped from a height of $2\varepsilon h$.

23.2.1 Engineering Formula

In general λ is a function of the Reynolds number, Re, and the relative magnitude of the roughness,

$$\delta = \hat{z}/h,$$

where $\hat{z}$ is the mean height of protrusions on the channel bottom, as illustrated in Figure 23.5.[10] Typically λ is graphed versus Re, for various values of δ, on a *Moody diagram*.[11] For laminar flow with $\text{Re} = \bar{u}h/\nu$, $\lambda = 24/\text{Re}$. This laminar formula is valid

[5] For a brief critique of a few of these approaches, see Appendix G.3.
[6] Introduced in § 22.1.1.
[7] Introduced in § 19.1.2 In the literature, f is often used in place of λ.
[8] The drag coefficient was introduced in § 11.5.1 and is investigated in § 23.7.
[9] See § 22.1.
[10] The factor $\hat{z}$ is equivalent to the factor k that is used in the literature on flow in rough pipes to denote the characteristic roughness height.
[11] By rights, this should be called a Nikuradse-Moody diagram, because Nikuradse produced the crucial experimental data on which the diagram is based.

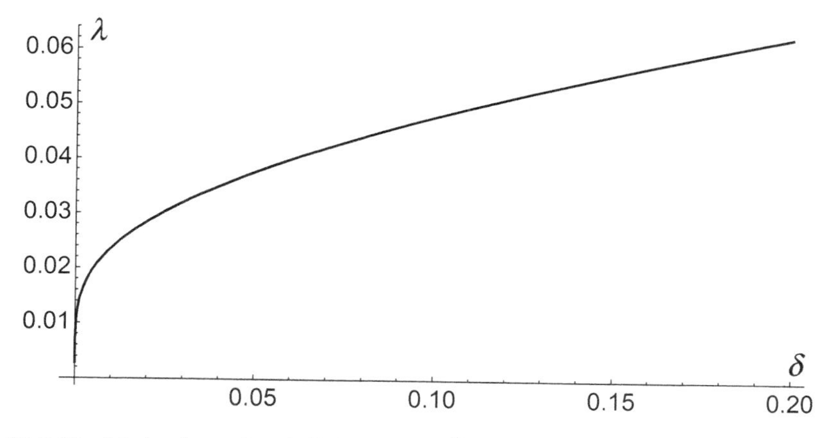

Figure 23.2 The friction factor $\lambda = 1.42/(3.15 - \ln \delta)^2$ plotted versus the bottom roughness $\delta = \hat{z}/h$, where $\hat{z}$ is the thickness of the rough bottom and h is the water depth.

for Re smaller than about 2000 if the channel bottom is smooth. For turbulent flow in a smooth-walled channel with $3000 < \text{Re} < 10^5$,[12] λ is given by the *Blasius formula*:[13] $\lambda = 0.266/\text{Re}^{1/4}$. These results are presented in the interest of completeness, but we will focus our attention on flow at large Reynolds number in a channel with a bottom sufficiently rough that the friction factor depends only on the roughness. The engineering formula for flow in a rough-walled channel or pipe is

$$\frac{\bar{u}}{\hat{u}} = \sqrt{\frac{8}{\lambda}} = B - \frac{1}{\kappa} \ln \delta ,$$

where κ and B are "universal" dimensionless constants; κ is called the *von Kármán constant*. The values of κ and B must be determined by experimentation; no theory is available. The best available estimates of these parameters, obtained from experiments in pipes, are $\kappa = 0.421 \pm 0.002$ and $B \approx 7.48$.[14]

Now we see that the analytic formula for the friction factor is

$$\lambda = \frac{8\kappa^2}{(B\kappa - \ln \delta)^2} \approx \frac{1.42}{(3.15 - \ln \delta)^2} .$$

The parameter λ is plotted versus the relative boundary roughness $\delta = \hat{z}/h$ in Figure 23.2.[15] With λ known for a given channel, we may use the Moody diagram to determine the range

[12] In the range $2000 < \text{Re} < 3000$, the flow is in transition.

[13] For pipe flow this formula is $\lambda \approx 0.3164/\text{Re}_D^{1/4}$, where Re_D is the Reynolds number based on the pipe diameter.

[14] See Shockling et al. (2006) and Allen et al. (2007).

[15] When comparing this plot with the Moody diagram, be aware that the Moody roughness is based on a pipe diameter, which is equivalent to twice the depth of water in a channel.

of Re for which the fully rough theory is valid. As the water depth varies, so does the value of λ. In particularly, as h decreases in magnitude, δ and λ both increase.

Substituting the formula for λ as a function of δ into the flow-depth relation $\bar{u} = \sqrt{8\varepsilon g h/\lambda}$, we obtain a formula relating the local mean flow speed to the dimensionless depth, $h^* = h/\hat{z} = 1/\delta$:

$$\bar{u}^* \approx 7.48(1 + 0.317 \ln h^*)\sqrt{h^*}, \qquad \text{where} \qquad \bar{u}^* = \bar{u}/\sqrt{\varepsilon g \hat{z}}$$

is the dimensionless mean speed, made so using $\sqrt{\varepsilon g \hat{z}}$, a characteristic speed based on the roughness height, $\hat{z}$. Typically h^* is much greater than unity and the profile of $\bar{u}^*$ versus h^* is very nearly parabolic.

In the following section, we will derive the engineering formula starting from the continuity and x-momentum equations governing flow in a channel and investigate the structure of turbulent channel flow.

23.3 Mean-Flow Equations

Turbulent flow is characterized by velocity perturbations (eddies), varying on relatively short time and length scales, that are superposed on a mean flow. To simplify the subsequent analysis, let's assume that the mean flow and the water depth are independent of downstream position and time and divide the velocity, $\mathbf{v}$, into a z-dependent mean speed in the x direction and a turbulent deviation, denoted by a prime: $\mathbf{v} = u\mathbf{1}_x + \mathbf{v}'$, with $\mathbf{v}' = u'\mathbf{1}_x + v\mathbf{1}_y + w\mathbf{1}_z$. The variable u depends only on z, while the primed variables are functions of position (x, y, z) and time (t). With this decomposition, the x momentum equation is[16]

$$\frac{\partial u'}{\partial t} + (u + u')\frac{\partial u'}{\partial x} + v\frac{\partial u'}{\partial y} + w\frac{\partial (u + u')}{\partial z} = \varepsilon g + \nu \frac{\partial^2 (u + u')}{\partial z^2}.$$

where ε is the (small) slope in the x direction.[17] In writing this, we have adopted a Newtonian viscous rheology, but, as we shall see below, the viscous term is unimportant. Also, we have not included the Reynolds stress term $(1/\rho)\mathrm{d}\tau/\mathrm{d}z$; the purpose of this exercise is to derive this term from the equation above. Using the continuity equation,

$$\frac{\partial u'}{\partial x} + \frac{\partial v}{\partial y} + \frac{\partial w}{\partial z} = 0,$$

the momentum equation may be written as

$$\frac{\partial u'}{\partial t} + \boldsymbol{\nabla} \bullet \left((u + u')\mathbf{v}'\right) = \varepsilon g + \nu \frac{\partial^2 (u + u')}{\partial z^2}.$$

[16] From § 6.2.2 with $\boldsymbol{\Omega} = \mathbf{0}$ and ν constant.
[17] We are reverting to εg to represent the gravitational force in the x direction, rather than g' as employed in Chapter 22, in order avoid the appearance of primes having differing meanings.

The turbulent deviations fluctuate rapidly in time and space and their time average at any point is zero. However, the time averages of their products may be non-zero. (This is reminiscent of the harmonic functions sine and cosine; their averages are zero, but the averages of their squares are not.) The statistical properties of the turbulent flow depend on distance from the bottom of the channel, but not on downstream or cross-stream distance. That is, the time-averaged products depend on z, but not on x or y. Time-averaging the momentum equation, we obtain

$$\frac{\mathrm{d}}{\mathrm{d}z}\left(\widetilde{u'w}\right) = \varepsilon g + \nu \frac{\mathrm{d}^2 u}{\mathrm{d}z^2},$$

where the tilde denotes the time average. This equation can also be expressed as

$$0 = \varepsilon \rho g + \frac{\mathrm{d}}{\mathrm{d}z}\left(\eta \frac{\mathrm{d}u}{\mathrm{d}z} - \rho \widetilde{u'w}\right).$$

In this latter form, we see that the factor $\rho\widetilde{u'w}$ is an effective stress, called the *Reynolds stress*. Assuming that the water surface at $z = h$ is stress-free, the integral of this equation is

$$\widetilde{u'w} - \nu \frac{\mathrm{d}u}{\mathrm{d}z} = \varepsilon g(z-h) = -\hat{u}^2\left(1 - \frac{z}{h}\right),$$

where we have used $\varepsilon g h = \hat{u}^2$.[18]

The Reynolds stress acts to redistribute momentum in the vertical direction, but is incapable of slowing the flow all by itself. This stress acts to transfer the momentum imparted to the fluid by the gravitational term (on the right-hand side of the momentum equation above) toward the channel bottom, but we need another mechanism to come into play to transfer the momentum to the bottom. One mechanism is obvious in the equation above: viscous drag. A second mechanism, not apparent in that equation, is boundary roughness. We will evaluate the relative importance of these two mechanisms in § 23.4. But before addressing that issue, in the following subsection we discuss and roughly quantify the Reynolds stress and the terms that comprise it.

23.3.1 Discussion of Reynolds Stress

In the following section we will relate the Reynolds stress, $\widetilde{u'w}$, to the mean flow, u, using mixing-length theory. But before doing so, let's try to develop an understanding of this term. We begin by noting that the perturbations of vertical and horizontal speed are strongly correlated as illustrated in Figure 23.3. The curve represents the mean profile $u(z)$ having a positive gradient. A positive vertical velocity (denoted by the vertical arrow on the left) moves a parcel initially near the boundary (represented by the lower circle) to a

[18] See § 22.1.1.

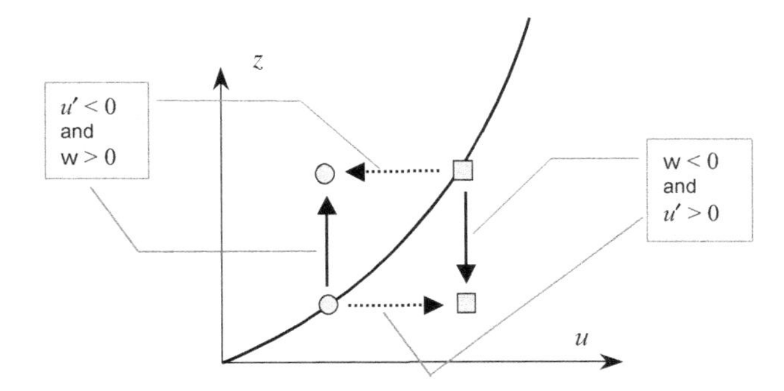

Figure 23.3 As parcels of fluid move off the mean profile (solid curve), the velocity perturbations u' and w are negatively correlated.

higher position (represented by the upper circle). If the parcel preserves its momentum, it is moving slower than the surrounding fluid; it has a negative horizontal velocity perturbation (represented by the upper dotted line). The velocity perturbations are correlated and their product $u'w$ is negative. Similarly a negative vertical velocity (represented by the vertical arrow on the right) moves a parcel initially far from the boundary (represented by the upper square) closer to the boundary (represented by the lower square) where it has a positive horizontal velocity perturbation (represented by the lower dotted line). Again the velocity perturbations are correlated and their product is negative.

The correlated velocity perturbations u' and w are instrumental in maintaining the fluid in steady flow in spite of being constantly accelerated by the downslope pull of gravity. The sizes of these perturbations necessary to accomplish this task may be estimated as follows. A flow speed u can be achieved by letting a parcel of fluid fall a vertical distance $u^2/2g$. If the parcel falls parallel to the bed, it moves a horizontal distance $L = u^2/2\varepsilon g$. If the parcel is not to flow faster than this speed, it must lose its excess momentum and energy to the bottom. It does so by moving from the main stream to the bottom (a distance of order h), on average once in each downstream distance L. The associated vertical speed is of order $(h/L)u$; that is[19]

$$w \sim \hat{u}^2/u\,.$$

The turbulent motions are commonly thought of as an ensemble of eddies, each having vertical and horizontal scales of comparable magnitude. It follows from conservation of mass that $u' \sim w$ and, since the flows are correlated,

$$\widetilde{u'w} \sim \hat{u}^4/u^2\,.$$

[19] The symbol $\sim$ means "of the order of," implying both a functional dependence and a rough numerical equality; see Table 2.1. Numerical factors such as 2 and π are ignored. It is different from the tilde over a symbol, which means the time average.

If the Reynolds stress term is larger than the viscous stress term in the momentum equation, then it must balance the downslope pull of gravity. That is, the momentum equation requires

$$\widetilde{u'w} \sim \hat{u}^2 .$$

These two estimates of the velocity correlation are compatible provided

$$u^2 \sim \hat{u}^2 .$$

The mean flow speed is roughly the same magnitude as the friction velocity. This speed is achieved by a parcel "falling" a distance h parallel to the bottom boundary; that is, $L \sim h$. In order to shed this excess speed, a parcel must visit the bottom boundary roughly once each time it moves a distance h downstream. This requires that the velocity perturbations be the same order of magnitude as the mean speed; turbulence is vigorous motion. This crude estimate does not take into account any variations of flow and structure in the z direction. This deficiency will be corrected in § 23.5, where we develop simple theory of the vertical structure of the Reynolds stress. But before that, let's take one more digression and discuss the nature of boundary roughness and its role in turbulence.

23.4 Boundary Roughness

The nature of the flow close to the bottom differs depending whether the bottom is smooth or rough. If it is smooth, turbulent perturbations become very small as the bottom is approached and the force balance is between the downslope component of gravity and viscous drag. Given the small value of viscosity, this requires an extraordinarily large value of the gradient of the average speed. On the other hand, if the bottom is rough, the associated perturbations of flow generate a Reynolds stress of sufficient magnitude to balance the slope force. The parameter measuring the relative importance of roughness and shear stress is a Reynolds number based on the roughness height $\hat{z}$ and friction velocity, $\hat{u}$:

$$\hat{R} = \hat{z}\hat{u}/\nu .$$

We shall consider flows for which the depth of water is much greater than the roughness scale: $\hat{z} \ll h$.[20] Three flow regimes have been identified, as follows (see p. 579–80 of Schlichting, 1968):

- if $0 \leq \hat{R} \leq 5$ the flow is hydraulically smooth, and roughness is unimportant. The turbulent flow is tied to the bottom by viscosity;
- if $5 < \hat{R} \leq 70$ the flow is in a transition regime and both viscous stress and roughness are important near the bottom; and
- if $70 < \hat{R}$ the flow is *hydraulically rough* and viscous stress is unimportant everywhere.

[20] That is, we are not considering flows in streams that have rocks poking out the surface, or even rising close to the surface.

Water $\{0.1, 1, 10\}$ m deep flowing down a slope of 0.1% ($\varepsilon = 0.001$) is in the hydraulically rough regime if the bottom roughness is greater than $\{20, 6, 2\} \times 10^{-4}$ m. It is safe to say that all natural flows of water are hydraulically rough, and the viscous term is unimportant.

Resistance to flow in a channel with a rough bottom is provided by *form drag*. Form drag is due to a pressure difference between the upstream and downstream faces of a protuberance; see Figure 11.3. Also, vigorous flow impinging on the front face of a protuberance can be strongly redirected back upward into the streaming flow, but with a much-diminished downstream momentum.[21]

We have seen that dynamic instabilities cause the interior of a flowing stream to be turbulent, with momentum within the fluid being redistributed vertically by turbulent eddies and transferred to the ground through interaction with rough protuberances on the bottom boundary. Now let's develop a simple theory that quantifies the vertical distribution of momentum.

23.5 Mixing-Length Theory

In order to develop an equation governing the variation of the mean flow speed with depth, i.e., $u(z)$, we need to develop an expression relating the Reynolds stress, $\widetilde{u'w}$, to this mean flow. Quantification of these perturbations and their effect on the mean flow is one of the most difficult challenges of mathematical physics. A rigorous analysis leads into a mathematical morass. One heuristic theory that is reasonably simple and accurate is *Prandtl's mixing-length theory*. To develop this theory, let's begin by considering a steady mean flow in the vicinity of some specified elevation, z_0, with a positive gradient: $\mathrm{d}u/\mathrm{d}z > 0$. Near that point, the truncated Taylor representation of the mean flow is

$$u(z) = u(z_0) + \left.\frac{\mathrm{d}u}{\mathrm{d}z}\right|_{z=z_0} (z - z_0) + \ldots.$$

Suppose that a parcel of fluid originally at $z = z_0$ with a downstream speed deviation of zero is displaced upward or downward a distance $\Delta z = z - z_0$ by a vertical speed w acting for an interval of time Δt, that is

$$w = \Delta z / \Delta t.$$

If the parcel's downstream momentum is conserved, it will have a downstream speed deviation of

$$u' = -\left.\frac{\mathrm{d}u}{\mathrm{d}z}\right|_{z=z_0} \Delta z.$$

Note that these two velocity perturbations are anticorrelated, with one positive and the other negative, and their magnitudes are both proportional to the distance Δz. In § 23.3.1

[21] Think of the spray produced as a storm wave strikes a boulder on the shore.

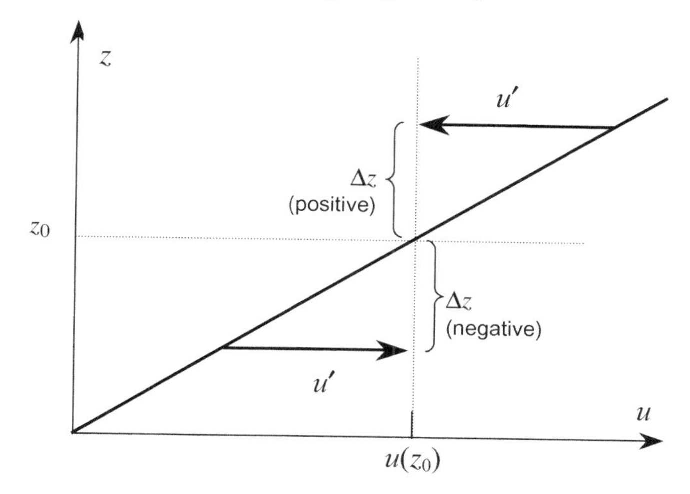

Figure 23.4 Schematic plot of the downstream velocity perturbations, u', due to vertical displacement Δz of a fluid parcel having downstream speed u.

we saw that $w \sim -u'$, on average. It follows that

$$\frac{1}{\Delta t} \sim \left.\frac{\mathrm{d}u}{\mathrm{d}z}\right|_{z=z_0} .$$

Since we are developing a heuristic theory, we may replace $\sim$ with $=$ in this expression. Combining these expressions, taking the short-time average and dropping the subscript on z_0, we have that

$$\widetilde{u'w} = -\nu_T \frac{\mathrm{d}u}{\mathrm{d}z}, \qquad \text{where} \qquad \nu_T \equiv l^2 \left|\frac{\mathrm{d}u}{\mathrm{d}z}\right|$$

is the *turbulent kinematic viscosity* and $l = |\Delta z|$ is the *mixing length*. We have added the absolute value sign to the definition of ν_T so that it can apply to flows having $\mathrm{d}u/\mathrm{d}z < 0$.

Ignoring the laminar viscous term, the momentum equation developed in §23.3 becomes

$$\nu_T \frac{\mathrm{d}u}{\mathrm{d}z} = \varepsilon g(h - z) = \hat{u}\left(1 - \frac{z}{h}\right) .$$

To make further progress toward our goal of developing an expression for $\nu_T(z)$, we need to determine the variation of the mixing length with vertical distance: $l(z)$. This is accomplished in the following section.

23.5.1 Vertical Variation of Mixing Length

The mixing length is the vertical distance a parcel moves before it merges with its surroundings. It is analogous to the mean free path of molecules of gas or the distance you can walk unimpeded across a crowded room. It is clear that this distance is limited by

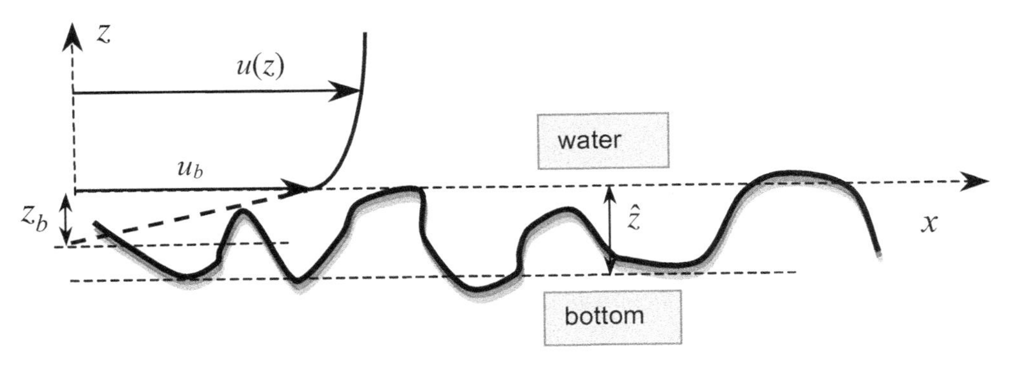

Figure 23.5 Magnified view of the channel bottom, showing the rough boundary of thickness $\hat{z}$ and its location in reference to the level $z=0$ marking the top of the rough boundary. The extrapolated logarithmic speed becomes zero at a distance z_b below the top of the rough boundary and u_b is the apparent slippage speed at the outer edge of the rough boundary.

the distance to the bottom boundary: $z > l$. This suggests "natural" scaling for l with $l \propto z$ near the bottom of the layer. We need to allow for the possibility that the channel bottom is rough, with l being small but non-zero there. With this in mind, we can set

$$l = \kappa(\hat{z} + z),$$

where $\hat{z}$ is the vertical height of irregularities in the channel bottom, as illustrated in Figure 23.5, and the constant of proportionality, $\kappa \approx 0.421$, is the von Kármán constant introduced in § 23.2. Note that the vertical distance z is measured from the top of the rough wall, but neither the horizontal nor the vertical velocity is zero there; water can and does penetrate into the rough layer.

23.5.2 Turbulent Diffusion of Momentum

Using the scaling for l presented above, the turbulent kinematic viscosity is

$$\nu_T = \kappa \hat{u}(z + \hat{z}).$$

Away from the bottom boundary, $\hat{z} \ll z$ and $\nu_T \approx \kappa \hat{u} z$.

Now the momentum equation becomes

$$\kappa(z + \hat{z})\frac{\mathrm{d}u}{\mathrm{d}z} = \hat{u}\left(1 - \frac{z}{h}\right).$$

Note that this formulation has ν_T tending to infinity as $z \to \infty$. This singular behavior can be avoided by specifying that ν_T is constant for z greater than some cut-off distance. But implementing this refinement would only complicate the mathematics when using this model (as in § 24.1 or § 25.1.2, for example) and would not result in any discernible change of flow structure.

The friction velocity is defined[22] as $\hat{u} = \sqrt{\tau/\rho}$ and the stress is commonly related to the flow speed by $\tau = C_D \rho U^2$, where U is the change of speed across the boundary layer and C_D is the drag coefficient.[23] These formulas provide an alternate representation for the turbulent kinematic viscosity

$$\nu_T = \hat{\varepsilon} U (z + \hat{z}), \qquad \text{where} \qquad \hat{\varepsilon} \equiv \kappa \sqrt{C_D},$$

with $\kappa = 0.421$ and $C_D \approx 0.0013$, $\hat{\varepsilon} \approx 0.02$. This scaling for ν_T is based on a flow parallel to a boundary, with U representing the change of speed across the layer.

It is interesting to note that, with this scaling and setting the lateral scale L equal to the vertical distance $z + \hat{z}$, the turbulent Reynolds number is

$$\mathrm{Re}_T = UL/\nu_T = 1/\hat{\varepsilon}.$$

If $\hat{\varepsilon} \approx 0.02$, $\mathrm{Re}_T \approx 50$; a constant value slightly smaller than the rough-bottom limit of 70 given by Schlichting (1968). This intriguing result suggests that the turbulent-velocity profile is at the margin of stability, in agreement with the arguments presented by Bak (1996).

The magnitude of our turbulent kinematic viscosity grows without bound with distance from the boundary. If the fluid is homogeneous and not rotating, it is well known that this parameterization yields a logarithmic velocity profile; see § 23.6, immediately below. It is less well known that if the fluid is stably stratified or rotating, this model gives uniform flow outside a viscous boundary layer near the wall; see § 24.1 and § 25.1.2. Our model of small-scale turbulent flow assumes that the Coriolis force does not affect the small-scale flow structures. This assumption is reasonable if the lifetime of an individual eddy is much less than one day, which is usually the case in the atmosphere and oceans.

23.6 Turbulent Velocity Profile Near the Bottom

Close to the bottom, with $z << h$, the momentum equation presented in § 23.5.2 is readily integrated to yield a logarithmic profile. Two versions of this profile are

$$\frac{u}{\hat{u}} = \frac{1}{\kappa} \ln\left(\frac{\hat{z} + z}{\hat{z} - z_b}\right) \quad \text{or equivalently} \quad \frac{u - u_b}{\hat{u}} = \frac{1}{\kappa} \ln\left(\frac{\hat{z} + z}{\hat{z}}\right),$$

where $z = z_b$ is the level at which u would equal zero if the logarithmic profile were extended into the rough bottom, and $u_b = u(0)$ is the flow speed at the outer edge of the rough region; see Figure 23.5. This logarithmic profile is known as the *law of the wall*.[24]

The values of the two parameters κ and either z_b or u_b must be determined by experimentation; there is no theory available. Values of equivalent parameters have been determined by measurements of turbulent flow in rough-walled pipes. In order to relate the

[22] See § 22.1.1.
[23] See § 11.5.1.
[24] Though it is not a law; it is just the result of a model.

present results to those experiments, we shall rewrite the two formulas for u as follows.

$$\frac{u}{\hat{u}} = \frac{1}{\kappa}\ln\left(\frac{z}{\hat{z}}\right) + \frac{1}{\kappa}\ln\left(1+\frac{\hat{z}}{z}\right) + C$$
$$\approx \frac{1}{\kappa}\ln\left(\frac{z}{\hat{z}}\right) + C,$$

where

$$C = \frac{1}{\kappa}\ln\left(\frac{\hat{z}}{\hat{z}-z_b}\right) = \frac{u_b}{\hat{u}}$$

is the *Nikuradse roughness function*. The constants κ and C are "universal" dimensionless parameters that have fixed values, provided molecular viscous effects are negligibly small: $\kappa = 0.421 \pm 0.002$ and $C = B + 1/\kappa \approx 9.86$.[25] Note that $z_b/\hat{z} = 1 - e^{-\kappa C} \approx 0.984$. The mean speed of flow at the top of the rough layer is about 10 times faster than the friction velocity and the level at which the extrapolated speed profile becomes zero is less than 2% above the bottom of the rough layer.

Integrating the profile from $z = 0$ to $z = h$, we find that the average downslope speed, $\bar{u}$, is given by

$$\frac{\bar{u}}{\hat{u}} = C - \frac{1}{\kappa} + \frac{1}{\kappa}\ln\left(\frac{h}{\hat{z}}\right) = B - \frac{1}{\kappa}\ln\delta.$$

This is the engineering formula introduced in § 23.2.

We have assumed that the magnitude of the roughness of the bottom is fixed and specified, independent of the flow. However, in real flows over erodible beds, the flow tends to deform the bed, modifying the roughness. This adds a significant layer of complexity to the physical and mathematical problem that is beyond the scope of this introductory text.

23.7 Drag Coefficient

The *drag coefficient* is a dimensionless factor representing the magnitude of the drag force exerted on (or by) a body due to fluid flow. This factor is employed in several contexts. In aerodynamics the drag force F_D exerted on a body having a frontal area[26] A_f by a fluid having density ρ and mean speed U relative to the body is given by

$$F_D = \tfrac{1}{2}C_A \rho A_f U^2,$$

where C_A is the *aerodynamic drag coefficient*.[27] In this formula the drag is *form drag* or *pressure drag* produced on a blunt body, principally due to flow separation, which produces a stagnant low-pressure region behind (in the lee of) the body.

[25] See § 23.2.1.
[26] Normal to the direction of flow.
[27] e.g., see Eq. (1.14) on p. 16 of Schlichting (1968).

Suppose we have a horizontal planar array of blunt bodies (simulating a rough boundary), with each body occupying a horizontal area A. Flow past this array exerts a mean shear stress

$$\tau_b = F_D/A = C_D \rho U^2, \qquad \text{where} \qquad C_D = C_A A_f / 2A$$

is our preferred form of the drag coefficient. We have encountered this formula previously in § 11.5.1, in the context of the stress exerted on the ocean surface by the prevailing winds. Commonly the stress-speed relation is expressed as

$$\hat{u} = \sqrt{C_D} U, \qquad \text{where} \qquad \hat{u} = \sqrt{\tau_b/\rho}$$

is the friction velocity, which we first encountered in § 22.1.1. As noted in § 23.2, $C_D = \lambda/8$ where λ is the Darcy friction factor.[28] A formula for the drag coefficient applicable to a rotating fluid is developed in Chapter 25.

23.8 Turbulent Diffusion of Heat

In the previous sections of this chapter we developed an expression for the turbulent diffusion of momentum, treating the downstream momentum as a conserved scalar quantity. According to *Reynolds analogy*, turbulent motions have an identical effect on the transport of other conserved scalar quantities, such as heat and chemical constituents. The turbulent diffusion of downstream momentum, $\widetilde{u'w}$, is represented by the turbulent diffusion coefficient, ν_T, according to $\widetilde{u'w} = -\nu_T \mathrm{d}u/\mathrm{d}z$, where $\nu_T = \hat{\varepsilon} U(z + \hat{z})$.[29] In a similar fashion, the turbulent diffusion of heat may be represented by

$$\widetilde{T'w} = -\nu_T \frac{\mathrm{d}T}{\mathrm{d}z},$$

where T is the temperature.

This parameterization is equivalent to setting the turbulent Prandtl number, which measures the strength of diffusion of momentum relative to heat, equal to unity. Reynolds analogy is a reasonable approximation provided buoyancy effects are unimportant.[30]

Transition

We have developed a simple theory of turbulence in shearing flows, based on Prandtl's mixing-length theory, in which the turbulent kinematic viscosity varies linearly with height above a rough bottom. This simple theory is used to investigate several non-rotating turbulent flows in the following chapter.

[28] Introduced in § 19.1.2
[29] See § 23.5.2.
[30] For an assessment of the validity of Reynolds analogy in atmospheric flows, see Li et al. (2015).

24
Some Non-Rotating Turbulent Flows

In this chapter we consider some non-rotating turbulent flows that are a bit more complicated than the flow of water down a slope considered in § 23.6. These include turbulent katabatic winds driven by thermal buoyancy (§ 24.1), avalanches driven by snow suspended in air (§ 24.2) and cumulonimbus clouds driven by the release of latent heat as water vapor condenses (§ 24.3).

24.1 Turbulent Katabatic Winds

In § 22.2.5, we investigate the katabatic wind down a slope in the case that the flow is laminar and, using typical parameter values, found that the flow is very likely turbulent rather than laminar. When flow is turbulent, the diffusivity coefficients for momentum and heat are not constant, but instead vary linearly with elevation. In this section, we will revisit the problem formulated in § 22.2.5, but with variable diffusivities, using Reynolds analogy to set the turbulent thermal diffusivity equal to the turbulent diffusivity of momentum.[1]

The governing equations now are

$$\frac{\mathrm{d}}{\mathrm{d}z}\left(\nu_T\frac{\mathrm{d}u}{\mathrm{d}z}\right) = -\frac{\Delta T g'}{T_0}T^* \qquad \text{and} \qquad \frac{\mathrm{d}}{\mathrm{d}z}\left(\nu_T\frac{\mathrm{d}T^*}{\mathrm{d}z}\right) = \frac{T_0\gamma_x}{\Delta T}u,$$

where z is elevation above the ground, u is the downslope speed, T^* is the dimensionless perturbation temperature, $g' = \varepsilon g$ is the reduced gravity, γ_x is the down-slope thermal gradient, ΔT is the temperature contrast, $\nu_T = \hat{\varepsilon}U(z+\hat{z})$ is the turbulent diffusivity, $\hat{\varepsilon}$ is a small dimensionless parameter (≈ 0.02), U is the velocity scale and $\hat{z}$ is the roughness scale. As before, these equations are to be solved on the domain $0 < z < \infty$ subject to the conditions $u(0) = T^*(0) - 1 = u(\infty) = T^*(\infty) = 0$. And as before, we can combine the two equations into a single complex equation, although the scalings are somewhat different. In the present case

$$U = \frac{\Delta T}{T_0}\sqrt{\frac{g'}{\gamma_x}} \qquad \text{and} \qquad h = \frac{2\hat{\varepsilon}\Delta T}{\gamma_x T_0},$$

[1] See § 23.8.

while the complex equation for $W = T^* - iu/U$ is

$$\frac{d}{dz^*}\left((z^* + \hat{z}^*)\frac{dW}{dz^*}\right) = 2iW,$$

where $z^* = z/h$ is the dimensionless vertical distance and $\hat{z}^* = \hat{z}/h$ is the scaled boundary roughness, subject to conditions $W(0) = 1$ and $W(\infty) = 0$.[2]

The problem for the turbulent katabatic winds is a bit more challenging than for the laminar winds because our complex ordinary differential equation now has a variable coefficient. We can get this equation in "standard" form by introducing a new independent variable; let

$$\xi = \sqrt{8(z^* + \hat{z}^*)} \qquad \text{with} \qquad \hat{\xi} = \sqrt{8\hat{z}^*}.$$

Now the governing equation becomes

$$\frac{d^2W}{d\xi^2} + \frac{1}{\xi}\frac{dW}{d\xi} - iW = 0;$$

this is to be solved in the domain $\hat{\xi} < \xi < \infty$, subject to conditions $W(\hat{\xi}) = 1$ and $W(\infty) = 0$. This differential equation is recognized as a modified Bessel equation with a complex coefficient.[3] The solution satisfying the boundary conditions may be expressed

$$W = \frac{K_1(\sqrt{i}\xi)}{K_1(\sqrt{i}\hat{\xi})} = \frac{\mathrm{ker}\,(\xi) + i\,\mathrm{kei}\,(\xi)}{\mathrm{ker}(\hat{\xi}) + i\,\mathrm{kei}(\hat{\xi})}$$

or equivalently

$$T^* = C \qquad \text{and} \qquad u = US,$$

where K_1 is the modified Bessel function of the second kind,[4] ker and kei are Kelvin functions,[5], and

$$C = \frac{\mathrm{ker}(\hat{\xi})\mathrm{ker}\,(\xi) + \mathrm{kei}(\hat{\xi})\mathrm{kei}\,(\xi)}{\mathrm{ker}^2(\hat{\xi}) + \mathrm{kei}^2(\hat{\xi})} \qquad \text{and} \qquad S = \frac{\mathrm{kei}(\hat{\xi})\mathrm{ker}\,(\xi) - \mathrm{ker}(\hat{\xi})\mathrm{kei}\,(\xi)}{\mathrm{ker}^2(\hat{\xi}) + \mathrm{kei}^2(\hat{\xi})}$$

are functions of ξ and $\hat{\xi}$ that represent the vertical structure of the turbulent katabatic wind. The laminar and turbulent profiles of u/U and T^* are plotted versus z/h in Figure 24.1.

24.1.1 Discussion of Turbulent Katabatic Winds

Let's estimate the magnitudes of the speed U and layer thickness h for turbulent katabatic flow, again estimating that $\Delta T/T_0 \approx 0.02$, $g' \approx 0.05$ m·s^{-2} and $\gamma_x \approx 5 \times 10^{-6}$ m^{-1}. If

[2] The factor 2 is introduced to simplify subsequent equations.
[3] See §10.25(i), p. 248 of Olver et al. (2010).
[4] See formula 10.31.1, p. 252 of Olver et al. (2010). A simple representation is $K_1(z) = \int_0^\infty \cos(zt)/\sqrt{t^2+1}\,dt$.
[5] See formula 10.61.2, p.268 of Olver et al. (2010).

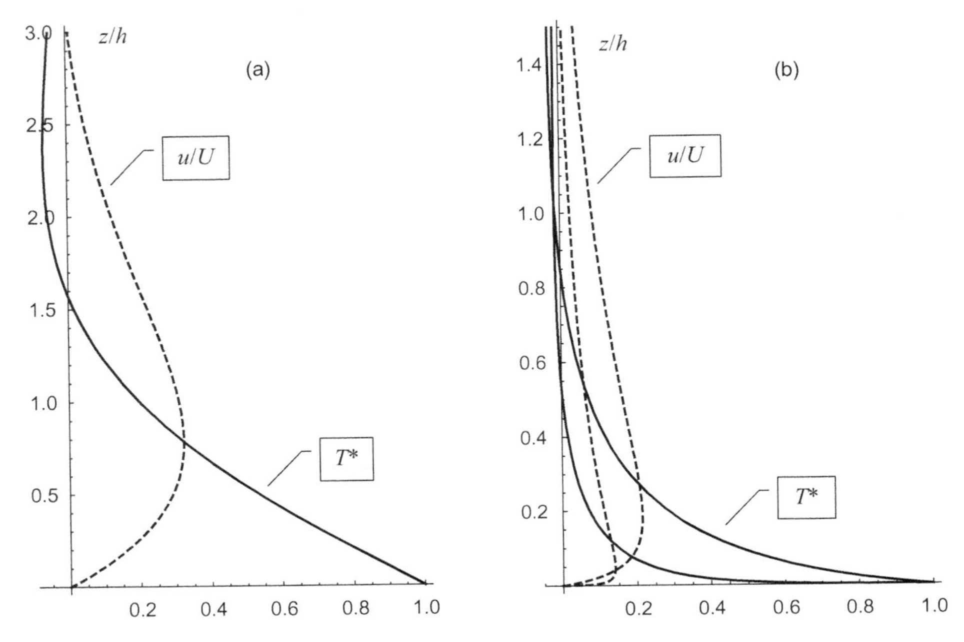

Figure 24.1 (a) Left panel: Laminar katabatic flow. Profiles of the functions T^* and u/U versus z/h. (b) Right panel: Turbulent katabatic flow. Profiles of the functions T^* and u/U versus z/h for $\hat{\xi} = 0.5$ (thick lines) and 0.05 (thin lines). Note the difference in vertical scale; the laminar flows decay more slowly with distance from the boundary than do the turbulent flows. These profiles also describe the variation of flow speeds in laminar and turbulent Ekman layers, as discussed in § 25.1, replacing u/U, T^* and z/h with u/U, v/U and z/L, where L is the appropriate Ekman layer thickness.

we estimate that $\hat{\varepsilon} \approx 0.02$,[6] we see that the velocity scale is again $U \approx 2$ m·s^{-1}, but the thickness of the flow is much greater: $h \approx 200$ m and reasonably close to the estimated thickness (300 m) of katabatic winds in Antarctica, but our speed estimate is much lower than the speeds observed on that continent.

With an increased thickness of the layer and the same characteristic speed, the Froude number is significantly less than unity, indicating sub-critical flow. But let's look at this more carefully. The Froude number for turbulent katabatic flow is

$$\mathrm{Fr} = \sqrt{\frac{\varepsilon \Delta T}{2\hat{\varepsilon} T_0}}.$$

Note that Fr is independent of the downslope temperature gradient, γ_x; it is an increasing function of the slope angle, ε, and the thermal contrast, $\Delta T/T_0$, and can become large if the drag coefficient $\hat{\varepsilon}$ is small.

6 From § 23.5.2.

In temperate latitudes katabatic winds are occasionally generated by nocturnal radiative cooling, while in Antarctica they are the dominant meteorologic phenomenon, being produced by radiative cooling combined with the proximity to the cold ice sheet that covers the continent to a depth of 4000 m. Katabatic winds in temperate latitudes are relatively mild, while those in Antarctica are very strong, often as great as 50 $m{\cdot}s^{-1}$, equivalent to a category 3 hurricane on the Saffir–Simpson hurricane scale (the strongest gust recorded was 84 $m{\cdot}s^{-1}$, which places it as category 5 on the Saffir–Simpson scale and category 4 on the modified Fujita scale for tornadoes); see Table 28.1.

Katabatic winds in Antarctica are often quite steady – blowing for days on end – but they can also be gusty. Similar winds in Patagonia, called williwaws, are notoriously gusty and sufficiently strong to knock an anchored sailboat on her beam ends. This erratic behavior can be explained in terms of a transient hydraulic shock perhaps occurring spontaneously,[7] but more likely induced as the wind encounters an air mass influenced by open waters, as explained by Ball (1956); also see Parish (1988).

24.2 Avalanches

An *avalanche* is the flow of granular material down a slope. All avalanches are transient; they begin abruptly and decay as the source of granular material is exhausted. Most commonly the material is snow, though it may be any loose or unconsolidated material, such as rocks, sand or soil. Broadly speaking, there are two types of snow avalanches: wet and dry. A wet snow avalanche is essentially a *slump*: a relatively smooth (often laminar) flow of material down a slope. On the other hand, a dry snow avalanche is a more violent affair, with a turbulent mix of air and snow flowing rapidly down a slope. Typically (Vilajosana et al., 2007) wet snow avalanches last between 50 and 80 s and have speeds less than 10 $m{\cdot}s^{-1}$, while dry snow avalanches last between 8 and 18 s and have speeds roughly 50 $m{\cdot}s^{-1}$. Let's focus on dry snow avalanches, as they are the most common, familiar and dangerous.

An avalanche has a complicated structure, as illustrated in Figure 3 of Sovilla et al. (2015), consisting of a head (labeled 1 and 2 in their Figure 3) followed by a main body (labeled 3 and 5) and a tail (labeled 4 and 5). Snow pack is entrained into the flow within the head region, is sustained by turbulence within the main body and settles to the ground in the decay region. In this section we will develop a simple model of steady flow within the main body of the avalanche.

24.2.1 Avalanche Mass

An avalanche is driven by the gravitational potential energy released as the mobilized particles of snow and ice flow downslope. The amount of snow in suspension may be determined using the particle conservation equation developed in § 6.4.3. Assuming no

[7] See § 21.3.1.

phase change, steady state and no downslope variation, this equation simplifies to

$$\frac{\mathrm{d}}{\mathrm{d}z}\left(\Phi w_s + \kappa_\Phi \frac{\mathrm{d}\Phi}{\mathrm{d}z}\right) = 0,$$

where w_s is the sedimentation speed, Φ is the volume fraction of particles and κ_Φ is the turbulent transport coefficient. This equation is readily integrated:

$$\kappa_\Phi \frac{\mathrm{d}\Phi}{\mathrm{d}z} = -\Phi w_s,$$

where the constant of integration has been set equal to zero since Φ tends to zero at great height. As with the thermal diffusivity, we can invoke Reynolds analogy set the particle diffusivity, κ_Φ, equal to the turbulent viscous diffusivity: $\kappa_\Phi = \hat{\varepsilon} U(z+\hat{z})$.[8] Now the particle conservation equation can be integrated again:

$$\Phi = \Phi_{max}\left(\frac{\hat{z}}{z+\hat{z}}\right)^{w_s/\hat{\varepsilon} U},$$

where Φ_{max} is the maximum particle concentration. The total volume Φ of suspended particles per unit area is

$$\Phi_{tot} = \int_0^\infty \Phi\, \mathrm{d}z = \hat{\varepsilon}\Phi_{max}\hat{z}U/w_s$$

and the mass per unit area of suspended particles is $\rho_p \Phi_{tot}$.

We are not yet in a position to quantify these results, because we don't yet know the avalanche speed, U; this is calculated in the following subsection.

24.2.2 Avalanche Speed

In the simplest model of steady uniform flow, the gravitational pull of the snow load, represented by $\rho_p \Phi_{tot} g \varepsilon$, where ε is the slope, is balanced by the basal drag, τ_b:

$$\rho_p \Phi_{tot} g \varepsilon = \tau_b .$$

If the snow were welded together, the flow would be simple sliding, as described in § 22.2.1. Actually, this is a reasonable approximation for the initial stages of a slab avalanche. We saw that sliding, if resisted only by dry friction, is a run-away process, with no steady state. However, if we account for the turbulent viscous resistance provided by the fluid component of the flow, which varies as the square of the speed, a steady balance is possible between downslope gravitational pull and drag. A relatively simple model that includes both types of resistance is the *Voellmy model* :

$$\tau_b = \mu_f \rho_p \Phi_{tot} g + (\lambda/8)\rho U^2,$$

[8] This ignores the effect of dense particles on the structure and strength of the turbulent motions. See Dyer and Soulsby (1988) for a discussion of other possible models of particle diffusivity.

where μ_f is the coefficient of sliding friction, λ is the Darcy friction factor, U is the mean speed of flow and ρ is the density of the mixture at the bottom: $\rho \approx \Phi_{max}\rho_p$. Substituting this drag formula into the force balance and noting that $\rho_f \ll \rho_p$, $\Phi_{tot} = \varepsilon\Phi_{max}\hat{z}U/w_s$ and $w_s \approx 2gR^2\rho_p/(9\eta)$,[9] gives a formula for the steady-state avalanche speed:

$$U = 36\left(\varepsilon - \mu_f\right)\hat{\varepsilon}\frac{\eta\hat{z}}{\lambda\rho_p R^2}.$$

Commonly avalanches begin on fairly steep slopes for which $\mu_f \ll \varepsilon$. Note that the speed is independent of the acceleration of gravity, even though gravity is an essential factor in avalanche dynamics and energetics. The speed of an avalanche is quite sensitive to the size of the particles. Since this parameter is quite poorly known, this formula is not very useful in predicting the avalanche speed.

24.3 Cumulonimbus Clouds

Cumulonimbus clouds are the "workhorse" of atmospheric motions, providing the motive force for nearly all tropical and mid-latitude motions – both large scale and small. These clouds are thermodynamic engines, converting heat to work.[10] In addition, the rain produced by these clouds is essential for all terrestrial life.

Although cumulonimbus clouds are time dependent with a definite life cycle, it is instructive to develop a simplified model of a steady-state cloud.[11] In this simple model a cloud is treated as a turbulent jet, directed upward, driven by the buoyancy produced as water vapor condenses, releasing its latent heat to the air. The turbulence of the jet entrains ambient air so that the vertical structure depends on the properties of the air within the jet and of the air being entrained. In this simple model the air within the cloud is assumed to be well mixed with uniform properties at any given level, while in a real cloud there will be a gradient of properties depending on distance to the lateral edge of the jet.

The goal of the model is to predict the magnitude and structure of the cloud as a function of height or equivalently pressure, p. Its magnitude is measured by the upward mass flow $\dot{M}$, while its structure depends on the temperature T and mass fraction of condensed water, ξ_c. The structure of the jet depends on the properties of the entrained air (denoted by a subscript r), specifically the temperature T_r and specific humidity (that is, vapor mass fraction; see Appendix E.9.3), ξ_r. The pressure within the cloud is the same as that outside: $p = p_r(z)$, while the mass fraction of vapor in the cloud is equal to the saturation value, ξ^*, as given in Appendix E.9.4.

To keep the analysis simple, let's follow Stommel (1947) and develop the evolution equations for the cloud as a vertical jet, based on conservation of energy and water, using

[9] From § 6.4.2.
[10] See Appendix E.11.
[11] Typically the life cycle of a cumulonimbus cloud is several hours, during which time the dynamics of the cloud is not strongly affected by rotation of the Earth.

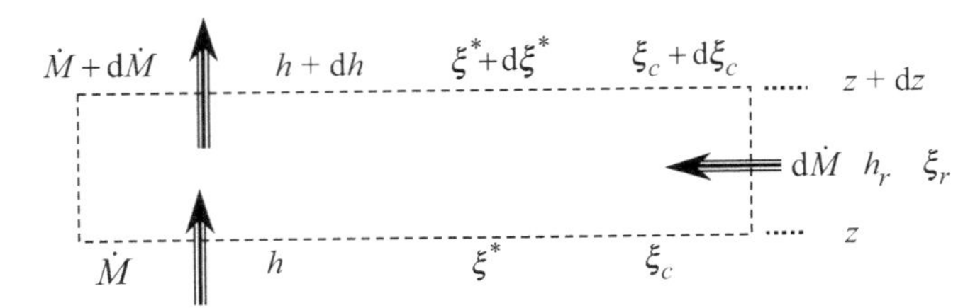

Figure 24.2 Illustration of a control volume for the balance within a cumulonimbus cloud, modeled as a vertical jet. $\dot{M}$ is the mass flow, h and h_r are the enthalpies within and outside the cloud, respectively, ξ^* is the saturated vapor mass fraction, ξ_c is the mass fraction of condensed water and ξ_r is the mass fraction of vapor in ambient air.

the *control-volume* approach illustrated in Figure 24.2.[12] The energy flow into and out of the control volume consists of

- internal energy (including sensible and latent heats);
- compressive energy (the product of pressure p and specific volume $v = 1/\rho$); and
- gravitational potential energy (represented by the gravitational potential, $\psi = gz$, where z is height).

The sum of internal and compressive energies is called the *enthalpy* and denoted by h. In steady state, the flows of energy into and out of the control volume are in balance; the energy flow through the lower control surface is $(h+\psi)\dot{M}$ and the flow out the top surface is increased due to the addition of entrained enthalpy and potential energy, $(h_r+\psi)\dot{M}$; that is,

$$\mathrm{d}\big((h+\psi)\dot{M}\big) = (h_r+\psi)\mathrm{d}\dot{M} \qquad \text{or} \qquad \dot{M}(\mathrm{d}h+g\mathrm{d}z) = (h_r-h)\mathrm{d}\dot{M}\,.$$

The differential relation for the enthalpy of moist air is[13]

$$\mathrm{d}h = T\mathrm{d}s + (1/\rho)\mathrm{d}p = c_p\mathrm{d}T + L\mathrm{d}\xi_v\,,$$

where s is specific entropy, ξ_v is the water-vapor content ($=\xi^*$ within and $=\xi_r$ outside the cloud) and L is the latent heat of vaporization. Treating c_p and L as constants, this is readily integrated to yield $h = c_pT + L\xi_v$ and the energy balance becomes

$$\dot{M}\left(\mathrm{d}T + T_c\mathrm{d}\xi^* - \gamma_d\mathrm{d}z\right) = \left(T_r - T + T_c\xi_r - T_c\xi^*\right)\mathrm{d}\dot{M}\,,$$

where[14]

$$T_c \equiv L/c_p \approx 2500\,\mathrm{K} \qquad \text{and} \qquad \gamma_d \equiv -g/c_p \approx -9.8\times10^{-3}\,\mathrm{K\cdot m^{-1}}$$

[12] A control volume is an arbitrarily specified portion of a system, used as an aid in developing the governing equations. Control volumes are commonly used in mechanical engineering.
[13] See Table E.1 and Appendix E.5.
[14] See Appendix E.4.2.

represent the influence of latent heat relative to the specific heat and the dry adiabatic gradient, respectively; see § 5.6 and § 7.2.1.[15]

Now let's consider the water content of the moist air. The variation of saturation vapor content is controlled by[16]

$$\mathrm{d}\xi^* = \frac{T_L\xi^*}{T^2}\mathrm{d}T - \frac{\xi^*}{p}\mathrm{d}p = \frac{T_L\xi^*}{T^2}\mathrm{d}T + \frac{g\xi^*}{R_dT}\mathrm{d}z,$$

where $T_L = L/R_v \approx 5400$ K.[17]

Substituting this into the energy balance and dividing by dz, we have

$$\left(1+\frac{T_cT_L}{T^2}\xi^*\right)\frac{\mathrm{d}T}{\mathrm{d}z} = \gamma_d\left(1+\frac{T_c}{T}\xi^*\right) + \left(T_r - T + T_c\xi_r - T_c\xi^*\right)\frac{1}{\dot{M}}\frac{\mathrm{d}\dot{M}}{\mathrm{d}z}.$$

If there were no entrainment ($\mathrm{d}\dot{M} = 0$) and no phase-change effects ($L = T_L = T_c = 0$), then this equation would simplify to the dry adiabat: $\mathrm{d}T/\mathrm{d}z = \gamma_d = -g/c_p$; see § 5.6 and § 7.2.1. With the inclusion of phase-change effects (but still without entrainment), this formula gives the moist adiabat; see Appendix E.9.4.

Assuming the condensed water remains in suspension (no rain yet), the water content is conserved provided the flows of water across the boundaries are in balance:

$$\left(\xi_r - \xi^* - \xi_c\right)\mathrm{d}\dot{M} = \dot{M}\left(\mathrm{d}\xi_c + \mathrm{d}\xi^*\right).$$

This equation serves to determine ξ_c, once the other variables have been determined.

With c_p, g, L and T_L constant, T_r and ξ_r being known functions of z and ξ^* being a known function of temperature,[18] the energy balance is a single relation involving two unknown quantities: M and T. A second relation can be obtained from the momentum equation, but this is a daunting task. An alternate approach is to rely on observation and experimentation to parameterize the vertical rate of change of M; de Rooy et al. (2013) suggests

$$\frac{\mathrm{d}\dot{M}}{\mathrm{d}z} \approx \frac{0.2}{D}\dot{M},$$

where D is the diameter of the cloud. Using this, the energy balance becomes

$$\left(1+\frac{T_cT_L}{T^2}\xi^*\right)\frac{\mathrm{d}T}{\mathrm{d}z} = -\frac{g}{c_p}\left(1+\frac{T_L}{T}\xi^*\right) + \frac{0.2}{D}\left(T_r - T + T_c\xi_r - T_c\xi^*\right).$$

The vertical evolution of the cloud depends on the temperature and humidity of the entrained air. If that air has the same properties as that within the cloud (that is, $T_r = T$

[15] $L \approx 2.5\times10^6$ J·Kg^{-1}, $c_p \approx 1005$ J·Kg^{-1}·K^{-1} and $g \approx 9.8$ m·s^{-2}.

[16] See Appendix E.9.4. We have used the hydrostatic equation $\mathrm{d}p = -\rho g\mathrm{d}z$ and the ideal gas equation $p \approx \rho R_dT$ to obtain the second version of this equation. This is an approximate relation because we are assuming that the gas constant for air is equal to that for dry air.

[17] $R_v \approx 461.5$ J·Kg^{-1}·K^{-1} is the specific gas constant for water vapor. Note that $T_L = c_pT_c/R_v \approx (7/2)T_c$.

[18] See Appendix E.9.4.

and $\xi_r = \xi$) then entrainment has no effect and the thermal structure of the cloud is the ideal moist adiabat. However, commonly $T > T_r$ and $\xi > \xi_r$, so that the entrainment term is negative and the actual adiabatic lapse rate is less than the ideal moist adiabatic lapse rate.

Transition

In this chapter we have investigated several one-dimensional turbulent flows including katabatic winds, avalanches and cumulonimbus clouds. This completes our study of non-rotating flows in the atmosphere and oceans. In the next part, we will begin our investigations of flows in rotating fluids.

Part VI

Flows in Rotating Fluids

In Part IV we investigated a number of types of waves in the atmosphere and oceans that are affected by rotation. We complement those studies in the present part, by investigating a number of flows that are influenced by rotation. We begin this investigation in the following chapter by investigating the Ekman boundary layer in some detail. This is followed in Chapter 26, with investigation of a number of large-scale flows in the atmosphere, including the general circulation, thermal wind and jet stream. Similarly in Chapter 27 we investigate oceanic currents, including Sverdrup currents, boundary currents and thermohaline circulation. This part concludes with a survey of various atmospheric vortices in Chapter 28, with focus on tornadoes and hurricanes.

25

Ekman Layers

An Ekman layer is a viscous boundary layer that occurs at a boundary of a large body of rotating fluid: at the bottom of the oceans and atmosphere and at the top of the oceans. This layer is an agent that communicates information regarding velocity (or stress) at the boundary to the fluid outside the layer. The communication is accomplished by means of an *Ekman pumping*: a vertical flow into (or from) the boundary layer. This vertical flow is determined by integration of the continuity equation, once the horizontal velocity within the layer has been determined by solving the momentum equation.

Ekman layers are encountered

- at the bottom or top of a fluid body;
- in either hemisphere; and
- adjoining a solid or fluid boundary,

giving us eight physical cases to consider. In the following analysis we will keep track of the position of the Ekman layer relative to the boundary and equator is categorized by the factor s which has values 1 or -1, as follows:

- $s = 1$ for bottom layers in the northern hemisphere;
- $s = -1$ for bottom layers in the southern hemisphere;
- $s = -1$ for top layers in the northern hemisphere; and
- $s = 1$ for top layers in the southern hemisphere.

There is yet another categorization of Ekman layers, dealing with the nature of the model employed. We will explore two models of Ekman layer flow and structure – having either constant or variable turbulent kinematic viscosity – and compare the results of these two models.

The equations governing the Ekman layer are presented and solved in § 25.1 and the horizontal Ekman transport is investigated in § 25.2 for both models. Then useful relations between the transport and basal velocity gradient are developed in § 25.3 and Ekman pumping is investigated in § 25.4. Finally, the Ekman-layer theory is applied to the atmosphere and oceans in § 25.5.

25.1 Formulation and Solution

The Ekman-layer equations are the horizontal momentum equation on an f-plane,[1]

$$\frac{\partial}{\partial z}\left(\nu\frac{\partial \mathbf{v}}{\partial z}\right) = s|f|\,\mathbf{1}_z \times \mathbf{v},$$

and the incompressible continuity equation,

$$\nabla_H \bullet \mathbf{v} + \frac{\partial w}{\partial z} = 0,$$

where $\mathbf{v}$ is the horizontal velocity in the Ekman layer,[2] w is the normal component of velocity, z is the normal coordinate with $z = 0$ at the base of the boundary layer and z increasing into the fluid[3] and $\mathbf{1}_z$ is the unit vector pointing in the direction of increasing z, $|f|$ is the magnitude of the Coriolis parameter and $s = \pm 1$ as discussed in the introduction to this chapter. For either a top or bottom layer, the Coriolis force acts to deflect flow to the {right, left} in the {northern, southern} hemisphere.

These equations are to be solved in the domain $0 < z < \infty$. The boundary conditions to be applied to this differential equation depend whether the boundary is rigid or fluid and whether we are viewing the layer as a floating or fixed observer, giving four cases in all.
(1) For a floating observer and rigid boundary, the conditions are

$$\text{at} \quad z=0: \quad \mathbf{v} = -\mathbf{v}_0 \qquad \text{and as} \quad z \to \infty: \quad \mathbf{v} \to \mathbf{0}.$$

(2) For a floating observer and fluid boundary, the conditions are

$$\text{at} \quad z=0: \quad \nu\partial\mathbf{v}/\partial z = \tau/\rho \qquad \text{and as} \quad z \to \infty: \quad \mathbf{v} \to \mathbf{0}.$$

where τ is the stress applied to the fluid.
(3) For a fixed observer and rigid boundary, the conditions are

$$\text{at} \quad z=0: \quad \mathbf{v} = \mathbf{0} \qquad \text{and as} \quad z \to \infty: \quad \mathbf{v} \to \mathbf{v}_0.$$

It is not advisable to consider the Ekman layer on a fluid boundary from a point of view fixed to that boundary because the velocity of this boundary (and the observer) is not specified; we will ignore the fourth possibility. Note that $\mathbf{v}_0$ is the horizontal velocity of the fluid outside the Ekman layer, as viewed by an observer fixed to the boundary.

The Coriolis and viscous forces in the Navier–Stokes equation are in balance within this layer provided the horizontal velocity has significant variation on the (vertical) *Ekman length*: $L_E = \sqrt{\nu/|f|}$. This scale is quite small in the atmosphere and oceans. With $|f| \approx 10^{-4}$ s^{-1} and laminar kinematic viscosities of $\nu \approx 2 \times 10^{-5}$ m^2·s^{-1} for air and

[1] See § 8.1.5. Since the Ekman layer is a local balance, not involving lateral derivatives, Ekman-layer solutions we will be developing are valid even if the Coriolis parameter, f, varies with horizontal position.
[2] Relative to an observer floating in the fluid outside the layer. Subsequently we will put subscript c or v on the velocity vector and its components depending whether the viscosity is constant or variable.
[3] Previously z has been the upward coordinate. Here z may point either upward or downward.

$\nu \approx 10^{-6}$ m^2·s^{-1} for water, the Ekman lengths are less than one meter. These thicknesses are unrealistically small; the scale of boundary roughness can be much greater than this. Boundary roughness induces turbulent motions, so we need to consider turbulent flow.

To get a more realistic estimate of the Ekman-layer thickness, we need to introduce a turbulent viscosity. In the following, we will consider two approaches to modeling turbulent flow in the Ekman layer. The traditional approach is to treat the viscosity as constant and write

$$\nu_c = L_c^2 |f|/2,$$

where L_c is a "realistic" estimate of the turbulent Ekman layer thickness. As we shall see, a superior approach is to introduce a variable viscosity:[4]

$$\nu_v = \hat{\varepsilon} U(z + \hat{z}),$$

where $\hat{z}$ is the boundary roughness scale,[5] U is the speed difference across the layer and $\hat{\varepsilon}$ is a small (≈ 0.02) dimensionless number. Note that we are using subscripts c and v to denote the constant- and variable-viscosity models, respectively.

In the following sections we will solve for the Ekman-layer structure and auxiliary quantities such as Ekman transport and Ekman pumping using these two models in sequence, and compare and contrast the results.

25.1.1 Constant-Viscosity Solution

With $\nu = \nu_c = L_c^2 |f|/2$, the vector form of the horizontal momentum equation is

$$L_c^2 \frac{\partial^2 \mathbf{v}_c}{\partial z^2} = 2s\, \mathbf{1}_z \times \mathbf{v}_c,$$

where $\mathbf{v}_c$ is the velocity within the constant-viscosity Ekman layer. This equation is to be solved in the interval $0 < z < \infty$, subject to one condition a $z = 0$ and one at $z = \infty$. We are viewing the boundary layer from a point fixed in the fluid outside the boundary layer so that the asymptotic condition is $\mathbf{v}_c \to \mathbf{0}$ as $z \to \infty$. The form of this second boundary condition depends whether the boundary is solid or fluid. We will consider these two possibilities subsequently. However, the structure of the solution is independent of that condition.

The solution satisfying conditions (1) is

$$\mathbf{v}_c = -C_c \mathbf{v}_0 + s S_c \mathbf{1}_z \times \mathbf{v}_0, \qquad \text{where}$$

$$C_c = e^{-z/L_c} \cos(z/L_c), \qquad S_c = e^{-z/L_c} \sin(z/L_c)$$

[4] Developed in § 23.5.2. This generalizes the model introduced by Ellison (1956) in the context of the atmospheric boundary layer and investigated by Madsen (1976) for oceanographic boundary layers. Use of this parameterization assumes that the lifetime of eddies is much shorter than one day; see § 23.5.2.

[5] Introduced in § 23.2. The effect of laminar viscosity may be included by changing $\hat{z}$ to $\hat{z} + 2\nu/(|f| L_v)$ everywhere.

and $\mathbf{v}_0$ is an arbitrary horizontal vector representing the velocity of the far fluid, relative to the boundary. This solution determines the flow within the Ekman layer as viewed by an observer floating in the fluid far from the boundary. Flow within the layer decays exponentially with distance from the boundary and is in the form of a {left, right} handed spiral (called the *Ekman spiral*) in the {northern, southern} hemisphere; that is, flow veers to the {right, left} in the {northern, southern} hemisphere with increasing distance from the boundary, as noted in § 8.1. The horizontal Ekman-layer flow speed normal to $\mathbf{v}_0$, represented by S_c, is a maximum of $e^{-\pi/4}/\sqrt{2} \approx 0.3224$ at $z/L_c = \pi/4$.

It is interesting to note that the solution for the laminar Ekman layer is mathematically identical to that for the laminar katabatic winds presented in § 22.2.5, with the parallel component of Ekman velocity equivalent to the katabatic temperature and the normal component of Ekman velocity equivalent to the downslope katabatic flow. In fact, profiles of $u/U = S_c$ and $v/U = C_c$ are plotted in panel (a) of Figure 24.1. We shall see that this analogy carries over to the turbulent cases.

It is well known that the constant-viscosity model is deficient in a number of aspects. In particular, it does not reproduce the logarithmic velocity profile observed in the atmospheric boundary layer. One way to improve this model is to patch an outer Ekman spiral to an inner logarithmic profile (see for example, §5.3 of Holton, 2004). However, this patch is quite unnecessary, as the variable-viscosity problem can be solved analytically in terms of known functions. This is demonstrated in the following section.

25.1.2 Variable-Viscosity Solution

Setting $\nu = \hat{\varepsilon} U(z+\hat{z})$, where $U = \|\mathbf{v}_0\|$ and $\hat{\varepsilon} \approx 0.02$, the vector form of the horizontal momentum equation becomes

$$L_v \frac{\partial}{\partial z}\left((z+\hat{z})\frac{\partial \mathbf{v}_v}{\partial z}\right) = 2s\,\mathbf{1}_z \times \mathbf{v}_v\,, \qquad \text{where} \qquad L_v = 2\hat{\varepsilon}\frac{U}{|f|}$$

is the turbulent Ekman-layer thickness.[6] Note that this thickness varies with U, in contrast to the thickness of the constant-viscosity (i.e., laminar) Ekman layer, whose thickness is independent of dynamic conditions.

The turbulent Ekman-layer scales for the atmosphere and oceans are readily estimated. These scales differ because the atmosphere and oceans have differing characteristic speeds. For the atmosphere with $|f| \approx 10^{-4}\ \mathrm{s}^{-1}$ and $U \approx 15\ \mathrm{m{\cdot}s}^{-1}$, $L_v \approx 2800$ m, while for the oceans with $|f| \approx 10^{-4}\ \mathrm{s}^{-1}$ and $U \approx 0.5\ \mathrm{m{\cdot}s}^{-1}$, $L_v \approx 100$ m. These are much more plausible values than obtained using the molecular viscosity. The neglect of molecular viscosity is justified provided $\nu \ll \hat{\varepsilon} U\hat{z} = |f| L_v \hat{z}/2$.

[6] The factor 2 is introduced to simplify subsequent equations.

Table 25.1. *Estimates of surface roughness $\hat{z}$ and the magnitude of the parameter $\hat{\xi}$ for Ekman layers in various locations.*

Location	$\hat{z}$ in m	$\hat{\xi}$
Atmosphere over land	1 to 10	0.053 to 0.17
Atmosphere over water	2 to 5	0.076 to 0.12
Top and bottom of ocean	2 to 5	0.4 to 0.63
Under fresh sea ice	0.2 to 0.5	0.13 to 0.2

The analysis is facilitated by employing horizontal unit vectors $\mathbf{1}_0$ pointing parallel to $\mathbf{v}_0$ and $\mathbf{1}_n = \mathbf{1}_0 \times \mathbf{1}_z$. Writing

$$\mathbf{v}_v = u_v\,\mathbf{1}_n + v_v\,\mathbf{1}_0 \qquad \text{and} \qquad \mathbf{v}_0 = U\mathbf{1}_0\,,$$

the components of the momentum equation are

$$L_v \frac{\partial}{\partial z}\left((z+\hat{z})\frac{\partial u_v}{\partial z}\right) = -2sv_v \quad \text{and} \quad L_v \frac{\mathrm{d}}{\mathrm{d}z}\left((z+\hat{z})\frac{\partial v_v}{\partial z}\right) = 2su_v\,.$$

As with the equations governing turbulent katabatic winds (see § 24.1), these components can be combined into a single complex equation:

$$L_v \frac{\partial}{\partial z}\left((z+\hat{z})\frac{\partial \mathrm{W}_v}{\partial z}\right) = 2\mathrm{i}s\mathrm{W}_v\,, \qquad \text{where} \qquad \mathrm{W}_v = v_v - \mathrm{i}u_v\,.$$

The complex form of boundary condition set (1) is

$$\text{at} \quad z=0: \quad \mathrm{W} = -U \qquad \text{and as} \qquad z \to \infty: \quad \mathrm{W} \to 0\,.$$

Note that the solution for the turbulent Ekman layer is mathematically identical to that for the turbulent katabatic winds presented in § 24.1, with the parallel component of Ekman velocity equivalent to the katabatic temperature and the normal component of Ekman velocity equivalent to the downslope katabatic flow.

We can transform the complex equation into a standard form by introducing a new independent variable:

$$\xi \equiv \sqrt{8(z+\hat{z})/L_v} \qquad \text{with} \qquad \hat{\xi} = \sqrt{8\hat{z}/L_v}\,.$$

In the following analysis, $\hat{\xi}$ is an important dimensionless parameter, representing the rough-boundary thickness, $\hat{z}$, relative to the turbulent Ekman scale, L_v. Let's take a moment to estimate the values of $\hat{\xi}$ in various locations. We already have estimated L_v to be 2800 m for the atmosphere and 100 m for the ocean. The roughness of Earth's surface ranges roughly from 1 to 10 m in various locations, resulting the range of values of $\hat{\xi}$ given in Table 25.1.

Changing the independent variable from z to ξ, complex equation and conditions become[7]

$$\frac{d^2 W_v}{d\xi^2} + \frac{1}{\xi}\frac{dW_v}{d\xi} - isW_v = 0;$$

$$\text{at} \quad \xi = \hat{\xi}: \quad W = -U \qquad \text{and as} \qquad \xi \to \infty: \quad W \to 0.$$

Now the differential equation is recognized as a modified Bessel equation with a complex coefficient.[8] The solution satisfying the boundary conditions may be expressed

- in complex form as

$$\begin{aligned} W_v &= -U\frac{K_1(\sqrt{is}\xi)}{K_1(\sqrt{is}\hat{\xi})} \\ &= -U\frac{\ker(\xi) + is\,\mathrm{kei}(\xi)}{\ker(\hat{\xi}) + is\,\mathrm{kei}(\hat{\xi})} \\ &= -U(C_v - isS_v), \end{aligned}$$

- in component form as

$$u_v = -UsS_v \qquad \text{and} \qquad v_v = -UC_v$$

- or in vector form as

$$\mathbf{v}_v = -C_v\mathbf{v}_0 + sS_v\mathbf{1}_z \times \mathbf{v}_0,$$

where K_1 is the modified Bessel function of the second kind,[9] ker and kei are Kelvin functions,[10]

$$C_v = \frac{\ker(\hat{\xi})\ker(\xi) + \mathrm{kei}(\hat{\xi})\mathrm{kei}(\xi)}{D_v(\hat{\xi})} \qquad \text{and} \qquad S_v = \frac{\mathrm{kei}(\hat{\xi})\ker(\xi) - \ker(\hat{\xi})\mathrm{kei}(\xi)}{D_v(\hat{\xi})},$$

are functions of ξ and $\hat{\xi}$ that determine the vertical structure of the turbulent Ekman layer parallel to and perpendicular to the direction of motion,

$$D_v(\xi) = \ker^2(\xi) + \mathrm{kei}^2(\xi)$$

and $\mathbf{v}_0$ is an arbitrary constant vector representing the velocity of the fluid outside the layer, relative to the boundary. The speed v of flow in the turbulent Ekman layer (as measured by an observer floating in the far fluid) varies with height according to

$$v = UD_v(\xi)/D_v(\hat{\xi}).$$

[7] With subscript v for "variable".
[8] See formula 10.25.1, p. 248 of Olver et al. (2010).
[9] See formula 10.31.1, p. 252 of Olver et al. (2010).
[10] See formula 10.61.2, p.268 of Olver et al. (2010).

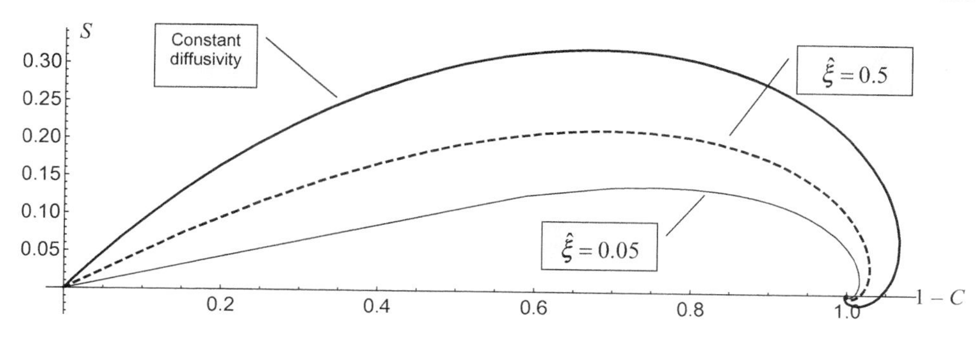

Figure 25.1 Hodographs of the constant-viscosity Ekman-layer spiral (thick solid line), together with two hodographs of the variable-viscosity solution for $\hat{\xi} = 0.5$ (thick dashed line) and 0.05 (thin solid line).

The solutions found here are identical to those for a turbulent katabatic wind in § 24.1. In fact, profiles of $u/U = S_v$ and $v/U = C_v$ versus z/L_v are plotted in panel (b) of Figure 24.1 for $\hat{\xi} = 0.5$ and 0.05. The variable viscosity is small in magnitude near the boundary, requiring a large velocity gradient to transmit the stress, while farther from the boundary, the viscosity is large, endowing the fluid with a stiffness that depresses the gradient and magnitude of the boundary-layer flow. The areas "under the curves" for variable-viscosity flow are clearly smaller than those for constant-viscosity flow; these areas are represented by the factors R_C and R_S, which will be introduced in the next subsection. However, if the Ekman-layer flow is driven by a boundary stress, rather than velocity, we will see in § 25.3 that both models predict the same Ekman transport.

In contrast to the Ekman spiral of the constant-viscosity flow, the variable-viscosity flow veers but does not exhibit a spiral, as can be seen in Figure 25.1, which displays a hodograph[11] of the constant-viscosity solution together with the variable-viscosity hodograph for two different values of $\hat{\xi}$. The latter hodographs are closer to the typical mean wind hodographs observed in the atmosphere than is the constant-viscosity hodograph; e.g., see Figure 5.5, p. 131 in Holton (2004).

For atmospheric flows it is likely that the vertical scale of surface roughness is much less than the turbulent Ekman scale: $\hat{z} \ll L_v$ (and equivalently $\hat{\xi} \ll 1$); see Table 25.1. In this case the velocity profile near the boundary becomes logarithmic:

$$\mathrm{W}_v \to U \log(\sqrt{\mathrm{i}s}\xi)/\log(\sqrt{\mathrm{i}s}\hat{\xi}),$$

like the non-rotating turbulent boundary of § 23.6 and like the observed structure of the atmospheric boundary layer near the surface.

We now are in a position to quantify the Ekman transport (in § 25.2), to relate the boundary stress to the Ekman transport (in § 25.3) and quantify the Ekman pumping (in § 25.4) for both models.

[11] A *hodograph* is a parametric plot of the variation of two velocity components versus the normal coordinate, showing how speed (radius) and direction (angle) vary.

25.2 Ekman Transport

The *Ekman transport*[12] $\mathbf{q}$ at the top of the ocean is found by integration of the horizontal velocity $\mathbf{v}$ from $z = 0$ to ∞:

$$\begin{aligned}\mathbf{q} &= \int_0^\infty \mathbf{v}\,\mathrm{d}z \\ &= \int_0^\infty (-C\mathbf{v}_0 + Ss\,\mathbf{1}_z \times \mathbf{v}_0)\,\mathrm{d}z \\ &= -q_C\mathbf{1}_0 + q_S s\mathbf{1}_z \times \mathbf{1}_0\,,\end{aligned}$$

where

$$q_C = U\int_0^\infty C\,\mathrm{d}z \qquad \text{and} \qquad q_S = U\int_0^\infty S\,\mathrm{d}z$$

are the transports parallel and normal to the direction of boundary motion, as seen by the floating observer. In the following, we will append subscripts to c and v on $\mathbf{q}$, q_C, q_S, C and S to denote constant or variable viscosity respectively.

25.2.1 Constant-Viscosity Ekman Transport

With constant viscosity

$$\int_0^\infty C_c\,\mathrm{d}z = \int_0^\infty S_c\,\mathrm{d}z = \frac{L_c}{2}\,,$$

and the transport may be expressed in vector form as

$$\mathbf{q}_c = \frac{L_c}{2}(-\mathbf{v}_0 + s\,\mathbf{1}_z \times \mathbf{v}_0) = q_c\mathbf{1}_c\,, \qquad \text{where}$$

$$q_c = L_c U/\sqrt{2} \qquad \text{and} \qquad \mathbf{1}_c = (\mathbf{1}_0 - s\,\mathbf{1}_z \times \mathbf{1}_0)/\sqrt{2}$$

are the magnitude and direction of the constant-viscosity transport. The Ekman transports in the direction of and normal to the boundary velocity have equal magnitudes.

25.2.2 Variable-Viscosity Ekman Transport

To calculate the Ekman transport with variable viscosity we need to transform the transport integrals from z to ξ:

$$\int_0^\infty C_v\,\mathrm{d}z = \frac{L_v}{4}\int_{\hat{\xi}}^\infty C_v\,\xi\,\mathrm{d}\xi = \frac{L_v}{2}R_C$$

and

$$\int_0^\infty S_v\,\mathrm{d}z = \frac{L_v}{4}\int_{\hat{\xi}}^\infty S_v\,\xi\,\mathrm{d}\xi = \frac{L_v}{2}R_S\,,$$

12 This is a specific volume flow: volume flow per unit of lateral distance.

where the factors R_C and R_S are given by

$$R_C \equiv \frac{1}{2}\int_{\hat{\xi}}^{\infty} C_v\,\xi\,\mathrm{d}\xi$$
$$= \frac{\hat{\xi}}{2\sqrt{2}D_v(\hat{\xi})}\left[\ker(\hat{\xi})\left(\ker_1(\hat{\xi})-\mathrm{kei}_1(\hat{\xi})\right)+\mathrm{kei}(\hat{\xi})\left(\ker_1(\hat{\xi})+\mathrm{kei}_1(\hat{\xi})\right)\right].$$

and

$$R_S \equiv \frac{1}{2}\int_{\hat{\xi}}^{\infty} S_v\,\xi\,\mathrm{d}\xi$$
$$= \frac{\hat{\xi}}{2\sqrt{2}D_v(\hat{\xi})}\left[\mathrm{kei}(\hat{\xi})\left(\ker_1(\hat{\xi})-\mathrm{kei}_1(\hat{\xi})\right)\ker(\hat{\xi})\left(\ker_1(\hat{\xi})+\mathrm{kei}_1(\hat{\xi})\right)\right]$$

The factors R_C and R_S quantify the relative magnitudes of transport in the direction of and normal to the boundary velocity, respectively. These are functions of the single variable $\hat{\xi}$ and are plotted in part (a) of Figure 25.2. The Ekman-layer models with constant and variable viscosity would give the same result if $R_C = R_S = 1$ for both models. But this is not possible; as is readily seen in part (a) of Figure 25.2, $R_C < R_S$ for all values of $\hat{\xi}$.

In the following, it will be convenient to write

$$R_C = R_T \sin\theta_T \qquad \text{and} \qquad R_S = R_T \cos\theta_T\,,$$

where

$$R_T \equiv \sqrt{R_C^2+R_S^2} = \frac{\hat{\xi}}{2}\sqrt{\frac{\ker_1^2(\hat{\xi})+\mathrm{kei}_1^2(\hat{\xi})}{\ker^2(\hat{\xi})+\mathrm{kei}^2(\hat{\xi})}}$$

and

$$\tan\theta_T \equiv \frac{R_C}{R_S}$$
$$= \frac{\mathrm{kei}(\hat{\xi})\left(\ker_1(\hat{\xi})+\mathrm{kei}_1(\hat{\xi})\right)+\ker(\hat{\xi})\left(\ker_1(\hat{\xi})-\mathrm{kei}_1(\hat{\xi})\right)}{\mathrm{kei}(\hat{\xi})\left(\ker_1(\hat{\xi})-\mathrm{kei}_1(\hat{\xi})\right)-\ker(\hat{\xi})\left(\ker_1(\hat{\xi})+\mathrm{kei}_1(\hat{\xi})\right)},$$

where R_T represents the total Ekman transport and θ_T is the angle between the transport vector, $\mathbf{q}$, and the forcing velocity vector, $\mathbf{v}_0$. The functions R_T and θ_T are plotted in part (b) of Figure 25.2. (The angle θ_T is re-plotted in degrees, rather than radians, in Figure 25.4.)

Now the transports may be expressed in component form as

$$q_{Cv} = \tfrac{1}{2}L_v U R_T \cos\theta_T \qquad \text{and} \qquad q_{Sv} = \tfrac{1}{2}L_v U R_T \sin\theta_T\,,$$

or in vector form as

$$\mathbf{q}_v = \tfrac{1}{2}R_T L_v\left(-\cos\theta_T \mathbf{v}_0 + \sin\theta_T s\,\mathbf{1}_z \times \mathbf{v}_0\right) = q_v \mathbf{1}_v\,,$$

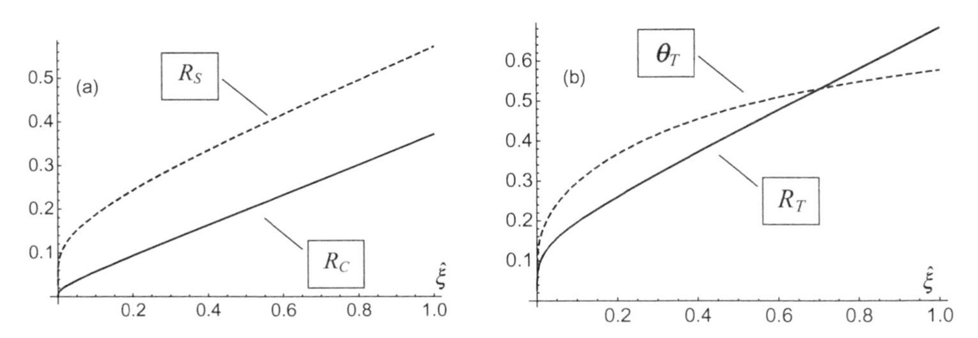

Figure 25.2 (a) Plots of the transport ratios R_C (solid line) and R_S (dashed line) versus the dimensionless roughness factor $\hat{\xi} = \sqrt{8\hat{z}/L_v}$. (b) Plots of the ratio magnitude R_T (solid line) and angle θ_T (dashed line) versus $\hat{\xi}$. These solutions behave logarithmically for small $\hat{\xi}$.

where

$$q_v = \tfrac{1}{2} R_T L_v U \qquad \text{and} \qquad \mathbf{1}_v = \cos\theta_T \mathbf{1}_0 - \sin\theta_T s\, \mathbf{1}_z \times \mathbf{1}_0$$

are the magnitude and direction of the Ekman transport in the variable-viscosity model. This result is identical to the constant-viscosity transport if $R_T = \sqrt{2}$. Otherwise, the magnitude and orientation of the transport relative to the boundary velocity differ from those predicted by the constant-viscosity model.

25.3 Relations Between Transport and Stress

An interesting feature of the Ekman layer is that the horizontal transport within the layer is directly related to the stress at the base. This expression is developed in the following two subsections for the constant- and variable-viscosity models.

25.3.1 Constant-Viscosity Transport and Stress

The integral of the constant-viscosity solution developed in § 25.1.1 from $z = 0$ to $z = \infty$ may be expressed in component form as

$$-\frac{dC_c}{dz}\bigg|_{z=0} = \frac{dS_c}{dz}\bigg|_{z=0} = \frac{1}{L_c}$$

and in vector form as

$$\frac{d\mathbf{v}_c}{dz}\bigg|_{z=0} = -\frac{2}{L_c^2} s\, \mathbf{1}_z \times \mathbf{q}_c = \frac{1}{L_c} \mathbf{1}_z \times (s\mathbf{v}_0 - \mathbf{1}_z \times \mathbf{v}_0)\,.$$

The corresponding basal shear stress (acting on the fluid) is

$$\boldsymbol{\tau}_c = -\rho \nu_c \frac{d\mathbf{v}_c}{dz}\bigg|_{z=0} = \frac{2\rho\nu_c}{L_c^2} s\, \mathbf{1}_z \times \mathbf{q}_c = \tau_c s\, \mathbf{1}_z \times \mathbf{1}_c\,,$$

where $\mathbf{1}_c = (\mathbf{1}_0 - s\,\mathbf{1}_z \times \mathbf{1}_0)/\sqrt{2}$ is the orientation of the Ekman transport and $\tau_c = 2\rho \nu_c q_c / L_c^2$ is the magnitude of the basal stress. Using $\nu_c = L_c^2 |f|/2$, this simplifies to

$$\tau_c = \rho |f| q_c .$$

The expression for $\boldsymbol{\tau}_c$ is readily inverted to give

$$\mathbf{v}_0 = \frac{U}{\sqrt{2}\tau_c}(-\boldsymbol{\tau}_c + s\mathbf{1} \times \boldsymbol{\tau}) .$$

In summary, the constant-viscosity model predicts that the drag exerted on the fluid, $\boldsymbol{\tau}_c$, may be expressed in terms of the Ekman transport, $\mathbf{q}_c$, and far-fluid velocity, $\mathbf{v}_0$, as

$$\begin{aligned} \boldsymbol{\tau}_c &= \rho |f| s\,\mathbf{1}_z \times \mathbf{q}_c \\ &= -\rho \sqrt{\nu_c |f|/2}\,(\mathbf{v}_0 + s\,\mathbf{1}_z \times \mathbf{v}_0) . \end{aligned}$$

Conversely, the Ekman transport may be expressed in terms of the stress acting on the fluid:

$$\mathbf{q}_c = -\frac{s}{|f|\rho}\,\mathbf{1}_z \times \boldsymbol{\tau}_c .$$

Note that the basal shear stress is normal to the Ekman transport.

25.3.2 Variable-Viscosity Transport and Stress

The complex equation for the variable-viscosity model may be written as

$$\frac{\mathrm{d}}{\mathrm{d}\xi}\left(\xi \frac{\mathrm{dW}_v}{\mathrm{d}\xi}\right) - \mathrm{i}s\xi \mathrm{W}_v = 0 .$$

Integrating this from $\hat{\xi}$ to ∞, we have that

$$\hat{\xi} \frac{\mathrm{dW}_v}{\mathrm{d}\xi}\bigg|_{\xi=\hat{\xi}} = -\mathrm{i}s \int_{\hat{\xi}}^{\infty} \mathrm{W}_v\,\xi\,\mathrm{d}\xi = -\mathrm{i}s \frac{4}{L_v} \mathrm{q}_v .$$

This may be expressed in component form as

$$\hat{\xi} \frac{\mathrm{d}C_v}{\mathrm{d}\xi}\bigg|_{\xi=\hat{\xi}} = -2R_T \sin\theta_T \qquad \text{and} \qquad \hat{\xi} \frac{\mathrm{d}S_v}{\mathrm{d}\xi}\bigg|_{\xi=\hat{\xi}} = 2R_T \cos\theta_T$$

or in vector form as

$$\begin{aligned} \hat{\xi} \frac{\mathrm{d}\mathbf{v}_v}{\mathrm{d}\xi}\bigg|_{\xi=\hat{\xi}} &= -\frac{4}{L_v} s\,\mathbf{1}_z \times \mathbf{q}_v \\ &= 2R_T\,\mathbf{1}_z \times (s\cos\theta_T \mathbf{v}_0 - \sin\theta_T\,\mathbf{1}_z \times \mathbf{v}_0) . \end{aligned}$$

The corresponding basal shear stress is

$$\begin{aligned}
\boldsymbol{\tau}_v &= -\rho \nu_T \left.\frac{\mathrm{d}\mathbf{v}_v}{\mathrm{d}z}\right|_{z=0} = -\rho(\hat{\varepsilon}U\hat{z})\frac{4}{L_v\hat{\xi}}\left.\frac{\mathrm{d}\mathbf{v}_E}{\mathrm{d}\xi}\right|_{\xi=\hat{\xi}} \\
&= \rho(\hat{\varepsilon}U\hat{z})\frac{8R_T}{L_v\hat{\xi}^2}\,\mathbf{1}_z \times (-s\cos\theta_T \mathbf{v}_0 + \sin\theta_T\,\mathbf{1}_z \times \mathbf{v}_0) \\
&= \tau_v s\,\mathbf{1}_z \times \mathbf{1}_v\,,
\end{aligned}$$

where $\mathbf{1}_v = \cos\theta_T \mathbf{1}_0 - \sin\theta_T s\,\mathbf{1}_z \times \mathbf{1}_0$ is the direction of the Ekman transport and $\tau_v = \hat{\varepsilon}R_T\rho U^2$ is the magnitude of the basal shear stress. The stress may be expressed as

$$\tau_v = C_{Dv}\rho U^2\,, \qquad \text{where} \qquad C_{Dv} = \hat{\varepsilon}R_T\,.$$

The components of basal stress now differ in magnitude. The transport may be expressed variously as

$$q_v = \frac{R_T L_v U}{2} = \frac{C_{Dv}U^2}{|f|} = \frac{\tau_v}{\rho|f|}\,.$$

The expression for $\boldsymbol{\tau}_v$ is readily inverted to give

$$\mathbf{v}_0 = \frac{U}{\tau_v}(-\sin\theta_T \boldsymbol{\tau}_v + s\cos\theta_T \mathbf{1} \times \boldsymbol{\tau}_v)\,.$$

In summary, the variable-viscosity model predicts that the drag exerted on the fluid, $\boldsymbol{\tau}_v$, may be expressed in terms of the Ekman transport, $\mathbf{q}_v$, and far-fluid velocity, $\mathbf{v}_0$, as

$$\begin{aligned}
\boldsymbol{\tau}_v &= \rho|f|s\,\mathbf{1}_z \times \mathbf{q}_v \\
&= -C_{Dv}\rho U(\sin\theta_T \mathbf{v}_0 + \cos\theta_T s\,\mathbf{1}_z \times \mathbf{v}_0)\,.
\end{aligned}$$

Conversely, the Ekman transport may be expressed in terms of the stress:

$$\mathbf{q}_v = -\frac{s}{|f|\rho}\,\mathbf{1}_z \times \boldsymbol{\tau}_v\,.$$

As with the constant-viscosity model, the transport is normal to the stress. Note that the relation between stress and transport is the same for both models, but that between the stress and boundary velocity differs for the two models.

25.4 Ekman Pumping

So far in this chapter we have focused on the flow within the Ekman layer. Now in this section we will see how this flow, here denoted by a subscript E, is related and connected to the remainder of the body of fluid, which we will call the "interior" and denote by a subscript I. The horizontal velocity (as seen by an observer fixed to the boundary) is the sum of the boundary-layer and interior parts:

$$\mathbf{v}(\mathbf{x}_H, z, t) = \mathbf{v}_I(\mathbf{x}_H, z, t) + \mathbf{v}_E(\mathbf{x}_H, z, t)\,,$$

where $\mathbf{x}_H$ is the horizontal position vector and t is time. These two parts are distinguished by the rapid variation of $\mathbf{v}_E$ with z, compared with that of $\mathbf{v}_I$, $\|\partial\mathbf{v}_I/\partial z\| \ll \|\partial\mathbf{v}_E/\partial z\|$, and by the fact that variables with subscript E decay exponentially with distance from the boundary. According to the fixed observer, the flow within the Ekman layer is given by $\mathbf{v}_E = -C_v \mathbf{v}_{I0} + S_v s\, \mathbf{1}_z \times \mathbf{v}_{I0}$, where $\mathbf{v}_{I0} = \mathbf{v}_0$ is the interior velocity evaluated at $z = 0$.

The continuity equation is

$$\nabla_H \bullet (\mathbf{v}_I + \mathbf{v}_E) + \frac{\partial (w_I + w_E)}{\partial z} = 0 .$$

The bottom boundary is impenetrable, requiring $w_I(0) = -w_E(0) \equiv w_\infty$, where w_∞ is called the *Ekman pumping*.

The continuity equation is readily separated into two, with the boundary-layer part satisfying

$$\begin{aligned} \frac{\partial w_E}{\partial z} &= -\nabla_H \bullet \mathbf{v}_E \\ &= C_v (\nabla_H \bullet \mathbf{v}_{I0}) + S_v s \zeta_I , \end{aligned}$$

where $\zeta_I = -\nabla_H \bullet (\mathbf{1}_z \times \mathbf{v}_{I0}) = \mathbf{1}_z \bullet (\nabla_H \times \mathbf{v}_{I0})$ is the vorticity of the geostrophic flow outside the Ekman layer.[13] The integral of this equation gives

$$\begin{aligned} \int_0^\infty \frac{\partial w_E}{\partial z} \mathrm{d}z &= -w_E(0) = w_\infty \\ &= (\nabla_H \bullet \mathbf{v}_{I0}) \int_0^\infty C_v \,\mathrm{d}z + s\zeta_I \int_0^\infty S_v \,\mathrm{d}z \\ &= \tfrac{1}{2} L_v \left(R_C \nabla_H \bullet \mathbf{v}_{I0} + R_S s \zeta_I \right) . \end{aligned}$$

The solution for constant viscosity is recovered by setting $R_C = R_S = 1$; otherwise, these factors are determined by the solution for variable viscosity as illustrated in Figure 25.2. We see that Ekman pumping is driven by both the horizontal divergence and vorticity of the interior flow. Commonly this flow is geostrophic, with $\nabla_H \bullet \mathbf{v}_I = 0$, and the Ekman pumping is driven by the vorticity of the interior flow:

$$w_\infty = \tfrac{1}{2} L_v R_S s \zeta_I .$$

Barring a steady state balance,[14] outside the Ekman layer (that is, in the interior of the fluid body) the magnitude of this vertical velocity decreases linearly with distance from the Ekman layer and conservation of mass is accommodated by a horizontal convergence or divergence in the interior that acts to drive the vorticity to zero, in a process called *spin up* if $s\zeta_I < 0$ and *spin down* if $s\zeta_I > 0$.

25.4.1 Discussion

Two deficiencies of the constant-viscosity model are that it produces a value for the drag coefficient that is unrealistically large and predicts that floating objects drift at a 45

[13] More specifically, it is the vertical component of the vorticity vector. See § 8.2.1.
[14] To be considered in § 27.2.

degree angle from the wind direction. However, the variable-viscosity model is capable of producing realistic values of drag and realistic drift angles.

If the drag coefficient C_D is known (e.g., from observations), then we may determine $\hat{\xi}$ from Figure 25.2 using[15] $R_T = C_D/\hat{\varepsilon}$ if $\hat{\xi}$ is reasonably large (greater than 0.1, say) or by using the asymptotic formula

$$\hat{\xi} \approx 2\exp\left(-\sqrt{\frac{\hat{\varepsilon}^2}{4C_D^2} - \frac{\pi^2}{16}} - 0.577\right)$$

if $\hat{\xi} < 0.1$.

25.5 Ekman Layers on Specific Boundaries

In this section, we investigate the form and consequences of the Ekman-layer theory developed in the preceding sections to specific boundaries, including

- the atmosphere over land;
- the atmosphere over the oceans;
- the top of the oceans;
- the bottom of the oceans; and
- the thermocline.

25.5.1 Atmospheric Ekman Layer over Land

Resistance to the wind on the land surface, represented by the roughness length, $\hat{z}$, or equivalently by $\hat{\xi}$, varies widely depending on the terrain and vegetation (primarily trees). Roughness is greatest in cities and least on grassy plains. One difficulty in applying the Ekman-layer theory to the land surface is that it assumes the boundary roughness to be constant, independent of flow conditions. However, trees bend and deform in the wind, so that their resistance to flow is poorly represented by the quadratic $\tau - U$ relation (see Cullen, 2005). In this case roughness is a dynamic – rather than static – parameter.

Dynamic roughness opens a can of worms, and forces us to think in terms of dimensional analysis.[16] Dimensionally, the factors τ, ρ and U combine to form a dimensionless parameter in only one way: $\tau/(\rho U^2)$, which we have set equal to a dimensionless constant, called C_D. A mathematical formula with any other structure requires the introduction of another physical quantity that we have not considered previously. One possible quantity is the elastic strength, σ, of trees, which controls the variation of canopy area with wind speed. Since σ has the same dimensions as τ, it is readily apparent that a possible second dimensionless parameter is $\Pi = \rho U^2/\sigma$. with C_D

[15] From § 25.3.2.
[16] See Appendix B.1.

being a function of Π. Note that $\Pi = 0$ and roughness becomes static if the trees are infinitely strong. Now what is the form of the functional relation $C_D(\Pi)$? It needs to be a finite value when $\Pi = 0$ and to decay to a smaller finite value as $\Pi \to \infty$. These conditions are satisfied by a negative exponential, such as $C_D = C_1 + C_2 e^{-\Pi}$, where C_1 is the drag provided by rigid obstacles (rocks, buildings, etc.) and C_2 is the variable drag provided by vegetation. But we are in uncharted territory here and are getting well beyond the scope of the book. However, before moving on, there are several "take-away" messages worth mentioning. First, models often have implicit assumptions that are not immediately obvious. In the present case, this was the assumption of constant roughness, independent of dynamic conditions. The second message is that dimensional analysis is a valuable guide when venturing into unknown territory.

25.5.2 Atmospheric Ekman Layer over the Oceans

Drag on atmospheric flow over the oceans originates in the boundary roughness provided by waves. Like vegetation, this roughness is a dynamic variable, but it does not vary nearly as much as on land because the prevailing winds (described in § 26.1), and the waves they induce, are fairly uniform. We have a reasonably good estimate of the drag coefficient: $C_{D10} \approx 0.0013 \pm 0.002$, based on the mean wind speed observed at 10 m height.[17]

The mean velocity $\mathbf{U}_{10}$ of the winds (at 10 m elevation) over the oceans are relatively easily measured and can be taken as a known quantity. The stress on the ocean surface due to the winds is given by

$$\boldsymbol{\tau}_D = C_D \rho U_{10} (\cos\theta_T \mathbf{U}_{10} + s \sin\theta_T \mathbf{1}_z \times \mathbf{U}_{10}),$$

where $U_{10} = \|\mathbf{U}_{10}\|$. The stress is primarily in the direction of the prevailing wind, with a smaller response in the normal direction.

25.5.3 Ekman Layer at the Bottom of Oceans

The Ekman layer at the bottom of the oceans is similar to the atmospheric Ekman layer over land, discussed in § 25.5.1, but with several differences worth noting. The primary difference is that the typical value of $\hat{\xi} = \sqrt{8\hat{z}/L_v}$, where $\hat{z}$ is the boundary-roughness scale and $L_v = 2\hat{\varepsilon} U/|f|$ is the turbulent Ekman-layer scale, is much larger in the ocean than in the atmosphere, primarily because oceanic flow speeds are far smaller than atmospheric. Another difference is the lack of vegetation on the ocean bottom, so that the boundary roughness is rigid and the quadratic stress-speed relation is valid.

Flow within the bottom Ekman layer produces a drag stress

$$\boldsymbol{\tau}_D = C_D \rho U (-\cos\theta_T \mathbf{v}_I + s \sin\theta_T \mathbf{1}_z \times \mathbf{v}_I),$$

[17] From § 11.5.1.

where U is the magnitude of velocity outside the layer: $U = \|\mathbf{v}_I\|$.

The magnitude of C_D at the bottom of the ocean is not well constrained. From § 25.3.2 we have that $C_D = \hat{\varepsilon} R_T$, with $\hat{\xi} \approx 0.02$, but the value of $R_T(\hat{\xi})$ is poorly constrained. We see from panel (b) of Figure 25.2 that a likely range for R_T is 0.2 to 0.4, giving a range of values for C_D from 0.004 to 0.01.

25.5.4 Ekman Layer at the Top of Oceans

The Ekman layer at the top of the ocean exists in response to a surface stress, $\boldsymbol{\tau}_W$, applied by the prevailing winds. The horizontal velocity within this layer is given by

$$\mathbf{v}_E = \frac{U}{\sqrt{T_W}} \left(C_\tau \boldsymbol{\tau}_W - s S_\tau \mathbf{1}_z \times \boldsymbol{\tau}_W \right),$$

where

$$C_\tau = \frac{\cos\theta_T}{R_T} C_v - \frac{\sin\theta_T}{R_T} S_v \quad \text{and} \quad S_\tau = \frac{\sin\theta_T}{R_T} C_v + \frac{\cos\theta_T}{R_T} S_v .$$

Profiles of C_τ and S_τ, representing flow in an Ekman layer having variable viscosity are compared with those in a layer having constant viscosity in Figure 25.3 with $\hat{\xi} = 0.5$.[18] Note that $C_\tau \approx 2.08$ and $S_\tau \approx 1.09$ for $\xi = \hat{\xi} = 0.5$. The variable viscosity is small near the boundary, and a larger velocity is needed to transmit a given shear stress. Again the constant-viscosity flow exhibits an Ekman spiral, while the variable-viscosity flow veers but does not exhibit a spiral.

Ekman Transport Revisited

Inverting the expressions given at the end of § 25.3.2 for $\boldsymbol{\tau}_v$ in terms of $\mathbf{q}_v$ and $\mathbf{v}_0$ and dropping the subscripts v, we have

$$\mathbf{q} = -\frac{\mathbf{1}_z \times \boldsymbol{\tau}}{\rho f}$$

and

$$\mathbf{v}_0 = \frac{1}{C_{Dv} \rho v_0} \left(\cos\theta_T \boldsymbol{\tau} + \sin\theta_T s \, \mathbf{1}_z \times \boldsymbol{\tau} \right),$$

where we have used $s = -\mathrm{sgn}(f)$. Recalling that in the Ekman layer at the top of the ocean $\mathbf{1}_z$ points downward and $s = \{-1, 1\}$ in the {northern, southern} hemisphere, we see that the Ekman transport is to the {right, left} of the wind stress in the {northern, southern} hemisphere.

[18] In contrast to a solid boundary, where turbulent motions are induced by the mean flow impinging on the rough bottom, turbulent motions are induced at the top of the oceans by breaking waves. In this case, the roughness scale, $\hat{z}$, may be equated to the mean wave height, which typically varies from 1 to 10 m in the oceans. Note that $\hat{z} \ll L_v$.

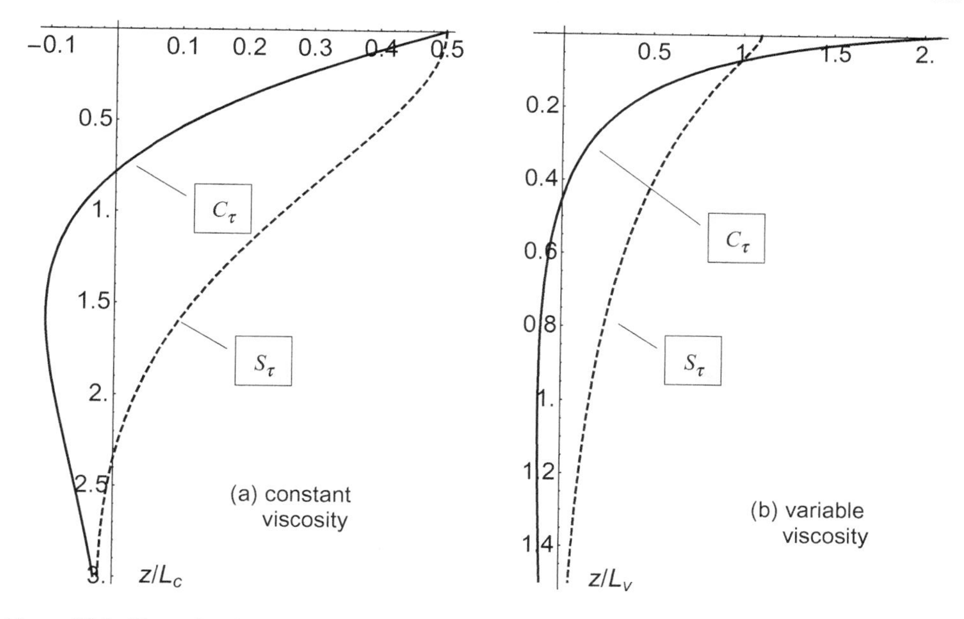

Figure 25.3 Plots of scaled speeds $C_\mathcal{T}$ and $S_\mathcal{T}$ versus dimensionless depth z/L for wind-driven motion at the top of the oceans. (a): constant viscosity (b): variable viscosity with $\hat{\xi} = 0.5$. Note that $z = 0$ is the top of the ocean, with z increasing downward. Also note the difference in vertical scale; the variable-viscosity flows decay more rapidly than the constant-viscosity flows.

We can readily estimate the magnitude of the Ekman transport; using $\rho \approx 10^3$ kg·m^{-3}, $|f| \approx 10^{-4}$ s^{-1}, $\|\boldsymbol{\tau}\| \approx 0.1$ Pa and $\|\mathbf{q}\| \approx 1$ m^2·s^{-1}. In § 27.2.2 we will compare the magnitude this transport to that of the Sverdrup transport in the mixed layer of the oceans.

The formulas above give the immediate response of the upper ocean to an applied wind stress, but this is only part of the story. The actual currents in the mixed layer are a combination of the response found above plus a geostrophic part driven by variations in the ocean surface, which act to redistribute the flow, as described in Chapter 27.[19]

25.5.5 Ekman Layers at the Thermocline

The thermocline may be modeled as a sharp horizontal interface separating a thin, turbulent, well-mixed top layer from the stably stratified bulk of the ocean beneath; see § 7.2.2. There are two Ekman layers associated with the thermocline: a turbulent layer above and a laminar layer beneath. We shall assume that the fluid at depth is quiescent, and

[19] This dual response is very similar to the particular and homogeneous solutions of a linear differential equation with a forcing term.

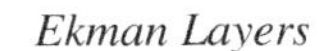

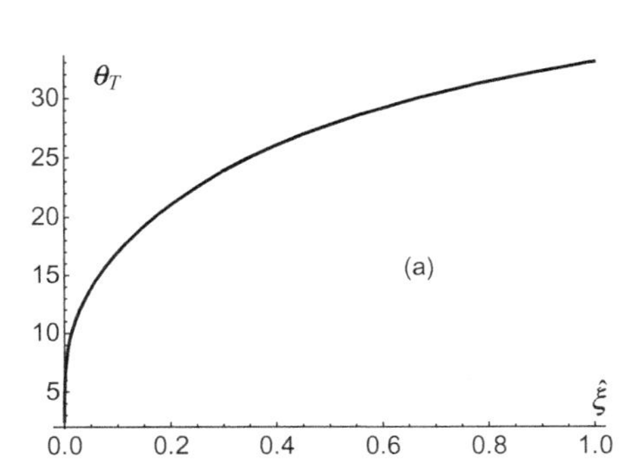

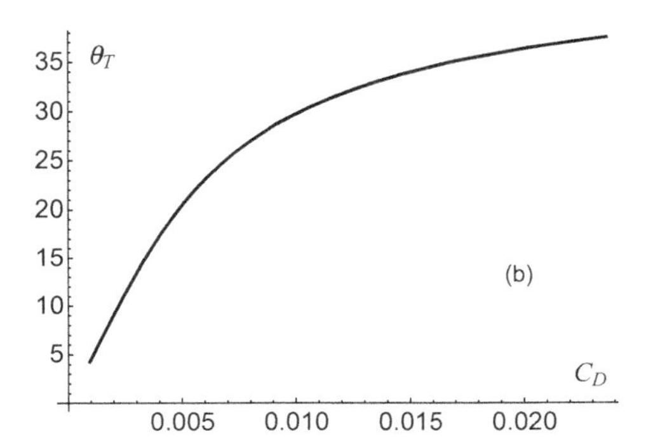

Figure 25.4 (a) A plot of the angle, θ_T, of surface flow, relative to the applied wind stress – in degrees – versus the roughness parameter $\hat{\xi}$. (b) A plot of θ_T – in degrees – versus the drag coefficient C_D. This graph may be used to determine C_D from the observed drift angle.

an observer there sees the fluid at the thermocline moving horizontally with velocity $\mathbf{v}_t$ and the fluid in the well-mixed layer moving with velocity $\mathbf{v}_{wm}$.

The Ekman layer underneath the thermocline lies within the stably stratified fluid and is given by the constant-viscosity solution presented in previous sections, with the laminar – rather than turbulent – viscosity. The Ekman layer atop the thermocline is turbulent, but we need to be careful about applying the variable-viscosity solution developed in previous sections. That solution has been developed assuming that the turbulence is generated by irregularities of the boundary and varies across the layer. However, the thermocline has no such irregularities; instead the turbulence is generated at the ocean surface and penetrates to the thermocline. The situation is similar to that of turbulent channel flow (studied in § 23.5.1), but turned upside down. At the thermocline, the viscosity of the mixed layer is roughly constant and the appropriate solution is the constant-viscosity solution with the viscosity having the turbulent value $\nu_{wm} = \kappa\hat{u}D$, where $\hat{u}$ is the roughness velocity for the ocean surface and D is the depth of the well-mixed layer.

We found in § 25.3.1 that the stress exerted on a constant-viscosity Ekman layer is $\boldsymbol{\tau}_c = \rho\sqrt{\nu_c|f|/2}\,(\mathbf{v}_0 + s\,\mathbf{1}_z \times \mathbf{v}_0)$. This formula applies to both Ekman layers, with appropriate values of ν_c. That is,

$$\tau_t = \rho\sqrt{\nu_t|f|}v_t \qquad \text{and} \qquad \tau_{wm} = \rho\sqrt{\nu_{wm}|f|}v_{wm}$$

are the magnitudes of the stress exerted across the thermocline, where v_t and v_{wm} are the magnitudes of the velocity differences across the upper and lower layers, respectively. As these stresses are equal in magnitude; it follows that

$$\sqrt{\nu_t}v_t = \sqrt{\nu_{wm}}v_{wm}\,.$$

The turbulent viscosity ν_{wm}, is far greater in magnitude than the laminar, ν_t. It follows that $v_{wm} \ll v_t$; the velocity difference across the turbulent boundary layer is relatively very small. By the same token, the stress exerted across the thermocline is also small compared with that applied at the ocean surface by the winds. This result can be understood as follows. Turbulence provides the fluid in the well-mixed layer with a certain stiffness, and it tends to move as a rigid unit, sliding on the slippery stable fluid beneath. Saying the same thing the other way around, the laminar Ekman layer beneath the thermocline is incapable of exerting a significant stress on the well-mixed layer because the laminar value of its viscosity is so small.

25.6 Comments on Ekman Layers

The transport produced by the variable-viscosity model is independent of $\hat{\xi}$ and identical to the constant-viscosity results. But this should be no surprise to us, because we have already seen in § 25.3 that the transports are directly related to the stresses, and these conditions must be satisfied by both models. So it turns out that for a wind-driven Ekman layer, the constant-viscosity model gets the transport right. But this model fails to predict the surface drift angle, the structure of the flow with depth and the magnitude of the Ekman pumping.

The constant-viscosity model predicts a drift angle of 45 degrees, which is much larger than typically observed. In the variable-viscosity model the drift angle θ_T is a function of $\hat{\xi}$; this relation is graphed in panel (a) of Figure 25.2 with θ_T in radians, and re-graphed in panel (a) of Figure 25.4 with θ_T in degrees. We previously found that $C_D = \hat{\varepsilon} R_T$, with $\hat{\varepsilon} \approx 0.02$.[20] It follows that the coefficient of drag and drift angle are not independent, but in fact are related, as shown in panel (b) of Figure 25.4.

The variable-viscosity model may be generalized to include molecular diffusion by replacing $\hat{z}$ with $\hat{z} + 2\nu/(|f|L_v)$. With this modification, the relation between L_v and $\hat{\xi}$ changes from $L_v = 8\hat{z}/\hat{\xi}^2$ to

$$L_v = \frac{4}{\hat{\xi}^2}\left(\hat{z} + \sqrt{\hat{z}^2 + \hat{\xi}^2 L_E^2}\right).$$

where $L_E = \sqrt{\nu/|f|}$ is the laminar Ekman scale. Molecular diffusion is negligible provided $\hat{\xi} L_E \ll \hat{z}$. With $\hat{\xi}$ typically small, this condition is almost always satisfied.

If the wind is blowing parallel to a coastline, the lateral Ekman transport must be accommodated by a vertical flow in the ocean below the Ekman layer. More specifically, a wind blowing toward the equator at the eastern edge of an ocean basin induces an Ekman transport away from the coast that requires a compensatory *upwelling*. This upwelling can bring to the surface nutrient-laden water that fuels a productive fishery – as happens, for example, off the coast of Peru.

The model of an Ekman layer presented in this chapter ignores a number of effects that can modify this simple solution, including density stratification, convective motions, gravity waves and time-dependent effects.[21] In particular, the analytic solutions for the

[20] In § 25.3.2.

[21] These can be determined using the *Laplace transform*, but this is beyond the scope of this book.

Ekman layer at the top of the ocean are valid only in the well-mixed layer above the thermocline. This provides us with another reason to prefer the variable-viscosity solution, since its velocities decay more rapidly with depth than do the velocities in the constant-viscosity solution.

Transition

This completes our investigation of the structure of both laminar and turbulent Ekman layers and a survey of the nature of these layers in the atmosphere and oceans. We now turn our attention to large-scale atmospheric flows that are influenced by rotation.

26

Atmospheric Flows

The predominant flow within the atmosphere is the *general circulation*, described in § 26.1. An associated flow, the *jet stream*, whose existence relies on the thermal wind (considered in § 26.2) is investigated in § 26.3. These flows are powered by the "workhorse" of atmospheric flow: the cumulonimbus rain cloud, which has been investigated in § 24.3.

26.1 General Circulation

If the Earth were not rotating, the general circulation would consist of a *Hadley cell* in each hemisphere, with rising motion near the equator, descending motion near the poles, poleward motion at high elevation and equatorward motion near the surface. However, due to the action of the Coriolis force, the general circulation consists of three cells in each hemisphere, a Hadley cell near the equator, a mid-latitude *Ferrel cell* and the aptly named *polar cell* at high latitude, as illustrated in Figure 26.1 for the northern hemisphere. The flow in the southern hemisphere is the mirror image, reflected about the equator.

Circulation of the atmosphere is driven primarily by convective motions in the tropics. These motions drive a circulation of roughly 5 to 8 $\times 10^{10}$ kg·s^{-1} in each of the Hadley cells (Liu et al., 2012) and these motions convey poleward a heat flow of roughly 4 PW.[1] The strength of circulation in these cells has been increasing in the past 150 years, possibly due to global warming. The trade winds within the Hadley cells on either side of the equator blow toward the equator and tend to concentrate convective motions in a band close to the equator called the *inter-tropical convergence zone* (ITCZ) by meteorologists and the *doldrums* by sailors. Descending motions at the boundary between the Hadley and Ferrel cells produce a band of weak surface winds called the *horse latitudes* by sailors. These descending motions inhibit convection, so that rainfall is low near the horse latitudes. Most of the world's deserts occur near this latitude, both north and south of the equator.

The moist adiabatic lapse rate is smaller in magnitude than the dry adiabatic lapse rate.[2] Consequently, strong updrafts in the ITCZ, accompanied by condensation and release of latent heat, produce a smaller adiabatic gradient, resulting in warm dry air at high

[1] 1 PW = 1 petawatt = 10^{15} W.
[2] See Appendix E.9.4.

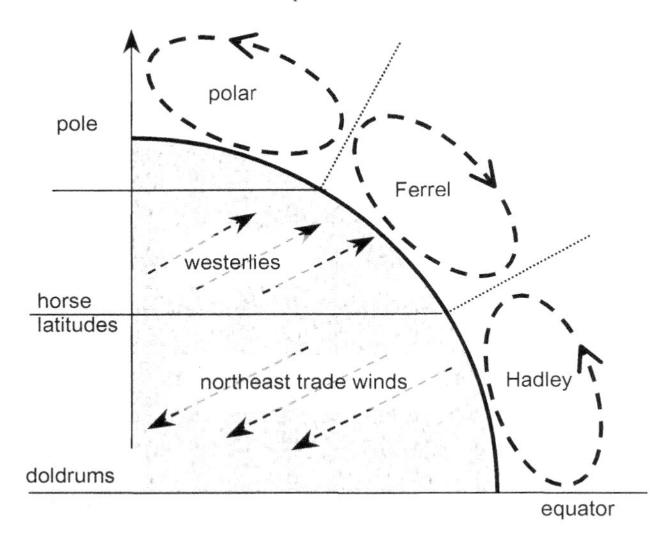

Figure 26.1 General circulation of the atmosphere showing the three cells of circulation. This pattern is symmetric about the equator. Surface winds, denoted by the dashed arrows within the stippled area, are identified by the direction from which they blow. The vertical scale of the atmosphere is greatly exaggerated in this cartoon.

altitude (near the top of the troposphere) close to the equator. This warm air moves away from the equator in the two Hadley cells, one on either side of the equator. The Coriolis effect deflects these high-altitude motions to the {right, left} in the {northern, southern} hemisphere – that is, to the east in both hemispheres. This deflection is understood in terms of conservation of angular momentum.[3] The northern extent of the Hadley cells is determined by a fluid instability, the analysis of which is beyond the scope of this book. Suffice it to say that around 30 degrees latitude, poleward motions at high altitudes give way to sinking within the downward legs of these cells.

As the sinking air approaches the surface, a portion flows equatorward. The Coriolis force deflects this low-latitude flow toward the west, creating the *northeast trade winds* in the northern hemisphere and the *southeast trade winds* in the southern hemisphere. The remainder of this sinking air is deflected away from the equator, forming part of the mid-latitude Ferrel cells. The Coriolis force deflects these near-surface flows toward the east, producing *westerlies* that play a large part in the propagation of weather systems. The Ferrel cells extend to about 60 degrees latitude, where they encounter the low-altitude equatorward flow of the polar cells. Upward motions near the boundary between the Ferrel and polar cells facilitates the formation and growth of rain-bearing clouds.

[3] See § 8.1.

Circulation in the polar cells is driven primarily by radiative cooling at high latitudes.[4] This cooling increases the density of the air, which sinks near the poles and then flows equator-ward, with a deflection to the west due to the Coriolis force. Since cooling is a weaker driving force than convection and since the area of the polar regions is much smaller than the tropics, the polar cells are relatively weak.

The thermodynamic efficiency of the Hadley cell, as a heat engine, is about 2.6%. This seems a low figure, but we need to realize that the energy provided by solar insulation in the tropics is very large.

26.2 Thermal Winds

The atmosphere is nearly in the geostrophic balance $\mathbf{v}_H = (g/f)\mathbf{1}_z \times \nabla_p Z$ developed in § 8.5. If the atmosphere were barotropic, then the form of Z would be independent of the value of p and the geostrophic balance would be the same at all levels. However, our atmosphere is baroclinic and the form of Z depends on p. Consequently, the geostrophic balance and the associated winds change with elevation. This change is called the *thermal wind*, though it is not really a wind; it is a wind shear.

Differentiating the geostrophic balance with respect to p and using the hydrostatic relation ($\mathrm{d}p = -\rho g \mathrm{d}Z$), we have

$$\frac{\partial \mathbf{v}_H}{\partial p} = \frac{g}{f}\mathbf{1}_z \times \nabla_p \frac{\partial Z}{\partial p} = \frac{1}{f\rho^2}\mathbf{1}_z \times \nabla_p \rho \,.$$

Using the ideal-gas equation ($p = \rho R_s T$) and the hydrostatic relation written as $\mathrm{d}p = -\rho g \mathrm{d}z$, this becomes the *thermal wind* equation:

$$\frac{\partial \mathbf{v}_H}{\partial z} = \frac{g}{fT}\mathbf{1}_z \times \nabla_p T \approx \frac{g}{fT}\mathbf{1}_z \times \nabla_H T \,.$$

This equation describes the change in geostrophic flow with elevation that is driven by horizontal changes in the structure of the temperature field. It applies to both the atmosphere and oceans.

To understand the thermal wind, let's consider a layer of incompressible fluid (e.g., sea water) with no vertical temperature gradient. Suppose the temperature decreases in the x direction: $T = T_0 - \gamma_x x$, where γ_x is a prescribed constant. Due to thermal expansion, the density increases in the x direction: $\rho = \rho_0(1 + \alpha\gamma_x x)$, where α is the coefficient of thermal expansion. Assuming a uniform pressure at the surface ($p = p_0$ at $z = h(x)$), the hydrostatic balance is $p = p_0 + \rho_0 g(h-z)(1+\alpha\gamma_x x)$ and the pressure at the base of the layer is $p_b = p_0 + \rho_0 g h(1+\alpha\gamma_x x)$. If the water were immobile with constant bottom pressure, the layer would decrease in thickness with increasing x and at a given elevation, the pressure

[4] The katabatic winds studied in § 24.1 form part of the Antarctic polar cell.

would decrease with x. In order for the fluid to remain immobile in geostrophic balance, this horizontal pressure gradient must be balanced by a geostrophic wind in the y direction: assuming no flow at the base, $\mathbf{v}_H = -(g\gamma_x z/fT)\mathbf{1}_y$. This flow increases in magnitude with elevation, with the rate of increase proportional to the horizontal temperature gradient. The principal source of lateral temperature variations in the atmosphere is the variation of solar heating with latitude requiring a compensatory geostrophic flow (that is, a thermal wind) from west to east, known as the *jet stream*.

26.3 Jet Streams

Jet streams are bands of high-altitude westerlies that encircle the globe. Flow is strongest at an elevation roughly equal to the height of the tropopause, marking the top of the troposphere. Two such bands occur in each hemisphere, roughly at the boundaries between the Hadley and Ferrel cells (called the *sub-tropical jet*) and Ferrel and polar cells (called the *polar jet*); see Figure 26.1. The polar jet is the stronger of the two. Typical speeds and elevations of the polar jet stream are 50 $\text{m}\cdot\text{s}^{-1}$ and 10 km. These values are consistent with $\partial v/\partial z \approx 5 \times 10^{-3}\ \text{s}^{-1}$. This vertical gradient is driven by a horizontal temperature gradient of magnitude 1 K per 100 km.

These flows are rarely steady or axially symmetric; they typically have four to six lobes, consisting of Rossby waves that have westward phase speeds.[5] However, these waves are carried eastward by the mean motion of the jet stream. These lobes have a dominant influence on surface weather. In particular, vacillation in the amplitude of the Rossby waves are associated with the movement of cold fronts on Earth's surface.

Transition

In this chapter, we have investigated three types of large-scale atmospheric flows that are affected by rotation of Earth: the general circulation, the thermal wind and the jet stream. In the following chapter, we investigate large-scale oceanic flows that are similarly affected by rotation of Earth.

[5] See § 15.3.

27

Oceanic Currents

As we have previously noted,[1] a major dynamic feature of the oceans is the thermocline, which separates a thin well-mixed top layer from a deep stably stratified layer beneath. The oceans have three types of large-scale motion that involve one or both of these layers, as summarized in Table 27.1. We have previously investigated the tides; in this chapter we will look into barotropic currents in the upper ocean (above the thermocline) driven by wind stress at the ocean surface and deep currents driven by subtle differences in temperature and salt content, beginning with the former.

Figure 27.1 shows the major oceanic currents in the upper ocean, consisting primarily of gyres having {clockwise, counterclockwise} flow in the {northern, southern} hemisphere. There are five main gyres, occurring in the five major ocean basins:[2] the northern Pacific, southern Pacific, northern Atlantic, southern Atlantic and (southern) Indian Oceans. In addition there are several minor gyres and currents near the equator and at high latitudes.

The near-surface flows within an ocean basin are driven by the prevailing winds illustrated in Figure 26.1, but in a somewhat indirect manner. The prevailing winds are predominantly along lines of constant latitude, being easterly (from west to east) in the tropics and westerly (from east to west) in mid-latitudes. As we saw in § 25.5.4, these winds drive near-surface flows (Ekman transport) in the normal direction: north-south, along lines of constant longitude. These potential north-south flows have no dynamic outlet and instead the wind stresses result in deformation of the ocean surface which sets up a counterbalancing north-south pressure gradient. Near-surface flows arise due to a latitudinal variation of this dynamic balance, quantified by the curl of the wind stress, forming the great oceanic gyres illustrated in Figure 27.1. These flows are predominantly in the east-west direction in the open oceans. With the exception of the southern ocean surrounding Antarctica, these east-west flows are blocked by continents and are diverted in the north-south direction in thin boundary layers, predominantly on the western boundaries of the oceans, forming the western boundary currents.

[1] See § 7.2.2.
[2] Note that the ocean basins are bounded by continents and the equator.

Table 27.1. *The three major types of oceanic flows, summarizing the involvement of the two oceanic layers (the well-mixed layer above the thermocline and the stratified ocean beneath), the forcing, the nature of the response and the section in which the flow is investigated.*

Motion	Involvement	Forcing	Nature	Section
Tides	Both layers	Gravitation	Barotropic	§ 12.2
Upper-ocean currents	Top layer	Wind stress	Barotropic	§ 27.2 and § 27.3
Thermohaline	Bottom layer	Density differences	Baroclinic	§ 27.4

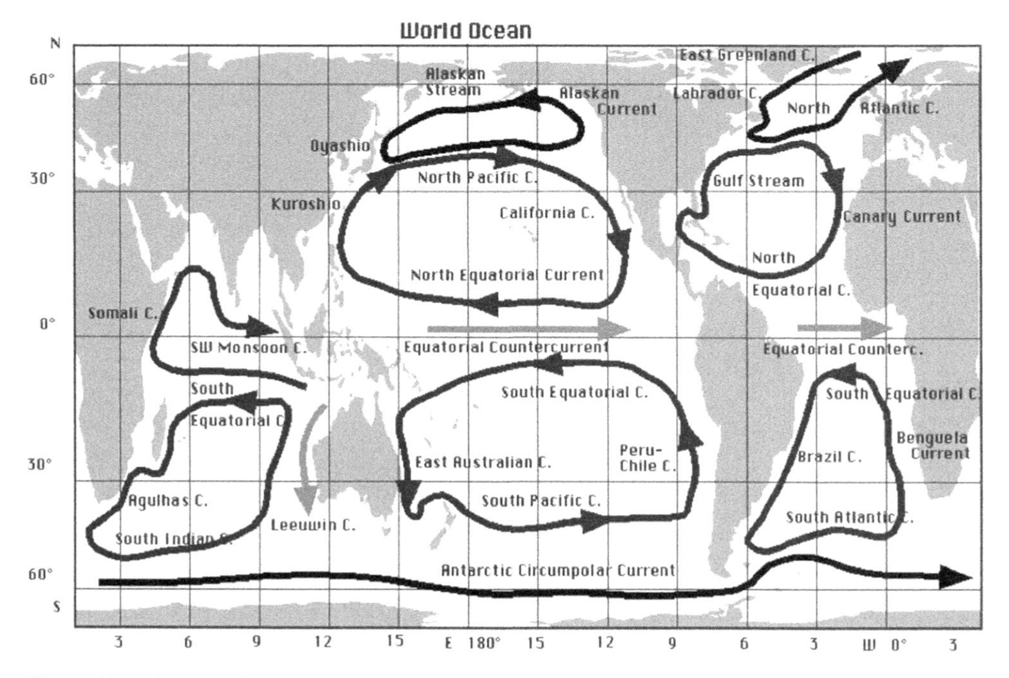

Figure 27.1 Geographical distribution of the major ocean surface currents. Figure by Matthias Tomszak; copied with permission from http://mt-oceanography.info/IntroOc/lecture02.html.

27.1 Formulation

To quantify these great oceanic currents, let's start with the form of the Navier–Stokes equation valid for a thin layer of rotating fluid presented in § 8.2.1. Ignoring the time-dependent and nonlinear terms, this becomes

$$\rho f \mathbf{1}_z \times \mathbf{v}_H = -\boldsymbol{\nabla}_H p + \frac{\partial}{\partial z}\left(\eta_T \frac{\partial \mathbf{v}_H}{\partial z}\right) + \eta_T \nabla_H^2 \mathbf{v}_H,$$

where we have added a subscript T to the dynamic viscosity η to remind us that it is a turbulent – not laminar – viscosity. This equation is to be applied in a layer of fluid extending from $z=-D+h_b$ to $z=h$ where h and h_b are functions of horizontal position. The density may be taken as constant within the well-mixed layer; deviations are small and dynamically unimportant.

The transport[3] $\mathbf{q}$ is given by the depth integral

$$\mathbf{q}=\int_{-D+h_b}^{h}\mathbf{v}_H\,\mathrm{d}z$$

and the vertical integral of the horizontal momentum equation is

$$\rho f\mathbf{1}_z\times\mathbf{q}=-\int_{-D+h_b}^{h}\nabla_H p\,\mathrm{d}z+\boldsymbol{\tau}_W+\eta_T\nabla_H^2\mathbf{q},$$

where

$$\boldsymbol{\tau}_W=\eta_T\left.\frac{\partial\mathbf{v}_H}{\partial z}\right|_{z=h}$$

is the known wind stress. In writing this, we have neglected to include a bottom stress term because we have seen in § 25.5.5 that the stress at the thermocline is negligibly small.[4]

The momentum equation is a two-dimensional vector equation involving a two-dimensional unknown vector, $\mathbf{q}$, and an unknown scalar p. A third scalar equation is provided by the continuity equation. For two-dimensional flow of an incompressible fluid, this is simply

$$\nabla_H\bullet\mathbf{q}=0.$$

Note that the Coriolis and stress terms in the momentum equation combine to give the Ekman transport investigated in § 25.5.4. However, the actual current in the mixed layer differs from the Ekman transport, primarily due to the action of the pressure gradient. Although this gradient plays an important role in steering the currents, it can be eliminated from the formulation and can be found after the currents have been determined. In order to eliminate the pressure-gradient term, it must be expressed as the horizontal gradient of a scalar; in this form, it will disappear when we take the curl of the horizontal momentum equation. We shall assume hydrostatic equilibrium with the boundary deflections satisfying $\rho h=(\rho-\rho_b)h_b$, where ρ_b is the density of the fluid lying beneath the well-mixed layer. The top pressure is atmospheric, $p(h)=p_a$ and the bottom pressure is

$$p(-D+h_b)=p_a+\rho g\,(D+h-h_b)=p_a+\rho gD+\rho gh\rho_b/(\rho_b-\rho).$$

[3] That is, the specific volume flow. Alternatively, the following formulation could have been carried through using the specific mass flow $\dot{\mathbf{M}}=\rho\mathbf{q}$.

[4] This simplification precludes consideration of Stommel's model of western boundary currents.

Using these together with *Leibniz's rule* the pressure term may be expressed as

$$\int_{-D+h_b}^{h} \nabla_H p \, \mathrm{d}z = \nabla_H P,$$

where

$$P = \int_{-D+h_b}^{h} p \, \mathrm{d}z - (p_a + \rho g D)\frac{\rho}{\rho_b - \rho} h - \frac{\rho^2}{2(\rho_b - \rho)^2} \rho_b g h^2 .$$

Now the vertically integrated horizontal momentum equation becomes

$$\rho f \mathbf{1}_z \times \mathbf{q} + \nabla_H P = \boldsymbol{\tau}_W + \eta_T \nabla_H^2 \mathbf{q}.$$

The stress term $\boldsymbol{\tau}_W$ is a prescribed forcing. This equation admits a static balance in which the applied wind stress is balanced by the horizontal pressure gradient. This balance is permissible provided the curl of the surface wind stress is zero. As it happens, the wind-stress curl is not zero and the stress drives surface currents. To quantify these currents, we may take the curl of the horizontal momentum equation, thereby eliminating the pressure term. Using the continuity equation and vector identities found in Appendix A.2.6 and writing[5] $\nabla_H f = \beta \mathbf{1}_y$, the vertical component of the curl of the horizontal momentum equation gives us

$$\beta \rho \, q_y = \mathbf{1}_z \bullet \nabla_H \times \left(\boldsymbol{\tau}_W + \eta_T \nabla_H^2 \mathbf{q} \right),$$

where $q_y = \mathbf{1}_y \bullet \mathbf{q}$ is the northward transport.

The momentum equation has been reduced to a single scalar equation involving a two-dimensional vector unknown, $\mathbf{q}$; this plus the continuity equation form our set of governing equations. These may be reduced to a single equation for a scalar unknown by expressing the transport in terms of a stream function;[6] let

$$\mathbf{q} = \mathbf{1}_z \times \nabla_H \Psi_q = -\frac{\partial \Psi_q}{\partial y} \mathbf{1}_x + \frac{\partial \Psi_q}{\partial x} \mathbf{1}_y .$$

Now the continuity equation is identically satisfied and the momentum equation becomes

$$\beta \frac{\partial \Psi_q}{\partial x} - \nu_T \nabla_H^4 \Psi_q = \frac{1}{\rho} \mathbf{1}_z \bullet \nabla_H \times \boldsymbol{\tau}_W = \frac{1}{\rho} \left(\frac{\partial \tau_y}{\partial x} - \frac{\partial \tau_x}{\partial y} \right).$$

where $\nu_T = \eta_T / \rho$ is the turbulent kinematic viscosity and we have written $\boldsymbol{\tau}_W = \tau_x \mathbf{1}_x + \tau_y \mathbf{1}_y$. It is clear from this equation that surface currents are driven by the curl of the wind stress.

It is time to pause and catch our breath. We have arrived at a single linear, non-homogeneous partial differential equation with a variable coefficient ($\beta(y)$) involving a single dependent variable, $\Psi_q(x, y)$. This equation applies on the horizontal surface of an

[5] See § 8.1.5.
[6] See Appendix C.4.

ocean basin and needs to be supplemented by two conditions at each point on the boundary of the basin. Physically reasonable boundary conditions are no flow normal or parallel to the boundary; that is,

$$\Psi_q = \mathbf{1}_n \bullet \nabla_H \Psi_q = 0$$

at each point on the boundary, where $\mathbf{1}_n$ is a unit vector pointing normal to the boundary. The Coriolis term with its x derivative singles out this coordinate as the most dynamically important and also imparts an east-west asymmetry to the structure of the oceanic currents. With this in mind, let's suppose that the ocean basin lies within $x_w(y) < x < x_e(y)$ where the functions $x_w(y)$ and $x_e(y)$ describe the geographic locations of its western and eastern boundaries, respectively.

27.1.1 Boundary-Layer Scaling

The problem formulated immediately above is a tough nut, but we can crack it with the use of dimensionless local coordinates on each boundary; let's introduce local coordinates near the western and eastern boundaries:

$$n_w = (\cos\varphi_w x + \sin\varphi_w y)/W \qquad \text{and} \qquad n_e = (\cos\varphi_e x + \sin\varphi_e y)/W,$$

where

$$\varphi_w = -\arctan\left(\frac{\mathrm{d}x_w}{\mathrm{d}y}\right) \qquad \text{and} \qquad \varphi_e = \arctan\left(\frac{\mathrm{d}x_e}{\mathrm{d}y}\right)$$

are the angles between the x and n axes on the western and eastern boundaries, respectively, as illustrated in Figure 27.2 and

$$W \equiv \left(\frac{\nu_T}{\beta|\cos\varphi|}\right)^{1/3}$$

is – as we shall see – the width of the boundary layer. A basic assumption of the boundary-layer approach to modeling surface currents in the oceans is that the width of the boundary layer is much smaller than the typical scale of variation of ocean winds, so that the viscous term in the momentum equation can be neglected in the interior region. An approximate magnitude of W is readily obtained from observations of western boundary currents; the Gulf Stream and Kuroshio currents are typically less than 100 km wide,[7] whereas the oceans are thousands of kilometers wide. Consequently, the equations governing oceanic surface currents can be simplified using boundary layer theory, as explained in the remainder of this section and in § 27.3.

[7] This observation may be used to estimate ν_T; see § 27.3.3.

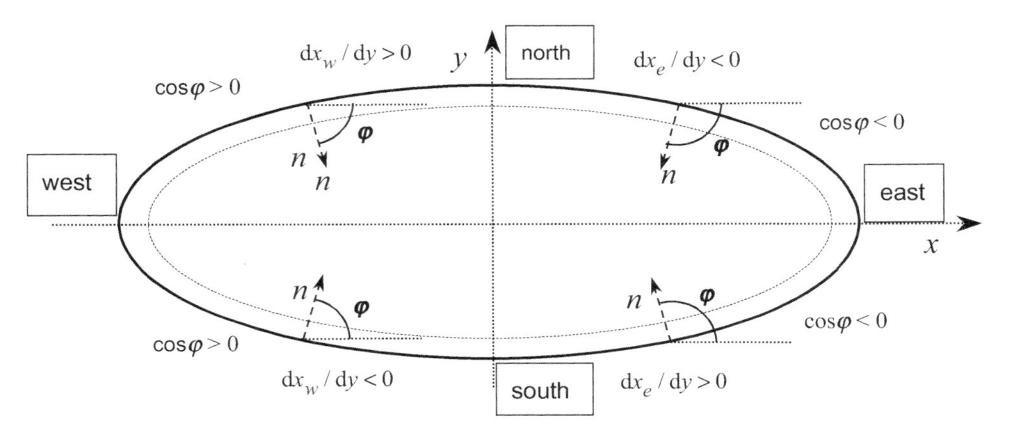

Figure 27.2 Cartoon of an oval ocean basin showing the relations among x, y, n and φ. Note that $\cos\varphi$ is positive on the western boundary and negative on the eastern.

Now the boundary conditions may be expressed as follows:

$$\text{at} \quad x = x_e: \quad \Psi_q = \frac{\partial \Psi_q}{\partial n_e} = 0;$$

$$\text{at} \quad x = x_w: \quad \Psi_q = \frac{\partial \Psi_q}{\partial n_w} = 0.$$

We may take advantage of the fact that the viscous term is small everywhere within the ocean basin except within narrow boundary layers near the ocean perimeter to separate Ψ_q into interior and boundary layer parts; let

$$\Psi_q = \Psi_i + \Psi_b,$$

where[8]

$$\beta \frac{\partial \Psi_i}{\partial x} = \frac{1}{\rho}\left(\frac{\partial \tau_y}{\partial x} - \frac{\partial \tau_x}{\partial y}\right) \qquad \text{and} \qquad \beta \frac{\partial \Psi_b}{\partial x} - \nu_T \nabla_H^4 \Psi_b = 0.$$

By dividing the equations in this manner, we have introduced a fifth derivative, and we apparently can impose a fifth boundary condition.[9] Let's prescribe

$$\text{at} \quad x = x_e: \quad \Psi_i = 0, \quad \Psi_b = 0, \quad \frac{\partial \Psi_b}{\partial n_e} = -\frac{\partial \Psi_i}{\partial n_e};$$

$$\text{at} \quad x = x_w: \quad \Psi_b = -\Psi_i, \quad \frac{\partial \Psi_b}{\partial n_w} = -\frac{\partial \Psi_i}{\partial n_w}.$$

The boundary-layer solutions are investigated in § 27.3, following an investigation of the Sverdrup balance and transport in the interior of the ocean basin.

[8] Since β and ν_T are positive at all points on the boundary, there is no need to place absolute-value signs on these variables.

[9] Actually this line of reasoning leads us a bit astray, but we will recover the correct path in § 27.3.1.

27.2 Sverdrup Balance and Transport

The interior equation describes the *Sverdrup balance*:

$$q_{iy} = \frac{\partial \Psi_i}{\partial x} = \frac{1}{\beta \rho} \left(\frac{\partial \tau_y}{\partial x} - \frac{\partial \tau_x}{\partial y} \right).$$

This balance is a combination of the Ekman transport investigated in § 25.5.4 and a geostrophic flow induced by the pressure gradient. The corresponding *Sverdrup transport* is easily obtained by integrating the equation with respect to x and satisfying the condition $\Psi_i = 0$ on the eastern boundary:

$$\Psi_i = \frac{1}{\beta \rho} \left(\tau_y - \tau_{ye} + \int_x^{x_e} \frac{\partial \tau_x}{\partial y} \, \mathrm{d}\grave{x} \right),$$

where $\tau_{ye}(y) = \tau_y(x_e(y), y)$ is the magnitude of the northern wind stress at the eastern boundary. At the western boundary the Sverdrup transport is not zero:

$$\begin{aligned} \Psi_{iw}(y) &\equiv \Psi_i(x_w(y), y) \\ &= \frac{1}{\beta \rho} \left(\tau_y(x_w(y), y) - \tau_{ye}(y) + \int_{x_w(y)}^{x_e(y)} \frac{\partial \tau_x}{\partial y} \, \mathrm{d}x \right). \end{aligned}$$

The Sverdrup transport, q_{iy}, is (if positive) a specific northward flow in the middle of an ocean basin. The stream function, Ψ_i, quantifies the total northward flow in the portion of the ocean basin lying to the east. The total northward flow across a line of latitude within an ocean basin must be zero. Since Ψ_i is not zero close to the western boundary, there must be a strong north-south flow in the western boundary layer, as we will see in § 27.3.2.

27.2.1 Sverdrup Simplification

The prevailing winds over the oceans are predominantly east-west, with relatively little latitudinal variation. If we assume that $\partial \tau_x / \partial x = 0$ and $\tau_y = 0$, the interior solution simplifies to

$$\Psi_i = \frac{1}{\beta \rho} \frac{\partial \tau_x}{\partial y} (x_e - x),$$

and the associated specific transport is

$$q_{iy} = -\frac{1}{\beta \rho} \frac{\partial \tau_x}{\partial y}.$$

In mid-latitude regions where $\partial \tau_x / \partial y > 0$, such as near the boundary between the Hadley and Ferrel cells, the specific transport is negative, indicating a southward flow.

27.2.2 Sverdrup Discussion

In § 25.5.4, we investigated the horizontal transport within the Ekman layer, $\mathbf{q}_E = -\mathbf{1}_z \times \boldsymbol{\tau}_W/\rho f$. This local transport has morphed into the Sverdrup transport, $q_{iy} = -(1/\beta\rho)\partial\tau_x/\partial y$. To understand how this transformation occurs, consider an initially quiescent ocean. At some instant of time, a uniform wind begins to blow, driving an Ekman transport perpendicular to the direction of the wind. If there is no horizontal outlet available for this flow, the water piles up, modifying the surface elevation of the ocean. This changed elevation creates a depth-independent geostrophic flow that adds to (or subtracts from) the Ekman transport, with the sum of the two being the Sverdrup transport. The magnitude of the surface elevation capable of modifying the Ekman transport can be estimated from a balance between the pressure and wind stresses: $h \approx L\tau_W/(\rho g D)$, where L is the horizontal scale. With $g \approx 10$ m·s^{-2}, $D \approx 10^2$ m, $L \approx 10^6$ m, $\rho \approx 10^3$ kg·m^{-3} and $\tau_W \approx 0.1$ Pa, the elevation change is a modest 0.1 m.

In order to estimate the magnitude of the Sverdrup transport, we need to estimate the north-south scale of variation of wind stress. We see from Figure 26.1 that the wind stress completely changes orientation over a latitude $\pi/6$ radians (30 degrees). The corresponding distance on Earth's surface is equal to half Earth's radius: $\Delta y \approx 3 \times 10^6$ m. Using this and $\rho \approx 10^3$ kg·m^{-3}, $\beta \approx 2 \times 10^{-11}$ m^{-1}·s^{-1} and $\tau_W \approx 0.1$ Pa, $\|\mathbf{q}_i\| \approx 1.6$ m^2·s^{-1}. This estimated magnitude is slightly larger than, but consistent with, that for the Ekman transport; see § 25.5.4.

The prevailing winds are primarily east-west, so that the total volume flow (or stream-function difference) across a line of latitude is the specific volume flow times the width of the ocean. For example, a specific flow of 1 m^2·s^{-1} across a line of latitude 1000 km wide produces a volume flow of 10^6 m^3·s^{-1}. This magnitude of flow is called *one Sverdrup*, in honor of Harald Sverdrup who first identified the balance and transport bearing his name (Sverdrup, 1947). Oceanic currents are typically several tens of Sverdrups in magnitude; see Table 27.2. In comparison, the flows of all the rivers in the world sum to about one Sverdrup. With $\boldsymbol{\tau}_W$ known, it is a relatively simple matter to determine Ψ_i and the corresponding Sverdrup transport $\mathbf{q}_i$, as illustrated in Figure 27.3.[10] As can be seen from this figure, Sverdrup's theory is only part of the story; Sverdrup transport is not zero at the western boundaries of the ocean basins. This is where the boundary-layer portion of the flow plays a dominant role, as explained in the following section.

27.3 Boundary Currents

Within the boundary layers at the perimeter of the ocean, normal derivatives are dominant. Noting that $\partial/\partial x = (1/W)\cos\varphi\,\partial/\partial n$,[11] the boundary-layer equation may be expressed as

$$s_\varphi \frac{\partial \Psi_b}{\partial n} - \frac{\partial^4 \Psi_b}{\partial n^4} = 0,$$

[10] Using wind stress data from Hellerman and Rosenstein (1983).
[11] From § 27.1.1. Recall that φ is the angle relative to the positive x axis.

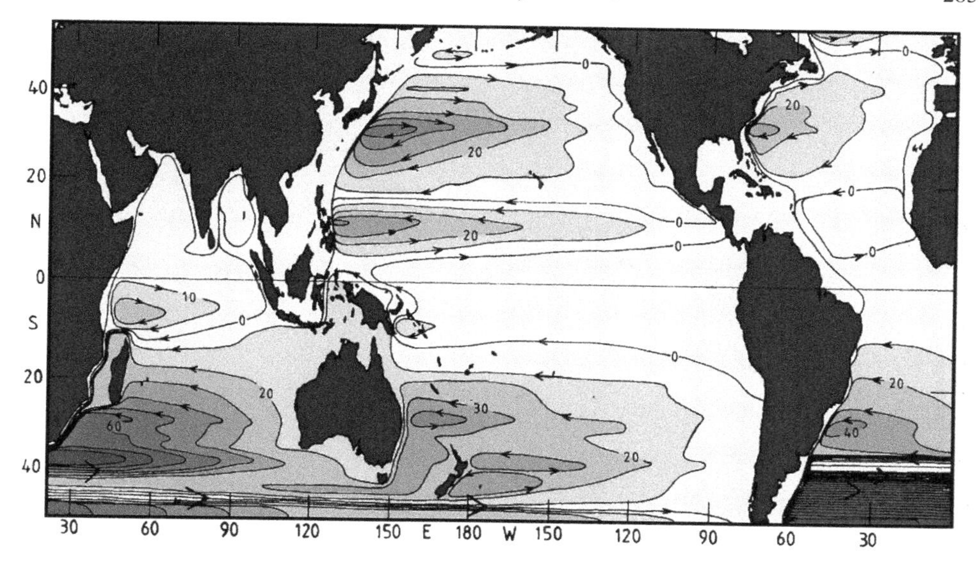

Figure 27.3 Sverdrup transport: near-surface ocean currents driven by wind stress. Flux units are Sverdrups. Figure by Matthias Tomszak; shown with permission.

where s_φ is the sign of $\cos\varphi$. This is a local balance, so we may treat $\beta(y)$ and $\varphi(y)$ as constants. Note that our boundary-layer scaling fails at the northern and southern extremities of the basin where the boundary layer thickness is no longer small; our simple analysis does not address the flow structure near these extremities. Assuming exponential behavior, $\Psi_b \propto e^{-\lambda n}$, the boundary-layer equation reduces to

$$s_\varphi \lambda + \lambda^4 = 0.$$

The mode corresponding to $\lambda = 0$ is the ghost of the Sverdrup mode and must be discarded. That leaves us with three boundary-layer modes having

$$\lambda = \left(-s_\varphi\right)^{1/3}.$$

We need the real part of λ to be positive. On the western boundary $\cos\varphi$ is positive. The cube root of -1 has roots $\{-1, (1 + \sqrt{3}\mathrm{i})/2, (1 - \sqrt{3}\mathrm{i})/2\}$, two of which have a positive real part. It follows that there are two boundary-layer modes on the western boundary; the boundary-layer modes can satisfy both boundary conditions on the western boundary. On the eastern boundary, $\cos\varphi$ is negative. The cube root of $+1$ has roots $\{1, (-1 + \sqrt{3}\mathrm{i})/2, (-1 - \sqrt{3}\mathrm{i})/2\}$ one of which has a positive real part. It follows that there is one boundary-layer mode on the eastern boundary; the conditions on the eastern boundary are satisfied by one boundary-layer mode and one interior mode.

27.3.1 Eastern Boundary Currents

We now see that it was premature to specify that the interior mode satisfies the no-flow condition on the eastern boundary. We need to be more careful. On the eastern boundary,

$$\Psi_b = \Psi_{be}(y)e^{-n} \quad \text{and} \quad \Psi_i = \frac{1}{\beta\rho}\left(\tau_y + \int_x^{x_e} \frac{\partial \tau_x}{\partial y}\,\mathrm{d}\grave{x}\right) + \Psi_{ie}(y),$$

where $\Psi_{be}(y)$ and $\Psi_{ie}(y)$ are arbitrary functions of integration. These functions are determined by the eastern boundary conditions:

$$\text{at} \quad x = x_e(y): \qquad \Psi_i + \Psi_b = 0 \quad \text{and} \quad \frac{\partial \Psi_i}{\partial n} + \frac{\partial \Psi_b}{\partial n} = 0.$$

Using the solutions above, these require

$$\Psi_{be}(y) + \Psi_{ie}(y) = -\tau_{ye}/(\beta\rho)$$

and

$$\Psi_{be}(y) = \frac{\partial \Psi_i}{\partial n} = W\left(\frac{\partial \Psi_i}{\partial x}\cos\varphi_e + \frac{\partial \Psi_i}{\partial y}\sin\varphi_e\right).$$

As we noted when W was introduced in § 27.1.1, a basic assumption of the boundary-layer formulation is that $W\partial/\partial x$ and $W\partial/\partial y$ are both small in magnitude. It follows that to dominant order in this small parameter, $\Psi_{be} = 0$ and $\Psi_{ie} = -\tau_y/(\beta\rho)$. That is, the eastern boundary-layer mode is weak and we recover the Sverdrup solution to dominant order.

27.3.2 Western Boundary Currents

The western boundary-layer has two modes:

$$\Psi_b = e^{-n/2}\left(\Psi_{b1}\cos\left(\tfrac{1}{2}n\right) + \Psi_{b2}\sin\left(\tfrac{\sqrt{3}}{2}n\right)\right)$$

and the western boundary conditions require

$$\Psi_{b1} + \Psi_{iw} = 0 \qquad \text{and} \qquad -\frac{1}{2}\Psi_{b1} + \frac{\sqrt{3}}{2}\Psi_{b2} + \frac{\partial \Psi_{iw}}{\partial n} = 0,$$

where Ψ_{iw} is the Sverdrup transport evaluated at the western boundary; see § 27.2. Since the Sverdrup transport has no boundary-layer structure, $\partial\Psi_{iw}/\partial n$ is small and may be neglected at dominant order. It follows that

$$\Psi_{b1} = -\Psi_{iw} \qquad \text{and} \qquad \Psi_{b2} = -\tfrac{1}{\sqrt{3}}\Psi_{iw},$$

and the flow at close to western boundary is given by

$$\Psi_q \approx \Psi_{iw} - \Psi_{iw}e^{-n/2}\left(\cos\left(\tfrac{1}{2}n\right) + \tfrac{1}{\sqrt{3}}\sin\left(\tfrac{\sqrt{3}}{2}n\right)\right).$$

The magnitude of the specific flow parallel to the coast, denoted by q_w, is given by the normal gradient:

$$q_w = \tfrac{1}{2}\Psi_{iw}e^{-n/2}\left(\cos\left(\tfrac{1}{2}n\right) - \cos\left(\tfrac{\sqrt{3}}{2}n\right) + \sin\left(\tfrac{1}{2}n\right) + \sin\left(\tfrac{\sqrt{3}}{2}n\right)\right).$$

The normalized boundary stream function (solid) and current (dashed) are plotted versus n in Figure 27.4. Note the oscillation of the functions; there is a weak off-shore counter-current.

27.3.3 Western Boundary Current Discussion

Western boundary currents are near-surface currents that occur – as the name implies – near the western boundaries of ocean basins. These currents are necessary to close the current loops of Sverdrup transport, as illustrated in Figure 27.3. As can be seen from this figure, currents within the western boundary layers flow toward the equator at low latitudes (<15 degrees) and away from the equator at mid-latitudes. There are five prominent mid-latitude currents, as summarized in Table 27.2, each of which is the western leg of an ocean gyre. These currents play an important role in conveying heat from the tropics toward the poles and in moderating climates in northern latitudes.

This western-boundary-layer solution, coupled with a key observation of maximum flow speed, permits us to estimate the magnitude of the turbulent viscosity, ν_T, in the oceans. The theoretical maximum speed of a western boundary current occurs for $n_{\max} \approx 1.63$. The corresponding off-shore distance of the maximum is $L = n_{\max} W$. This distance is readily measured from surface observations of the western currents. For example at the Straits of Florida, where $\varphi \approx 0$, the maximum current is about 25 km offshore. It follows from this and the definition $W = (\nu_T/\beta \cos\varphi)^{1/3}$ (with $\varphi \approx 0$) that

$$\nu_T \approx \beta(L/n_{\max})^3.$$

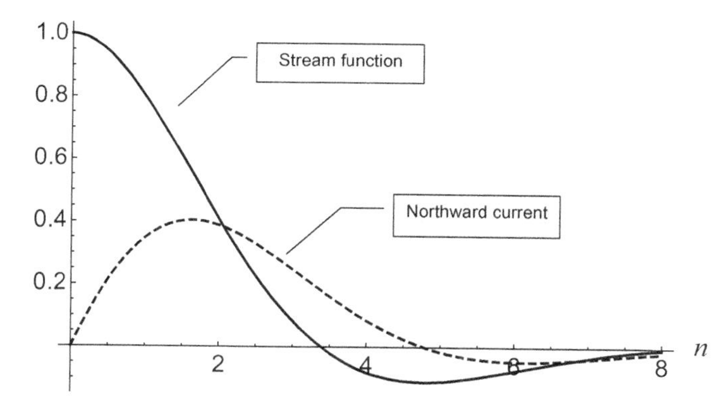

Figure 27.4 Plots of the normalized boundary stream function (solid line) and current (dashed line) versus the dimensionless normal boundary coordinate, n.

Table 27.2. *Names of the five western boundary currents and their associated gyres. Strengths are approximate.*

Name	Gyre	Strength (10^6 $m^3 \cdot s^{-1}$)
Kurishiro	Northern Pacific	50
East Australian	Southern Pacific	40
Gulf Stream	Northern Atlantic	30
Brazil	Southern Atlantic	40
Agulhas	(Southern) Indian Ocean	60

The Straits of Florida is at a latitude of 25 degrees north, giving $\beta \approx 2 \times 10^{-11}$ $m^{-1} \cdot s^{-1}$; altogether $\nu_T \approx 70$ $m^2 \cdot s^{-1}$.

The lateral (off-shore) thickness of the western boundary current is sensitive to the orientation of the coastline. The current is strongest and thinest off a north-south shore, such as at the Strait of Florida. As the shore trends toward east-west, the boundary layer thickens and weakens. This behavior contributes to the detachment of the Gulf Stream from the coastline of North America near Cape Hatteras and the detachment of the Kurishiro from Japan near the middle of Honshu.

27.4 Thermohaline Circulation

The oceanic gyres transport heat from equatorial regions toward the poles, with the heat flow $\dot{Q}$ related to the volume flow of the currents Ψ by

$$\dot{Q} = \rho c_p (\Delta T) \Psi ,$$

where $\rho \approx 10^3$ $kg \cdot m^{-3}$ is the density of water, $c_p \approx 4 \times 10^3$ $J \cdot kg^{-1} \cdot K^{-1}$ is the specific heat of water and ΔT is the temperature contrast.[12] The flows listed in Table 27.2 sum to 220 Sverdrups (2.2×10^8 $m^3 \cdot s^{-1}$). This current is able to transport the reported 2 PW of heat poleward provided the temperature excess is $\Delta T \approx 2.3$ °C.[13]

These warm currents flow poleward on the western edges of ocean basins at low latitudes (in the tropics), then turn and flow eastward in mid and sub-polar latitudes, gradually transferring their excess heat to the atmosphere, primarily through evaporation. Since the prevailing winds at these latitudes are westerly,[14] this excess heat acts to moderate the weather on the western edges of the continents, particularly North America and Europe.

[12] See Table B.5 for parameter values.

[13] By comparison, the atmospheric circulation conveys about 4 PW of heat poleward. 1 PW = 1 petawatt = 10^{15} W.

[14] That is, blowing from west to east.

The western boundary currents consist of warm (due to solar heating) salty (due to evaporation) water, with the excess heat tending to make the water in these currents buoyant and the excess salt tending to make them dense. Evaporation selectively removes heat, with the salts remaining in the liquid water. Consequently the water becomes progressively denser as the currents move poleward then eastward. The fate of these currents depends on the delicate balance between the two competing effects: buoyancy due to warmth and density due to salt. In the northern Pacific Ocean, warmth wins and the water in northern Pacific gyre remains on the surface, turning southward at this ocean's eastern edge. However, in the northern Atlantic Ocean, saltiness wins and some of the water in the northern Atlantic gyre sinks in the Greenland and Labrador Seas, becoming *north Atlantic deep water* (or NADW).

This sinking of salty water is a key component of the *Atlantic meridional overturning circulation* (or AMOC) and has profound implications for the structure of the deep ocean. There is an additional input of water to the deep ocean in the Ross and Weddell Seas near the Antarctic Peninsula. The formation of Antarctic deep water is driven by the formation of sea ice, which is composed predominantly of fresh water;[15] the sea water that remains unfrozen is enriched in salt and sufficiently dense to sink.

The volume flows of these two sources of deep water are seasonal and highly variable on decadal timescales; on average the deep ocean overturns roughly every 1000 years ($\approx 3 \times 10^{10}$ s). Since the volume of the oceans is about 1.3×10^{18} m^3, this translates to a mean overturning flow of about 40 Sverdrups (4×10^7 m$^3\cdot$s^{-1}). Heat and salt diffuse a distance given by the square root of diffusivity times time. The diffusivities of heat ($\approx 1.4 \times 10^{-7}$ m$^2\cdot$s^{-1}) and salt ($\approx 2 \times 10^{-9}$ m$^2\cdot$s^{-1}) are sufficiently small that molecular diffusion is not important; with heat and salt diffusing about 70 m and 8 m, respectively, in 1000 years. It follows that the density of a parcel of water will change appreciably only by turbulent mixing with adjacent waters.

The deep ocean, below the thermocline, is stably stratified, with density increasing with depth, as noted in § 7.2.2. Newly created deep water sinks until it reaches its equilibrium depth where its density is equal to the surroundings; North Atlantic deep water tends to sink to mid-ocean depths (about 2000 m), whereas Antarctic deep water (called *Antarctic bottom water*), being more dense, sinks right to the bottom. The newly arriving deep water spreads out and displaces upward the more buoyant water lying at shallower depths. Eventually a layer of deep water gets pushed up to the thermocline and entrained into the upper ocean, completing the circuit. Of course, this description is highly simplified, as there are flows from basin to basin, with waters exchanging and mixing.

There is one important ingredient missing from this simple picture: how the deep ocean water becomes more buoyant over time. This evolution is the result of downward transport of heat due to turbulent mixing. Turbulent mixing in the deep ocean is induced in part as waters having differing properties move laterally across sills from one basin to another. Also, internal waves generated by various mechanisms can break, causing

[15] See Appendix E.10.2.

turbulent mixing. The power (estimated to be 0.4 TW) for these turbulent motions is provided by surface wind stresses and tidal motions.

The volume and location of formation of NADW, and the associated northward transport of heat, is the result of a delicate balance between temperature and salinity of Gulf Stream waters, with salinity driving the formation. It follows that an input of fresh surface water to the north Atlantic would throw a monkey wrench into this process, potentially stopping the formation entirely. Stommel (1961) showed that a thermohaline circulation can be bi-stable; as illustrated in Figure 27.5. That is, for a given input of fresh water, the rate of formation of NADW can be appreciable (solid curve) or zero (solid horizontal line). As the input of fresh water increases the system migrates to the right along the top curve as indicated by the dotted arrow labeled a, with the rate of deep-water formation decreasing gradually until reaching the critical point labeled p_1 where the solid and dotted curves meet. A further increase in fresh-water input causes the rate of formation to drop abruptly to zero, as indicated by the dotted arrow labeled b. A decrease in fresh-water input below the critical does not get us back in business immediately. In order to get the formation of NADW to turn on, the input of fresh water must decrease to some low level, as indicated by the dotted arrow labeled c. Once this input drops below the threshold at p_2, then formation of NADW turns back on, as indicated by the dotted arrow labeled d.

An obvious source of fresh water in the north Atlantic is melting of the Greenland ice sheet. There is observational evidence in ice cores from Greenland and the sedimentary record in the north Atlantic that the formation of NADW has been severely disrupted in the past, with either formation being either much reduced (a *Dansgaard–Oeschger* event) or entirely absent (a *Heinrich* event). A Dansgaard–Oeschger event is characterized by rapid warming ($\approx 10°C$ in Greenland) followed by a gradual cooling, while a Heinrich event is

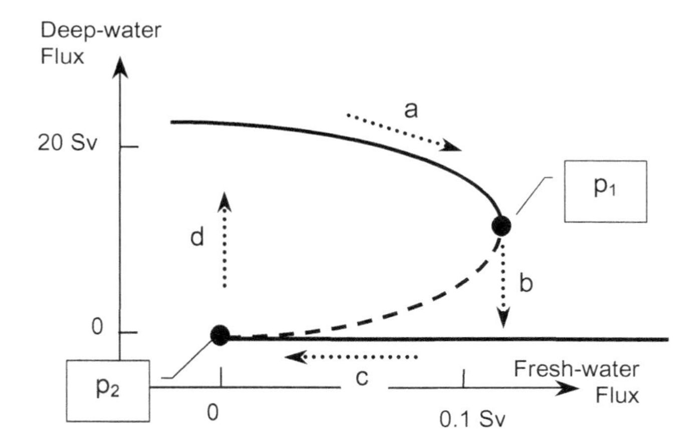

Figure 27.5 Schematic of a bi-stable regime for the formation of Atlantic deep water. The two stable branches are shown as solid lines and the unstable branch is dashed. Fluxes are in Sverdrups (10^6 $m^3 \cdot s^{-1}$); magnitudes are only suggestive.

a cold period, during which icebergs are prevalent in the north Atlantic. The Greenland ice cores show that Dansgaard–Oeschger events occur fairly regularly, about once every 1470 years, with the duration of individual events being somewhat variable, ranging from several hundred to 1000 years or more; see Figure 1 of Clement and Peterson (2008); also see Rahmstorf (2003). However, for the past 10,000 years, during which time modern civilization arose, the climate has been exceptionally steady.

The weakening or cessation of NADW formation leads to a number of climatic changes, including:

- cooling of the north Atlantic (by up to 10°C);
- warming of the southern hemisphere;
- less precipitation in northern-hemisphere mid-latitudes;
- southward shift of the inter-tropical convergence zone;
- a shift of the Gulf Stream close to the shore of New England and the Canadian maritime provinces with:
  - an immediate increase (of up to 1 m) of sea level in the north Atlantic; and
  - stronger storms over northern Europe.

Cooling of the north Atlantic will act to increase the formation of glacial ice in Greenland and reduce the input of fresh water into the north Atlantic. This will not immediately reinstate NADW formation; the input of fresh water must be reduced substantially before formation of NADW commences, as indicated in Figure 27.5; the system must follow the dotted arrow labeled c in Figure 27.5 all the way to the second critical point (p_2) before the formation of NADW is re-established.

There is evidence in data currently being collected that the strength of AMOC and rate of formation of NADW are currently waning and the Greenland ice sheet is melting; it has lost about 10^4 km^3 of ice in the past century (the rate of loss is less than 0.01 Sv).[16] This abbreviated summary is highly simplified; the atmosphere-ocean climate system is highly complex. For a more detailed description of deep-ocean structure and flows, see Rahmstorf (2006) and Schmittner et al. (2007) and for a discussion of variability of the thermohaline circulation and AMOC, see Clement and Peterson (2008).

Transition

In this chapter we have surveyed oceanic currents, which consist of broad, near-surface wind-driven currents, western-boundary currents and thermohaline circulation. These currents have been surveyed separately, but in fact they are parts of a complicated flow system that is strongly coupled to atmospheric flows and affects weather and climate.

In the next chapter, we will shift our attention to atmospheric vortices.

[16] See Rhamstorf et al. (2015).

28

Vortices

Vortices are important because they are an intrinsic feature of the most damaging storms that occur on earth: hurricanes[1] and tornadoes. Also, vortices are an interesting example of nonlinear interaction between a rotating flow and the structure that spontaneously develops due to the dynamic constraint imposed by the Proudman–Taylor theorem.[2]

This chapter begins in the following section with a survey of various types of vortices and a brief discussion of their dynamics. The equations governing axisymmetric vortices are introduced in § 28.2, then two simple vortices are investigated in § 28.3. These simple vortices are the basis for the more realistic models that are developed in § 28.4. Finally in § 28.5 we briefly discuss hurricanes and consider their efficiency in converting heat to kinetic energy.

28.1 Survey of Vortices

A *vortex* is a swirling mass of fluid; it is a three-dimensional structure in which fluid flows roughly symmetrically about an axis.[3] There are many kinds of atmospheric vortices, ranging in size and strength from the flow at a street corner that rustles leaves around in a circle to hurricanes that devastate coastlines. On the other hand, only one type of vortex occurs in water: the *whirlpool*.

Vortices may be categorized as either barotropic or baroclinic. These two types are discussed in the following subsections.

28.1.1 Barotropic Vortices

Barotropic vortices arise as a nonlinear consequence of the instability of a basic-state shear. They often are abetted by topography and are distinguished from baroclinic vortices by the lack of a source of energy other than the kinetic energy of the flow. A familiar example of a barotropic vortex is that generated as wind blows past a building. A down-wind corner

[1] These large storms have differing names - hurricane, typhoon, cyclone - depending on the geographic region in which they occur. To simplify the presentation, all are referred to as hurricanes.

[2] See Appendix D.9.

[3] Circular fluid motion lacking a well-defined axis is called an *eddy* or *gyre*.

of the building is a singular point in the flow-field, where *flow separation* occurs,[4] with a weak vortex forming in the lee of the building.

Whirlpools – vortices in water – commonly form in rivers and near narrow channels that have strong tidal flows by the same mechanism: flow separation at an irregular boundary, with the whirlpool being a region of strong re-circulation in the lee of the obstacle;[5] they rarely occur in open waters. Whirlpools are observed as a depression of the water surface. Whirlpools are dynamically different from most atmospheric vortices, which do not require topography for their generation or maintenance.

Barotropic vortices, which focus the existing kinetic energy of a flow, are typically less vigorous and destructive than baroclinic vortices that gain kinetic energy from the release of gravitational potential energy. Consequently our studies will focus on baroclinic vortices.

28.1.2 Baroclinic Vortices

Baroclinic vortices occur in the atmosphere due to the nonlinear interaction between a swirling flow field, a source of buoyancy and ambient vorticity. The source of buoyancy is typically the moist air that fuels convective clouds, but it also can simply be very hot air (perhaps laced with hot dust particles). As convective motion is initiated, a localized patch of air rises and is replaced by horizontal convergence of air near the ground. This convergence does two things. First, it replenishes the supply of moist or hot air. Second, it advects vorticity toward the updraft, forming a vortex that provides structure to the flow. The rotational structure of the vortex controls and reinforces the inflow, particularly within the turbulent boundary layer near the ground. Typically, the buoyancy force that drives the flow is fairly large-scale and diffuse, while the strong vertical flow occurs on a smaller scale. On the scale of the vortex itself, the buoyancy force is small and can be ignored in a first approximation.

Baroclinic vortices[6] in the atmosphere include – in order of increasing size and strength:

- dust devils;
- water spouts;
- tornadoes; and
- hurricanes.

The last two of these are the most damaging, and the dangers they pose are categorized by two similar rating systems: the *Fujita scale* for tornadoes and the *Saffir–Simpson* scale for hurricanes, as shown in Table 28.1. Winds slower than those included in these scales are rated using the Beaufort scale; see Table 11.1. A *tropical storm* has an organized circulation

[4] See Appendix C.5.3.
[5] Small small-scale whirlpools are readily produced by the movement of an oar or paddle through the water.
[6] This discussion is limited to vortices having vertical - or nearly vertical - axes and excludes horizontal rollers. It also does not include vortices produced by the presence of topography, buildings, etc.

Table 28.1. *Enhanced Fujita and Saffir–Simpson scales. The Fujita scale is used to rate tornadoes and the Saffir–Simpson scale is used to rate hurricanes.*

Rating	Enhanced Fujita speed in m·s^{-1} (miles per hour)	Saffir–Simpson speed in m·s^{-1} (miles per hour)
0	29–37 (65–85)	—
1	38–49 (86–110)	33–42 (74–95)
2	50–60 (111–135)	43–49 (96–110)
3	61–73 (136–165)	50–58 (111–129)
4	74–90 (166–200)	59–69 (130–156)
5	90– (200–)	70– (157–)

and sustained winds of $17-32$ m·s^{-1} ($38-73$ miles per hour, $33-62$ knots, Beaufort force $8-11$), while a *tropical depression* has an organized circulation and sustained winds < 17 m·s^{-1} (< 38 miles per hour, < 33 knots, Beaufort force 7 or less). Tropical depressions are identified by numerals, while tropical storms and hurricanes are identified by names.

Atmospheric vortices may be categorized by their aspect ratio, that is, lateral extent divided by height. In this classification hurricanes, having large aspect ratio (roughly 10 to 200), are in a class by themselves, while tornadoes, waterspouts and dust devils all have small aspect ratios (roughly 1/5 to 1/50). All vortices interact with the ground and it is this interaction that produces the damage seen in the wake of a tornado or hurricane. The importance of ground interaction to the dynamics of an atmospheric vortex depends in large part on its aspect ratio. Small-scale vortices having small aspect ratio are not as strongly affected by interaction with the ground as hurricanes are.

In what follows, we will focus our attention on the two most important atmospheric vortices: tornadoes and hurricanes. From a dynamic point of view, the smaller fry (dust devils and waterspouts) may be considered to be weak tornadoes, while tornadoes and hurricanes are two very distinct types of storms; as noted by Lewellen (1993), "The hurricane is a complete, large storm system, while the tornado is an appendage to a thunderstorm."

28.1.3 Vortex Dynamics

Vortices are complex dynamic structures that are created and maintained by the interaction of five ingredients:

- buoyancy;
- ambient vorticity;
- meridional flow;
- viscous drag near the axis; and
- viscous drag near the ground.

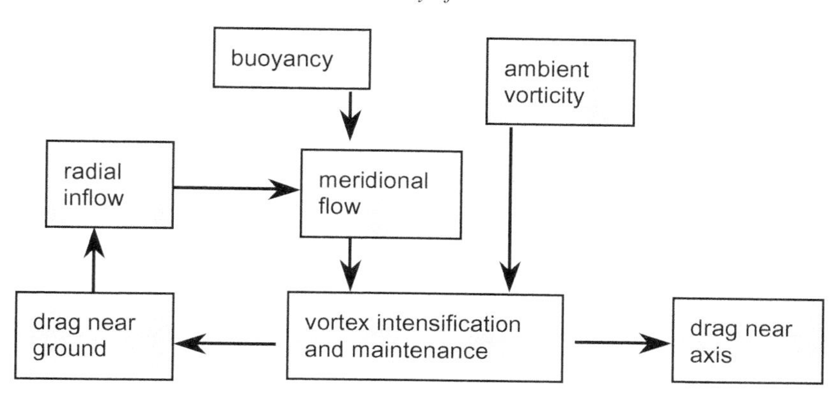

Figure 28.1 Schematic of the ingredients that create, maintain and degrade an atmospheric vortex. Energy is supplied by buoyant instability and degraded by drag near the ground and near the vortex axis.

The meridional flow consists of two components: a large-scale flow driven by buoyancy and a flow driven by a pressure imbalance within the boundary layer near the ground. The interaction among these components is illustrated in Figure 28.1, with the arrows schematically indicating influence and transfer of energy. Vortex formation is due largely to vigorous large-scale buoyancy-driven meridional flow that concentrates the ambient vorticity. Once the vortex is established, meridional flow is dominated by the radial inflow associated with the boundary layer near the ground.

Commonly vortices arise from instability of a shearing flow; this process is analyzed in Appendix G.1. In order to develop, a vortex needs a mechanism whereby the ambient vorticity is intensified. Since the atmosphere is inviscid to a good approximation, the only mechanism available is vortex stretching, which consists of vertical extension and horizontal convergence of flow near the surface. Typically the vertical extension is induced by convection. The convective motions associated with hurricanes and tornadoes are very evident due to the presence of impressive cumulonimbus clouds, while the clouds associated with water spouts often look very innocuous and dust devils rarely are accompanied by clouds. The energy for convection is supplied by the release of latent heat if clouds are present (as discussed in § 24.3) or by strong heating of the ground if clouds are absent. We will not quantify development of meridional flow by the buoyancy forces, but will simply take that flow as given, and focus our attention on fully developed, steady vortices. By the same token, in this preliminary study of vortices we will ignore the large-scale buoyancy force.

Vortices are characterized by a symmetric (or nearly symmetric) circulation[7] about a vertical (or nearly vertical) axis. This circulation is the result of the advection of ambient vorticity toward the axis of the vortex by the meridional flow. The ambient vorticity for hurricanes is due to the rotation of Earth and represented by the Coriolis parameter, while

[7] Circulation is introduced in § 8.1.3.

the other vortices utilize local vorticity that is enhanced by some mesoscale (that is, larger than the scale of the vortex but smaller than the scale of weather systems) mechanism. Mesoscale vorticity may be enhanced, for example, along a weather front or squall line or within a rotating supercell (large-scale cumulonimbus cloud). It follows that hurricanes always rotate *cyclonically*: counterclockwise in the northern hemisphere and clockwise in the southern when viewed from above, while other atmospheric vortices can rotate either way. Tornadoes commonly form near a cold front that has a strong shear that is {counterclockwise, clockwise} in the {northern, southern} hemisphere, so that they rotate cyclonically. In the following analysis, we will limit our attention to cyclonic vortices in the northern hemisphere, having a positive Coriolis parameter.[8]

Two features common to all atmospheric vortices are a two-cell structure and strong updrafts near their center. We will see in § 28.4.2 how viscous drag near the vortex axis causes the two-cell structure and in § 28.4.3 and § 28.4.4 we will see that the strong updraft arises due to the dynamics of the boundary layer near the ground. Another feature of vortices is a low central pressure. The central pressure in a tornado is usually sufficiently low to induce condensation of water vapor, making the eye of a tornado very visible. However, a tornado eye can exist without condensation and a common misconception is that a tornado is not present if an eye is not visible.

This completes our qualitative discussion of atmospheric vortices. Now we will turn our attention to the components that make up these intriguing flows. We begin in the following section with a summary of the relevant governing equations.

28.2 Vortex Equations

We begin our investigation of vortex dynamics making the following simplifying assumptions concerning the flow structure and dynamics:

- the fluid is incompressible with uniform density;
- flow is independent of time;
- the vortex is axisymmetric with all variables independent of the azimuthal angle ϕ; and
- turbulent viscosity affects
  - the radial flow only near the ground;
  - the azimuthal flow near the vortex axis and near the ground; and
  - the axial flow not at all.

The assumption of incompressibility is made to simplify the mathematical development. It introduces quantitative errors in our analysis, but compressibility is not an essential physical feature of strong vortices. The viscosity has been included only in those regions – and for those components – having strong shear.

Due to their geometry, vortices are best investigated using cylindrical coordinates ϖ, ϕ and z with the z axis being an axis of symmetry. Writing the velocity as $\mathbf{v} = u\mathbf{1}_{\varpi} + v\mathbf{1}_{\phi} +$

[8] This limitation is for convenience, to avoid cluttering the equations with absolute value signs.

$w\mathbf{1}_z$, the Navier–Stokes and continuity equations[9] are

$$u\frac{\partial u}{\partial \varpi}+w\frac{\partial u}{\partial z}-\frac{v^2}{\varpi}-f_0 v=-\frac{1}{\rho_r}\frac{\partial p'}{\partial \varpi}+\frac{\partial}{\partial z}\left(\nu_z\frac{\partial u}{\partial z}\right),$$
$$u\frac{\partial v}{\partial \varpi}+w\frac{\partial v}{\partial z}+\frac{uv}{\varpi}+f_0 u=\frac{\partial}{\partial \varpi}\left(\frac{\nu_\varpi}{\varpi}\frac{\partial(\varpi v)}{\partial \varpi}\right)+\frac{\partial}{\partial z}\left(\nu_z\frac{\partial v}{\partial z}\right),$$
$$u\frac{\partial w}{\partial \varpi}+w\frac{\partial w}{\partial z}=-\frac{1}{\rho_r}\frac{\partial p'}{\partial z}$$

and

$$\frac{\partial(\varpi u)}{\partial \varpi}+\frac{\partial(\varpi w)}{\partial z}=0,$$

where a prime denotes a perturbation of the hydrostatic state. Note that we are using the f-plane approximation – treating the Coriolis parameter as a positive constant.[10]

28.2.1 Parameterization of Turbulent Viscosity

The parameters ν_ϖ and ν_z are the turbulent viscosities near the vortex axis and ground, respectively. We saw in § 23.5.2 that the turbulent viscosity near a horizontal boundary may be parameterized by

$$\nu_z=\hat{\varepsilon}U(z+\hat{z}),$$

where $\hat{\varepsilon}$ is a small (roughly 0.02) dimensionless parameter, U is azimuthal speed outside the boundary layer,[11] z is the distance from the boundary and $\hat{z}$ is the scale of boundary roughness. By analogy, the turbulent viscosity near the axis may be expressed as the product of the speed and the distance from the axis:

$$\nu_\varpi=\hat{\varepsilon}U\varpi\,.$$

If the flow outside the boundary layer is approximately a potential vortex[12] with $U=\Gamma_0/(2\pi\varpi)$, where Γ_0 is a prescribed constant, then the radial turbulent viscosity is constant:

$$\nu_\varpi=\hat{\varepsilon}\Gamma_0/(2\pi)\,.$$

Recall that we are limiting our attention to cyclonic vortices in the northern hemisphere, for which f and v have the same sign.

[9] From § 7.3.2.
[10] See § 8.1.5.
[11] This speed may be a function of horizontal position.
[12] See § 28.3.1.

28.3 Simple Vortices

The governing equations possess two surprisingly simple solutions if the ambient vorticity has been concentrated on – or close to – the axis of the vortex. If we assume

- no background rotation ($f = 0$);
- turbulent viscosity near the vortex axis is constant;
- no meridional flow ($u = w = 0$); and
- v and p' are independent of z and t,

the continuity and axial momentum equations are identically satisfied and the radial and zonal components simplify to

$$\frac{v^2}{\varpi} = \frac{1}{\rho_r}\frac{\partial p'}{\partial \varpi} \qquad \text{and} \qquad 0 = \frac{\partial}{\partial \varpi}\left(\frac{\partial(\varpi v)}{\varpi\, \partial \varpi}\right).$$

Note that the factor v^2/ϖ is just the centripetal acceleration first encountered in § 4.6.2. The pressure associated with a simple vortex is readily obtained by integration

$$p' = -\rho_r \int_{\varpi}^{\infty} \frac{v^2(\grave{\varpi})}{\grave{\varpi}}\, \mathrm{d}\grave{\varpi}\,.$$

If the fluid were inviscid, the zonal momentum equation would be satisfied for any function $v(\varpi)$. If the fluid is viscous, this equation is satisfied provided v varies with either ϖ or $1/\varpi$. The first choice is rigid-body motion of the fluid, while the second is a potential vortex. This latter solution is investigated in the following subsection, while the two are combined in § 28.3.2 to form a compound vortex.

28.3.1 Potential Vortex

The simplest vortex is the *potential vortex* having velocity

$$\mathbf{v} = v\mathbf{1}_\phi = \frac{\Gamma}{2\pi\,\varpi}\mathbf{1}_\phi\,,$$

where the circulation[13] Γ is a constant, ϖ is the distance from the axis and $\mathbf{1}_\phi$ is a unit vector pointing in the angular direction about that axis. This solution satisfies all our governing equations, even with arbitrary radial flow and non-zero (constant) viscosity.

An interesting feature of the potential vortex is that it has non-zero circulation, but the vorticity of the fluid flow is zero everywhere except at $\varpi = 0$, where the velocity and vorticity are infinite. A second feature of interest is that parcels of fluid in this flow have constant angular momentum. It follows that, if two parcels exchange positions (say, during turbulent motion) without the application of torques, then they will be dynamically indistinguishable compared with their new surroundings. This explains in part why the

[13] Circulation was introduced in § 8.1.3.

solution is valid for any magnitude of viscosity. But that brings up a worrisome point. If the fluid is viscous, then dissipation must degrade the fluid motions and slow the vortex. Another unphysical feature of the potential vortex is the singularity on the axis. In the following section, we will investigate a vortex that lacks this singularity, but this does not get us very far in our search for a realistic model of a vortex.

28.3.2 Rankine Vortex

We have noted that with constant viscosity the momentum equation is satisfied provided v is proportional to ϖ or $1/\varpi$. With this in mind, we can modify the potential vortex by adding a rigidly rotating core; the result, called the *Rankine vortex*, may again be expressed as $v = \Gamma/(2\pi\varpi)$, but now with

$$\Gamma = \Gamma_0 \begin{cases} \varpi^2/\varpi_0^2 & \text{for} \quad 0 \le \varpi < \varpi_0 \\ 1 & \text{for} \quad \varpi_0 \le \varpi \,, \end{cases}$$

where ϖ_0 is a specified value of the (cylindrical) radius and Γ_0 represents the vortex strength. This vortex is a crude approximation to real vortices, which always have an *eye*: a central region that is relatively calm and clear.

The pressure deficit at the center of a Rankine vortex (the pressure at $\varpi = \infty$ minus that at $\varpi = 0$) is

$$\Delta p = \rho_r v_{max}^2, \qquad \text{where} \qquad v_{max} = \Gamma_0/(2\pi\varpi_0)$$

is the maximum wind speed. Typically the lower pressure at the center of a vortex is induced and maintained by the buoyancy force; a (negative) density deficit ρ' acting over a vertical distance L induces a pressure deficit $\Delta p = \rho' g L$. A thermally induced density deficit is related to the temperature excess T' by the density equation of state.[14] With $\alpha = 1/T_r$ for air,[15] this relation is $\rho'/\rho_r = -T'/T_r$ and

$$v_{max} = \sqrt{(T'/T_r)\,gL}\,.$$

With $g \approx 10$ m·s^{-2}, $L \approx 4$ km and $T'/T_r \approx 0.06$, say, $v_{max} \approx 50$ m·s^{-1}, which is a weak category 2 tornado; see Table 28.1.

The Rankine vortex is an improvement on the potential vortex, but it suffers a similar problem: singular structure at $\varpi = \varpi_0$. We can clearly see that the assumption of steady flow at the rim of the core (at $\varpi = \varpi_0$) is physically untenable, because the fluid in either side is moving more slowly. This flow cannot be maintained in the presence of dissipation without the addition of a mechanism that acts to maintain the structure. That mechanism must involve flow in the meridional ($\varpi - z$) plane. This idea, along with others, is pursued in the following section.

[14] See § 5.2 and Appendix E.5.
[15] See Appendix E.8.5.

28.4 Approach to Realistic Vortices

In this section we investigate the structure of vortex flow with the addition of the remaining ingredients of real vortices: meridional flow (in § 28.4.1), drag near the axis (in § 28.4.2) and drag near the ground (in § 28.4.3). The latter induces a strong radial inflow within the boundary layer near the ground that is balanced by outflow within the corner region (see § 28.4.4). These flow elements are illustrated in Figure 28.2.

28.4.1 Meridional Flow

Let's start with an axisymmetric vortex having circulation $\Gamma(\varpi)$ within an inviscid fluid ($\nu_T = 0$) and add meridional flow in the form of axisymmetric *stagnation-point flow* with stream function $\Psi = \varpi^2 z/t_0$, where t_0 is a specified interval of time representing the strength of the meridional flow, so that

$$u = u_m \equiv -\varpi/t_0 \qquad \text{and} \qquad w = w_m \equiv 2z/t_0\,.$$

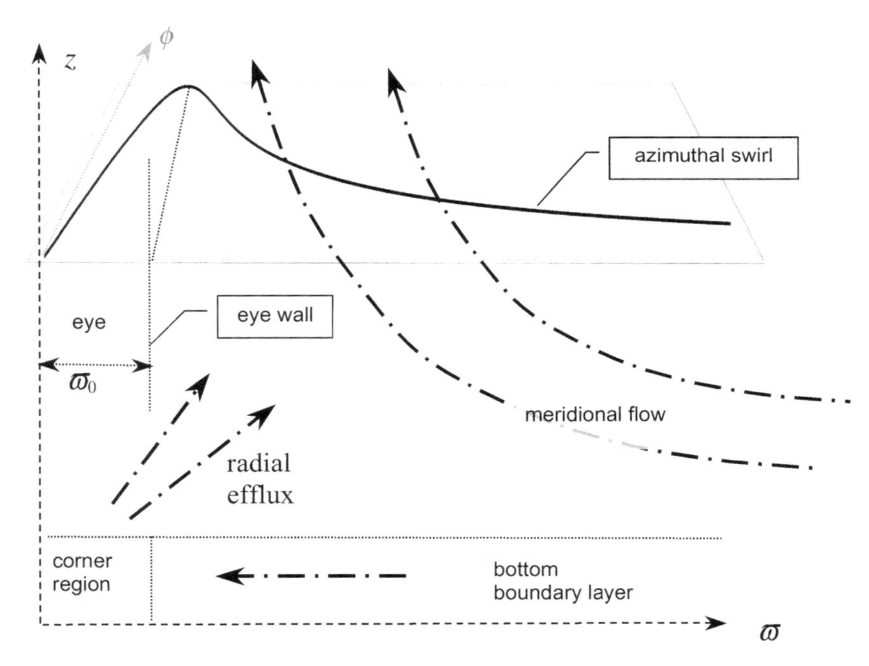

Figure 28.2 A schematic drawing of the elements of an atmospheric vortex. The solid curve shows a perspective representation of the Burgers–Rott vortex (azimuthal speed – in the ϕ direction), investigated in § 28.4.2. The dash-dotted lines denote meridional flows; the curves depict the stagnation-point flow investigated in § 28.4.1, the arrow pointing toward the origin depicts the flow in the bottom boundary layer investigated in § 28.4.3 and the arrows pointing away from the origin depict the outflow from the corner flow investigated in § 28.4.4.

This assumed form simulates the radial inflow and updraft associated with the formation of tornadoes and other vortices. Note that this flow is well behaved at $\varpi = 0$.

The continuity equation is satisfied and the components of the momentum equation[16] simplify to

$$\frac{\varpi}{t_0^2} - \frac{\Gamma^2}{\varpi^3} = -\frac{1}{\rho_0}\frac{\partial p'}{\partial \varpi}, \qquad 4\frac{z}{t_0^2} = -\frac{1}{\rho_0}\frac{\partial p'}{\partial z}$$

and

$$\frac{\partial \Gamma}{\partial \varpi} = 0.$$

The addition of meridional circulation does not affect the potential vortex (having Γ constant), but renders the rigid-body portion of the Rankine vortex (having $\Gamma \propto \varpi$ constant) untenable. The radial inflow tends to collapse the Rankine core, reinstating the potential vortex. We will find in the next section that this tendency for collapse is resisted by the azimuthal viscous force.

28.4.2 Burgers–Rott Vortex

Let's add the next ingredient – the azimuthal viscous drag near the axis of the vortex – to our recipe for a realistic vortex. Assuming that the circulation Γ is a function only of the cylindrical radius ϖ, the azimuthal momentum equation[17] becomes

$$\frac{\mathrm{d}}{\mathrm{d}\varpi}\left(\frac{\mathrm{d}\Gamma}{\varpi\,\mathrm{d}\varpi}\right) + \frac{2\pi}{\hat{\varepsilon}\Gamma_0 t_0}\frac{\mathrm{d}\Gamma}{\mathrm{d}\varpi} = 0.$$

This equation is to be solved in the domain $0 < \varpi < \infty$ subject to the conditions that far from the axis the solution is a potential vortex ($\Gamma \to \Gamma_0$ as $\varpi \to \infty$) and the swirl is zero on the axis ($\Gamma = 0$ at $\varpi = 0$). This problem is readily solved:

$$\Gamma = \Gamma_0\left(1 - e^{-\varpi^2/\varpi_0^2}\right), \qquad \text{where} \qquad \varpi_0 = \sqrt{\hat{\varepsilon}\Gamma_0 t_0/\pi}$$

is a characteristic radius. This is known as the *Burgers–Rott vortex*. As ϖ/ϖ_0 becomes large, the exponential term becomes very small and the solution tends rapidly (exponentially) to the potential-vortex solution, but as ϖ/ϖ_0 tends to zero, the viscous force becomes dominant and the solution tends toward rigid rotation. Note that this solution has a boundary-layer structure,[18] with $\Gamma = \Gamma_0$ being the interior solution and the exponential $-\Gamma_0 e^{-\varpi^2/\varpi_0^2}$ being the boundary-layer portion.

[16] Given in § 28.2.

[17] From § 28.2, with the radial turbulent viscosity parameterized by $\nu_\varpi = \hat{\varepsilon}\Gamma_0/(2\pi)$ (see § 28.2.1) and $u = -\varpi/t_0$ (see § 28.4.1).

[18] See Appendix G.2.

The pressure associated with this flow consists of three parts: due to meridional flow, swirl and back-ground rotation. The pressure deficit at the center of this vortex due to the swirl is

$$\Delta p = \ln(2)\rho_r \Gamma_0^2/\varpi_0^2 .$$

The central pressure of a vortex increases as the square of the circulation and decreases as the inverse square of the eye radius. This explains why the central pressure is so low and wind speed so high within compact hurricanes. Hurricane Wilma (2005) had both the lowest eye pressure (882 millibar) and smallest eye ($\varpi_0 = 1.85$ km) measured in an Atlantic hurricane, along with a wind speed of 295 km/hr (82 m·s^{-1}), only slightly less than the record speed for Atlantic hurricanes held by Hurricane Camille (1969).[19]

The low pressure in the center of the vortex induces a downward flow of air within the central cell (that is, the eye) of the vortex. Due to adiabatic compression this air is typically warm and dry.[20]

The incorporation of azimuthal viscous drag – coupled with meridional flow – into our model of a vortex is a big step forward; we have found a vortex solution having the appearance of a smoothed Rankine vortex – the singularity is gone. The characteristic radii of atmospheric vortices range from about one meter for dust devils, through 5–75 m for water spouts and 100–1000 m for tornadoes to 20–30 km for hurricanes. The next ingredient to add to our model is drag near the ground.

28.4.3 Vortex Bottom Boundary Layer

Finally, let's add the last ingredient: the bottom boundary layer. All atmospheric vortices interact strongly with the bottom surface (ground or water) across a turbulent viscous boundary layer. This layer is an important component in the dynamics of strong vortices because there is a large radial inflow within it and this flow typically conveys air having a large potential temperature and a lot of moisture. When expelled from the boundary layer close to the vortex axis, this air feeds the convective motions that power the vortex.

This boundary layer may be thought of as a nonlinear version of the Ekman layer studied in Chapter 25. A characteristic feature of this boundary layer is that it is thin, with axial derivatives much larger than radial: $|\partial/\partial\varpi| \ll |\partial/\partial z|$. This feature permits the axial momentum equation to be reduced to $\partial p'/\partial z = 0$ so that the perturbation pressure depends only on ϖ. (The validity of this approximation may be verified after the solution has been obtained.) This pressure is established by the *gradient-wind balance* outside the boundary layer:

$$\frac{1}{\rho_r}\frac{\mathrm{d}p'}{\mathrm{d}\varpi} = \frac{U^2}{\varpi} + f_0 U ,$$

[19] The highest recorded speed for a hurricane is 345 km/hr (96 m·s^{-1}), jointly held by Typhoon Nancy (1961) in the western Pacific and Hurricane Patricia (2015) in the eastern Pacific and the lowest pressure recorded was 870 millibar in Typhoon Tip (1979).

[20] See § 7.2.1.

where v_a is the azimuthal speed outside the layer. Within the boundary layer, viscous forces squelch the swirl, but not the pressure gradient and this imbalanced pressure drives a strong radial inflow - much stronger than the Ekman-layer flows investigated in Chapter 25 or the meridional flow of § 28.4.1. A primary goal of the analysis of this section is to determine this radial inflow and the rate with which this flow is expelled from the boundary layer, but due to the complexity of the mathematical model of these flows, we only get part way to our goal.

Ignoring the radial viscous term, the radial and azimuthal momentum equations from § 28.2 may be expressed as

$$\frac{\partial}{\partial z}\left(\nu_z \frac{\partial u}{\partial z}\right) = -f_{\varpi}\,(v-U) + u\frac{\partial u}{\partial \varpi} + w\frac{\partial u}{\partial z},$$
$$\frac{\partial}{\partial z}\left(\nu_z \frac{\partial v}{\partial z}\right) = f_{\phi} u + w\frac{\partial v}{\partial z} \quad \text{and} \quad \frac{\partial(\varpi u)}{\varpi\,\partial \varpi} + \frac{\partial w}{\partial z} = 0,$$

where

$$f_{\varpi} = f_0 + \frac{U+v}{\varpi} \quad \text{and} \quad f_{\phi} = f_0 + \frac{\partial(\varpi v)}{\varpi\,\partial \varpi}$$

are the effective (nonlinear) Coriolis parameters in the radial and azimuthal directions, respectively. With the flow being turbulent, our preferred parameterization of viscosity is $\nu = \hat{\varepsilon} U(z+\hat{z})$.[21] These equations are to be solved on the domain $0 < \varpi < \infty$ and $0 < z < \infty$, subject to the conditions that

$$\begin{aligned}
&\text{at} \quad \varpi = 0: \quad && u = 0; \\
&\text{at} \quad z = 0: \quad && u = u_b, \quad v = v_b, \quad w = w_b; \quad \text{and} \\
&\text{as} \quad z \to \infty: \quad && v \to U \quad \text{and} \quad u \to 0.
\end{aligned}$$

These conditions deserve comment. It is important to recall that the level $z = 0$ is at the top of the rough layer on the ground (see Figure 23.5) and realize that there can be appreciable flow within that layer. That is, all three velocity components are small – but non-zero – at the top of the layer. These possible flows (denoted by subscript b) are included in the formulation out of caution. In most cases, they may be set equal to zero. The last condition states that the radial speed tends to zero outside the boundary layer. This is an approximate condition, since there will be a non-zero radial flow outside the layer, but it is of smaller order than that within the layer.

This is one of the most challenging problems we have encountered; we need to unravel a set of coupled nonlinear partial differential equations. Our goal is reasonably modest: to quantify the flow within the boundary layer, so we do not need to solve for the entire flow structure; a relatively crude model of the flow should be sufficient. But at this point, we know almost nothing of the structure of the boundary layer. In order to get a peek behind the curtain, let's investigate a linearized version of this problem.

[21] See § 28.2.1.

Linearized Solution

If we ignore the inertia terms completely, the vortex boundary layer becomes the atmospheric Ekman layer, summarized in § 25.5.1. We can use this as a guide to developing a linearized solution that includes the effect of inertia. The linearization consists of writing $v = U + v'$ and linearizing the momentum equations by neglecting squares and products of u, v' and w. Note that U may depend on ϖ, but not on z. The linearized momentum equations are

$$\hat{\varepsilon} U \frac{\partial}{\partial z}\left((z+\hat{z})\frac{\partial u}{\partial z}\right) = -f_{\varpi} v' \quad \text{and} \quad \hat{\varepsilon} U \frac{\partial}{\partial z}\left((z+\hat{z})\frac{\partial v'}{\partial z}\right) = f_{\phi} u$$

where now

$$f_{\varpi} = f_0 + \frac{2U}{\varpi} \qquad \text{and} \qquad f_{\phi} = f_0 + \frac{\mathrm{d}(\varpi U)}{\varpi \,\mathrm{d}\varpi}.$$

For a wide range of swirl profiles, $f_0 \le f_{\phi} < f_{\varpi}$. For example with a potential vortex ($U = \Gamma_0/2\pi\varpi$), $f_{\phi} = f_0$ and $f_{\varpi} = f_0 + \Gamma_0/\pi\varpi^2$.

The linearization preserves the qualitative structure of the nonlinear equations, and the solution of the linearized equations should provide some insights into the structure of and flow within the vortex boundary layer. A distinct advantage of the linearized equations is their structural similarity to the Ekman layers studied in § 25.1.2, except that the Coriolis factors are no longer equal, with f_{ϖ} greater than f_{ϕ}. We can anticipate that, in compensation, u will be larger or v' smaller – most likely the former.

We can combine the linearized momentum equations into a single complex equation by multiplying the radial equation by $\sqrt{f_{\phi}/f_{\varpi}}\,\mathrm{i}$ and subtracting it from the azimuthal equation:

$$L_v \frac{\partial}{\partial z}\left((z+\hat{z})\frac{\partial \mathrm{W}}{\partial z}\right) = 2\mathrm{i}\mathrm{W},$$

where

$$\mathrm{W} \equiv v - U - \mathrm{i}\sqrt{f_{\phi}/f_{\varpi}}\,u, \qquad \text{and} \qquad L_v \equiv 2\hat{\varepsilon} U/\sqrt{f_{\varpi} f_{\phi}}$$

This equation is to be solved for $\mathrm{W}(z)$ on the interval $0 < z < \infty$ subject to the conditions[22]

$$\text{at} \quad z = 0: \quad \mathrm{W} = -U \qquad \text{and as} \quad z \to \infty: \quad \mathrm{W} \to 0.$$

This problem is nearly identical to that governing turbulent katabatic winds, which we developed and solved in § 24.1. With minor modification, we can employ that solution here:

$$\mathrm{W} = -U(C - \mathrm{i}S),$$

[22] Ignoring the rough-layer flow by setting $u_b = v_b = w_b = 0$.

where again

$$C = \frac{\ker(\hat{\xi})\ker(\xi) + \mathrm{kei}(\hat{\xi})\mathrm{kei}(\xi)}{\ker^2(\hat{\xi}) + \mathrm{kei}^2(\hat{\xi})}, \qquad S = \frac{\mathrm{kei}(\hat{\xi})\ker(\xi) - \ker(\hat{\xi})\mathrm{kei}(\xi)}{\ker^2(\hat{\xi}) + \mathrm{kei}^2(\hat{\xi})},$$

$$\xi = \sqrt{8(z+\hat{z})/L_v} \qquad \text{and} \qquad \hat{\xi} = \sqrt{8\hat{z}/L_v}\,.$$

The graphs of the functions C and S versus z/L_v are as seen in Figure 24.1 panel (b). The velocity components are

$$u = -\sqrt{f_\varpi/f_\phi}\,US \qquad \text{and} \qquad v = U(1-C)\,.$$

The minus sign on u indicates that the radial flow is directed inward, toward the vortex axis. With $f_\phi < f_\varpi$, the linearized inertial terms cause an increase in the radial inflow within the vortex boundary layer, as we had anticipated.

The linearized radial transport within the boundary layer may be calculated as in § 25.2.2:

$$q_\varpi = \int_0^\infty u\,\mathrm{d}z = -S_r U L_v\,, \qquad \text{where} \qquad S_r = \sqrt{\frac{f_\varpi}{f_\phi}}\frac{R_S}{2}$$

is the swirl ratio and the factor R_S is again given by

$$\begin{aligned} R_S &= \frac{1}{2}\int_{\hat{\xi}}^\infty S\,\xi\,\mathrm{d}\xi \\ &= \frac{\hat{\xi}\,\mathrm{kei}(\hat{\xi})\left(\ker_1(\hat{\xi}) - \mathrm{kei}_1(\hat{\xi})\right) - \hat{\xi}\ker(\hat{\xi})\left(\ker_1(\hat{\xi}) + \mathrm{kei}_1(\hat{\xi})\right)}{2\sqrt{2}\left(\ker^2(\hat{\xi}) + \mathrm{kei}^2(\hat{\xi})\right)}\,. \end{aligned}$$

The factor $R_S(\hat{\xi})$ is plotted in panel (a) of Figure 25.2. Typically R_S is small relative to unity. Using the definition $L_v \equiv 2\hat{\varepsilon}U/\sqrt{f_\varpi f_\phi}$, we have

$$q_\varpi = -R_S\hat{\varepsilon}U^2/f_\phi\,.$$

The specific radial transport within the bottom boundary layer of a vortex varies as the square of the azimuthal speed. If the vortex has the structure of a Burgers–Rott vortex, this formula predicts the speed and radial inflow to be greatest at the eye wall, but remember that this is only a linearized approximation.

Discussion

An important factor controlling the structure of the boundary layer is the *swirl ratio* S_r, introduced above; it is defined as the azimuthal speed outside the boundary layer, U, divided by the mean radial speed, $|q_\varpi|/L_v$:

$$S_r \equiv UL_v/|q_\varpi|\,.$$

For the linearized problem we found that $S_r = (2/R_S)\sqrt{f_\phi / f_\varpi}$. This ratio is large (roughly 20) in rotating layers, such as the Ekman layer, that are dominated by ambient rotation and described by linear dynamics, but inertia causes a decreases in its magnitude.

One drawback of the linearized solution is that it is local, with the variables dependent only on distance from the ground, z; when the full effect of inertia is taken into account, the flow depends on both ϖ and z. However, we can use the linearized solution to gain some insight into the radial variation of the boundary layer. Specifically we see that

- inertia causes the radial inflow within the boundary layer to increase;
- the increase in the mean Coriolis term $\sqrt{f_\varpi f_\phi}$ with increasing inertia tends to make the boundary layer thinner; and
- the swirl ratio, S_r, decreases as ϖ decreases.

To see this last point more clearly, let's assume the outer flow is a potential vortex, with $U = \Gamma_0 / 2\pi\varpi$ and Γ_0 constant, then $f_\varpi = f_0 + \Gamma_0/\pi\varpi^2$, $f_\phi = f_0$ and

$$S_r = \frac{2}{R_S}\sqrt{\frac{\pi f_0 \varpi^2}{\pi f_0 \varpi^2 + \Gamma_0}}.$$

As we move radially inward toward the vortex axis, the swirl ratio (initially much greater than unity because $R_S \ll 1$) decreases toward zero. The radial inflow speed increases as we move radially inward due to two effects: the inflow of air into the boundary layer (like an enhanced Ekman pumping) and geometric concentration. Close to the vortex axis, where the swirl ratio is small, the boundary layer acts as an annular jet of air flowing radially inward, with the swirl being an afterthought. The speed of this jet is the greatest within a vortex and, being close to the ground, is capable of causing the greatest damage.

If we are to avoid a singularity at $\varpi = 0$, the radial inflow must be diverted upward – out of the boundary layer. This occurs in the corner-region, described in the following section. Once the swirl ratio reaches roughly unity, the nature of the boundary-layer changes dramatically, with axial flow reversing direction from a boundary-layer inflow where $S_r > 1$ to out flow where $S_r < 1$.

28.4.4 Corner Region

The corner region of a vortex is that portion of the bottom boundary layer (close to the axis) in which the swirl ratio is small and flow is predominantly radially inward toward the axis. The intensity of flow in this region is seen in photographs and videos of tornadoes, with a cloud of debris being flung up and out from the corner region at an angle to the vertical. Of course, the air still swirling, so the trajectory of air, and debris sucked along with it, is a rising spiral. The structure of – and flow within – this region is investigated in this section.

Corner-Region Equations

Ignoring the Coriolis and radial viscous terms and dropping the subscript z on ν, the corner-region equations are

$$\frac{\partial(\varpi u)}{\partial \varpi}+\frac{\partial(\varpi w)}{\partial z}=0, \qquad \frac{\partial}{\partial z}\left(\nu\frac{\partial v}{\partial z}\right)=u\frac{\partial(\varpi v)}{\varpi\,\partial \varpi}+w\frac{\partial v}{\partial z}$$

and

$$\frac{\partial}{\partial z}\left(\nu\frac{\partial u}{\partial z}\right)=\frac{U^2}{\varpi}-\frac{v^2}{\varpi}+u\frac{\partial u}{\partial \varpi}+w\frac{\partial u}{\partial z},$$

with $U(\varpi)$ being a forcing function.[23] These equations are to be solved on the domain $0<\varpi<\infty$ and $0<z<\infty$ subject to the conditions listed in § 28.4.3. Since the flow speeds within this region are very large, we can ignore the small boundary speeds u_b, v_b and w_b and write the conditions as

$$\begin{aligned}
&\text{at} \quad && \varpi=0: \quad && u=0;\\
&\text{at} \quad && z=0: \quad && u=v=w=0;\\
&\text{as} \quad && z\to\infty: \quad && v\to U \quad \text{and} \quad u\to 0.
\end{aligned}$$

We are faced with a difficult problem, involving two coupled PDEs[24] containing an arbitrary forcing function $U(\varpi)$, subject to conditions at $\varpi=0$ and $z=0$ and as $z\to\infty$. These equations describe a boundary layer near the ground, having rapid variation in the z direction, with the ϖ variation being complicating factor. Since the boundary layer is thin, with $z\ll\varpi$, we can try to look for a local boundary-layer solution, governed by ODEs, with ϖ acting as a parameter.

In order to reduce the PDEs to ODEs, we need to unravel the z and ϖ dependencies. This process is guided by a scale analysis of the PDEs,[25] which shows us that the terms in the governing equations are in balance provided

$$z\sim\sqrt{\nu\varpi/v}, \qquad u\sim v\sim U \qquad \text{and} \qquad w\sim\sqrt{\nu v/\varpi}\,.$$

To make further progress, we need to decide on the parameterization of the viscosity, $\nu(z)$, and the radial structure of the outer flow, $U(\varpi)$. One helpful way to approach a complicated problem such as this is to generalize a known solution. In the limiting case that the viscosity is constant and the outer flow is a rigid body motion, the solution is known as Bödewadt flow. This solution would prevail in the eye of a vortex (close to the axis) provided the flow were laminar.

The Bödewadt solution is summarized in the following subsection. With this as a starting point, in § 28.4.4 we develop the similarity equations in the case that the flow is turbulent and the swirl velocity varies arbitrarily with ϖ.

[23] From § 28.4.3.
[24] PDE and ODE (used in the next paragraph) are acronyms for partial and ordinary differential equations, respectively.
[25] Introduced in § 2.4.5.

Bödewadt Flow

If the fluid is in rigid-body rotation outside the boundary layer and flow within the boundary layer is laminar, then U is constant and $U = \Omega\varpi$ with Ω being a known constant. The PDEs can be reduce to ODEs, and the continuity equation automatically satisfied, by writing

$$z = \sqrt{\nu/\Omega}\,\zeta \qquad \text{and} \qquad \{u, v, w\} = \{\Omega\varpi\,\tilde{u}, \Omega\varpi\,\tilde{v}, \sqrt{\nu\Omega}\tilde{w}\},$$

where the tilded variables are dimensionless functions of ζ. Now the problem simplifies to

$$\begin{aligned}
&\text{at} \quad \zeta = 0: && \tilde{u} = \tilde{v} = \tilde{w} = 0; \\
&\text{for} \quad 0 < \zeta < \infty: && \tilde{w}' = -2\tilde{u}, \qquad \tilde{v}'' = 2\tilde{u}\tilde{v} + \tilde{w}\tilde{v}' \\
&\text{and} \quad \tilde{u}'' = 1 - \tilde{v}^2 + \tilde{u}^2 + \tilde{w}\tilde{u}'; && \\
&\text{as} \quad \zeta \to \infty: && \tilde{u} \to 0 \quad \text{and} \quad \tilde{v} \to 1,
\end{aligned}$$

where a prime denotes differentiation with respect to ζ (and no longer denotes a perturbation). This problem was solved by U. T. Bödewadt in 1940; the flow profiles shown in Figure 28.3 have been produced using the data from Table 11.1 of Schlichting (1968). The asymptotic value of $\tilde{w}$ is 1.38 and the specific volume flow within the boundary layer

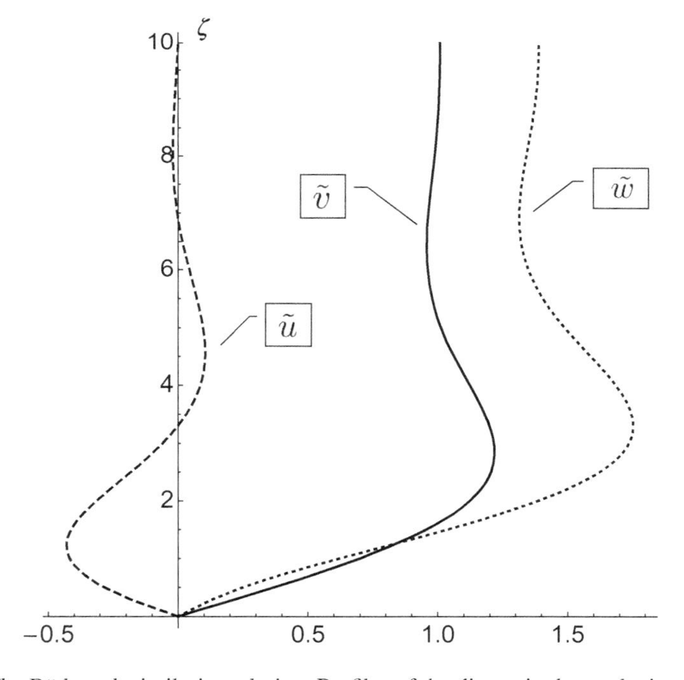

Figure 28.3 The Bödewadt similarity solution. Profiles of the dimensionless velocity components $\tilde{u}$, $\tilde{v}$ and $\tilde{w}$ versus the dimensionless distance from the boundary, ζ.

is $q = -1.38\sqrt{\Omega\nu}\varpi/2\pi$. The minus sign indicates that the radial flow is inward, toward the axis of rotation.

The boundary layer has a constant thickness similar to that of a laminar Ekman layer (see § 25.1), except that the relevant rate of rotation is that within the vortex eye (Ω), rather than the Coriolis term ($|f|$). Note that the normal (vertical) flow exits the boundary layer in Bödewadt flow. Usually vertical flow enters a boundary layer, with this inward advection countering the tendency for viscous diffusion to thicken the boundary layer. That is, the typical balance is $w\partial u/\partial z = \nu\partial^2 u/\partial z^2$ with $w < 0$. In Bödewadt flow, the radial advective term $u\partial(\varpi v)/\varpi\,\partial\varpi$ is negative and sufficiently large to balance the tendency for both the advective term $w\partial u/\partial z$ and diffusive term $\nu\partial^2 u/\partial z^2$ to thicken the layer. The dominance of this radial advection term is seen by the swirl ratio $S_r = 1.38/2\pi \approx 0.22$, which is small compared with unity. This feature of Bödewadt flow highlights the non-local nature of the vortex boundary layer. However, the non-local nature of the flow is encapsulated in the ansatz and we are able to quantify this flow by solving a set of ODEs. With these insights as encouragement, we now turn our attention to the problem with variable viscosity and rotation.

Variable Viscosity and Variable Rotation

With the kinematic viscosity parameterized[26] by $\nu = \hat{\varepsilon}Uz$ and assuming that the swirl speed within the boundary layer is comparable in magnitude to that outside, $v \sim U$, the boundary-layer scaling is $z \sim \sqrt{\nu\varpi/v} \sim \sqrt{\hat{\varepsilon}\varpi z}$ or equivalently

$$z \sim \hat{\varepsilon}\varpi\,.$$

This is a simple and interesting result: the boundary-layer thickness varies linearly with distance from the vortex axis, with the constant of proportionality being the small roughness parameter, $\hat{\varepsilon}$. This is a robust result, independent of the magnitude and structure of the azimuthal swirl speed, U. This flow structure avoids the problem of singularity on the vortex axis by simply vanishing at that axis. The small magnitude of $\hat{\varepsilon}$ assures validity of the assumption $|\partial/\partial\varpi| \ll |\partial/\partial z|$ and also assures that the vertical component of velocity, now scaled with $\hat{\varepsilon}U$, is small compared with the radial and azimuthal components.

The present scaling is valid provided this layer is much thinner than that of the turbulent Ekman layer. Using the scaling $L_v = 2\hat{\varepsilon}U/|f|$ from § 25.1.2, we see that the present model should be valid provided $\varpi \ll U/|f|$. Typically in mid-latitudes $|f| \approx 10^{-4}$ s^{-1}; if $U \approx 10$ m·s^{-1}, say, the present scaling is valid provided $\varpi \ll 100$ km.

Guided by this scaling, let's consider the ansatz

$$z = \hat{\varepsilon}\varpi\,\zeta \qquad \text{and} \qquad \{u, v, w\} = U\{\tilde{u}, \tilde{v}, \hat{\varepsilon}\tilde{w}\}\,.$$

[26] See § 23.5.2.

Now the problem becomes

$$\begin{array}{llll} \text{at} & \zeta = 0: & \tilde{u} = \tilde{v} = \tilde{w} = 0; & \\ \text{for} & 0 < \zeta < \infty: & \tilde{w}' = \zeta \tilde{u}' - \Xi \tilde{u}, & (\zeta \tilde{v}')' = \Xi \tilde{u}\tilde{v} + (\tilde{w} - \zeta \tilde{u})\tilde{v}' \\ \text{and} & (\zeta \tilde{u}')' = 1 - \tilde{v}^2 + (\Xi - 1)\tilde{u}^2 + (\tilde{w} - \zeta \tilde{u})\tilde{u}'; & & \\ \text{as} & \zeta \to \infty: & \tilde{u} \to 0 \quad \text{and} & \tilde{v} \to 1, \end{array}$$

where

$$\Xi \equiv \frac{1}{U} \frac{\mathrm{d}}{\mathrm{d}\varpi} (\varpi U)$$

quantifies the radial variation of swirl.

A general rotation may be expressed locally by a power-law relation: $U \sim \varpi^n$, with rigid rotation characterized by $n = 1$, a potential vortex by $n = -1$ and $-1 \leq n \leq 1$. For the Burgers–Rott vortex for example, Ξ is a monotonic decreasing function of ϖ, having a value of 2 on the vortex axis and decreasing asymptotically to 0 as $\varpi \to \infty$. Since the boundary layer is thin, Ξ may be treated as a specified (constant) parameter.

Sorry to let you down, but we have reached the end of this path; this problem has not been solved. It is time to drop this matter and turn our attention to hurricanes.

28.5 Hurricanes

Hurricanes are large, long-lived, intense rotating storms that involve the nonlinear interaction of several factors including ambient vorticity, convectively unstable air generated by evaporation of warm ocean water and an organized flow pattern. These factors combine and interact as follows. As air is drawn radially inward toward the eye of the hurricane,[27] its ambient vorticity is concentrated, causing the air to swirl. This swirling motion does two things. First, the Proudman–Taylor effect inhibits vertical variations, giving the hurricane a vertically uniform structure.[28] Second, the centrifugal force associated with the swirl is balanced by a pressure field that has a gradient directed radially outward. As with tornadoes, this swirl is squelched within the viscous boundary layer near the ground, and the unbalanced pressure field drives strong radial inflow within the layer, as described in § 28.4.3.

Ambient fluid is drawn into the boundary layers outside the eye regions of both tornadoes and hurricanes. However, in contrast to the tornado, the hurricane boundary layer has isolated patches of outflow that is organized into spiral bands of cumulonimbus convection. The vorticity of this outflowing air is greater than the air above the layer, causing the azimuthal velocity profile of the hurricane to have a variation with radius less than that of a potential vortex.

[27] By meridional circulation such as that investigated in § 28.4.1.
[28] See Appendix D.9 .

We found in § 28.4.2, the eye of a tornado results from the effect of radial viscous drag. The eye of a hurricane is too large and the air within too quiescent for it to be caused by the same mechanism. The factors determining the radius of the eye of a hurricane are not well understood; one possibility is that the boundary-layer instability responsible for producing the spiral bands becomes so extreme that the boundary layer completely breaks down at the eye-wall radius, preventing further radial inflow.

Hurricanes, like virtually all atmospheric flows, are driven by convective motions involving condensation of water vapor and release of latent heat of vaporization. Normally these motions involve a great deal of turbulent entrainment that tends to dilute the buoyant air and weaken the convection.[29] However, the relatively large vorticity associated with a hurricane inhibits turbulent entrainment and permits the convection to proceed more efficiently. As we shall see in the following section, in the extreme, this efficiency approaches the Carnot efficiency.

28.5.1 Hurricane Efficiency

Convection in the atmosphere may be thought of as a form of thermodynamic heat engine, converting solar heat into the kinetic energy of fluid motions. The efficiency of conversion of heat into mechanical energy by any means is limited by the *Carnot efficiency*, defined as $(T_{\text{hot}} - T_{\text{cold}})/T_{\text{hot}}$ where T_{hot} is the temperature of the reservoir supplying heat and T_{cold} is the temperature of the reservoir absorbing heat; see Appendix E.11. The efficiency of heat conversion of normal cumulonimbus clouds is low, just a few percent. This low efficiency is due to the large losses engendered by turbulent motions in cumulonimbus clouds; see § 24.3.

The destructive power of hurricanes is in part a result of a greatly enhanced efficiency in conversion of heat to kinetic energy. This enhancement arises due to the organizational effect of rotation, which constrains the large-scale convective motions and limits the dissipative influence of small-scale turbulence. In this section we will quantify the efficiency of the large-scale circulation of a hurricane, following the seminal work of Kerry Emanuel (Emanuel, 1986, 1991).

Idealized thermodynamic heat engines are *Carnot cycles* typically consisting of four steps, that leave the system in its original state. A hurricane operates in such a cycle as illustrated in Figure 28.4.[30] The cycle points labeled **a**, **b**, **c**, and **d** correspond to those points in parts (a) and (b) of Figure E.2.[31] The four steps of the cycle are as follows.

1. *Isothermal expansion* from state **a** to state **b**. With the temperature fixed at T_H, the system adds heat Q_H, principally in the form of latent heat as water vaporizes off the

[29] See § 24.3.
[30] See Appendix E.11.
[31] Note that the thermodynamic axes of Figure E.2 have been superposed on the physical axes in Figure 28.4.

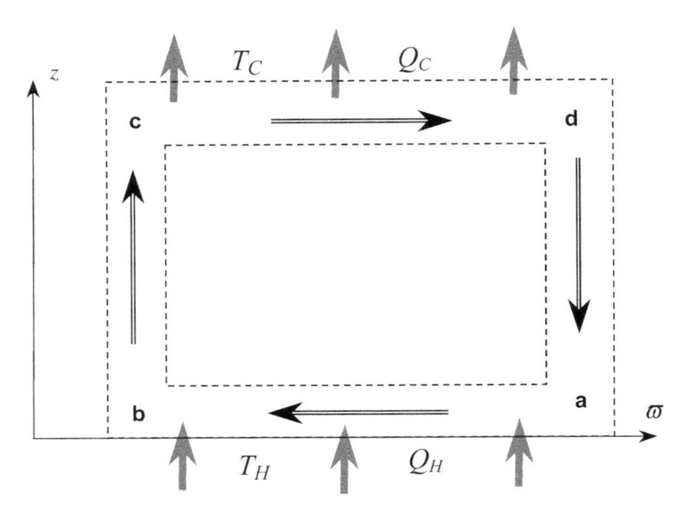

Figure 28.4 Schematic cross section of the circulation in a hurricane. The letters **a, b, c** and **d** mark states in the thermodynamic cycle.

surface of the warm ocean. During this step, the system gains an amount of entropy $S_H = Q_H/T_H$. (This is an expansion because the swirl creates a radial pressure gradient.)

2. *Isentropic expansion* from state **b** to state **c**. Rising air expands isentropically, doing work on its surroundings. Its entropy remains essentially constant and the temperature drops from T_H to T_C.[32]
3. *Isothermal contraction* from state **c** to **d**. With the temperature fixed at T_C, the system casts off heat Q_C by radiation to outer space. During this step, it loses an amount of entropy $S_C = Q_C/T_C$.
4. *Isentropic contraction* from state **d** to **a**. Barring exchanges of heat, descending air contracts isentropically. Its entropy remains constant and the temperature increases from T_C to T_H.

Quantification of hurricane thermodynamics involves a blend of dynamic and thermodynamic equations. The relevant thermodynamic equation is the entropy equation of state for a moist atmosphere[33]

$$T\mathrm{d}s = c_p\,\mathrm{d}T - (1/\rho)\,\mathrm{d}p + L\mathrm{d}\xi_v\,.$$

This equation relates the change of specific entropy (s) to changes in temperature (T) pressure (p) and water-vapor content (ξ_v), with the relative magnitude of the changes quantified by the specific heat at constant pressure (c_p), the specific volume ($v = 1/\rho$) and

[32] This is the shakiest part of the model because turbulence will cause some increases of entropy.
[33] See Appendix E.9.3.

the latent heat of vaporization (L). The relevant dynamic equation is the differential form of Newton's second law expressed as a statement of conservation of mechanical energy,[34]

$$\mathrm{d}\left(gz - v^2/2\right) = (1/\rho)\mathrm{d}p + \mathbf{f}_D \bullet \mathrm{d}\mathbf{l},$$

relating changes in the specific kinetic ($v^2/2$) and potential (gz) energies to the work provided by pressure and resistive loss provided by friction drag; $\mathrm{d}\mathbf{l} = \mathbf{v}\mathrm{d}t$ is a small step in the direction of flow and the factor $\mathbf{f}_D \bullet \mathrm{d}\mathbf{l}$ is always negative – drag resists motion and flow.

These two equations have a common term $\mathrm{d}p/\rho$, representing the work done by pressure. Eliminating this term between the two equations and treating c_p and L as constant, we have

$$\mathrm{d}u_T = \mathrm{d}q - \mathbf{f}_D \bullet \mathrm{d}\mathbf{l}, \qquad \text{where} \qquad \mathrm{d}q = T\mathrm{d}s$$

is the theoretical maximum amount of heat added (per unit mass) to a parcel of air,[35] and

$$u_T = \tfrac{1}{2}v^2 - gz + c_pT + L\xi_v$$

is the total specific energy in that parcel. Note the similarity of the present equation to the specific form of the first law of thermodynamics given in Appendix E.3.3. Two differences are that the compositional term ($L\mathrm{d}\xi_v$) have been folded into the specific internal energy and the pressure work term has been replaced by the viscous drag.

It is important to note that the internal energy u is a thermodynamic state variable,[36] whereas neither q nor the drag term is. This means that as a parcel of air completes a cycle as depicted in Figure 28.4, there is no net change of u. It follows that over a complete cycle,

$$\oint \mathrm{d}q = \oint T\mathrm{d}s = \oint \mathbf{f}_D \bullet \mathrm{d}\mathbf{l}.$$

This equation is in specific form; the corresponding extensive form is

$$\oint \mathrm{d}Q = \oint T\mathrm{d}S = W, \qquad \text{where} \qquad W = \oint \mathbf{F}_D \bullet \mathrm{d}\mathbf{l}$$

is the work done by the drag force, $\mathbf{F}_D$, during a cycle. Applying this equation to the cycle depicted in Figure 28.4 and noting that Q_H is added and Q_C is subtracted from the system, we have

$$W = Q_H - Q_C.$$

Further, if the system operates at Carnot efficiency, then $Q_C = Q_H T_C/T_H$ and

$$W = Q_H(T_H - T_C)/T_H.$$

[34] From § 4.6.3.
[35] Assuming heat is added reversibly.
[36] Whose value depends on the state of a thermodynamic system, independent of the processes leading to the establishment of that state.

Let's pause and take stock of this result. Normally convective motions are disorganized and operate at a low thermodynamic efficiency (several percent). Here we have shown that if instead the motions are organized and heat transfers occur reversibly – or nearly so – then a much greater percentage of heat is converted to work, which in this case is mainly the kinetic energy of the motions of air. This helps to explain the destructive power of rotating storms. The rotation severely constrains the turbulent motions, making them more efficient thermodynamic engines.

Since our model hurricane is in steady state, the kinetic energy must be degraded back to heat at the same rate it is being produced. Normally this degradation occurs by turbulent cascade of energy within the clouds, but this mechanism is suppressed by the constraint on turbulent motions imposed by rapid rotation. The only other mechanism capable of degrading kinetic energy significantly is turbulent shearing within the bottom boundary layer, which we studied in § 28.4.3 and § 28.4.4. This intense shearing motion requires a large shear stress at the ground, which involves fluid motions impinging on features that contribute to the boundary roughness, as described in § 23.4. It almost goes without saying that ground features responsible for this roughness – such as trees and buildings – take a real beating during a hurricane.

Transition

This chapter has been concerned with vortices, with an emphasis on atmospheric vortices. After surveying and discussing types of vortices and presenting the equations governing steady, axisymmetric vortices, we investigated several idealized vortices (the potential and Rankine vortex), then added realistic effects including meridional flow and radial and vertical diffusion of momentum, ending with a section devoted to the dynamics of hurricanes.

Now we switch gears and in the following part we investigate flows of silicates.

Part VII

Silicate Flows

We now turn our attention to flows in silicates, including mantle convection, volcanic eruptions and sheet flow of lava on the surface. Like flows in the atmosphere and oceans, the star of these flows is the Navier–Stokes equation, with the continuity equation remaining a co-star, but we now have a much different cast of supporting characters. Rotation, inertia and turbulence, which affect most atmospheric and oceanic flows, are completely negligible in the force balances for silicates. This is good news, because much of the difficulty in analyzing those flows stems from the nonlinear inertia terms; on the face of it the laminar, linearized Navier–Stokes equation would seem to be much easier to solve. However, two new characters that intrude on this idyllic scene are

- non-Newtonian creep; and
- the strong dependence of creep rate on temperature.

Each of these brings nonlinearity into the formulation; the first introduces a velocity nonlinearity while second couples the Navier–Stokes and energy equations. The energy equation contains, among other factors, a nonlinear advective term, presenting us once again with a set of challenging equations to analyze and solve.

The set of equations governing silicate flows is summarized in the following chapter. Following this, the equations are applied to two rather disparate flow problems: mantle convection and volcanic flows. Earth's mantle convects in order to cool both itself and core. The mantle is cooled by the introduction of cold lithospheric slabs, as explained in Chapter 31, while heat from the core is conveyed upward by discrete plumes, as explained in Chapter 32. Changing gears, Chapter 33 presents an overview of volcanic flows. Then flow in a volcanic conduit is briefly investigated in Chapter 34 and lava sheet flow on the surface is investigated in Chapter 35.

29

Equations Governing Silicate Flows

In this chapter, we summarize the equations governing flows of silicates in the mantle, in volcanic vents and on Earth's surface. With inertial terms negligibly small, the perturbation Navier–Stokes equation is[1]

$$\mathbf{0} = \rho'\mathbf{g} - \nabla p' + \nabla \bullet \mathbb{S}' \qquad \text{or equivalently} \qquad 0_i = \rho' g_i - \frac{\partial p'}{\partial x_i} + \frac{\partial \tau_{ij}}{\partial x_j},$$

where $\mathbf{g}$ is the acceleration of gravity, p' is the perturbation pressure,[2] ρ' is the perturbation density and $\mathbb{S}'$ is the deviatoric stress tensor, having elements τ_{ij}. Ignoring seismic waves, we may adopt the anelastic approximation,[3] and write the continuity equation as

$$\nabla \bullet (\rho_r \mathbf{v}) = 0,$$

where $\mathbf{v}$ is the (Eulerian) fluid velocity and ρ_r is the reference-state density.[4] These equations are supplemented by the equation of state for density. Within the mantle variations of density due to pressure and composition are dynamically unimportant; it is sufficient for our purposes to consider only the variation of density with temperature:

$$\rho' = \rho_r \alpha (T - T_r),$$

where T_r is a reference temperature profile.[5]

Let's pause and take stock. So far we have a set of five scalar equations involving three scalar variables (p', ρ' and T), one vector ($\mathbf{v}$) and one tensor ($\mathbb{S}'$ or equivalently τ_{ij}), containing a number of parameters – many of which may be functions of spherical radius or elevation.[6] To complete the set of equations we need one additional scalar equation

[1] From § 7.3.2.

[2] Recall that p and ρ are expressed as the sum of a reference-state, denoted by a subscript r, and a perturbation, denoted by a prime.

[3] See § 3.4.1.

[4] Recall that the reference-state quantities may vary with position.

[5] See § 5.2.

[6] ρ_r, T_0, α, and $\mathbf{g}$ are evident in the equations above, while c_p, k, n, T_m, β, $\dot{\epsilon}'_{ij}$, η, μ and Ψ_R have not yet made their appearances.

(the energy equation) and one tensor equation (the constitutive relation), plus a number of auxiliary equations governing the parameters.

The temperature dependence of density and viscosity introduces T as a dependent variable and requires us to consider the energy equation:[7]

$$\rho_r c_p \frac{\mathrm{D}T}{\mathrm{D}t} = \alpha T \rho_r (\mathbf{v} \bullet \mathbf{g}) + \rho_r \Psi_R + \dot{\epsilon}'_{ij} \tau_{ij} + k \nabla^2 T,$$

where c_p is the specific heat at constant pressure, α is the coefficient of thermal expansion, k is the thermal conductivity, Ψ_R is the radioactive heating (having dimensions power/mass) and $\dot{\epsilon}'_{ij}\tau_{ij}$ represents the viscous dissipation of mechanical energy. The radioactive heating is a function of depth and geological time. That is, if we are concerned with mantle convection in the present geological epoch, we may assume Ψ_R is independent of time, but if we want to explore the history of mantle convection over times of order 10^9 years, say, then Ψ_R must be treated as a function of time.

The deviatoric rate of strain, $\dot{\epsilon}'_{ij}$, is given in terms of velocity gradients in § 3.4.2. Complications of modeling silicate flows arise from the constitutive relation between $\mathbb{S}'$ and $\mathbf{v}$ or equivalently τ_{ij} and v_i. Deformation and flow of silicates is quantified reasonably well by the power-law creep formula introduced in § 6.2.3:

$$\tau_{ij} = \mu \left(\frac{\eta}{\mu} \frac{\partial v_i}{\partial x_j} \right)^{1/n}.$$

where the shear modulus μ is a known function of depth,[8] n is a positive integer (typically 3) and the viscosity η is a function of temperature and pressure:[9]

$$\eta(p,T) = \eta_0 \exp\left(\beta T_m(p) \frac{T_0 - T}{T_0 T} \right),$$

where $T_m(p)$ is the melting (eutectic) temperature,[10] β is a dimensionless constant ($\beta \approx$ 25 to 30) and a subscript 0 denotes a constant reference value. Note that the relative change in η is approximately (setting $T = T_m$) related to the relative change in T by $\Delta\eta/\eta = -\beta\Delta T/T$; the large magnitude of β implies that the magnitude of η is sensitive to changes in temperature; a few percent increase of T results in an e-fold decrease in the magnitude of η.

There is an additional complication in modeling silicate flows: the extreme sensitivity of the eutectic temperature, and hence the viscosity η, on trace amounts of dissolved volatiles, particularly water and carbon dioxide. These components lower the viscosity significantly and their presence in the upper mantle is an essential ingredient in the plate-tectonic mode of convection. (In effect, the dissolved volatiles lubricate the tectonic plates, permitting

7 See § 5.4 and Appendix E.7. Since thermal conduction is important only in thin regions, we have assumed that the thermal conductivity, k, is constant.
8 See Figure 7.2.
9 See § 6.2.4.
10 See § 6.2.4 and Appendix E.10.2.

them to slide past each other at subduction zones. If these lubricants were absent, the plates would seize up and plate-tectonic convection would cease.)

These equations apply to silicate flows, including:

- sub-solidus convection within the mantle (studied in Chapters 30 – 32); and
- flows of molten silicates:
  - as magmas within volcanic vents (Chapter 34); and
  - as lavas on the surface (Chapter 35).

29.1 Mantle Parameters

The dimensional parameters appearing in the governing equations for mantle silicates may be summarized as follows:

- c_p – the specific heat: may be considered constant, with an approximate value 1250 $\mathrm{J{\cdot}kg^{-1}{\cdot}K^{-1}}$;
- $\mathbf{g}$ – the gravitational vector: points downward and has mean magnitude 9.81 $\mathrm{m^2{\cdot}s^{-1}}$ on Earth's surface and magnitude close to 10 $\mathrm{m^2{\cdot}s^{-1}}$ throughout the mantle (see Figure 7.2);
- k – the thermal conductivity: varies with depth in the mantle, from roughly 5 $\mathrm{W{\cdot}m^{-1}{\cdot}K^{-1}}$ at the top to 10 $\mathrm{W{\cdot}m^{-1}{\cdot}K^{-1}}$ at the bottom;
- T_m – the melting temperature profile: varies with pressure (or equivalently depth or radius) in the mantle (see § 6.2.4, Figure 7.3 and Appendix E.10.2);
- T_r – the thermal profile of Earth: apart from boundary layers at the top and bottom of the mantle, varies roughly linearly from 1700 K at the surface to 2800 K at the core–mantle boundary (see § 30.7 and Figure 7.3);
- α – the coefficient of thermal expansion: varies moderately within the mantle from roughly 3.5×10^{-5} $\mathrm{K^{-1}}$ at the top and 1.1×10^{-5} $\mathrm{K^{-1}}$ at the bottom, with an approximate mean value of 2×10^{-5} $\mathrm{K^{-1}}$;
- η – the dynamic viscosity: varies with pressure and temperature; the strong variation with temperature is a dominant factor in mantle dynamics;
- ρ_r – the reference-state density: increases steadily with depth due to compression, from a value of 3.3×10^3 $\mathrm{kg{\cdot}m^{-3}}$ at the top of the mantle to 5.5×10^3 $\mathrm{kg{\cdot}m^{-3}}$ at the bottom (see Figure 7.2. We will develop an approximate representation for ρ_r in § 29.2. More accurate polynomial approximations for this parameter with depth are given in Table 1 of Dziewonski and Anderson (1981));
- μ – the shear modulus: is known from seismic inversion; at the base of the mantle $\mu \approx 3 \times 10^{11}$ Pa. (It is apparent from Figure 7.2 that μ is strongly correlated with ρ_r); and
- Ψ_R – the radioactive heating per unit mass: estimated to be $(3.25 \pm 0.3) \times 10^{-12}$ $\mathrm{W{\cdot}kg^{-1}}$ for the mantle on average,[11] but can be somewhat less in slabs due to geochemical fractionation.

[11] 13 TW of total heating (see Figure 30.2) divided by the mass of the mantle, 4×10^{24} kg.

For more detail on the structure of the mantle, see Stacey and Davis (2008).

The Grüneisen parameter, defined as

$$\gamma_G \equiv \frac{\alpha}{\rho\beta_T c_v} = \frac{\alpha}{\rho\beta_s c_p},$$

is useful in modeling mantle structure because it varies less with depth than its constituent parameters, varying from 1.15 to 1.3 in the lower mantle.

The ratio of dynamic viscosity to elastic modulus gives us a *viscoelastic relaxation timescale* $t_r = \eta/\mu$. Estimating $\eta \approx 3 \times 10^{22}$ Pa·s and $\mu \approx 3 \times 10^{11}$ Pa, $t_r \approx 10^{11}$ s, or roughly 3000 years, for the lower mantle. For phenomena with timescales significantly shorter than t_r (such as earthquakes), the mantle acts as an elastic solid, while it behaves as a fluid for phenomena (such as mantle convection) having timescales much greater than t_r.

29.2 Parameterized Lower-Mantle Structure

As an aid to understanding the dynamic structure of the mantle – particularly the lower mantle – let's develop some simple approximations to the reference density, ρ_r, and the adiabatic, melting and actual temperatures, denoted by T_A, T_m and T_r, respectively.

The observation that ρ varies linearly with radius[12] in the lower mantle leads us to a linear representation for the reference-state density:

$$\rho_r \approx \frac{(\rho_1 - \rho_2)r + \rho_2 r_1 - \rho_1 r_2}{r_1 - r_2},$$

where ρ_i is the density at radius r_i. If, for example, we select (from PREM) $r_1 = 3.48 \times 10^6$ m, $\rho_1 = 5566$ kg·m^{-3}, $r_2 = 5.6 \times 10^6$ m and $\rho_2 = 4443$ kg·m^{-3}, we have a linear parameterization of density for the lower mantle:

$$\rho_r \approx \rho_c S, \qquad \text{where} \qquad S \equiv 1 - 0.2492\, r/r_c,$$

$r_c = 3.48 \times 10^6$ m is the radius of Earth's core and $\rho_c = 5566$ kg·m^{-3} is the density at the base of the mantle. When extrapolated to the upper mantle, this parameterization predicts a greater density than actual (because it does not account for the phase changes), but is useful for exploring the basic behavior of the mantle.

The profiles of T_r and T_m in the lower mantle seen in Figure 7.3 are nearly linear, and we can adapt the linear representation for density to these temperatures. If, for example, we pick from Figure 7.3 $r_1 = 4.0 \times 10^6$ m, $T_{r1} = 2600$ K, $T_{m1} = 3772$ K, $r_2 = 5.5 \times 10^6$ m, $T_{r2} = 2000$ K and $T_{m2} = 2770$ K, we have linear parameterizations of temperatures for the lower mantle:

$$T_r \approx (4200 - 1390\, r/r_c)\,\text{K} \qquad \text{and} \qquad T_m \approx (6440 - 2320\, r/r_c)\,\text{K}.$$

[12] See Figure 7.2.

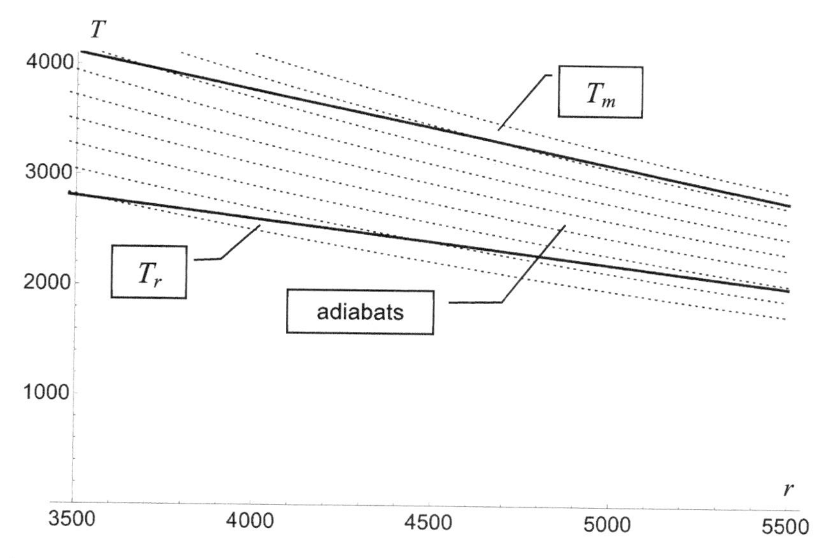

Figure 29.1 Plots of the linearized melting (T_m) and reference (T_r) temperatures for the lower mantle (solid lines), together with a family of adiabatic profiles (dotted lines).

Extrapolated to the surface the first approximation gives $T_r \approx 1650$ K, which approximates the potential temperature of the lower mantle. The approximation for T_r is not valid in the D″ layer, close to the core.

The equation governing the mantle adiabat,[13] $\mathrm{d}T_A/\mathrm{d}r = -T_A/r_A$, is readily integrated to yield

$$T_A(r) = T_0 e^{-r/r_A},$$

where T_0 is a constant of integration and $r_A \equiv c_p/\alpha g$ is the thermal scale height. We previously estimated that $r_A \approx 4.2 \times 10^6$ m for the mantle.[14] Adiabatic temperature profiles for the lower mantle are plotted in Figure 29.1, along with the linear approximations to T_r and T_m. Note that the temperature within the lower mantle has a shallower profile than the adiabat (warmer above and cooler below), indicating that the lower mantle is slightly stably stratified, and convecting passively in response to the forcing provided by slabs and plumes.

Substituting the parameterized forms for T_r and T_m into the expression for viscosity,[15] $\eta = \eta_0 e^{-\beta T_r/T_m}$, we readily see that the viscosity of the lower mantle increases with depth, increasing by about a factor 5 from top to bottom. There is observational evidence that the

[13] From § 5.5; assuming r_A = constant.
[14] In § 5.5.
[15] From § 6.2.4.

viscosity jumps modestly in magnitude across the 660 km phase transition, with the lower mantle as a whole being 2 to 3 orders of magnitude more viscous than the upper mantle.[16]

Transition

This completes our review of the equations governing silicates, our survey of the parameter values in the mantle and our rough parameterization of the density and temperatures within the mantle. In the following chapter we begin in earnest to understand why and how the mantle is in convective motion.

[16] E.g., see Davies and Richards (1992).

30
Cooling the Earth

Earth cools its mantle and core by casting off heat to outer space, with the heat being transported upward through the mantle by a combination of convective motions in the interior and conduction near the top and bottom surfaces. The top thermal boundary layer is the lithosphere,[1] and the bottom layer is called the D″ layer.[2] A thermal boundary layer has existed at the top of the mantle since the Earth accreted as a hot planetary body some 4.53×10^9 years ago, but a thermal layer exists at the bottom of the mantle now only because the mantle has cooled more rapidly than the core since accretion. In what follows, we will be interested in the current convective state of the mantle, and will not investigate the thermal history of the mantle.

The mantle can directly transfer its heat to the surface, but heat from the core must be transmitted upward through the mantle. This means that the mantle is convecting in two distinct modes, with *slab* convection acting to cool the mantle and *plume* convection acting to cool the core; these two modes of cooling are investigated and discussed in Chapters 31 and 32, respectively. But before getting into that, let's briefly discuss plate tectonics (in the following section), slabs and plumes (in § 30.2), volcanism (in § 30.3) and radioactivity (in § 30.5) and quantify the heat flows within the mantle (§ 30.4), the boundary conditions at the core–mantle boundary (§ 30.6) and the temperature profile in the bulk of the mantle (§ 30.7). A comprehensive summary of the structure of the convecting mantle may be found in Davies and Richards (1992).

30.1 Plate Tectonics

Although it took people a long while to recognize because of the slow speeds (typically a few centimeters per year) involved, it is now clear that Earth's surface is divided into a set of nearly rigid tectonic plates that move relative to each other. The current configuration of the plates is illustrated in Figure 30.1 and the current motions of Earth's 56 lithospheric plates are given in Argus et al. (2011). The tectonic plates behave as coherent entities because of

[1] See § 7.2.3.
[2] This curious name arose from its identification as a seismic anomaly at the base of the lower mantle (originally labeled the D' layer).

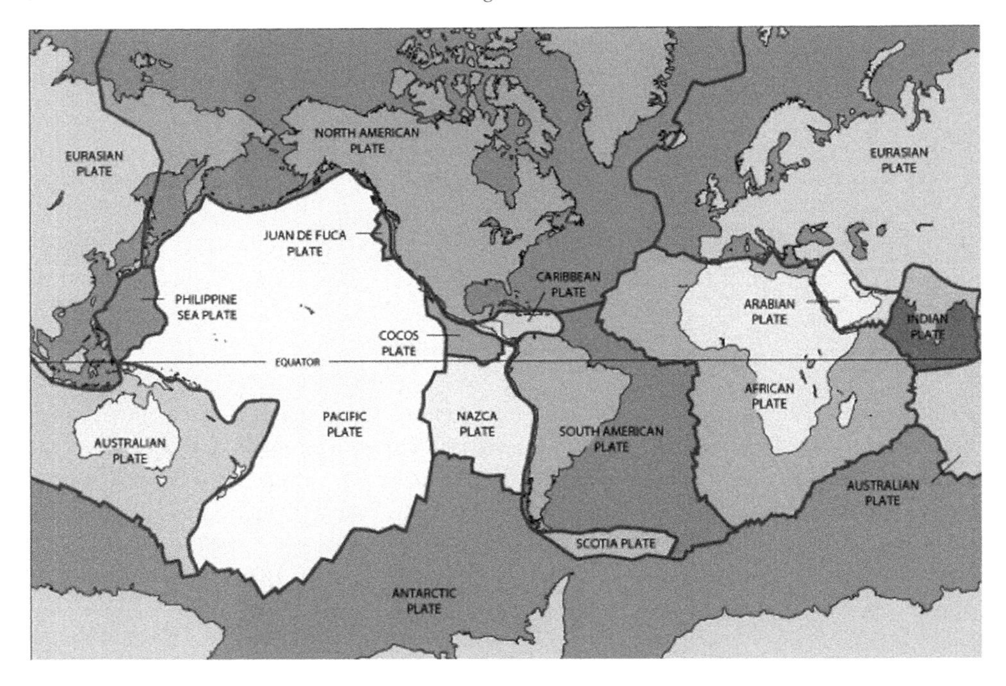

Figure 30.1 The surface of Earth is composed of a dozen or so major plates – and somewhat more minor ones - that move nearly rigidly. Courtesy of the USGS: http://pubs.usgs.gov/publications/text/slabs.html.

the strong dependence of viscosity on temperature.[3] Plates are "born" at mid-ocean ridges, where upwelling molten silicate cools and solidifies, slide past each other along strike-slip faults and dive back into the mantle at subduction zones beneath oceanic trenches. The key aspect of plate tectonics is the lubrication provided by water and carbon dioxide that permits the subsiding plate to slide beneath its neighbor. This lubrication permits a plate-tectonic planet such as Earth to cool faster than a planet lacking these lubricants and the plate-tectonic mode of convection; see Figure 31.1.

As it trundles from ridge to subduction zone, a plate cools and contracts, as described in § 7.2.3. This contraction increases the density of the plate relative to the underlying mantle. After the dense plate turns the corner at a subduction zone (and becomes a slab) and begins to plunge down into the mantle, the gravitational pull on this denser material provides the motive force for the convective motions within the bulk of the mantle.

30.2 Slabs and Mantle Plumes

The mantle is in vigorous convective motion due both to the input of heat at the bottom boundary and removal of heat at the top.[4] The strong dependence of viscosity

[3] See § 6.2.4.
[4] See § 31.3.

on temperature causes the mantle to convect differently in response to these two thermal forcings. A general feature of systems that are in vigorous convective motion is narrow rapid flows away from horizontal boundaries and broad, slow return flows, with the return flow having a neutral or stable density profile.

Cooling at the top produces cold rigid lithospheric plates, as described in § 7.2.3. The narrow rapid flows away from the top boundary are these cold plates, which are called slabs once they turn the corner at subduction zones. These serve to cool the interior of the mantle by mingling cold slab material with the ambient mantle. This process is somewhat clumsy and complicated by the fact that the slab material is much more viscous than its surroundings, causing the slabs to act as coherent entities.

On the other hand, heating of the mantle by the core produces a thin layer of hot mobile material within the D'' layer. The narrow rapid flows away from the bottom boundary are (roughly) cylindrical plumes arising from this layer. As we shall see in § 32.2, plumes entrain ambient mantle as they convey heat from the bottom to the top of the mantle, so that their thermal contrast with the adjacent mantle weakens with increasing radius. Some of the plume material is expelled as lava via shield volcanoes located above hotspots and some diverges beneath the lithospheric plates, forming *hotspot swells* and contributing to the low-viscosity layer called the *asthenosphere*.

30.3 Volcanism

Although the vast bulk of the mantle is below the solidus temperature, there remain localized pockets of melt (that is, molten silicate) near the surface that are source regions for volcanism. Broadly speaking, there are three types of volcanism associated with mantle convection: mid-ocean-ridge volcanism, arc volcanism and hotspot volcanism.

Mid-ocean-ridge volcanism is the result of adiabatic decompression of normal mantle material. The melting-temperature gradient is steeper than the adiabatic gradient.[5] Consequently as mantle material rises rapidly (and nearly adiabatically) to the surface at and near mid-ocean ridges, it becomes partially molten. The molten fraction percolates to the surface at mid-ocean ridges, where it quietly solidifies and forms the lithospheric plates.

Arc volcanism occurs above subduction zones where oceanic lithospheric plates dive into the mantle. As these plates rub against adjacent non-subducting plates, frictional heating causes hydrated minerals to decompose, releasing water that facilitates melting. This molten material rises through the overlying (usually continental) plate to the surface via a series of conduits topped by arc volcanoes. The volatile content of these magmas causes violent eruptions, as described in Chapter 34. These volcanoes occur along an arc if the subducting plate remains competent (without tearing or buckling laterally). (You can produce a similar arc by depressing the surface of a hollow ball.) If the subducting plate tears, there is a kink in the arc (such as that between Honshu and Hokkaido).

[5] See Figure 7.3 and Appendix E.10.2.

In addition to the volcanism occurring at mid-ocean ridges and above subduction zones, a third type, called *shield volcanism*, occurs in association with *hotspots*: regions of persistent volcanism not directly associated with plate boundaries. The heat for this type of volcanism originates in the core and rises through the mantle via plumes, as we noted in § 30.2 and will quantify in Chapter 32.

30.3.1 Geochemical Distillation

The volcanism associated with plate tectonics (that is, mid-ocean ridge and arc volcanism) acts as a two-step geochemical distillery. The first step occurs at the mid-ocean ridges where *mafic* magma[6] rising toward the mid-ocean ridge is partially distilled to produce basalt, which is the dominant type of rock on the sea floor. The second distillation step occurs during subduction, when magma heated by friction and mobilized by water rises to the surface as arc volcanism.

Each step of the process involves geochemical fractionation, also called *igneous differentiation* or *fractional crystallization*. The solidifying phases are composed preferentially of so-called compatible elements, having similarly sized atoms and ions, with the incompatible elements, having disparately sized (typically larger) atoms, preferentially remaining in the liquid phase. The end product of this two-step process is a suite of incompatible buoyant minerals that make up the continental crust. The radioactive elements are highly incompatible, so that the distillation process results in a lithospheric slab that is depleted in the heat-producing radioactive elements. Radioactive atoms preferentially end up rocks and minerals that solidify from late-stage magmas, particularly rhyolites and granites.

30.4 Global Heat Flow

The current heat balance for the mantle is somewhat uncertain (Dye, 2012), but several lines of evidence strongly indicate that the heat lost is greater than the amount of radioactive heating, with the result that Earth is gradually cooling. In particular, the geo-dynamo operating in Earth's core, maintaining the geomagnetic field, requires Earth to be cooling (e.g., see Olson, 2016).

Heat escapes Earth's interior by means of

- conduction through the rigid crust (particularly the oceanic crust; see § 7.2.3);
- hydrothermal circulation (particularly near mid-ocean ridges); and
- ejection of molten magma via volcanic vents.

The total rate of heat loss is estimated to be 46 ± 3 TW.[7] The primary sources of heat can be categorized as mantle radioactivity, mantle secular cooling and heat from the core; see Figure 30.2. This accounting ignores the (negligibly small) heat produced by tidal dissipation and the (potentially significant) heat escaping via geoneutrinos; see Dye (2012).

[6] Mafic magma is rich in magnesium and iron.
[7] 1 TW = 1 terawatt = 10^{12} W.

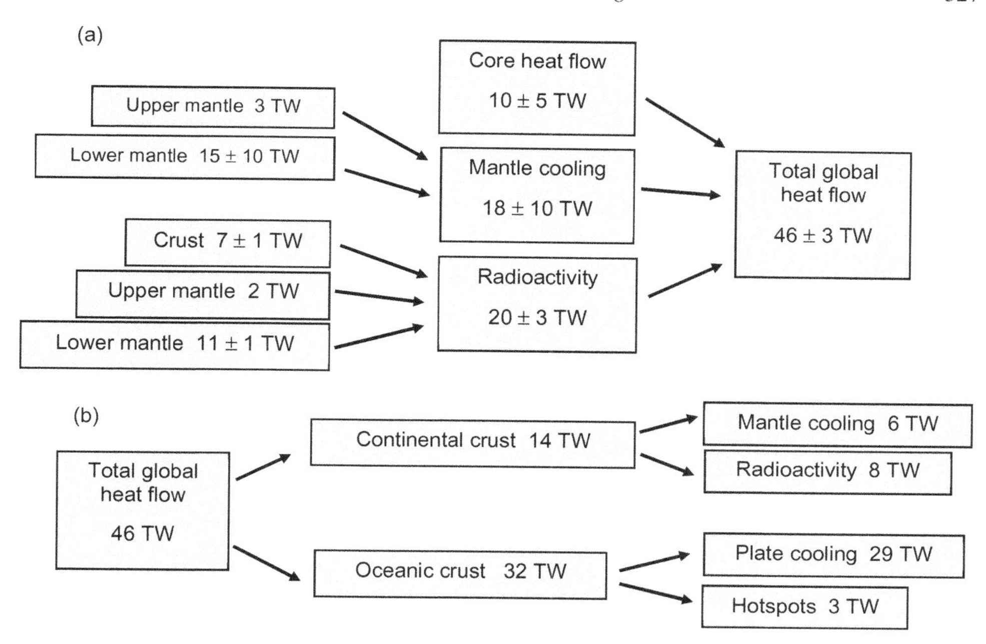

Figure 30.2 Contemporary estimates of the global heat flow from Earth: (a) Sources of heat within the mantle (including error estimates) and (b) means of heat escape from the mantle (excluding error estimates). 1 TW = 10^{12} W. By comparison humanity employs about 17 TW, whereas solar insolation is about 17×10^4 TW. See Lay et al. (2008), Furlong and Chapman (2013) and Jaupart et al. (2015).

The heat flow due to mantle cooling (18 TW) requires the mantle to cool currently at a rate of approximately 100 K per 10^9 years. (The mantle must have cooled significantly more rapidly in the past in order to have built up a temperature contrast of roughly 800 K across the D$''$ layer.) Heat flow from the core includes secular cooling of the core, latent heat and gravitational potential energy released as the inner core solidifies and compressive heating. Mantle cooling and radioactivity together act as a volumetric heat source within the mantle, while heat from the core enters the bottom of the mantle. In the following chapters, we will focus on flows within the mantle in response to these two heating sources, dealing with cooling the mantle in Chapter 31 and with mantle flows that act to cool the core in Chapter 32.

30.5 Radioactive Heating

Earth's effort to cool has been – and is – complicated by radioactive heating within the mantle.[8] The mantle is composed principally of silicates, but it contains small amounts of nearly every element in the periodic table, some of which are radioactive. The value of specific radioactive heating quoted in § 29.1 (3.25×10^{-12} W·kg^{-1}) is equivalent to

[8] Radioactive heating in the crust easily escapes to the surface and the rate of radioactive heating in the core is relatively small compared with that in the mantle.

roughly 13 TW of heating in the entire mantle; see Figure 30.2. If Earth were insulated, radioactivity would warm the mantle at an approximate current rate of 2.4×10^{-15} K·s^{-1} ≈ 77 K per 10^9 years.[9] Radioactive heating tends to decrease mantle viscosity, promoting convective motions.

At first glance one might think that radioactive heating within Earth's mantle drives convection. We can readily show that this is not the case because uniform internal heating (that is, $\Psi_R =$ constant, so that heating is proportional to density) leads to stable stratification. To demonstrate this, suppose the mantle is initially in a neutrally stable state with an adiabatic profile,[10] having $T \propto \rho^{\gamma_G}$ where γ_G is the Grüneisen parameter,[11] with T increasing with depth. It follows from the energy equation that, in response to uniform heating, $\partial T/\partial t \propto c_p^{-1}$. The specific heat in the mantle varies little with depth, so the rate of heating is nearly independent of depth, and the relative rate of heating is greatest at the top of the mantle, where the temperature is lowest. This causes the mantle to evolve to a stably stratified temperature profile. However, lateral variations in heat production can cause convection in the mantle, but the resulting motions are relatively weak and serve to equilibrate the internal temperature, rather than to cool the mantle.

It follows that the mantle convects principally in response to forcing at its top and bottom boundaries. Cooling at the top produces cold, dense, high-viscosity material (lithospheric slabs) and heating at the bottom produces hot, buoyant, low-viscosity material (plumes). These two modes of convection are discussed in the following two chapters.

30.6 Conditions at the Core–Mantle Boundary

Flows of mass and heat are induced within Earth's mantle and core as heat is radiated from its surface to outer space and these two bodies are in dynamic and thermal contact at their common boundary, called the *core–mantle boundary* (often referred to in the literature as the CMB). Due to the strong density contrast between silicate ($\rho = 5566$ kg·m^{-3}) and iron alloy ($\rho = 9903$ kg·m^{-3}) at the CMB,[12] this boundary is very close to spherical,[13] located at mean radial distance 3480 km from Earth's center.

The momentum and heat equations each are governed by equations containing the Laplacian, so that we need to specify one (vector) momentum condition and one (scalar) thermal condition[14] for both mantle and core at the CMB. These four conditions are continuity of tangential velocity, tangential stress, temperature and radial heat flux. Our attention in the following chapters will be focused on the flows of mass and heat within the mantle. In order to have a well-posed mathematical problem for flow of mass and heat

[9] The strength of radioactive heating within the mantle has decreased over geological time due to two effects: geochemical distillation that selectively removed radioactive elements and the finite half life of these elements.
[10] See § 5.5.
[11] See Appendix E.5.1.
[12] Density values from PREM; see Dziewonski and Anderson (1981).
[13] The equatorial radius of the core is about 9 km greater than the polar radius due to rotation. Given the large density contrast between the silicate mantle and metallic core, convective motions do not appreciably perturb the elevation of the CMB.
[14] See Appendix A.6.

within the mantle, we need to uncouple these conditions and obtain a single velocity and a single thermal condition for the mantle. Fortunately, the disparate transport properties of silicate and metal permit this decoupling.

The dynamic conditions are readily decoupled by noting the vast disparity of the kinematic viscosities of mantle silicates ($\approx 10^{19}$ $m^2 \cdot s^{-1}$) and molten iron of the core (< 10 $m^2 \cdot s^{-1}$). This disparity means that the tangential stress impressed on the mantle by the core is negligibly small compared with stresses within the mantle and the appropriate dynamic condition at the base of the mantle is that the tangential stress is zero. This readily translates to the condition that the radial gradient of velocity in the mantle is zero at the CMB. The difference in viscosity also means that flow speeds in the core are typically much larger than those in the mantle. (This difference is apparent in the rapidity of secular variation of the geomagnetic field.) From the point of view of the core, the mantle is convecting very slowly and is essentially immobile on time scales of interest in core dynamics. The velocity of core fluid must be zero at the CMB (as seen by an observer rotating with the mantle). Note that these conditions are homogeneous for both the mantle and core; each is satisfied if there is no flow.

In order to determine the thermal conditions applied to the mantle at the CMB, we must consider the dynamic and thermal state of the outer core. Evidence of this state is provided by the mere existence of the geomagnetic field. The only plausible explanation for the existence of Earth's magnetic field is the action of a dynamo within the outer core, involving convective motions of the outer-core liquid sufficiently vigorous to create an adiabatic thermal profile. If this adiabatic profile extends to the CMB, this profile dictates the temperature at the CMB. However, given the relatively large value of thermal conductivity in the core, it is possible that the amount of heat conducted down the adiabat (as much as 15 TW, see Pozzo et al. (2012)) within the outer core exceeds that conducted from core to mantle (across the CMB). In this case the convective state would be confined to the lower reaches of the outer core with the top of the outer core being stably stratified.

In either case, the flux of heat across the CMB is proportional to the radial thermal gradient within the mantle just above the CMB. This gradient and the associated flux are very likely to vary laterally. If the top of the core is convecting, the variations of heat flux are conveyed by vigorous radial motions of fluid parcels having small departures of temperature (less than 10^{-4} K)[15] from the adiabat and the mantle sees a laterally uniform temperature at the CMB. If the top of the core is stably stratified, the lateral variations of heat flux and temperature are smoothed by lateral advection of heat within a thin layer at the top of the outer core (similar to a thermal wind; see § 26.2). Again, due to the mobility of core fluid, the associated temperature variations are small so that the mantle sees a laterally uniform temperature at the CMB. The current value of the temperature at the CMB is the result of the long-time evolution of the mantle and core as a thermally coupled system.[16] The thermal evolution of Earth is beyond the scope of this book.

[15] See Stacey (1991).
[16] See Olson (2016).

We will see in § 32.1 how the mantle adjusts to the thermal and dynamic conditions at the CMB across the D″ layer, but before delving into that, we will investigate the overall mass and heat balances in the mantle in the following section.

30.7 Mantle Mass and Heat Balances

Let's investigate the mass and heat balances associated with mantle convection, using a simple model of mantle convection based on Figure 30.3. The total heat balance for the mantle may be expressed as

$$\frac{dQ_M}{dt} = \dot{Q}_R + \dot{Q}_p - \dot{Q}_t$$

where Q_M is the heat content of the mantle, $\dot{Q}_p$ is the rate that heat is transferred to the mantle from the core, $\dot{Q}_t$ is the rate that heat leaves the mantle at the top,

$$\dot{Q}_R = \int_M \Psi_R dM = \bar{\Psi}_R M$$

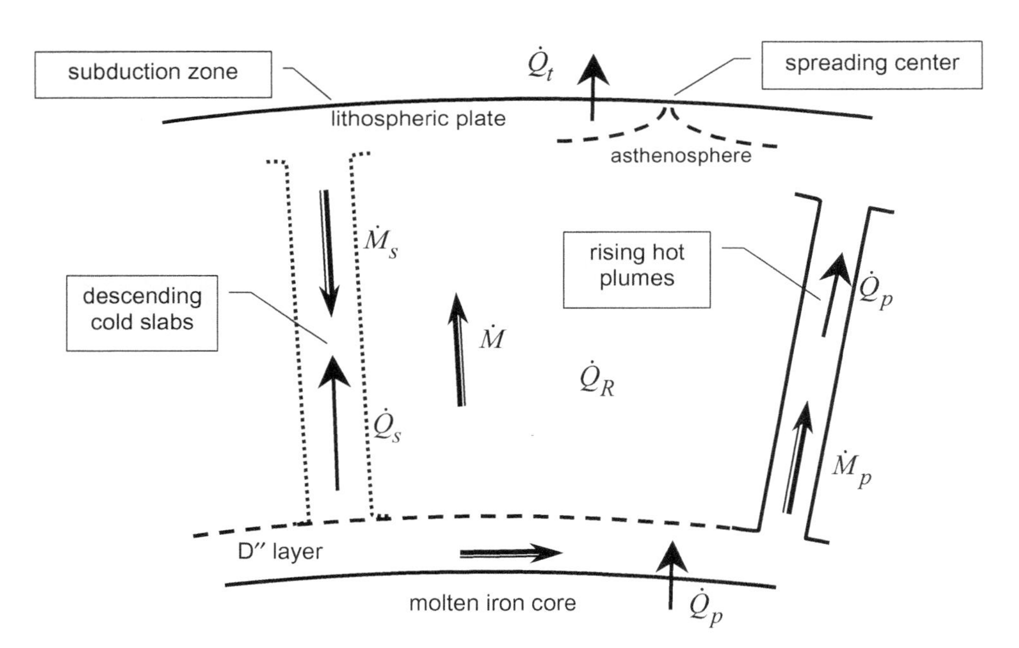

Figure 30.3 A cartoon of mantle mass and heat flows. The vertical mass flow in the bulk of the mantle, $\dot{M}$, is a passive response to the slab and plume flows, $\dot{M}_s$ and $\dot{M}_p$. A heat flow $\dot{Q}_p$ enters the mantle at the bottom and a flow $\dot{Q}_t$ leaves at the top. Cold slabs having (negative) mass flow $\dot{M}_s$ convey a positive upward heat flow $\dot{Q}_s$. Also, hot plumes having (positive) mass flow $\dot{M}_p$ convey a positive upward heat flow $\dot{Q}_p$. Radioactive heating is denoted by $\dot{Q}_R$.

is the rate of radioactive heating, M is the mass of the mantle and an overbar denotes a mass average.[17] Cooling of the mantle and core is controlled by $\dot{Q}_t$.

As the mantle cooled over geological time, a thermal boundary layer (called the D″ layer) developed at the bottom and now heat is drawn out of the core at rate $\dot{Q}_p$; this process is investigated in Chapter 32 and particularly § 32.1. This core heat is conveyed upward through the mantle by plumes having mass flow $\dot{M}_p$ and temperature excess (relative to the adjacent mantle) ΔT_p. Similarly the slab heat flow, $\dot{Q}_s$, is conveyed by a slab mass flow $\dot{M}_s$ having a temperature deficit ΔT_s:

$$\dot{Q}_p = c_p \Delta T_p \dot{M}_p \qquad \text{and} \qquad \dot{Q}_s = c_p \Delta T_s \dot{M}_s \,.$$

Note that $\Delta T_s < 0$ and $\dot{M}_s < 0$.

Since plumes serve to convey heat from the core to the top of the mantle, $\dot{Q}_p$ is independent of depth in the mantle and equal to the heat transferred to the mantle from the core. Unfortunately, this flow is rather difficult to estimate, with plausible values ranging from 5 to 15 TW.[18] Due to entrainment of ambient mantle material into plumes $\dot{M}_p$ increases with radius in the mantle, while ΔT_p decreases. The mantle heat flow,[19] $\dot{Q}_s$, is a maximum of roughly 32 TW at the top of the mantle and decreases with depth, reaching zero at the core–mantle boundary.

Expressing the heat content of the mantle as

$$Q_M = M c_p \bar{T}_r \Theta(t)\,,$$

the heat balance becomes

$$\bar{T}\frac{\mathrm{d}\Theta}{\mathrm{d}t} = \frac{\bar{\Psi}_R}{c_p} + \frac{\dot{Q}_p - \dot{Q}_t}{M c_p}$$

If we have reliable parameterizations of $\bar{\Psi}_R(t)$, $\dot{Q}_p(t)$ and $\dot{Q}_t(t)$, this equation governs the thermal history of the mantle, $\Theta(t)$, but we won't pursue this avenue of investigation.

30.7.1 Mantle Turn-over Times

Convection within the mantle can be characterized by two turn-over times: the time it would take the entire mantle to cycle through the lithosphere or the D″ layer if the mass flows were constant, independent of time: let's introduce

$$t_b \equiv M/\dot{M}_{b0}\,,$$

for $b = s$ or p, where M is the mass of the mantle, $\dot{M}_{s0}$ is the slab flow at the top of the mantle and $\dot{M}_{p0}$ is plume flow at the bottom. These mass flows can be estimated from

[17] This balance ignores heat flow (roughly 7 TW) due to radioactivity in the crust.

[18] See Stacey and Loper (2007), Pozzo et al. (2012) and Jaupart et al. (2015). The recent large estimate of core-to-mantle heat flow has profound implications for convection in Earth's core and the geodynamo, but these topics are beyond the scope of this volume.

[19] Due to secular cooling and radioactive heating; see Figure 30.2.

the heat fluxes out the oceanic floor and across the core–mantle boundary, together with estimates of the temperature differences and specific heats:

$$\dot{M}_{b0} = \dot{Q}_{b0}/(c_p \Delta T_{b0}) \,.$$

We can use these two formulas to estimate the mass flows and turn-over times for slab and plume convection. To a good approximation, $c_p = 1250$ J·kg^{-1}·K^{-1} throughout the mantle and $M \approx 4 \times 10^{24}$ kg.

First, for slabs with[20] $\dot{Q}_s \approx 32$ TW and $\Delta T_s \approx 1100$ K, it follows that $\dot{M}_{s0} \approx 3.2 \times 10^7$ kg·s^{-1}, and $t_s \approx 2 \times 10^{17}$ s $\approx 6 \times 10^9$ y (remarkably close to the age of the Earth). At this rate, the entire mantle would cycle through the oceanic lithosphere in 6 billion years. Of course, the heat flow and temperature difference have not been uniform over billions of years, so this estimated turn-over time is only a rough guide. With a hotter early Earth, the slab turn-over time likely was shorter earlier in Earth's history. However, it is virtually certain that some mantle material (particularly that in the upper mantle) has cycled through the lithosphere more than once, with portions of the mantle (mostly in the lower mantle) not having undergone that experience and so remaining in their primitive geochemical state.

Now for plumes; with $\dot{Q}_{p0} \approx 10^{13}$ W and $\Delta T_{p0} \approx 800$ K, we estimate that $\dot{M}_{p0} \approx 10^7$ kg·s^{-1} and $t_p \approx 6.4 \times 10^{17}$ s $\approx 20 \times 10^9$ y. The current plume turn-over time is significantly longer than the age of Earth. And it is likely that the plume mass flow was smaller and turn-over time was larger, earlier in Earth's history. It follows that only a small portion of the mantle has ever participated in plume convection.

30.7.2 Mantle Vertical Speed

The vertical speed in the bulk of the mantle may be expressed as

$$w = \dot{M}/(A\rho_r) \,,$$

where $\dot{M}$ is the mass flow within the bulk of the mantle (outside plumes and slabs), ρ_r is the reference-state density parameterized in § 29.2 and[21] $A \approx 4\pi r^2$ is the horizontal area at radius r. Each of the factors in this equation are functions of radius, r. The overall mass-flow balance is

$$\dot{M} + \dot{M}_s + \dot{M}_p = 0 \,,$$

where the plume mass flow $\dot{M}_p$ is positive (and increasing with radius due to entrainment), the slab mass flow $\dot{M}_s$ is zero at the base of the mantle and decreases with increasing radius, reaching a maximum (negative) magnitude at subduction zones; see Figure 30.3. It follows that $\dot{M}$ is negative at the bottom of the mantle, positive at the top and increases uniformly

[20] See Figure 30.2 and § 7.2.3.

[21] Since mantle convection occurs at high Rayleigh number (see § 31.3), the horizontal area occupied by plumes and slabs is relatively small.

with radius. Now the upward speed in the bulk of the mantle is

$$w = -(\dot{M}_s + \dot{M}_p)/(A\rho_r) .$$

In order to determine w in the bulk of the mantle we need to determine how $\dot{M}_s$ and $\dot{M}_p$ vary with r. These flows are investigated in Chapters 31 and 32, respectively.

Transition

This chapter has given us an overview of the cooling of Earth, focusing on the distinctive features associated with mantle convection: plates, slabs and plumes, and the gross heat balance within the mantle. It also identified the two functions of mantle convection: cooling of the mantle and core. In the following chapter, we investigate the first of these, cooling of the mantle, in more detail.

31
Cooling the Mantle

Earth's mantle is cooled by the introduction of cold slabs. This plate-tectonic mode of cooling is somewhat unusual, as mantle convection in the other large terrestrial bodies in our solar system operates differently, as explained in § 31.1. Mantle convection acts to convey heat toward the surface at a rate greater than can be achieved by conduction alone. This rate is parameterized in § 31.3. This chapter concludes in § 31.4 with a simple model of the mantle thermal structure and a discussion in § 31.4.1 of slab divergence.

31.1 Modes of Mantle Convection

All planetary bodies in our solar system (and likely elsewhere) formed by accretion relatively quickly ($\approx$ 10 million years or possibly less) and the associated gravitational energy released by this process led to a set of hot young planets. Ever since, these planets have been doing their best to cast this heat off to outer space as fast as they can. Convection of molten silicate is an efficient mode of cooling for a terrestrial planet; upwelling brings hot molten silicate material to the surface, where its heat can readily be radiated to outer space. Young planets with this mode of convection are in the *magma-ocean* phase. This state is fairly short-lived (several million years), as the surface temperature soon falls below the solidus temperature, and the mode of convection that ensues depend whether there is significant amounts of liquid water present or not. If little or no water is present, the planet transitions through three cooling modes having progressively weaker convection, weaker heat flow and lower temperatures. Fortuitously, since the rate of transitioning varying inversely with the size of the planet, we have good examples of each of these three modes within our own solar system:

- sluggish-lid convection (Venus): the surface is hot enough and convective motions are sufficiently vigorous to cause the surface to deform;
- rigid-lid convection (Mercury):[1] the surface is sufficiently cold that convective motions are unable to cause it to deform and convection must occur beneath this rigid lid; and

[1] Mercury has evolved toward a conductive state over its lifetime and possibly is now conductive rather than convective.

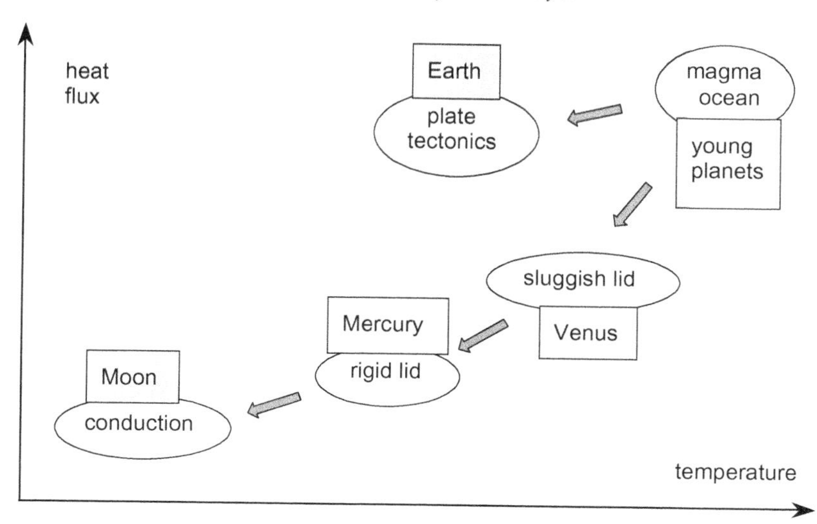

Figure 31.1 The five modes of planetary cooling. All young planets are in the magma-ocean mode, with a large heat flow. This mode does not last long and the planets evolve with time as indicated by the arrows, following the lower branch if liquid water is not present and the upper branch if water is present.

- conduction (Moon): The rate of internal heating and global cooling are sufficiently small that heat is transferred entirely by conduction.

On the other hand, if there is enough liquid water present to lubricate the plates, the planet is able to simulate the rapid cooling of the magma-ocean phase – even though the material at the surface has cooled well below the melting temperature – by operating in a *plate-tectonic* mode. These modes of cooling are illustrated in Figure 31.1. Note that on Earth, we experience the last small vestige of the magma-ocean phase at the mid-ocean ridges.

Mars presents us with an interesting tectonic history. There is strong evidence that it had both liquid water and plate-tectonic activity early in its history, but currently it has no liquid water and plate-tectonic activity ceased long ago. The loss of water and change of tectonic regime are tied to the loss of a dynamo creating and sustaining its magnetic field.[2]

31.2 Formation of the D″ Layer

Earth was hot immediately following accretion, with the mantle being partially molten (the magma-ocean phase shown in Figure 31.1). Initially the mantle and core cooled relatively rapidly – and in step – via counter-flow convection (rising liquid and descending solid) within the mantle. Once the bulk of the mantle cooled below the solidus,[3] the mantle switched to sub-solidus convection, which is a much less efficient mode of cooling.

[2] See Loper (2009).
[3] See Appendix E.10.

Following this switch, the rate of core cooling decreased dramatically, with the rate subsequently increasing as a thermal boundary layer (the D″ layer) developed at the base of the mantle as the mantle cooled more rapidly than the core over geological time.[4]

31.3 Parameterization of Mantle Convection

The strength of convective cooling of a horizontal layer of fluid is represented by the *Nusselt number*, commonly denoted by Nu, which is the ratio of the total heat flow divided by the conductive heat flow. Necessarily, $\mathrm{Nu} \geq 1$. The Nusselt number is a function of the strength of convective motions, which is represented by the Rayleigh number, commonly denoted by Ra.[5] The mantle Rayleigh number is very large, roughly 10^8, indicating that convection in the mantle is strong, with conduction important only in thin boundary layers (i.e., the lithosphere and D″ layer). At high Rayleigh number the functional relationship is a power law: $\mathrm{Nu} \approx b\mathrm{Ra}^c$, where b and c are constants. The key parameter is c. Roberts (1979) found that $c = 1/5$ if the top and bottom boundaries are fixed and $c = 1/3$ if the boundaries are mobile. This difference means that a planet with a mobile top surface (plate tectonics) cools more rapidly than one whose top surface is immobile, as indicated in Figure 31.1. This makes sense; a young mobile top boundary layer is thinner than an old static one and a body cools more rapidly under a thinner blanket.

31.4 Mantle Thermal Structure

In this section we take a stab at determining the radial thermal structure within the bulk of the lower mantle, outside slabs and plumes. Our simple model of the convecting mantle assumes that there are thermal boundary layers at the top and bottom, plus isolated descending cold slabs and rising hot plumes, with the bulk of the mantle being in uniform vertical motion with a uniform temperature. This is a crude model that ignores a lot of details of mantle convection, but permits us to describe the broad features of the thermal structure of the mantle. In particular, this simple model ignores the dynamic effects of the phase transitions responsible for the seismic discontinuities at 220, 400 and 660 km depth.[6]

In slabs and plumes, gradients of temperature and velocity are large and conduction of heat (represented by the term $k\nabla^2 T$) and viscous dissipation (represented by the term $\dot{\epsilon}'_{ij}\tau_{ij}$) are important. However, in the bulk of the mantle, these terms are negligibly small and, writing $\mathbf{v} = \mathbf{v}_H + w\mathbf{1}_r$, the energy equation (from Chapter 29) reduces to

$$\frac{\partial T}{\partial t} + w\frac{\partial T}{\partial r} + H = \frac{\alpha T g}{c_p} w + \frac{1}{c_p}\Psi_R, \qquad \text{where} \qquad H \equiv \mathbf{v}_H \bullet \nabla_H T$$

[4] For more detail, see Loper (1991).

[5] $\mathrm{Ra} \equiv g\alpha(\Delta T)D^3/\nu\kappa$, where ν is the kinematic viscosity, κ is the thermal diffusivity, D is the mantle thickness and ΔT is decrease of adiabatic temperature at the top boundary, compared with the bottom; see § 7.5.

[6] See Figure 7.2.

quantifies horizontal advection of heat. This equation balances advection of heat with adiabatic compression and radioactive heating. Motions in the mantle are driven principally by thermally induced density differences (in slabs and plumes) and are strongly modulated by the viscosity structure.

Assuming that the radioactive heating per unit mass is uniform within the lower mantle, so that $\Psi_R \propto \rho_r$ and using the parameterization for density given in § 29.2, we can write

$$\Psi_R(\mathbf{x},t) = \psi(t)S, \qquad \text{where} \qquad S \equiv 1 - 0.2492\, r/r_c\,.$$

The temporal variation of ψ is governed by the half lives of the radioactive elements (principally K, Th and U) that are present and the rate of extraction of radioactive elements by geochemical fractionation, as described in § 30.3.1.

Given the strong coupling among temperature, flow and viscosity, a reasonable ansatz is that the mantle temperature maintains its zonally averaged radial profile as it slowly cools; that is, $T(\mathbf{x},t) = T_r(r)\Theta(t)$, where T_r is known and has been parameterized for the lower mantle in § 29.2 and Θ is the potential temperature of the mantle. Now the energy equation becomes

$$T_r \frac{\mathrm{d}\Theta}{\mathrm{d}t} + w\left(\frac{\mathrm{d}T_r}{\mathrm{d}r} - \frac{\alpha g}{c_p} T_r\right)\Theta = \frac{1}{c_p}\psi S - \langle H \rangle\,,$$

where $\langle H \rangle$, the zonal average of $H = \mathbf{v}_H \bullet \nabla_H T$, represents the effect of horizontal divergence in the bulk of the mantle. This equation shows us that three factors cause the temperature to vary with time. Vertical advection of the non-adiabatic portion of the thermal profile can cause either local heating or cooling, while radioactivity causes heating and horizontal divergence causes cooling.

Horizontal divergence is caused by two factors: entrainment of mantle material into plumes and the introduction of cold slab material. Divergence due to plume entrainment acts to smooth lateral temperature gradients and hence contributes little to $\langle H \rangle$. On the other hand, cold slab material introduces large lateral thermal gradients and provides the dominant contribution to $\langle H \rangle$. Possible modes of slab divergence are discussed in the following subsection.

31.4.1 Slab Divergence

Slabs are the downward extension of cold lithospheric plates and serve to cool the bulk of the mantle by diverging and mingling with the warmer adjacent mantle. Slab divergence can take several forms. One of the most obvious – and least effective – forms is thermal diffusion; the bulk of the mantle can cool as its heat diffuses into the cold slabs. The distance that heat can diffuse in a time t is roughly $\sqrt{\kappa t}$.[7] With a silicate thermal diffusivity $\kappa \approx 10^{-6}\ \mathrm{m^2{\cdot}s^{-1}}$,[8] heat can diffuse a distance less than 180 km in 10^9 years.

[7] This relation can be obtained by dimensional analysis; see Appendix B.1.
[8] See Table B.5.

More effective modes of divergence are broadening and buckling. As a slab descends in the lower mantle, it must push through mantle material having a viscosity that increases with depth.[9] This increased resistance causes the slab to broaden and diverge and even buckle and fold. This mode of divergence is particularly pronounced near 660 km depth where there is an abrupt increase in viscosity with depth due to phase change. This behavior can be observed with seismic tomography; for an example, see Figure 11 (p. 165) of Davies and Richards (1992).

An additional mode of slab divergence that is not so obvious, but is quite effective, is temporal variation. The geographic locations of slabs varies with geological time as the plates move and reorganize. A newly descending slab must necessarily displace the ambient mantle material in a process very similar to broadening and buckling.

Slab divergence is clearly a highly complicated and heterogeneous process and we are sweeping a lot of difficult physics and mathematics up into a small package by representing its effect by the term $\langle H \rangle$. We are doing so because our goal is to introduce the important concepts and processes without undue mathematical complications. If our ansatz of a cooling mantle having a temperature profile given by $T = T_r(r)\Theta(t)$ is valid, then slab divergence, represented by $\langle H \rangle$, is determined by the energy equation given in Chapter 29.

Transition

In this chapter we have reviewed the possible modes of mantle convection, investigated the overall heat balance of the mantle and discussed the modes of slab divergence. In the next chapter, we investigate the structures and flows within the mantle that are associated with cooling the core.

[9] As we saw in § 29.2.

32
Cooling the Core

In this chapter we investigate the flows within the mantle that convey heat from Earth's metallic core upward to the surface. Heat is drawn out of the core by a thermal boundary layer (the D″ layer) at the base of the mantle, having a temperature contrast of roughly 800 K (see Figure 7.3). The large change in temperature within the D″ layer alters the properties of seismic waves and is believed to be the primary cause of subtle seismic anomalies at the base of the mantle.[1] Also, this large temperature increase causes a large decrease in viscosity (by as much as a factor of 1000; see § 6.2.4), making this layer very mobile compared with the adjacent mantle material. This mobility permits hot material within the D″ layer to migrate horizontally to the base of a set of plumes, where it feeds upward flow within the plumes, which act as chimneys to convey this hot material to the surface. Flow within the D″ layer is investigated in § 32.1 and plume flow is studied in § 32.2.

32.1 D″ Layer

The D″ layer is both a thermal and dynamic feature of the lower-most mantle. It draws heat from the core via thermal conduction and advects heated mantle material toward the bases of plumes. This layer is maintained in a steady state with thermal diffusive thickening being balanced by a slow downward flow of mantle material into the layer to replace the heated material that is advected away. Since the viscosity of mantle material decreases strongly with temperature,[2] the D″ layer has a double-layer structure, with the temperature adjusting across a thicker outer layer (roughly 150 km depth) and the viscosity and velocity adjusting in a thinner inner layer (roughly 20 km depth). In the following these two scales will be denoted by h with a suitable subscript: η for viscosity and T for temperature. In addition, we will encounter a third vertical scale, h_c, representing the temperature gradient at the core–mantle boundary.

Our analysis of temperature and flow the D″ layer will be highly idealized and intended to illustrate the basic processes involved. Since the D″ layer is thin, we can use cylindrical coordinates, ϖ, ϕ and z, where z is distance above the core–mantle boundary. Let's suppose

[1] The D″ seismic signature may also be due – in part – to lower-mantle phase transitions and compositional anomalies due to mass transport across the core–mantle boundary.
[2] See § 6.2.4 and Appendix D.2.2.

that there are a discrete number of plumes within the mantle, each having a catchment area in the D″ layer that supplies it with hot mobile material. Let's consider one idealized circular catchment area having radius ϖ_0 and assume that the variables are independent of ϕ. Within this area, hot silicate flows radially inward to the base of a plume located at $\varpi = 0$.

32.1.1 D″ Viscosity Structure

As we learned in § 6.2.4, the structure of flow within the D″ layer is dominated by the strong variation of viscosity. Setting the reference value of viscosity to be that at the core–mantle boundary (and replacing subscript r with c) and noting that the melting temperature T_m may be considered constant within this thin layer, the expression for viscosity becomes

$$\eta(p,T) = \eta_c \exp\left(\beta \frac{T_m}{T_c T}(T_c - T)\right).$$

It is sufficient for our purposes to expand $T(z)$ in a Taylor series about the level of the core–mantle boundary (that is, about $z = 0$) and keep only the linear term.[3] This may be expressed as

$$T \approx T_c(1 - z/h_c),$$

where h_c is a scale height representing the gradient of temperature at the core–mantle boundary. Substituting this expansion into the expression for η, we have

$$\eta = \eta_c e^{z/h_\eta}, \qquad \text{where} \qquad h_\eta = h_c T_c / \beta T_m$$

is a scale height representing the variation of viscosity. We don't know the magnitude of either h_η or h_c yet.

32.1.2 D″ Flow

Now let's consider the horizontal flows within the D″ layer. As with other boundary layers we have encountered, the vertical component of the momentum equation simplifies to $\partial p'/\partial z = 0$ to dominant order.[4] Writing $\mathbf{v} = u\mathbf{1}_\varpi + w\mathbf{1}_z$, the continuity equation and the radial component of the momentum equation simplify to

$$\frac{1}{\varpi}\frac{\partial(\varpi u)}{\partial \varpi} + \frac{\partial w}{\partial z} = 0 \qquad \text{and} \qquad \frac{\mathrm{d}p'}{\mathrm{d}\varpi} = \frac{\partial \tau_{z\varpi}}{\partial z},$$

with the shear stress[5] expressed as

$$\tau_{z\varpi} = \mu\left(\frac{\eta}{\mu}\frac{\partial u}{\partial z}\right)^{1/n} = \mu\left(\frac{\eta_c}{\mu}e^{\zeta}\frac{\partial u}{\partial z}\right)^{1/n}, \qquad \text{where} \qquad \zeta \equiv \frac{z}{h_\eta}$$

[3] Note that $z = r - r_c$ where r_c is the radius of the CMB.
[4] See § 24.1 and Chapters 25 and 28.
[5] From § 6.2.3.

is a dimensionless boundary-layer thickness and μ is the shear modulus. These equations are to be solved in the domain $0 < \varpi < \varpi_0$ and $0 < z$, subject to the conditions that

$$\begin{array}{lll} \text{at} & z=0: & w=\frac{\partial u}{\partial z}=0; \\ \text{as} & z\to\infty: & u\to 0 \quad \text{and} \quad w\to -W_b; \\ \text{at} & \varpi=\varpi_0: & u=0, \end{array}$$

where W_b is the downward speed just outside the D″ layer. Since the mantle above the D″ layer is highly viscous, it is reasonable to suppose that W_b is independent of ϖ, which implies that lower-mantle material moves downward into the D″ layer as close to rigidly as possible. The condition on u at $z = 0$ is a no-stress condition; as noted in § 30.6 the underlying core is essentially inviscid and cannot exert an appreciable shear stress on the mantle.

Note that the mass volume flow through this catchment basin is $\dot{M}_D = \pi\varpi_0^2 W_b$ and the volume flow through the entire D″ layer is $\dot{M}_{p0} = 4\pi r_c^2 W_b$, where r_c is the radius of the core–mantle boundary.

The continuity equation may be satisfied by postulating that w is independent of ϖ and writing

$$w = -W_b f_n(\zeta) \qquad \text{and} \qquad u = -\frac{W_b}{2h_\eta\varpi}(\varpi_0^2 - \varpi^2)f_n'(\zeta),$$

where a prime denotes differentiation with respect to ζ. The radial inflow ($u < 0$) within the D″ layer supplies the plume centered at $\varpi = 0$, much as the bottom boundary layer of a tornado or hurricane feeds upward flow in and near the eye region.[6] The D″ layer solution does not hold all the way to $\varpi = 0$; there is a corner region (which we will not quantify) in which the flow turns from radial to vertical. Substituting this ansatz into the momentum equation, we have

$$\frac{dp'}{d\varpi} = -\frac{\mu}{\varpi_0}\Pi_n\left(\frac{\varpi_0^2 - \varpi^2}{\varpi_0\varpi}\right)^{1/n}\left(\left(e^\zeta f_n''\right)^{1/n}\right)',$$

where

$$\Pi_n = \frac{\varpi_0}{h_\eta}\left(\frac{W_b\eta_c\varpi_0}{2h_\eta^2\mu}\right)^{1/n}$$

is a dimensionless parameter.[7] The condition[8] that $\partial p'/\partial z = 0$ requires

$$\left(\left(e^\zeta f_n''\right)^{1/n}\right)' = C_n^{1/n},$$

[6] See § 28.4.3 and § 28.4.4.
[7] The subscript n on Π reminds us that the value of Π_n depends on the chosen value of n.
[8] From the vertical component of the momentum equation.

where C_n is a dimensionless constant to be determined. This equation is to be solved on the interval $0 < \zeta < \infty$ subject to the conditions

$$f_n(0) = f_n''(0) = 0 \qquad \text{and} \qquad f_n(\infty) = 1\,.$$

A first integral satisfying the second-derivative condition at $\zeta = 0$ is

$$\left(e^{\zeta} f_n''\right)^{1/n} = C_n^{1/n} \zeta \qquad \text{or equivalently} \qquad f_n'' = C_n \zeta^n e^{-\zeta}\,.$$

The solution of this equation satisfying the boundary conditions is

$$f_n = 1 - \frac{\Gamma(n+2,\zeta) - \zeta\,\Gamma(n+1,\zeta)}{\Gamma(n+2)} \qquad \text{and} \qquad C_n = \frac{1}{\Gamma(n+2)}\,,$$

$$\text{where} \qquad \Gamma(\lambda, x) \equiv \int_x^{\infty} \grave{x}^{\lambda-1} e^{-\grave{x}} \mathrm{d}\grave{x}$$

is the incomplete gamma function[9] and $\Gamma(n) = \Gamma(n,0)$ is the gamma function.

This is rather remarkable; we have found an analytic solution for the non-Newtonian flow within the D$''$ layer! If we assume Newtonian rheology with $n = 1$, the solution of this equation satisfying the boundary conditions is

$$f_1 = 1 - (1 + \zeta/2)e^{-\zeta} \qquad \text{and} \qquad C_1 = 1/2\,.$$

With non-Newtonian rheology having $n = 3$, the solution is

$$f_3 = 1 - \left(1 + \frac{3}{4}\zeta + \frac{1}{4}\zeta^2 + \frac{1}{24}\zeta^3\right) e^{-\zeta} \qquad \text{and} \qquad C_3 = \frac{1}{24}\,.$$

The dimensionless radial-velocity profiles f' are plotted versus ζ (with the ζ axis vertical) for $n = 1$ and 3 in Figure 32.1. The total radial flow is the same in each case. Non-Newtonian rheology tends to broaden the radial flow, but the change is not dramatic because of the strong constraint provided by the exponential variation of viscosity. The decreased level of shear strain for non-Newtonian rheology is in accord with Figure 6.1, which shows that at low stress levels, deformation of a non-Newtonian shear-thinning fluid is less than that for a Newtonian fluid.

32.1.3 D$''$ Thermal Structure

Next, consider the thermal adjustment across the D$''$ layer. The temperature varies from T_c at the core–mantle boundary to T_∞ outside the layer. The dominant terms in the heat equation are vertical diffusion and vertical advection; radioactive heating, secular cooling, viscous dissipation and adiabatic compression are negligibly small.[10]

[9] e.g., see §8.2(i) of Olver et al. (2010).

[10] Viscous dissipation in the D$''$ layer is small because the flow speed and shear are much smaller than in a plume.

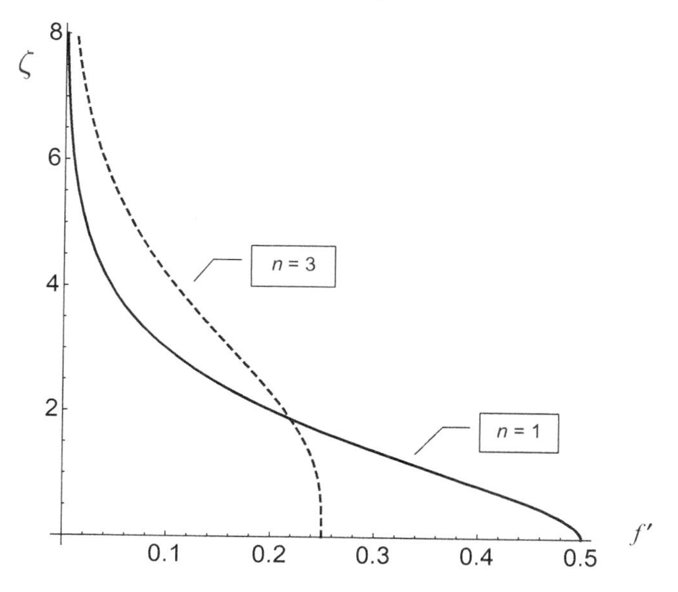

Figure 32.1 Plots of the dimensionless radial speed f' (horizontal axis) in the D″ layer versus dimensionless vertical coordinate $\zeta = z/h_\eta$ (vertical axis) for Newtonian (solid line) and non-Newtonian (dashed line) rheologies. The areas to the left of the curves are identical.

Noting that $w = -W_b f_n(\zeta)$ and writing $T = T_c - (T_c - T_\infty)F_n(\zeta)$ with $\zeta = z/h_\eta$, the heat equation becomes

$$-\lambda f_n F_n' = F_n'', \qquad \text{where} \qquad \lambda = W_b h_\eta \rho_c c_p / k$$

is a small dimensionless parameter. This equation is to be solved in the domain $0 < \zeta < \infty$ subject to the conditions that

$$F_n(0) = 0 \qquad \text{and} \qquad F_n(\infty) = 1\,.$$

If f_n were constant (equal to its asymptotic value 1) the thermal profile would simply be

$$F_n = e^{-\lambda\zeta} = e^{-z/h_T}, \qquad \text{where} \qquad h_T = k/(W_b \rho_c c_p)$$

is the vertical scale for the thermal boundary layer. Although the actual thermal profile is more complicated than this because f_n is not constant, this simple solution does give the correct thermal-layer scale. If we estimate[11] $c_p = 1250$ J·kg^{-1}·K^{-1}, $\rho_c = 5500$ kg·m^{-3}, $k \approx 10$ W·m^{-1}·K^{-1} and $W_b \approx 10^{-11}$ m·s^{-1}, then $h_T \approx 145 \times 10^3$ m $= 145$ km, which – given the many uncertainties in parameter values – is remarkably close to the seismically determined thickness (150 km) of the D″ layer.

[11] See § 29.1.

A first integral of the equation for F_n with f_n as given above is

$$F_n' = F_0' \exp(-\lambda\zeta - \lambda G + \lambda(n+2)/2),$$

where

$$G(\zeta, n) \equiv \frac{\Gamma(n+3,\zeta) - 2\zeta\Gamma(n+2,\zeta) + \zeta^2\Gamma(n+1,\zeta)}{2\Gamma(n+2)}$$

and $F_0' = F_n'(0)$ is a constant of integration. Note that $G(0,n) = (n+2)/2$. The boundary conditions on F_n at $\zeta = 0$ and ∞ require the solvability condition

$$1 = F_0' \int_0^\infty \exp(-\lambda\zeta - \lambda G + \lambda(n+2)/2)\, \mathrm{d}\zeta$$

be satisfied. This is an implicit equation for λ as function of n and F_0', and the scale for viscosity and velocity follows from $h_\eta = \lambda h_T$. The integral may be evaluated numerically using a standard numerical mathematics package. The functional relation λ versus F_0' is plotted in Figure 32.2 for representative values of n (specifically $n = 1$ and 3). In the limit $F_0' \ll 1$ the relation is linear: $\lambda = F_0'$.

32.1.4 D″ Thickness

In order to complete the solution for velocity and temperature within the D″ layer we need to determine the vertical length scales h_η and h_c. Recall that h_c was introduced as a scale for the temperature gradient at the core–mantle boundary, with $T = T_c(1 - z/h_c)$ close to the core–mantle boundary. A second representation of temperature is

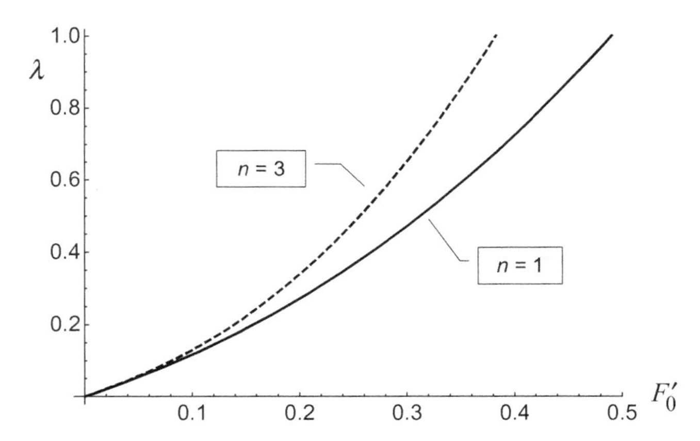

Figure 32.2 Plots of the thickness ratio $\lambda = h_\eta/h_T$ versus the dimensionless temperature gradient at the core–mantle boundary F_0' for Newtonian (solid line: $n = 1$) and non-Newtonian (dashed line: $n = 3$) rheologies.

$T = T_c - (T_c - T_\infty)F_n(\zeta)$, where $\zeta = z/h_\eta$ and $h_\eta = h_c T_c/\beta T_m$. These two representations give the same gradient at the core–mantle boundary provided

$$F_0' = \frac{T_c^2}{\beta T_m (T_c - T_\infty)}.$$

The terms on the right-hand side are known quantities. Due to the large magnitude of β, $F_0' \ll 1$. For example, typical values are $\beta \approx 27$, $T_m \approx 4100$ K, $T_c \approx 3700$ K and $T_c - T_\infty \approx 800$ K, giving $F_0' \approx 0.15$. The solvability condition[12] requires that with $F_0' = 0.15$, $\lambda = 0.188$ for $n = 1$ and 0.222 for $n = 3$ and with $h_T \approx 145 \times 10^3$ m, $h_\eta \approx 27 \times 10^3$ m for $n = 1$ and $h_\eta \approx 32 \times 10^3$ m for $n = 3$.

The thermal gradient at the base of the mantle is represented by the scale $h_c = \beta T_m h_\eta / T_c$, with $h_c \approx 808 \times 10^3$ m for $n = 1$ and $h_c \approx 957 \times 10^3$ m for $n = 3$. The thermal gradient at the base of the mantle is small because the vertical speed tends to zero at the core–mantle boundary.

32.1.5 D″ Discussion

We have considered a simple model of flow within the D″ layer at the base of the mantle, with the core–mantle boundary being divided into roughly 20 patches (like a soccer ball) and each patch being treated as a circular catchment area that feeds hot material horizontally inward toward the center of the patch where it turns upward and feeds a mantle plume. We consider plume structure in the following section.

The material within the D″ layer is significantly hotter than that above due to a flux of heat from the core. This heat reduces the viscosity of mantle material within the layer, making the bottom of the layer particularly mobile. The structure of this variable-viscosity, non-Newtonian flow has been expressed in terms of the incomplete gamma function in § 32.1.2.

Our model is highly idealized, having neglected, for example, the possibility that the flow has a spoke-like non-axisymmetric pattern. Nonetheless the basic structure will still consist of a thin region of rapid flow embedded within a thick thermal halo. Another issue that we have neglected in our simple model is the possibility that there may be "crypto-continents" at the base of the mantle: thin layers of silicates that are too dense to be entrained in the heated flow feeding mantle plumes; see Stacey (1991) and Loper and Lay (1995). If dense material resides at the base of the mantle, it can cause lateral variations in heat flux from the core and possibly choke a plume as this material gets rafted to the plume base.

32.2 Mantle Plumes

The heated mantle material that flows radially inward within the D″ layer, as investigated in the previous section, ascends through the mantle in a set of plumes. The structure of flow and temperature within one such plume is investigated in this section.

[12] The solvability condition is shown in § 32.1.3.

The structure of a plume is similar to that of the D″ layer: a broad thermal halo surrounding a narrow, low-viscosity conduit, although – due to geometric factors – the thermal and viscosity contrasts within the plume are somewhat less than those of the D″ layer. These differing lateral (cylindrical radial) structures will be quantified in the following analysis by various lateral scale factors, with the radius ϖ_T representing the lateral variation of temperature (determined in § 32.2.4), the radius ϖ_η representing the lateral variation of viscosity and the radius ϖ_μ representing the lateral variation of vertical speed. We anticipate that $\varpi_\eta \ll \varpi_T$. Each of these lateral factors is a function of z, but may be treated as (locally) constant while determining the lateral structure of a plume.

32.2.1 Plume Viscosity Structure

As with the D″ layer, the structure of flow within a plume is dominated by the strong variation of viscosity with temperature:

$$\eta(p,T) = \eta_a \exp\left(\beta \frac{T_m}{T_a T}(T_a - T)\right),$$

with subscript a denoting values on the plume axis. This formula is structurally identical to that introduced in § 32.1.1, except that the parameters T_a and η_a, as well as T_m, are now functions of p or equivalently distance, z, above the CMB. However, in determining the lateral (ϖ) structure of the plume, they may be treated as constant. The temperature is greatest, and the viscosity lowest, at the plume axis. Due to axisymmetry, the temperature close to the axis varies quadratically with ϖ,

$$T \approx T_a - \Delta T(\varpi/\varpi_T)^2, \qquad \text{where} \qquad \Delta T \equiv T_a - T_r$$

is the temperature excess on the plume axis, $T_r(z)$ is the temperature far from the plume and the radial scale ϖ_T represents the strength of the lateral variation of temperature close to the plume axis. Substituting this into the equation for viscosity and keeping only dominant terms in powers of ϖ/ϖ_T (with $T \approx T_a$), we have

$$\eta(p,T) \approx \eta_a e^{\varpi^{*2}},$$

where

$$\varpi^* \equiv \varpi/\varpi_\eta$$

is a dimensionless radial coordinate and

$$\varpi_\eta \equiv \varpi_T T_a/\sqrt{\beta T_m \Delta T}$$

is a radial scale representing variation of viscosity close to the plume axis. Due to the large magnitude of β (≈ 30), $\varpi_\eta \ll \varpi_T$.

32.2.2 Plume Vertical Flow

We will continue to use a cylindrical coordinate system ϖ, ϕ and z in modeling plume flow and again assume steady axisymmetry with $\mathbf{v}(\mathbf{x},t) = u(\varpi,z)\mathbf{1}_\varpi + w(\varpi,z)\mathbf{1}_z$. Let's focus our attention on the plume at some distance above the D″ layer, with $\varpi \ll z$, and stay well away from the corner region where hot material from the D″ layer is fed into the plume.

The plume acts as a chimney with flexible walls and as such is incapable of creating a draft. It follows that the pressure term in the vertical component of the momentum equation is negligibly small and the balance is between buoyancy provided by the hot material within the plume and lateral viscous drag:

$$0 = \rho' g + \frac{\partial \tau_{\varpi z}}{\partial \varpi}, \qquad \text{where} \qquad \tau_{\varpi z} = \mu \left(\frac{\eta}{\mu} \frac{\partial w}{\partial \varpi} \right)^{1/n}$$

is the non-Newtonian shear stress.[13]

The equation of state for the perturbation density[14] is given by $\rho' = -\rho_r \alpha (T - T_r)$, where $\rho_r(z)$ and $T_r(z)$ are the density and temperature outside the plume. Close to the plume axis the density may be expressed

$$\rho' \approx -\rho_r \alpha \Delta T,$$

where $\Delta T = T_a - T_r$. Using this, vertical momentum equation may be expressed as

$$\varpi_\mu \frac{\partial}{\partial \varpi} \left(\frac{\eta}{\mu} \frac{\partial w}{\partial \varpi} \right)^{1/n} = -1, \qquad \text{where} \qquad \varpi_\mu \equiv \frac{\mu}{\rho_r g \alpha \Delta T}$$

is a radial scale representing the lateral variation of vertical speed. This equation is to be solved in the domain $0 < \varpi < \infty$ subject to the conditions

$$\text{at} \quad \varpi = 0: \quad \frac{\partial w}{\partial \varpi} = 0; \qquad \text{and as} \quad \varpi \to \infty: \quad w \to 0.$$

This equation is readily integrated once with respect to ϖ:

$$\varpi_\mu \left(\frac{\eta}{\mu} \frac{\partial w}{\partial \varpi} \right)^{1/n} = -\varpi .$$

The constant of integration has been set equal to zero in order to satisfy the boundary condition at $\varpi = 0$. Using formulas $\eta \approx \eta_a e^{\varpi^{*2}}$ and $\varpi^* = \varpi / \varpi_\eta$ (from § 32.2.1), this equation may be expressed as

$$\frac{\partial w}{\partial \varpi^*} = -2 w_n \varpi^{*n} e^{-\varpi^{*2}}, \qquad \text{where} \qquad w_n \equiv \frac{\varpi_\eta}{2 t_a} \left(\frac{\varpi_\eta}{\varpi_\mu} \right)^n$$

[13] See § 6.2.3.
[14] See § 7.3.2.

is a velocity scale and $t_a = \eta_a/\mu$ is a characteristic timescale.[15] This equation is readily integrated; the solution satisfying the boundary conditions is

$$w = w_n \Gamma\left((n+1)/2, \varpi^{*2}\right),$$

where $\Gamma(a,z)$ is the incomplete gamma function.[16]

We have again obtained an analytic solution for flow with non-Newtonian rheology. However, this plume solution is not yet normalized to mass flow as our D″-layer solution was. In order to normalize this solution, we need to calculate the plume mass flow; this is given by

$$\begin{aligned}\dot{M}_p &= \rho_r \int_0^\infty w\, 2\pi\, \varpi\, \mathrm{d}\varpi \\ &= \pi\,\Gamma\left(\frac{n+3}{2}\right)\rho_r \varpi_\eta^2 w_n = \pi\,\Gamma\left(\frac{n+3}{2}\right)\frac{\rho_r}{2t_a \varpi_\mu^n}\varpi_n^{3+n}.\end{aligned}$$

This mass flow is an increasing function of z due to entrainment; the rate of entrainment is investigated in § 32.2.4. If we scale the mass flow with the mean vertical speed w_p within a cylinder of radius ϖ_η, by writing $\dot{M}_p = \rho_r \pi \varpi_\eta^2 w_p$, then we see that w_n is given by $w_n = w_p/\,\Gamma\left((n+3)/2\right)$ and our scaled solution is

$$w = w_p W_n, \qquad \text{where} \qquad W_n(\varpi^*) \equiv \frac{\Gamma\left((n+1)/2, \varpi^{*2}\right)}{\Gamma\left((n+3)/2\right)}.$$

Note that $W_n(0) = 2/(n+1)$. If we assume Newtonian rheology with $n = 1$, the solution is simply

$$w = w_p e^{-\varpi^{*2}}.$$

With non-Newtonian rheology having $n = 3$, the solution is

$$w = w_p \frac{1+\varpi^{*2}}{2} e^{-\varpi^{*2}}.$$

The normalized vertical velocity component w/w_p is plotted versus ϖ^* in panel (a) of Figure 32.3 for Newtonian ($n = 1$) and non-Newtonian ($n = 3$) rheologies.

Our solution for the vertical speed within a plume contains an arbitrary parameter, ϖ_T, representing the radial variation of temperature. This parameter may be found by considering the heat equation, but we first need to determine the plume radial flow.

[15] See § 29.1 and Appendix D.1.2.
[16] e.g., see §8.2(i) of Olver et al. (2010).

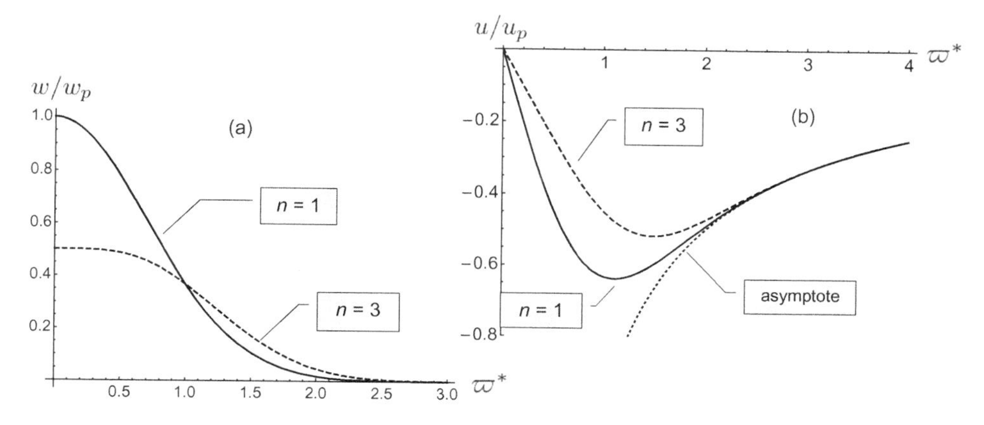

Figure 32.3 (a) Plots of the dimensionless vertical speed w/w_p in a plume versus dimensionless cylindrical radial coordinate $\varpi^* = \varpi/\varpi_\eta$ for Newtonian (solid line: $n = 1$) and non-Newtonian (dashed line: $n = 3$) rheologies. (b) Plots of the dimensionless radial speed u/u_p in a plume or Newtonian (solid line: $n = 1$) and non-Newtonian (dashed line: $n = 3$) rheologies. The asymptote is $-1/\varpi^*$.

32.2.3 Plume Radial Flow

With the reference density of the mantle, ρ_r, being a function of z only, the continuity equation may be expressed as

$$\frac{\partial(\varpi u)}{\varpi\,\partial\varpi} + \frac{1}{\rho_r}\frac{\partial(\rho_r w)}{\partial z} = 0.$$

The radial flow within the plume is determined by integrating this equation with $w = w_p(z)W_n(\varpi)$, $w_p = \dot{M}_p/(\rho_r \pi \varpi_\eta^2)$ and W_n known in terms of the incomplete gamma function. This equation can be solved analytically if we make the physically reasonable assumption that the viscosity scale ϖ_η is independent of z. Now it may be expressed as

$$\frac{\partial(\varpi^* u)}{\partial\varpi^*} = -u_p\frac{\varpi^*\Gamma\left((n+1)/2, \varpi^{*2}\right)}{\Gamma\left((n+3)/2\right)}, \quad \text{where} \quad u_p(z) \equiv \frac{1}{2\pi\rho_r\varpi_\eta}\frac{\mathrm{d}\dot{M}_p}{\mathrm{d}z}$$

is a radial velocity scale and $\varpi^* = \varpi/\varpi_\eta$. The integral satisfying the condition that $u = 0$ at $\varpi^* = 0$ is

$$\frac{u}{u_p} = U_n - \frac{1}{\varpi^*},$$

where

$$U_n(\varpi^*) \equiv \frac{\Gamma((n+3)/2, \varpi^{*2}) - \varpi^{*2}\Gamma((n+1)/2, \varpi^{*2})}{\varpi^*\Gamma\left((n+3)/2\right)}.$$

Note that close to the plume axis $u/u_p \approx -2\varpi^*/(n+1)$.

For a Newtonian rheology ($n = 1$)

$$\frac{u}{u_p} = \frac{1}{\varpi^*}\left(e^{-\varpi^{*2}} - 1\right),$$

while for $n = 3$

$$\frac{u}{u_p} = \frac{1}{\varpi^*}\left(\left(1 + \frac{1}{2}\varpi^{*2}\right)e^{-\varpi^{*2}} - 1\right).$$

The normalized radial velocity component u/u_p is plotted versus ϖ^* in panel (b) of Figure 32.3 for $n = 1$ and 3. Note that far from the plume axis, $U_n \to 0$ and the equation for u becomes a statement regarding the variation of $\dot{M}_p$ with z:

$$\frac{\mathrm{d}\dot{M}_p}{\mathrm{d}z} = -2\pi\rho_r(\varpi u)_{\varpi\to\infty}\,.$$

We have determined the structure of the vertical and radial flow in the plume, but we don't yet know the magnitudes of these flows and the variation of plume mass flow with elevation. These are determined by considering the energy equation on the plume axis in the following subsection.

32.2.4 Plume Thermal Structure

Each mantle plume conveys heat at a fixed rate, denoted by $\dot{Q}_e$, from the D″ layer to the top of the mantle. This heat flow is given by

$$\begin{aligned}\dot{Q}_e &= \rho_r c_p \int_0^\infty w\,(T - T_r)\,2\pi\varpi\,\mathrm{d}\varpi \\ &= c_p\dot{M}_p\Delta T + \rho_r c_p \int_0^\infty w\,(T - T_a)\,2\pi\varpi\,\mathrm{d}\varpi\,.\end{aligned}$$

We anticipate that the dynamic and thermal structure of the plume is similar to that of the D″ layer: a broad, relatively stagnant thermal halo surrounding a soft inner region with significant flow and nearly uniform temperature. With this ansatz, the integral involving $w(T - T_a)$ is small and to a good approximation,

$$\dot{Q}_e \approx c_p\dot{M}_p\Delta T = c_p\dot{M}_{p0}\Delta T_0\,,$$

where a subscript 0 denotes evaluation at the base of the mantle (at $z = 0$). The flow in an average plume is equal to the total flow of heat $\dot{Q}_p$ from the core, divided by the number of plumes. Taking $\dot{Q}_p \approx 5 \times 10^{12}$ W and 20 plumes, $\dot{Q}_e \approx 2.5 \times 10^{11}$ W. Also, with $\Delta T_0 \approx 800$ K at the base of the mantle, $\dot{M}_{p0} \approx 2.5 \times 10^5$ kg·s^{-1}. Although $\dot{Q}_e$ is independent of depth, we shall see that $\dot{M}_p$ increases with elevation while ΔT decreases.

Now let's turn to the details of the thermal structure of a plume, writing the temperature as

$$T = T_r(z,t) + \Theta(\varpi^*)\Delta T(z)\,,$$

where the reference temperature $T_r(z,t)$ has been parameterized for the present epoch in § 29.2 $\Delta T(z) = T_a - T_r$ is the temperature contrast across the plume and Θ represents the (positive) dimensionless radial temperature structure within the plume. Subtracting the reference energy equation given in § 31.4 from the full energy equation given in Chapter 29 and noting that cylindrical radial derivatives are dominant in the thermal conduction term, the perturbation energy equation may be expressed as

$$\lambda\left(U_n - \frac{1}{\varpi^*}\right)\frac{\mathrm{d}\Theta}{\mathrm{d}\varpi^*} - 2\lambda W_n\Theta = \frac{\mathrm{d}}{\varpi^*\mathrm{d}\varpi^*}\left(\varpi^*\frac{\mathrm{d}\Theta}{\mathrm{d}\varpi^*}\right) + \frac{\varpi_\eta^2}{k\Delta T}\dot{\epsilon}'_{ij}\tau_{ij},$$

where

$$\lambda(z) \equiv \frac{c_p}{2\pi k}\frac{\mathrm{d}\dot{M}_p}{\mathrm{d}z}$$

is a dimensionless parameter representing the relative strengths of diffusion and advection of heat. If $\lambda \ll 1$, the structure of the plume is similar to that of the D″ layer: a narrow region of rapid flow surrounded by a broad sluggish thermal halo.

Plume convection acts as a thermodynamic heat engine converting heat to mechanical work; this work is immediately re-converted to heat, as quantified by the viscous heating term, which is dominated by radial shear of the vertical flow:

$$\begin{aligned}\dot{\epsilon}'_{ij}\tau_{ij} &= \mu\frac{\partial w}{\partial\varpi}\left(\frac{\eta}{\mu}\frac{\partial w}{\partial\varpi}\right)^{1/n}\\ &= \mu\, t_a^{1/n}\left(2w_n/\varpi_\eta\right)^{(n+1)/n}\varpi^{*(n+1)}e^{-\varpi^{*2}}.\end{aligned}$$

The thermal equation holds in the interval $0 < \varpi^* < \infty$, subject to the conditions

$$\text{at} \quad \varpi^* = 0: \quad \Theta = 1; \qquad \text{and as} \quad \varpi^* \to \infty: \quad \Theta \to 0.$$

The energy equation is a second-order, linear, non-homogeneous ordinary differential equation with variable coefficients controlling the radial thermal profile: $T(\varpi^*)$. This is a difficult equation to solve, but in order to determine the rate of entrainment of material into a plume, we don't need to solve the whole equation. We need only consider the energy balance close to the plume axis.

Energy Balance Close to the Plume Axis

In developing the expression for the variation of viscosity in § 32.2.1, we assumed that $T \approx T_a - \Delta T(\varpi/\varpi_T)^2$ close to the plume axis. That formulation is compatible with the current ansatz provided

$$\Theta \approx 1 - \frac{T_a}{\Delta T}\left(\frac{\varpi_\eta}{\varpi_T}\right)^2\varpi^{*2}$$

when $\varpi^* \ll 1$. In this same limit, the viscous-dissipation term ($\dot{\epsilon}'_{ij}\tau_{ij} \sim \varpi^{*(n+1)}$) and radial-flow term ($U_n - 1/\varpi^* \approx -2\varpi^*/(n+1)$) in the energy equation are negligibly small; the dominant balance is between vertical advection and radial diffusion of heat. Noting that $W_n \approx 2/(n+1)$ close to the axis, using the definition of λ above and recalling that $\varpi_\eta \equiv \varpi_T T_a/\sqrt{\beta T_m \Delta T}$ and $\Delta T \approx \dot{Q}_e/c_p\dot{M}_p$, the energy balance close to the axis is

$$\frac{\lambda}{n+1} = \frac{T_a}{\Delta T}\left(\frac{\varpi_\eta}{\varpi_T}\right)^2$$

or

$$\frac{\mathrm{d}\dot{M}_p}{\mathrm{d}z} = \frac{\dot{M}_p^2}{\dot{M}_{p0}Z}, \qquad \text{where} \qquad Z(z) \equiv \frac{\beta \dot{Q}_e T_m \Delta T_0}{2\pi(n+1)kT_a^3}$$

is a parameter having dimensions of length, representing the rate of variation of $\dot{M}_p$ with elevation z.

Let's estimate the rate of variation of plume mass flow with elevation in the mantle, assuming the parameters are independent of z. With[17] $\dot{Q}_e \approx 2.5 \times 10^{11}$ W, $k \approx 5$ W·m^{-1}·K^{-1}, $T_a \approx 2500$ K, $T_m \approx 3500$ K, $\beta \approx 30$, $\Delta T_0 \approx 800$ K and $n = 3$, $Z \approx 10^7$ m. With the depth of the mantle being 2.9×10^6 m, we see that the mass flow of a plume increases moderately (by about 30%) as it rises through the mantle. Note that our parameter estimates are quite uncertain and the fractional increase in mass flow could easily be 50% or more.

Since plume heat flow is constant, the temperature contrast at the top of the mantle is reduced by a similar fraction; a plume having $\Delta T = 800$ K at the base of the mantle could have a temperature contrast of 200 to 400 K at the top, due to the entrainment of ambient mantle material. This is in accord with geochemical analyses (e.g., see Putirka et al. 2007) that find a plume typically has a temperature excess of several hundred degrees at the top of the mantle.

32.2.5 Plume Discussion

In this section we have focused our attention on steady-state mantle plumes that serve to convey heat, in the form of hot silicate material, from the core upward through the mantle. As it reaches the lithosphere, some of this heated material erupts on the surface as hotspot volcanism, as explained in the following chapter, while most diverges beneath the lithosphere, forming a hotspot swell and contributing to the low-viscosity asthenosphere. Plumes, which are roughly cylindrical structures extending upward from the D″ layer, move laterally, along with the bulk of the mantle, as cold slabs are introduced from above. Since the thickness of the lithosphere is much less than the thickness of the mantle, the lateral speed of plumes is much less than the rate of sea-floor spreading. That is, a plume is a relatively stationary source of magma that produces a sequence of volcanoes having ages that increase with distance from the hotspot. The most famous example of this is the

[17] See Table B.5 and Figure 7.3.

Hawaiian-Emperor chain of volcanoes, but many others can be seen on the World Ocean Floor Panorama, produced in 1977 by Bruce Heezen and Marie Tharp.[18]

Mantle plumes are not permanent and a plume may decay if, for example, its source of hot material is cut off by a subducting slab or a crypto-continent. On the other hand, a new plume may form if the hot material in the D″ layer has no ready outlet via an existing plume and becomes unstable. A newly forming plume, called a *starting plume*, has a hard row to hoe; it must push its way upward through the overlying mantle that is relatively cold and viscous, and unwilling to move aside. This is accomplished by the rise of a roughly spherical head containing hot silicate supplied from below by a trailing plume. When it reaches the surface, the plume head delivers a large amount of hot material in a relatively short period of time (several million years). This material erupts on the surface to produce a large igneous province, called a *flood basalt*. There are a number of flood basalts seen in the geological record, each of which has a trail of volcanoes of decreasing age culminating at an active hotspot.[19] Interestingly, the eruption times of flood basalts occurring in the past 3×10^8 y coincide with mass extinctions of species, strongly suggesting a causal link.[20]

Mantle plumes exist as distinct dynamic structures because the viscosity of mantle silicates is strongly dependent on temperature. Interestingly, there are several common fluids with this same property, including corn syrup and silicone oil. This permits the investigation of plume structure and dynamics within the laboratory; see Davaille and Limare (2009).

Transition

In this chapter we have developed analytical models of the mantle flows that serve to cool the core: flows in the D″ layer and in plumes. In the following chapters, we will investigate the nature of volcanism near Earth's surface associated with cooling the core and mantle, beginning in the next chapter with a brief overview of the types of volcanoes and volcanic eruptions.

[18] Available from the Library of Congress at https://loc.gov/resource/g9096c.ct003148/.
[19] See Richards et. al. (1989)
[20] See Table 1 and Figure 1 of Courtillot and Renne (2003).

33
Overview of Volcanic Flows

As we noted in § 30.3, there are three types of volcanic flows:

- at mid-ocean spreading centers;
- beneath subduction zones; and
- at hotspot volcanoes.

Volcanic flows at mid-ocean spreading centers are passive upwellings that fill the void created as the lithospheric slabs on either side are pulled away. Due to the parabolic shape[1] of the cooling slabs, the upwelling speed is greatest, the motion is closest to adiabatic and the temperature is hottest beneath the spreading center. As can be seen from Figure 7.3, the mantle adiabat, when extrapolated to the top of the mantle, is greater than the melting temperature. This means that the upwelling material becomes partially molten. The viscosity of molten silicates is far smaller than the viscosity characterizing sub-solidus creep[2] and the density of melt[3] is less than that of the solidified remainder. It follows that the molten silicate (now called *magma*),[4] percolates upward, collecting in *magma chambers*. The molten material in these chambers wells upward to the surface, where it forms basaltic *pillow lava*. Since this flow occurs in an expanding environment, the magma is not confined and out-gassing of volatiles is a quiet affair.

The situation regarding out-gassing is quite different for the *andesitic* volcanism that occurs above subduction zones. This volcanism originates at the top of the subducting slabs due to a combination of viscous-dissipative heating and the presence of water. Most andesitic volcanoes occur on the rim of the Pacific Ocean, forming the so-called *ring of fire*. As the lithospheric slabs trundle their way from spreading center to subduction zone, minerals in the top of the slab become hydrated; that is, they incorporate water into their crystalline structure.[5] After subduction, these hydrated minerals break down, releasing the water from its crystalline cage. This liberated water does two things. First, it lubricates the top of the slab, permitting it to slip beneath the adjoining crustal material. Second, it acts

[1] See § 7.2.3.
[2] See § 6.2.3.
[3] That is, molten silicate.
[4] Magma describes molten silicate containing volatile components. After the volatiles have de-gassed upon reaching the surface, the molten silicate is called *lava*.
[5] A common hydrated mineral is *serpentine*.

to reduce the melting temperature of the silicates, facilitating partial melting of the slab material. In contrast to the mid-ocean melting that occurs close to the surface, subduction melting occurs at depth (typically 100 to 200 km) and beneath an overburden of crust. The melt, which is less dense than the surrounding material, makes its way upward through the overlying crust via semi-permanent channels. Upon reaching the surface, these magmas over time build volcanic cones, exemplified by Mount Fuji in Japan.

As the andesitic magma rises through the crust, the reduction of pressure promotes melting, but conductive cooling induces freezing. Each of these processes is accompanied by *chemical fractionation* which changes the composition of the melt. Typically when it gets close to the surface, the melt accumulates in one or more magma chambers, where further fractionation occurs. In particular, as the non-volatile components of the magma within a magma chamber solidify, the remaining melt becomes increasingly enriched in volatile components. Since the solubility of volatiles in magma is a strong function of pressure, a reduction of pressure can cause the abrupt ex-solution of volatiles, leading to an explosive eruption. For example, the reduction of pressure due to a landslide led directly to the eruption of Mount St. Helens on May 18, 1980. The hot volatiles released by the explosion formed a *pyroclastic flow* or a *nuées ardentes*: a cloud of superheated gas and debris that flows rapidly downslope.

Hotspot volcanism occurs above mantle plumes. This type of volcanism is similar to mid-ocean volcanism, in that it tends to be effusive, rather than explosive. Also, since the viscosity of hotspot magma is much lower than that of andesitic magmas, the volcanic edifices formed by hotspot volcanism have much lower slopes, forming broad *shield volcanoes*.[6] However, there are exceptions to this benign behavior when a mantle plume impinges on continental crust as occurs, for example, at Yellowstone. In this case, the behavior is similar to that of andesitic volcanoes, but on a somewhat larger scale, leading to super-volcanic eruptions that can have a strong impact on climate and survivability; see Robock (2000) and Mason et al. (2004).

33.1 Types of Magmas

The physical characteristics of magmas (and lavas) vary principally with their silica (SiO_2) content, and they may be roughly divided into three categories, as summarized in Table 33.1. Basaltic magmas are the most primitive type and typically form directly from mantle silicates by decompression melting. Andesitic magmas are a secondary type, commonly form by the addition of volatiles (mainly water); this process is called *flux melting*. Rhyolitic magma is the most evolved type, typically formed by re-heating of cooled andesitic magma. Note the progression of increasing silica content, lowering melting temperature and increasing viscosity.

Basaltic magmas flow most easily, forming smooth pahoehoe lava on land and pillow lavas under water. Late-stage basaltic flows and andesitic flows form more jagged a'a lava,

[6] So named because the look like a circular shield.

Table 33.1. *Categorization of the types of magmas. MOR = mid-ocean ridge. Melting temperatures are in °C. Viscosities are kinematic viscosities in* $m^2 \cdot s^{-1}$*; these are rough estimates, as the viscosities are very sensitive to variations in temperature, silica content and volatile content.*

	Basaltic	Andesitic	Rhyolitic
Location	MOR and hotspots	Ring of fire	Ring of fire
Silica content	45–55%	55–65%	65–75%
Temperature	1000–1200	800–1000	650–800
Viscosity	0.1–1	100–1000	10^7–10^8

Table 33.2. *Categorization of volcanic eruptions using the Volcanic Eruption Index.*

VEI	Name	Volume (m^3)	Frequency per year
0	Hawaiian	10^4	Continuous
1	Hawaiian/Strombolian	10^5	25
2	Strombolian/Vulcanian	10^6	12
3	Vulcanian/Peléan	10^7	4
4	Peléan/Plinian	10^8	1–2
5	Peléan/Plinian	10^9	0.1
6	Plinian	10^{10}	0.01–0.02
7	Ultraplinian	10^{11}	0.001–0.002
8	Supervolcano	10^{12}	2×10^{-5}

which is dangerous to walk upon. Magmas with high silica content do not flow easily, and tend to form lava domes. In particular, in the late stage of an eruption, a volcanic vent often fills with high-viscosity rhyolitic magma, which solidifies to form a plug, setting the stage for the next eruption.

33.2 Categorizing Volcanic Eruptions

Volcanic eruptions are categorized by the *volcanic eruption index*, or VEI, which ranges from 0 to 8, depending on the magnitude of the erupted magma and the severity; see Table 33.2. The nature of a volcanic eruption depends on several factors, including volatile content and the size and depth of the magma chamber.

Volcanoes pose a number of risks to life and property. Explosive volcanoes lift material into the air, ranging from large blocks that are lobbed up like mortar shells to fine dust that threatens airplanes and even gases capable of altering the climate. For example, the eruption in 1815 of Mount Tambor in the Dutch East Indies led to the "year without a summer" in 1816. Volcanic hazards on the ground are associated with the rain of debris lofted by the volcano and flows of lava and other volcanic matter down the slopes. For example, stone and ash ejected from Mount Vesuvius (located on the Gulf of Naples in Italy) in AD 79 entombed Pompeii and Herculaneum. We are most familiar with the gentle lava flows from the Hawaiian volcanoes, but these are far less dangerous than the avalanche-like flows associated with andesitic (island-arc) volcanoes, which can be devastating. A notable example is the eruption of Mount Pelée on the island of Martinique on May 8, 1902, which produced a pyroclastic flow that flattened the town of St. Pierre, killing nearly all of its 30,000 residents.

Although supervolcanoes are rare, they pose a risk comparable to an all-out nuclear war or the impact of a large asteroid. For example, the eruption of Mount Toba in Sumatra about 75,000 years ago ejected some 3×10^{12} m^3 of magma and nearly led to the extinction of *Homo sapiens* (see Jones and Savino, 2015). The size of the Toba eruption relative to other notable eruptions is illustrated in Figure 33.1. In spite of their impressive sizes, these are not the largest eruptions to have occurred; for example, the Deccan Traps in India are volcanic deposits having an estimated volume 5×10^5 km^3. The Deccan eruptions may have played a role in the Cretaceous–Tertiary (K–T) mass extinction event that occurred some 66 My ago.

33.3 Triggering Volcanic Eruptions

It is well-known that volcanoes can and do erupt violently after remaining dormant for centuries. Eruptions are driven by gases that had been dissolved in the magma. The mass fraction of volatiles that can remain in solution is a function of pressure, with volatiles preferentially remaining in solution at high pressure. Exsolution occurs when the magma pressure becomes less than the saturation pressure or – saying it the other way round –

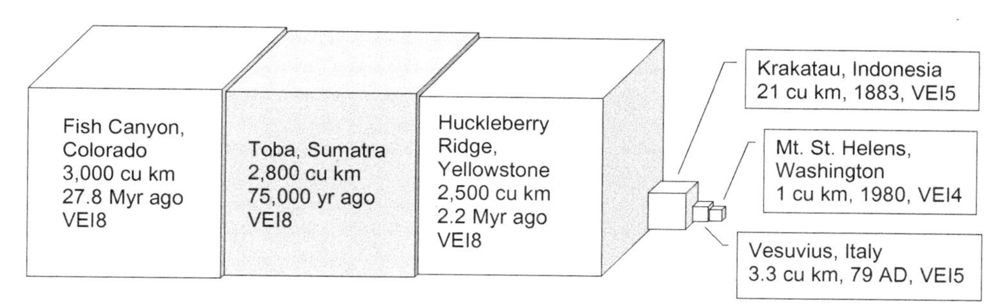

Figure 33.1 Illustration of the relative sizes (volume of erupted magma) of six notable volcanic eruptions. 1 cu km = 10^9 m^3.

when the mass fraction of volatiles in the melt becomes greater than can remain in solution at a given pressure. Eruptions are sustained by the first version – by the reduction in pressure due to the presence of volatiles in a volcanic conduit; this effect is quantified and discussed in § 34.3. A volcanic eruption can be triggered by a sudden reduction of the lithostatic pressure caused, for example, by an avalanche or slumping of the overburden above a magma chamber. The eruption of Mount St. Helens in 1980 (a VEI4 event; see Figure 33.1) was triggered by such an event.

It is less appreciated and understood that an eruption can be triggered simply by the slow cooling of a magma chamber. Cooling causes crystallization and during this process volatiles, being incompatible,[7] preferentially remain in the liquid phase, gradually increasing the mass fraction of volatiles in that phase. Eventually the volatiles will exsolve. If pathways do not exist for the gases to escape, they cause an increase the pressure within the magma chamber. Once this pressure exceeds the lithostatic pressure an eruption ensues. This effect can be replicated by placing a can of carbonated beverage in a freezer. (This is a dangerous experiment – do not try it at home.)

Normally crystallization and exsolution occur gradually, but exsolution can occur abruptly in a process similar to that known as "rollover" in the liquefied natural gas industry. Cooling can lead to the development of an unstable density profile and the ensuing convection can mix disparate liquids and trigger the rapid exsolution of volatiles.

Transition

This chapter has given us a brief overview of the three main types of magmas and the categorization of volcanic eruptions. In the following chapter we will investigate flow in volcanic conduits, with emphasis on the role of volatiles in driving volcanic eruptions.

[7] See § 30.3.1.

34
Flow in Volcanic Conduits

A *volcanic conduit*[1] is a roughly circular opening extending upward from a magma chamber, through which magma rises and erupts on the surface. Magma is a mixture of molten silicate and volatiles, both in solution and in gaseous form, with the total mass fraction of volatiles typically being several (2 to 6) percent. The dominant volatile component is water (with a typical mass fraction of 4%), with smaller amounts of CO_2, HS, SO_2, HCl, HF, etc.[2] To simplify the analysis and discussion in the following sections, we will assume the volatile is entirely water, and will take the mass fraction of water to be 0.04.

The behavior of the magma flowing up a conduit depends in large part on the magma viscosity. If the magma has low viscosity, the outgasing volatiles can ascend relatively quickly and vent to the surface. This behavior is typical of hotspot volcanoes, which erupt with mild burps of gas. However, if the magma viscosity is high, the gases tend to rise in tandem with the liquid phase. These gases expand dramatically during an eruption – with spectacular results.

In this chapter we will investigate and discuss the physical processes involved in the steady flow of magma up a conduit having a cross-sectional area that varies with elevation: $A(z)$. Let's assume that the variation of area is gradual so that we can treat the flow locally as entirely vertical, ignoring any horizontal flow within the conduit.

34.1 Conduit Equations

Flow of magma is characterized by five variables: the upward speed w, the pressure p, the density ρ, the mass fraction of volatiles (that is, water) in solution, ξ, and the mass fraction of vapor, ξ_v.[3] These variables are governed by five equations: continuity, vertical component of momentum for magma and vapor, equation of state and gas solubility.

The momentum equation for the gaseous phase determines how much of the vapor escapes upward. At one extreme (for example, in hotspot volcanism), the gas is very mobile and ξ_v is close to zero. At the other extreme, the gas rises so slowly relative to the liquid

[1] Also called a volcanic vent or pipe.
[2] See Plank et al. (2013) and Wallace et al. (2015).
[3] At the high temperatures and pressures within a volcanic vent, water is above its critical point and so technically is a fluid, rather than a gas.

phase that it may be considered immobile and conservation of water tells us that

$$\xi_v = \xi_0 - \xi\,,$$

where ξ_0 is the (prescribed) mass fraction of volatile in the magma prior to any exsolution. We are interested in modeling explosive eruptions that occur sufficiently rapidly that motion of the gas phase, relative to the liquid, is negligibly small. In this case, this equation determines ξ_v, and we don't need to consider the momentum equation for the vapor.

In steady state the speed depends on both vertical and horizontal position while p, ρ and ξ are functions only of z. We are interested in the gross features of the flow and will be content with determining the mean speed $\bar{w}(z)$ rather than the point-wise speed w.

34.1.1 Continuity

With steady flow within a conduit of area $A(z)$, conservation of mass requires that

$$\int_A \rho w \,\mathrm{d}A = \rho \bar{w} A = \dot{M}\,,$$

where the mass flow up the conduit, $\dot{M}$, is constant. The area is a known function of z, while ρ and $\bar{w}$ are functions of z to be determined. Note that

$$\frac{1}{\bar{w}}\frac{\mathrm{d}\bar{w}}{\mathrm{d}z} = -\frac{1}{\rho}\frac{\mathrm{d}\rho}{\mathrm{d}z} - \frac{1}{A}\frac{\mathrm{d}A}{\mathrm{d}z}\,.$$

34.1.2 Momentum

The vertical component of the momentum equation is[4]

$$w\frac{\partial w}{\partial z} + g = -\frac{1}{\rho}\frac{\mathrm{d}p}{\mathrm{d}z} + \frac{\partial}{\varpi\,\partial\varpi}\left(\nu\varpi\frac{\partial w}{\partial\varpi}\right),$$

where g is the local acceleration of gravity and ν is the molecular or turbulent viscosity as appropriate. We are using the full version of this equation, rather than the perturbation form,[5] because deviations from the hydrostatic balance are large within a conduit.

In taking the area average of this equation, we may express the vertical component of velocity as $w(\varpi, z) = \bar{w}(z)\left(1 + w'(\varpi)\right)$, where w' is the dimensionless deviation of the vertical speed from the average. Note that

$$\overline{w^2} = \frac{1}{A}\int_A w^2 \mathrm{d}A = \alpha\bar{w}^2\,, \qquad \text{where} \qquad \alpha \equiv \frac{1}{A}\int_A \left(1 + 2w' + w'^2\right)\mathrm{d}A$$

is the ratio of the average of the square to the square of the average. For laminar flow in a circular conduit of radius ϖ_0, $w = 2\bar{w}\left(1 - (\varpi/\varpi_0)^2\right)$ and $\alpha = 4/3$.[6] For turbulent flow,

[4] From § 6.2.2.
[5] See § 7.3.2.
[6] This is called *Poiseuille flow* (pronounced "pwa-sey").

α is much closer to unity. To simplify the subsequent analysis, we will assume that $\alpha = 1$, realizing that this introduces a small quantitative error.

The area average of the vertical component of the momentum equation gives us

$$\bar{w}\frac{\mathrm{d}\bar{w}}{\mathrm{d}z} + g = -\frac{1}{\rho}\frac{\mathrm{d}p}{\mathrm{d}z} + \frac{2}{\varpi_0}\frac{\tau_b}{\rho},$$

where $\varpi_0 = \sqrt{A/\pi}$ is the mean radius of the conduit (not necessarily circular) and

$$\tau_b = \rho\nu\left.\frac{\partial w}{\partial\varpi}\right|_{\varpi=\varpi_0}$$

is the stress on the magma at the boundary. Note that the velocity gradient at the wall is negative, so that τ_b is negative, indicating resistance to upward motion. Expressing the stress in terms of the drag coefficient,[7] $\tau_b = -C_D\rho\bar{w}^2$, the momentum equation becomes

$$\bar{w}\frac{\mathrm{d}\bar{w}}{\mathrm{d}z} + g = -\frac{1}{\rho}\frac{\mathrm{d}p}{\mathrm{d}z} - \frac{2}{\varpi_0}C_D\bar{w}^2.$$

If the magma is a Newtonian fluid flowing laminarly, $C_D \approx 3\nu/\varpi_0\bar{w}$, while for vigorous turbulent flow, C_D is a small constant, whose magnitude depends on the roughness of the boundary.

34.1.3 Equation of State and Solubility

Treating the magma as a mixture of liquid silicate (containing dissolved water) and exsolved gas (water vapor), having densities ρ_l and ρ_g, respectively, the density equation of state[8] is $(\rho)^{-1} = (1-\xi_v)(\rho_l)^{-1} + \xi_v(\rho_g)^{-1}$. The gas and liquid that comprise the magma are in intimate contact so that they have a common temperature. Given its much greater density and hence heat capacity, the liquid phase controls this temperature. Further, the compressibility of the liquid phase is small, so that the liquid and gas rise nearly isothermally. It follows from the ideal gas law[9] that $\rho_g = \rho_0 p/p_0$, where a subscript 0 denotes a constant value. (p_0 and ρ_0 are determined and discussed in the next paragraph.) Altogether,

$$\frac{1}{\rho} = \frac{1-\xi_v}{\rho_l} + \frac{\xi_v}{\rho_0 p^*}, \qquad \text{where} \qquad p^* \equiv \frac{p}{p_0}.$$

Though ξ_v is typically small (a few percent at most), the effect of gas on the total density is dramatic as p becomes small (near the surface) because the density of water vapor is far less than that of liquid silicate: $\rho_0 \ll \rho_l$.

[7] See § 23.7.
[8] From 6.4.1.
[9] See § 5.6.

Since we are free to select the reference pressure p_0, let's take it to be the lowest pressure at which all the water remains in solution. That is, if $p > p_0$ (or equivalently $p^* > 1$) all the water is dissolved in the magma ($\xi_v = 0$), while some water is exsolved (in a gaseous state, with $\xi_v > 0$) if $p < p_0$ (or equivalently $p^* < 1$). The magnitude of p_0 depends on the total mass fraction of water, ξ_0. Specifically, $p_0 \approx \xi_0^2\, 6 \times 10^{10}$ Pa.[10] If this is a lithostatic pressure given by $p_0 = \rho_l g h_0$, where h_0 is the depth below Earth's surface, then with $\rho_l \approx 2600$ kg·m^{-3} and $g = 9.8$ m·s^{-2}, $h_0 \approx \xi_0^2\, 2.4 \times 10^6$ m. For example, if $\xi_0 = 0.04$ (a typical mass fraction of water in magmas), then $p_0 \approx 10^8$ Pa and $h_0 \approx 4$ km. Now we see that ρ_0 is the density of the tiny fraction of water vapor that first forms as p drops below p_0. For example, the density of water vapor[11] at a temperature of 1600 K and pressure 10^8 Pa is about 135 kg·m^{-3}. This density is far less than that of the parent liquid: $\rho_0 \ll \rho_l$.

So far we have four equations (conservation of water ($\xi_v + \xi = \xi_0$), continuity in § 34.1.1, vertical component of momentum in § 34.1.2 and equation of state immediately above) for five unknowns: $\bar{w}$, p, ρ, ξ_v and ξ. We need a fifth equation, governing variation of the mass fraction of dissolved water, ξ. The solubility of water in magma is a function of pressure: $\xi(p)$. Ideally, the solubility is a linear function of pressure,[12] but the solubility of H_2O deviates considerably from this ideal (see Zhang et al., 2007), with its solubility varying approximately as $\sqrt{p}$, specifically $\xi \approx \xi_0 p^{*1/2}$, giving

$$\xi_v = \begin{cases} 0 & \text{if} \quad 1 < p^*, \\ \xi_0 - \xi_0 p^{*1/2} & \text{if} \quad p^* \leq 1 . \end{cases}$$

Altogether, the combined equation of state for density as a function of pressure, valid for $p^* \leq 1$, is

$$\begin{aligned} \frac{1}{\rho} &= \frac{1 - \xi_0 \left(1 - p^{*1/2}\right)}{\rho_l} + \frac{\xi_0}{\rho_0}\left(\frac{1}{p^*} - \frac{1}{p^{*1/2}}\right) \\ &\approx \frac{1}{\rho_l} + \frac{\xi_0}{\rho_0}\left(\frac{p_0}{p} - \frac{p_0^{1/2}}{p^{1/2}}\right). \end{aligned}$$

The simplified version of the equation of state (on the second line) is obtained assuming that $\xi_0 \ll 1$.

34.2 Speed of Sound

The speed of sound[13] in a molten (liquid) silicate is given by $U_l^2 = (\mathrm{d}\rho_l/\mathrm{d}p)^{-1/2} = (K/\rho_l)^{1/2}$, where K is the adiabatic incompressibility and[14] has a magnitude roughly several (3 to 4) km·s^{-1}. The presence of water vapor in magma causes a dramatic decrease

[10] See Zhang et al. (2007).
[11] From the ideal gas law, $\rho = p/R_s T$ with $R_s = 461.5$ J·kg^{-1}·K^{-1}.
[12] This is the essence of *Henry's law* and *Raoult's law*.
[13] See § 5.5.
[14] With $\rho_l \approx 2600$ kg·m^{-3} and $K \approx 3 \times 10^{10}$ Pa; see de Koker and Stixrude (2009).

in this speed. To quantify this decrease, let's calculate the speed of sound, U_s, in the magma using[15] $U_s = (\mathrm{d}\rho/\mathrm{d}p)^{-1/2}$ together with the simplified equation of state developed in the previous section. Noting that ξ_0 and ρ_0 are independent of p, we have that

$$U_s \approx \frac{\rho_l U_0 p^*}{\rho\left(2 - p^{*1/2} + (U_0/U_l)^2 p^{*2}\right)^{1/2}}, \qquad \text{where} \qquad U_0 \equiv \sqrt{\frac{2\rho_0 p_0}{\xi_0 \rho_l^2}}.$$

With previous parameter estimates ($\rho_0 \approx 135$ kg·m^{-3}, $p_0 \approx 10^8$ Pa, $\xi_0 \approx 0.04$ and $\rho_l \approx 2600$ kg·m^{-3}), we see that $U_0 \approx 316$ m·s^{-1}.

It is readily seen that $\mathrm{d}U_s/\mathrm{d}p^* > 0$, so that in the pressure range $0 < p^* < 1$ the speed of sound in magma is greatest when the fraction of vapor is vanishingly small (with $p^* = 1$ and $\rho = \rho_l$). In this limit the speed of sound in the magma is $U_s \approx U_0 U_l / \left(U_0^2 + U_l^2\right)^{1/2}$. With previous estimates ($U_l \approx 4000$ m·s^{-1} and $U_0 \approx 316$ m·s^{-1}), we find that $U_{s1} \approx 315$ m·s^{-1}, a value slightly smaller than U_0. This speed is roughly 1/10 of the speed of sound in silicate and also much less than the speed of sound (approximately[16] 1 km·s^{-1}) in water vapor.

Why is U_s so small? To understand this slow speed, let's recall that sound is a compressive wave, with a speed inversely proportional to the compressibility; the more compressible the material, the slower the speed of sound. Water vapor – by itself – is more compressible than silicates and so has a slower sound speed (roughly 1 km·s^{-1} versus 3 to 4). The thing is, the water vapor in a magma is not by itself; it is encased in molten silicate and reacts with it in such a way as to increase its effective compressibility. As pressure is increased during the passage of a sound wave, some of the water vapor condenses, giving the gas extra compressibility, which results in a slower speed of sound. The phase-change process is not instantaneous, so that there is a phase lag between compression and phase change, which translates into a loss of wave energy. This effect is responsible for the muffling of sounds in a heavy fog.

34.3 Exsolution and Fragmentation

At great depth ($p^* > 1$) all water is in solution. As the magma ascends above the level at which $p^* = 1$, water begins to exsolve. In order to estimate the pressure, or equivalently the depth h, at which exsolution begins in an open conduit, let's consider the hypothetical situation in which the magma is at rest and assume that the liquid magma and water vapor do not separate. The static pressure varies with depth according to

$$\left(1 + \xi_0 \frac{\rho_l}{\rho_0}\left(\frac{1}{p^*} - \frac{1}{\sqrt{p^*}}\right)\right)\frac{\mathrm{d}p^*}{\mathrm{d}z} = -\frac{1}{h_0}, \qquad \text{where} \qquad h_0 = \frac{p_0}{\rho_l g}$$

[15] See § 5.5.

[16] Using the last equation of § 5.6, $U_s \approx \sqrt{\gamma R_s T}$ with $\gamma = 4/3$, $R_s = 461.5$ J·kg^{-1}·K^{-1} and $T \approx 1600$ K. At this temperature and pressure, water is above the critical point and likely deviates from ideal-gas behavior, so this calculation is just a rough estimate.

is the exsolution depth in the absence of density changes due to the presence of the gaseous phase. The depth h may be determined by integrating this equation from atmospheric pressure $p = p_a$ or equivalently $p^* = p_a^* \equiv p_a/p_0$ to $p^* = 1$:

$$h \approx h_0 \left(1 - \xi_0 \frac{\rho_l}{\rho_0} \left(\log p_a^* - 2\right)\right).$$

With previous parameter estimates ($g \approx 9.8$ m^2·s^{-2}, $\rho_0 \approx 135$ kg·m^{-3}, $p_0 \approx 10^8$ Pa, $\xi_0 \approx 0.04$ and $\rho_l \approx 2600$ kg·m^{-3}) and[17] $p_a = 1.01325 \times 10^5$ Pa, $p_a^* \approx 10^{-3}$, $h_0 \approx 4$ km and $h \approx 6.7$ km. This increase in exsolution depth is due to the reduction of density and pressure in the conduit due to the presence of exsolved gas. This means that when an eruption starts, exsolution decreases the pressure, causing magma to begin to exsolve at greater depth. If there is a large magma chamber just below the static exsolution depth, a very large eruption could ensue.

34.3.1 Fragmentation

Initially, as the magma ascends above the level at which $p^* = 1$, exsolved water occurs as small isolated bubbles that are carried along with the upward flow. Due to their smaller density, these bubbles will tend to migrate upward relative to the magma, but this upward percolation is significant only if the bubbles are large and the magma has low viscosity and is fairly quiescent. For example, percolation in basaltic magmas is important if the bubble diameters exceed about 0.01 m and the ascent speed is less than 10^{-2} m·s^{-1}. In this situation, water can *outgas* and the eruption is effusive, rather than explosive. As the magma moves upward, the mass fraction of gas increases, as does the bubble size, until the bubbles begin to merge. At this level, called the *fragmentation level*, the character of the magma changes rather abruptly from a liquid containing bubbles to a gas containing liquid fragments, called *pyroclasts*. These fragments tend to fall downward relative to the gas, but usually this motion is negligibly small within the conduit, where the upward speed is large.

34.4 Magma Flow up a Conduit

Now let's consider magma moving steadily up a conduit of cross-sectional area $A(z)$. As the magma moves upward, the pressure decreases and more water exsolves and the water vapor expands as the pressure decreases. Since in steady state the mass flow is constant, the mean upward velocity, $\bar{w}$, increases to compensate for the reduction in density. This upward acceleration leads to dramatic effects, with the pryoclasts being lofted high into the air. In this section we will investigate the vertical variation in the mean upward speed, $\bar{w}$.

Substituting the derivative of the continuity equation ($\mathrm{d}(\rho\bar{w}A)/\mathrm{d}z = 0$ from § 34.1.1) and the definition of speed of sound ($U_s = (\mathrm{d}\rho/\mathrm{d}p)^{-1/2}$ from § 5.5) into the vertical

[17] See Table B.5.

component of the momentum equation ($\rho\bar{w}\mathrm{d}\bar{w}/\mathrm{d}z + \rho g = -\mathrm{d}p/\mathrm{d}z - 2C_D\rho\bar{w}^2/\varpi_0$ from § 34.1.2), we have

$$\left(1 - \frac{\bar{w}^2}{U_s^2}\right)\frac{\mathrm{d}p}{\mathrm{d}z} = \left(\frac{1}{A}\frac{\mathrm{d}A}{\mathrm{d}z} - \frac{2}{\varpi_0}C_D\right)\rho\bar{w}^2 - \rho g\,.$$

This equation, coupled with the continuity equation ($\rho\bar{w}A = \dot{M}$ from § 34.1.1) and the combined equation of state $\left(\rho^{-1} = \rho_l^{-1} + (\xi_0/\rho_0)\left((p^*)^{-1} - (p^*)^{-1/2}\right)\right)$ from § 34.1.3), form a set of three equations for three unknowns: p, $\bar{w}$ and ρ, with auxiliary equations for $U_s(p)$ and $A(z)$ prescribed. This flow problem is superficially similar to that encountered in the study of flow in a rocket nozzle,[18] but in fact is considerably more complicated, primarily because the speed of sound is not constant, but is a function of pressure.

34.4.1 Conduit Area and Roughness

We have formulated the problem assuming that the conduit area A is a specified function of z, independent of the flow and have implicitly assumed that the coefficient of drag, C_D, representing the retarding effect of small-scale irregularities in the wall, is known and is independent of the flow. However, in truth, neither of these assumptions is true; both A and C_D are functions of the flow. The walls of the conduit are made of cooled magma and, as an eruption proceeds, hot magma rising from depth can – and does – erode and reshape the walls of the conduit, changing both A and C_D.

This problem, in which flow and structure are coupled, is very difficult and has yet to be addressed. One way to side-step this difficult issue is to use the fact that the evolution of conduit shape is a slow process and over any short interval of time, we can treat A and C_D as given.

In a volcanic conduit with the wall being sculpted by the flow, it is likely that the wall is relatively smooth. Any protuberance is in a hotter-than-average environment, making it prone to ablation. Conversely, liquid magma within a cavity is in a cooler-than-average environment and prone to solidification. This suggests that C_D may be quite small; in what follows we will assume that wall drag is negligibly small and set $C_D = 0$. Now the momentum equation simplifies to

$$\left(1 - \frac{\bar{w}^2}{U_s^2}\right)\frac{\mathrm{d}p}{\mathrm{d}z} = \frac{1}{A}\frac{\mathrm{d}A}{\mathrm{d}z}\rho\bar{w}^2 - \rho g\,.$$

34.4.2 Flow Transition

When $\bar{w} = 0$, the momentum equation simplifies to the hydrostatic balance: $\mathrm{d}p/\mathrm{d}z = -\rho g$. Since the actual (lithostatic) pressure at depth exceeds the conduit pressure (due to the presence of exsolved gases), magma is forced upward with $\bar{w} > 0$. The momentum equation

[18] Called a *de Laval nozzle*.

is a first-order ordinary differential equation predicting the variation of pressure with elevation. However, this equation has a curious feature. As the fluid is accelerated upward, the coefficient multiplying the pressure derivative becomes smaller and reaches zero when the upward magma speed equals the speed of sound. The flow speed can actually reach the speed of sound only if the right-hand side of the momentum equation above is equal to zero; that is,

$$\mathrm{d}A/\mathrm{d}z = gA/U_s^2 .$$

If this condition is not met, the flow is said to be *choked*. If it is met, the level at which this equation is satisfied is called the *throat* and at that level the flow is said to be *critical*. At elevations greater than the throat, the flow can either return to sub-sonic, or become super-sonic. We have encountered a similar situation in our study of channel flow; see Figure 20.2. Note that the transition from sub- to super-sonic flow requires the conduit area to increase with height. If the conduit has uniform area, the flow is choked.

Typically explosive flow out a volcanic vent is supersonic. This means that flow erosion has been able to sculpt the walls of the vent into the shape of a de Laval nozzle. This distinctive shape, with a narrow throat, is evident, for example, in kimberlite pipes that have been excavated for diamonds.

34.4.3 Flow Solution

Although the momentum equation is only a first-order ordinary differential equation, the problem of the flow of magma up a conduit is too complicated for analytic solution; we must resort to numerical methods to determine the flow. A numerical scheme for solving this problem is available courtesy of the USGS; see Mastin and Ghiorso (2000). This scheme requires specification of either the conduit geometry ($A(z)$) or pressure profile ($p(z)$) and the program calculates the other.

Transition

In this chapter we have investigated the flow of magma up a volcanic conduit, first developing a set of equations for the mean flow, then using these to determine how the presence of exsolved gas lowers the speed of sound dramatically. Next, we investigated and discussed the depth at which exsolution begins and the transition to fragmented flow. Finally, we studied the nature of flow up a conduit and found that it can readily be supersonic. In the following chapter, we investigate the nature of lava flow on the surface, focusing on basaltic lavas that have relatively low viscosity.

35

Lava Flows

Commonly andesitic and rhyolitic magmas rise up discrete conduits, whereas basaltic magma often – at least during the early stages of an eruption phase – rises to the surface through a *magmatic dike* (also called a rift or fissure): a linear crack extending from a magma chamber at depth to the surface.[1] Typically as magma erupts from a dike, outgassed volatiles (mainly water) produce a linear *fire fountain* composed of incandescent blobs of lava.

After magma reaches the surface, its subsequent behavior depends largely whether the eruption is effusive or explosive. A discussion of the possible styles of lava flow is found in Griffiths (2000). Effusive eruptions lead to the production of lava which flows relatively gently downslope, initially in the form of a planar sheet. We will consider a simple model of lava sheet flow in the following section.

On the other hand, explosive eruptions display a variety of behaviors depending on the speed, mass flow and composition of the erupting magma. Some of these behaviors are explored in § 35.2.

35.1 Lava Sheet Flow

In this section we will develop a simple model of subaerial (that is, on land, as opposed to under the sea) lava sheet flow: a sheet of hot basaltic lava flowing steadily down a slope (in the x direction) from a magmatic dike.[2]

Our simple model will ignore

- variations in topography;
- horizontal variations in the lava surface;
- fluid inertia;
- loss of heat to the ground;[3]
- radiative cooling;[4]

[1] As a dike eruption proceeds, nonlinear interaction between the rising magma and the dike walls leads to the concentration of rising magma flow in a series of conduits.
[2] A discussion of this type of flow is found in Hon, et al. (1994). Particularly note their Figure 3A.
[3] If the lava flow has been proceeding for a while, the ground beneath will have been warmed sufficiently that bottom cooling is negligibly small.
[4] Radiative cooling would be significant if the surface temperature of the lava sheet were large.

- variations in the ambient temperature;
- effects of rain;
- the dynamic effect of crystals suspended in the liquid lava;
- buoyancy effects; and
- variations of thermal conductivity with temperature.

In order to simplify the analysis further, we will assume that the cool upper layer of lava is rafted along with the flow, rather than being anchored to the ground at the sides. The weight of this rafted layer is borne by the liquid beneath, so that there is a vertical hydrostatic balance. This rafted layer may deform as needed in order to adjust to the flow of the liquid beneath. Typically this adjustment is seen as a buckling of the surface of the sheet;[5] we will not model this small-scale deformation. However, we shall see in § 35.1.3 that the sheet tends to thicken as it cools.

35.1.1 Sheet-Flow Equations

Let's consider a sheet of lava having initial depth $z = h_I$ and initial uniform temperature T_I, exiting a dike at $x = 0$. The lava flows down a uniform plane sloping at an angle ε to the horizontal, with z measuring (almost vertical) distance above the plane. Writing $\mathbf{v} = u\mathbf{1}_x + w\mathbf{1}_z$, steady downslope flow is governed by the x momentum and continuity equations:

$$\frac{\partial}{\partial z}\left(\eta\frac{\partial u}{\partial z}\right) = -\varepsilon\rho g \qquad \text{and} \qquad \frac{\partial u}{\partial x} + \frac{\partial w}{\partial z} = 0,$$

where g is the local acceleration of gravity.[6]

Basaltic magma, such as that discharged by Hawaiian volcanoes, behaves as a liquid if $1340\,\mathrm{K} < T$, flows as a sub-solidus material if $1070\,\mathrm{K} < T < 1340\,\mathrm{K}$ and is essentially rigid if $T < 1070\,\mathrm{K}$.[7] The viscosity of the liquid is also a strong function of temperature:[8]

$$\eta(T) = \eta_I \exp\left(\beta\frac{T_I - T}{T_I}\right),$$

where a subscript I denotes initial values (at $x = 0$) and β is a dimensionless constant somewhat larger than unity. In modeling lava flows, this temperature dependence is the dominant dynamic feature; once the lava has cooled to 1340 K the viscosity is so large that deformation has effectively ceased and sub-solidus creep and crystallization are dynamically irrelevant.

[5] See, for example, the photos of lava flows in Hon et al. (1994).
[6] Omission of the pressure term is justified provided $\mathrm{d}h/\mathrm{d}x \ll \varepsilon$.
[7] Lava temperatures in the literature are commonly given in degrees Celsius, rather than kelvin. The difference between these two scales is discussed at the end of Appendix B.2.
[8] This is similar to the sub-solidus behavior of the silicate mantle, but with far smaller values.

Variation of heat within the sheet is governed by the energy equation:

$$u\frac{\partial T}{\partial x}+w\frac{\partial T}{\partial z}=\kappa\frac{\partial^2 T}{\partial z^2},$$

where $\kappa = k/\rho c_p$ is the thermal diffusivity. Note that we have assumed that vertical conduction of heat is much larger than horizontal and have omitted the latter term from the energy equation. (This is a good assumption provided $h_c \ll x$, where h_c is the vertical length scale introduced in boundary condition at $z=h$, presented below.)

These equations are to be solved in the interval $0<x<\infty$ and $0<z<h(x)$, where $h(x)$ is the top of the sheet. The boundary conditions at the bottom are

$$\text{at} \qquad z=0: \qquad u=w=\frac{\partial T}{\partial z}=0.$$

At the top of the sheet, the vertical velocity is determined by the kinematic condition found in § 3.5 and the downslope velocity satisfies the no-stress condition (zero vertical gradient). Our thermal boundary condition at the top of the sheet is *Newton's law of cooling*, which states that the heat flow from a body is proportional to the temperature difference between the body and its surroundings. Altogether

$$\text{at} \quad z=h(x): \quad \frac{\partial u}{\partial z}=0, \quad w=u\frac{\mathrm{d}h}{\mathrm{d}x} \quad \text{and} \quad h_c\frac{\partial T}{\partial z}=-(T-T_a),$$

where h_c is a length scale representing the rate of cooling and T_a is the specified ambient temperature.

The initial conditions are

$$\text{at} \qquad x=0: \qquad h=h_I, \qquad T=T_I \qquad \text{and} \qquad u=u_I\left(3\frac{z}{h_I}-\frac{3}{2}\frac{z^2}{h_I^2}\right),$$

$$\text{where} \qquad u_I\equiv\frac{\varepsilon\rho g h_I^2}{3\eta_I}$$

is the mean speed. The entry profile for u is the constant-viscosity creeping-flow profile we investigated in § 22.2.2. The specific volume flow[9] is simply $q=h_I u_I$. Since we are assuming steady state, the steady volume flow is an integral constraint on the flow:

$$\int_0^{h(x)} u\,\mathrm{d}z = q = h_I u_I\,.$$

This equation determines $h(x)$.

A first integral of the momentum equation that satisfied the no-stress condition is

$$\eta\frac{\partial u}{\partial z}=\varepsilon\rho g(h-z)\,.$$

[9] Volume flow per unit width, having SI units $\mathrm{m}^2\cdot\mathrm{s}^{-1}$.

Taken together, the two thermal conditions require that the lava sheet cools as it flows downslope. Since the volume flow and gravitational forcing are constant, the resistance to flow must remain constant even though the viscosity increases as the temperature decreases. This adjustment is accomplished by an increase in the thickness of the sheet as it flows downslope, but this variation is typically slow: $0 < \mathrm{d}h/\mathrm{d}x \ll 1$.

Typical parameters for basaltic lavas are:

- $\eta_I \approx 10 - 10^3$ Pa·s;
- $T_I = 1750$ K;
- $T_a \approx 300$ K;
- $\beta \approx 25$;
- $c_p \approx 1100$ J·kg^{-1}·K^{-1};
- $\rho \approx 2700$ kg·m^{-3}.
- $h_c \approx 0.02$ m;
- $\kappa \approx 10^{-6}$ m^2·s^{-1};
- $\varepsilon \approx 0.01$–0.1;
- $u \approx 0.1$–1 m·s^{-1};
- $h \approx 0.1$–1 m; and

Note that the momentum balance requires $g\varepsilon h^2 \approx \eta u$.

Lava sheet flow is governed by a set of three partial differential equations (x momentum, continuity and energy) involving four unknowns: $u(x,z)$, $w(x,z)$, $T(x,z)$ and $h(x)$ with η being a function of T. The extra equation comes from the boundary conditions; we have five conditions on a fourth-order set of equations. This is a free-boundary problem, with the upper surface at $z = h(x)$ to be determined as part of the solution. Since the problem contains a rather large number of dimensional parameters, analysis and solution can be facilitated by non-dimensionalizing the problem.

35.1.2 Non-Dimensionalization

As can be seen by the list of values above, the problem has a fairly large number of dimensional parameters. The number of parameters may be reduced to a minimum by judiciously non-dimensionalizing the problem. Specifically, we may write

$$h = h_I h^*, \qquad u = u_I u^*, \qquad w = \frac{\kappa}{h_I}\left(\frac{z^*}{h^*}\frac{\mathrm{d}h^*}{\mathrm{d}x^*}u^* + w^*\right)$$

$$\text{and} \qquad T = T_a + (T_I - T_a)T^*,$$

where u^*, w^* and T^* are functions of the dimensionless spatial coordinates

$$x^* \equiv \kappa x/(u_I h_I^2) \qquad \text{and} \qquad z^* \equiv z/h = z/(h_I h^*).$$

and h^* is a function of only x^*. Note the extreme anisotropy in scaling the vertical and horizontal coordinates; $z^* = O(1)$ corresponds to $z = O(h) \approx 0.1$–1 m, while $x^* = O(1)$ corresponds to $x = O(u_I h_I^2/\kappa) \approx 10^3$–$10^6$ m. Also, noting that

$$\frac{\partial}{\partial x} = \frac{\kappa}{u_I h_I^2}\left(\frac{\partial}{\partial x^*} - \frac{z^*}{h^*}\frac{\mathrm{d}h^*}{\mathrm{d}x^*}\frac{\partial}{\partial z^*}\right),$$

the dimensionless governing equations are

$$\frac{\partial u^*}{\partial z^*} = 2e^{-\beta^*(1-T^*)}(1-z^*),$$
$$\frac{\partial (h^* u^*)}{\partial x^*} + h^* \frac{\partial w^*}{\partial z^*} = 0 \qquad \text{and}$$
$$u^* \frac{\partial T^*}{\partial x^*} + w^* \frac{\partial T^*}{\partial z^*} = \frac{\partial^2 T^*}{\partial z^{*2}},$$

where

$$\beta^* = \beta(T_I - T_a)/T_I.$$

These equations are to be solved in the domain $0 < x^* < \infty$ and $0 < z^* < 1$, subject to the conditions

$$\text{at} \quad x^* = 0: \quad h^* = T^* = 1 \quad \text{and} \quad u^* = 2z^* - z^{*2};$$
$$\text{at} \quad z^* = 0: \quad u^* = w^* = \frac{\partial T^*}{\partial z^*} = 0;$$
$$\text{at} \quad z^* = 1: \quad w^* = 0 \quad \text{and} \quad \frac{\partial T^*}{\partial z^*} = -\mathrm{Bi}\, T^*,$$

where

$$\mathrm{Bi} = h_I/h_c$$

is the *Biot number*, representing the relative importance of conduction within the body and convection at the surface.

The integral constraint on the volume flow becomes

$$h^* = \frac{2}{3I_u}, \qquad \text{where} \qquad I_u \equiv \int_0^1 u^* \mathrm{d}z^*.$$

This volume-flow constraint can be used to verify that the integral of the continuity equation from $z^* = 0$ to $z^* = 1$ satisfies the boundary conditions on w^*. Note that, integrating by parts and using the momentum equation,

$$I_u = u^*|_{z^*=1} - \int_0^1 \frac{\partial u^*}{\partial z^*} z^* \mathrm{d}z^*.$$

The problem contains two large dimensionless parameters: β^* (≈ 25) and Bi (≈ 50). The large value of β^* means that viscosity varies strongly with temperature and the top of the sheet tends to flow rigidly. The large value of Bi means that thermal conduction is relatively weak; the lava can flow large distances before it cools significantly.

35.1.3 Approaching a Solution

Let's try to develop a solution by following progression of the sheet as it moves downslope (with x^* increasing from 0). The lava is cooled from above, which leads to an increase in the viscosity and decrease in the downslope speed. The volume flow remains constant provided the layer thickens.

Initial Temperature Adjustment

The point $\{x^*, z^*\} = \{0, 1\}$ is a singular point, where the convective cooling of the sheet abruptly begins. Due to the large value of Bi, cooling is initially confined to a thin layer near the top of the sheet, where the downslope speed is nearly uniform (with $u^* \approx 1$ and $w^* \approx 0$). The structure of the energy equation and the top thermal boundary condition require T^* to have exponential dependence on x^*. Assuming that

$$T^*(x^*, z^*) = e^{-\lambda^2 x^*} Z(z^*),$$

the energy equation simplifies to

$$Z'' + \lambda^2 Z = 0,$$

where a prime denotes differentiation with respect to z^*. This ordinary differential equation is to be solved for $Z(z^*)$ on the interval $0 < z^* < 1$, subject to the conditions

$$Z'(0) = 0 \qquad \text{and} \qquad Z'(1) = -\text{Bi}\, Z(1).$$

This is a homogeneous problem with an unknown parameter, λ. The problem admits the trivial solution $Z = 0$, and this is the only permissible solution if λ does not have a specific value, called an *eigenvalue*. This type of problem is called an *eigenvalue problem*; it has an infinite number of eigenvalues and a corresponding set of solutions, called *eigenfunctions*. The eigenfunctions form a complete set of functions, permitting representation of an arbitrary function as a linear combination of eigenfunctions. For more detail, see Dettman (1969).

The general solution of the differential equation is

$$Z = Z_C \cos(\lambda z^*) + Z_S \sin(\lambda z^*).$$

The thermal condition at $z^* = 0$ tells us that $Z_S = 0$, while that at $z^* = 1$ is satisfied provided

$$Z_C = 0 \qquad \text{or} \qquad \lambda \tan \lambda = \text{Bi}.$$

The former option gives only the trivial solution $T^* = 0$, so we must opt for the latter, which is a transcendental algebraic equation that gives us an infinite set of eigenvalues, λ_i, where j is a positive integer identifier, and a set of eigenfunctions, $\cos(\lambda_j z^*)$. To see that we have an infinite set of solutions, let's first note that $\tan \lambda$ is periodic with period π and takes on all real values in that interval; this equation has an infinity of solutions, where the hyperbola Bi/λ intersects the graph of $\tan \lambda$. Since $\lambda \tan \lambda$ is an even function of λ, we need to consider only solutions having $\lambda > 0$. For example, with $\text{Bi} = 50$, the first ten values of λ, ordered from smallest to largest, are 1.540, 4.620, 7.701, 10.783, 13.867, 16.952, 20.392, 23.129, 26.221 and 29.315.

The function

$$T^*(x^*,z^*) = \sum_{j=1}^{\infty} Z_j e^{-\lambda_j^2 x^*} \cos(\lambda_j z^*)$$

satisfies the heat equation and the conditions at $z^* = 0$ and 1. The magnitudes Z_j are determined by satisfaction of the condition at $x^* = 0$:

$$1 = \sum_{j=1}^{\infty} Z_j \cos(\lambda_j z^*) .$$

This equation may be solved for Z_j by multiplying by $\cos(\lambda_i z^*)$ and integrating from $z^* = 0$ to 1:

$$\int_0^1 \cos(\lambda_i z^*) \mathrm{d}z^* = \sum_{j=1}^{\infty} Z_j I_{ij} , \qquad \text{where} \qquad I_{ij} \equiv \int_0^1 \cos(\lambda_j z^*) \cos(\lambda_i z^*) \mathrm{d}z^* .$$

It is a routine exercise in trigonometry to show, with the use of $\lambda \tan \lambda = \mathrm{Bi}$ with either subscript, that $I_{ij} = 0$ if $i \neq j$. With $i = j$ we obtain

$$Z_j = \frac{4 \sin \lambda_j}{2\lambda_j + \sin(2\lambda_j)}$$

and the thermal profile is

$$T^*(x^*,z^*) = \sum_{j=1}^{\infty} \frac{4 \sin \lambda_j}{2\lambda_j + \sin(2\lambda_j)} e^{-\lambda_j^2 x^*} \cos(\lambda_j z^*) .$$

The vertical profile of T^* depends on two parameters, x^* and Bi. Reasonable results may be obtained by truncating the series at a finite value of j. To illustrate this, the profile of T^* versus z^* using the first 20 terms of the series is plotted in Figure 35.1 with Bi = 50 for several values of x^*. The profile for $x^* = 0$ is the best effort of the first 20 terms to replicate unity. The overshoot close to $z^* = 1$ is called the *Gibbs phenomenon*. This overshoot, and all other oscillations, are effectively squelched by the exponential term by the time $x^* = 0.002$, and the profile decreases smoothly as a function of z^* for $x^* > 0.002$.

Initial Velocity Adjustment

The decrease in temperature at the top of the sheet causes the viscosity near the top to increase, and this alters the downslope velocity. This alteration can be visualized by numerically integrating the momentum equation at various x^* positions with T^* as found in the previous subsection. These profiles are plotted in Figure 35.2 for x^* =0, 0.01, 0.02, 0.03, 0.04 and 0.05. Note that the speed at the top of the lava sheet initially decreases linearly with downslope distance. Of course this linear decrease cannot persist; the volume flow of lava within the sheet has been assumed to be constant, independent of x^*.

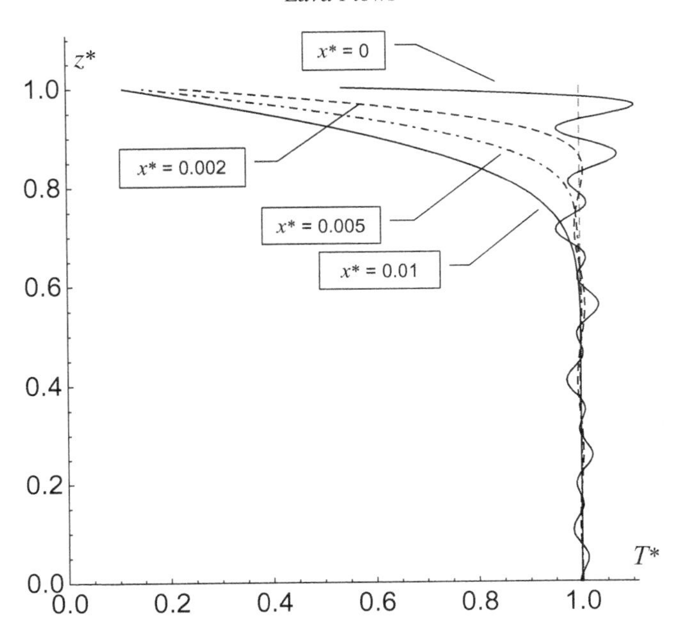

Figure 35.1 Plots of the dimensionless temperature T^* within a lava sheet versus dimensionless depth z^* with Bi = 50 for various values of the dimensionless downslope distance x^*, using the first 20 terms of the summation.

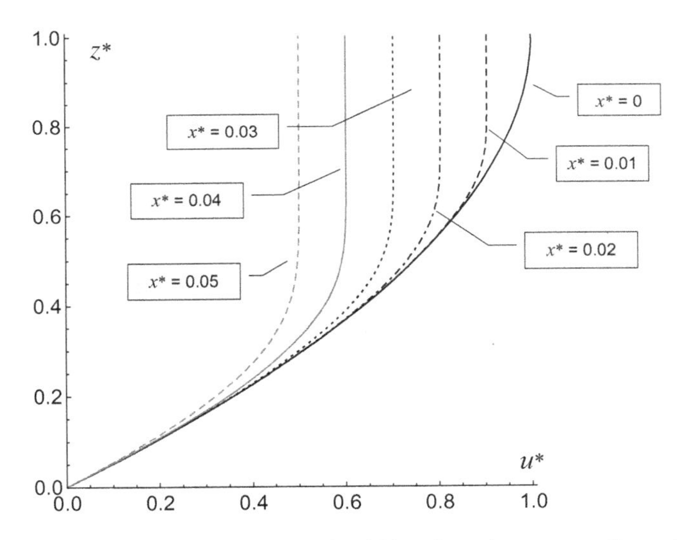

Figure 35.2 Plots of the dimensionless speed u^* within a lava sheet versus dimensionless depth z^* with Bi = 50 and $\beta^* = 25$ for various values of the dimensionless downslope distance x^*.

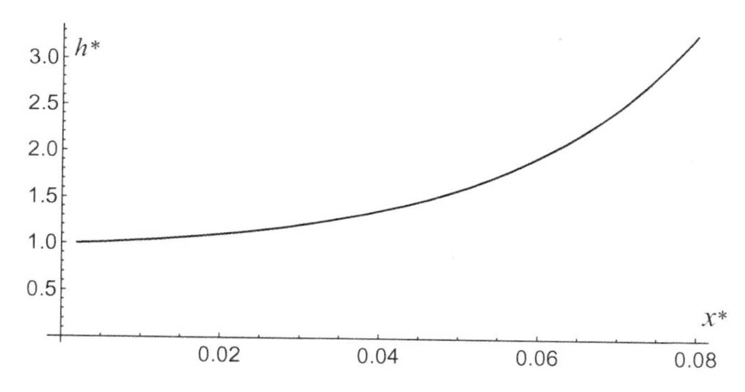

Figure 35.3 Plots of the dimensionless thickness h^* of a lava sheet versus dimensionless downslope distance x^* with $\mathrm{Bi} = 50$ and $\beta^* = 25$.

Thickening of Lava Sheet

Figure 35.2 shows that the scaled downslope speed u^* decreases in magnitude as x^* increases due to the effect of cooling. Since the mass flow of lava is independent of x^*, there must be a compensatory thickening of the lava sheet, determined by the integral constraint. The thickening of the sheet with downslope distance x^* is illustrated in Figure 35.3. Note that the rate of thickening is initially modest, but increases dramatically for $x^* > 0.05$.

Lava Sheet Flow Discussion

The problem of lava sheet flow solved in the previous sections is highly simplified and idealized and glosses over many complications. For example, we have assumed that the flow is uniform in the lateral direction. While this assumption may be reasonable close to the vent, it becomes suspect as the flow progresses downhill. The uniform plane beneath the lava is just cold lava that was deposited by previous sheet flows. The hot lava interacts thermally and dynamically with its bed, forming lateral undulations that develop into channeled flows that are contained laterally by dikes. As the top of a channeled flow cools, instead of rafting along the cold top tends to adhere to the sides of the channel, forming a *lava tube*. These tubes are rather efficient mode of lava transport, permitting flows to progress long distances.

In this section we have investigated the laminar flow of a homogeneous sheet of basaltic lava. In the following section, we investigate the opposite extreme: turbulent flow of a heterogeneous mix of gas and lava fragments.

35.2 Pyroclastic Flows

Explosive volcanic eruptions produce a variety of flows, consisting of varying amounts of hot gas, fragments of lava and water vapor, including:

- pyroclastic flow: an avalanche of volcanic debris and hot gases;[10]
- nuées ardente: a pyroclastic flow that is sufficiently hot to glow; and
- lahar: a muddy flow of water and volcanic debris. Lahars commonly form when hot volcanic debris melts glacier ice.

All these flows are gravity currents, similar to – but rather more complicated than – the channel flow investigated in § 23.6, katabatic winds investigated in § 22.2.5 and § 24.1 and avalanches investigated in § 24.2. In this section, we will focus our attention on pyroclastic flows.

A pyroclastic flow is a two-phase mixture of gas and debris. In contrast to the gravity currents we studied previously (river flows and katabatic winds), the structure of a pyroclastic flow evolves as it flows downslope. The dynamic regime of this mixture depends on the ratio of debris to gas. In the dilute regime (having a small ratio), the mixture can be considered as gravity current composed primarily of gas with debris particles in suspension, while in the concentrated regime (having a large ratio), the mixture can be considered as an avalanche similar to a rock or debris slide,[11] fluidized by the gas. As a pyroclastic flow progresses downslope, its components tend to separate, with the denser volcanic debris settling toward the ground, forming a concentrated layer and possibly settling out, and the hot gases rising to the top of the flow, forming a dilute layer and possibly escaping into the atmosphere above. The tendency to separate is inhibited – but not prevented – by strong turbulence within the flow.

Like their parent magma, pyroclastic flows and nuées ardents are phase-equilibrium mixtures that have a small speed of sound.[12] This property does not permit sounds that they generate internally to be propagated to the surrounding air. They flow silently downhill, and this makes them particularly dangerous. On June 3, 1991, during the eruption of Mount Unzen in Japan, a pryoclastic flow crept down upon and killed 41 people, including two experienced volcanologists, Katia and Maurice Krafft.

Transition

In this the final chapter, we have investigated two types of surface volcanic flows: lava sheet flows (in § 35.1) and pyroclastic flows (in § 35.2).

35.2.1 A Final Word

Any book such as this that covers a wide range of subjects is – almost by definition – incomplete. Choices must be made regarding what material to include and what to exclude, and how deeply to delve into each topic. There are a number of interesting geophysical

[10] The mix of debris and gas is called *tephra*. Often the debris contains trapped volatiles; when solidified it is called *pumice*. When a pyroclastic flow or nuées ardente comes to rest, the volcanic debris often is welded into an unconsolidated mass called an *ignimbrite* or *tuff*.

[11] That is, a *fluidized bed*.

[12] See § 34.2.

waves and flows that have not been investigated or even mentioned previously in this monograph, including, but not limited to:

- waves and flows in the upper atmosphere;
- waves and flows in Earth's core – particularly those related to the geodynamo;
- ground-water flows;
- hydrothermal circulations;
- dust and sand storms;
- breaking waves;
- swash zone and along-shore currents;
- cold fronts;
- river meanders; and
- turbidity currents.

In the present case, most of the choices have been dictated by the flow of the text and by the need to keep the book of a manageable size and modest cost.

The aim of this work has been to provide a solid introduction to a number of geophysical flows, with enough theoretical underpinning so that the interested reader has a good start on tackling the intricacies of a given subject area, and also to put a given topic in a broader perspective, illustrating the commonality of the fundamental laws governing waves and flows and a variety of mathematical approaches to solving the governing differential equations. Of course, in a wide-ranging book there is the danger that the result will be "a mile wide and an inch deep".[13] In this monograph, I have tried for a mile wide and a foot deep.

[13] Or in metric, "a kilometer wide and a centimeter deep" – but this doesn't have the same ring to it.

Part VIII

Fundaments

The fundaments contain material that is underpins the main text. They are intended to "fill the gaps" in a reader's background or else to provide generalizations of simplified presentations in the main text. The 47 fundaments are grouped into seven appendices according to general topic, including:

- A: mathematics;
- B: dimensions and units;
- C: kinematics;
- D: dynamics;
- E: thermodynamics;
- F: waves; and
- G: flows.

These fundaments are not meant to be read in sequence, but rather to be read or referred to as the need arises; most have pointers from the main text provided via footnotes. Most of the fundaments are "stand-alone" summaries, with the exception of those dealing with thermodynamics, which are necessarily somewhat interdependent.

Appendix A
Mathematics

The fundaments of mathematics include:

- A.1: a brief introduction to the algebra of vectors and tensors in three dimensions;
- A.2: a summary of the calculus of three-dimensional vectors;
- A.3: an introduction to curvilinear coordinate systems;
- A.4: an introduction to Taylor series;
- A.5: an introduction to Fourier series and integrals;
- A.6: classification of several simple types of linear, second-order partial differential equations;
- A.7: a listing of the Greek symbols that are commonly used in applied mathematics; and
- A.8: an introduction to scalar and vector potentials.

These mathematical fundaments will employ the orthogonal reference coordinates system introduced in § 2.1 (and visualized in Figure 2.1) and will use extensively the associated unit vectors $\mathbf{1}_1$, $\mathbf{1}_2$ and $\mathbf{1}_3$. These vectors are arranged in a right-hand configuration such that $\mathbf{1}_1 \times \mathbf{1}_2 = \mathbf{1}_3$. Any or all of these will be designated by $\mathbf{1}_i$, with $i = 1$, 2 and/or 3, as the situation requires. These vectors satisfy the orthogonality relation $\mathbf{1}_i \bullet \mathbf{1}_j = \delta_{ij}$, where δ_{ij} is the Kronecker delta.[1]

A.1 Vectors and Tensors

This fundament is not a general summary, but instead focuses on vectors and tensors in three-dimensional space, specifically vectors having three elements (or components) and tensors of rank two having nine elements.

A.1.1 Definitions, Notation and Representation

A *tensor* is an ordered set of n^k numbers, where k and n are non-negative integers, that obeys certain tensor transformation rules. The integer k is the *rank* of the tensor. The numbers comprising the tensor are called its *elements*.

[1] See Appendix A.1.4.

A tensor of rank 0 is a *scalar*: a number.

A tensor of rank 1 is a *vector*: an ordered set of n numbers. Each number is a *component* of the vector. (Typically $n = 3$)

Tensors may be thought of as square arrays in k dimensions. A *matrix* is a related two-dimensional array of size m by n, where m and n are positive integers. A matrix is *square* if $m = n$. A tensor of rank two behaves algebraically the same as a square matrix, and the tensor elements may be written out and manipulated in matrix-display form.

In this document, a *scalar* is denoted by a lower case letter in italics (e.g., x).

A *vector* is denoted by a lower or upper case letter in bold (e.g., **b**) or a letter in italics with a single subscript (b_i). The last of these in fact represents the set of scalar components comprising the vector. A vector having unit length (by design) is denoted by a bold **1** with suitable subscript, e. g., $\mathbf{1}_i$ and $\mathbf{1}_1$. Alternative representations of vectors (not employed in this text) include a letter having an arrow on top or an underline.

A *tensor* of rank two is denoted by an upper case letter in outline ($\mathbb{C}$) and its elements by an italicized letter (typically lower case) with two subscripts[2] (c_{ij}). Alternative representations of tensors (not used in this text) include a letter with two arrows on top or with a double underline.

Commonly vectors representing physical quantities such as position and velocity have three components. These may displayed in two forms: as row vectors or column vectors:

$$\mathbf{b} = \begin{bmatrix} b_1 & b_2 & b_3 \end{bmatrix} \qquad \mathbf{b} = \begin{bmatrix} b_1 \\ b_2 \\ b_3 \end{bmatrix}.$$

An alternate but equivalent representation is

$$\mathbf{b} = b_1\mathbf{1}_1 + b_2\mathbf{1}_2 + b_3\mathbf{1}_3 = b_i\mathbf{1}_i \qquad \text{(sum over } i\text{)},$$

where $\mathbf{1}_1$, $\mathbf{1}_2$ and $\mathbf{1}_3$ are vectors of unit length that are mutually perpendicular. The alternate form above employs the Einstein summation convention; see Appendix A.1.2. A third option is to express a vector in terms of a magnitude and direction: $\mathbf{b} = b\mathbf{1}_b$, where the magnitude of the vector is given by

$$b = \sqrt{b_1^2 + b_2^2 + b_3^2}.$$

The zero vector is that having all three components equal to zero:

$$\mathbf{0} = \begin{bmatrix} 0 & 0 & 0 \end{bmatrix}.$$

[2] The elements of a tensor of rank k are denoted a letter in italics with k subscript (e.g., for $k = 4$, c_{ijpq}).

Tensors of rank two with $n = 3$ contain nine components; these may be expressed in matrix-display form as follows:

$$\mathbb{C} = \left[c_{ij}\right] = \begin{bmatrix} c_{11} & c_{12} & c_{13} \\ c_{21} & c_{22} & c_{23} \\ c_{31} & c_{32} & c_{33} \end{bmatrix}.$$

A set of vectors is said to be linearly {independent, dependent} if {no, any} member of the set can be expressed as a linear combination of the others. Any set of linearly independent vectors may be used as a *basis set*, with all vectors being expressed as a linear combination of these. The most versatile basis sets consist of mutually orthogonal vectors of unit length. We represent these in the text as $\mathbf{1}_1$, $\mathbf{1}_2$ and $\mathbf{1}_3$.

A.1.2 Summation Rules

Tensor equations that are written with subscripts (called indices) follow the *Einstein summation convention*, summarized in three rules.

Rule 1. If an index (i.e., a particular letter) appears once in a term of an equation, that index must appear once in every other term. This is a *free index*.

Rule 2. If an index appears twice in a term, it is automatically summed from 1 to J. That is, $a_i b_i = \sum_{j=1}^{J} a_j b_j$ and (with $J = 3$) $a_i \mathbf{1}_i = a_1 \mathbf{1}_1 + a_2 \mathbf{1}_2 + a_3 \mathbf{1}_3$. This is a *dummy index*.

Rule 3. If an index appears three or more times in a term, you have made a mistake (by using the same letter to represent two types of indices). Use a different letter for one of them.

Vector and tensor equations may be expressed alternatively in symbolic or subscript notation, e.g.,

$$\mathbf{a} = \mathbb{C} \bullet \mathbf{b} \qquad \text{or} \qquad a_i = c_{ij} b_j .$$

Symbolic notation is better for conceptual understanding, while subscript notation is better for manipulation and calculation. In many places in the text, equations are presented in both forms.

A.1.3 Vector Algebra

There are four fundamental arithmetic operations: addition, subtraction, multiplication and division. Addition of vectors mirrors that of scalars, as does subtraction. Division of one vector by another is an undefined operation,[3] but that lack is compensated by the existence of three types of vector multiplication: multiplication by a scalar, the dot product and cross product. These operations are described in the following four subsections.

[3] But mathematical programs, such as *Mathematica*, permit operations that appear to be division by a vector.

Addition of Vectors

Given two vectors, $\mathbf{b} = b_i\mathbf{1}_i$ and $\mathbf{c} = c_i\mathbf{1}_i$, their sum is

$$\mathbf{b}+\mathbf{c} = (b_i + c_i)\mathbf{1}_i\,.$$

It follows from this definition that vectors obey certain algebraic rules:

- commutative law $\mathbf{c}+\mathbf{b} = \mathbf{b}+\mathbf{c}$; and
- associative law $(\mathbf{a}+\mathbf{b})+\mathbf{c} = \mathbf{a}+(\mathbf{b}+\mathbf{c})$.

Be aware that there are mathematical creatures which appear to be vectors, but in fact are not. We shall see in Appendix C.3.1 that finite rigid-body rotation is one such pseudo-vector; it does not obey the commutative law.

Multiplication by Scalar

If a vector $\mathbf{b} = b_i\mathbf{1}_i$ is multiplied by a scalar q, its orientation is preserved, while its magnitude is changed according to the usual rules of multiplication:

$$q\mathbf{b} = (qb_i)\mathbf{1}_i\,.$$

Dot Product

This is an operation (a contraction) whereby two vectors multiply to form a scalar; it is also called the *scalar product* or *inner product*. Given $\mathbf{b} = b_i\mathbf{1}_i$ and $\mathbf{c} = c_i\mathbf{1}_i$, the dot product is

$$\mathbf{b}\bullet\mathbf{c} = b_ic_i\,.$$

The scalar is equal to the product of the magnitudes of the vectors and the cosine of the angle between them. If the dot product is zero the two vectors are said to be *orthogonal*. Geometrically they are perpendicular.

The dot product is commutative:

$$\mathbf{c}\bullet\mathbf{b} = \mathbf{b}\bullet\mathbf{c}\,.$$

The vectors of our reference coordinate system are orthogonal, in that $\mathbf{1}_1\bullet\mathbf{1}_2 = \mathbf{1}_2\bullet\mathbf{1}_3 = \mathbf{1}_3\bullet\mathbf{1}_1 = 0$. Also they are normalized such that $\mathbf{1}_1\bullet\mathbf{1}_1 = \mathbf{1}_2\bullet\mathbf{1}_2 = \mathbf{1}_3\bullet\mathbf{1}_3 = 1$. These two sets of conditions may be combined into a single compact expression

$$\mathbf{1}_i\bullet\mathbf{1}_j = \delta_{ij}\,,$$

where δ_{ij} is the Kronecker delta; see Appendix A.1.4. In brief, the set of vectors $\mathbf{1}_1$, $\mathbf{1}_2$ and $\mathbf{1}_3$ is *orthonormal*.

Cross Product

The *cross product* is an operation whereby two vectors multiply to form a third vector which is orthogonal to each of the input vectors. Given $\mathbf{b} = b_i\mathbf{1}_i$ and $\mathbf{c} = c_i\mathbf{1}_i$, the cross product is defined mathematically

$$\mathbf{b} \times \mathbf{c} = (b_2c_3 - b_3c_2)\mathbf{1}_1 + (b_3c_1 - b_1c_3)\mathbf{1}_2 + (b_1c_2 - b_2c_1)\mathbf{1}_3 .$$

The cross product is not commutative, in fact

$$\mathbf{b} \times \mathbf{c} = -\mathbf{c} \times \mathbf{b} .$$

Note also that $\mathbf{b} \bullet (\mathbf{b} \times \mathbf{c}) = 0$ and $\mathbf{c} \bullet (\mathbf{b} \times \mathbf{c}) = 0$.

More complicated vector products are often encountered. The following formulas are of use in simplifying these.

$$\begin{aligned}
\mathbf{a} \bullet \mathbf{b} \times \mathbf{c} &= \mathbf{a} \times \mathbf{b} \bullet \mathbf{c} \\
\mathbf{a} \times (\mathbf{b} \times \mathbf{c}) &= (\mathbf{a} \bullet \mathbf{c})\mathbf{b} - (\mathbf{a} \bullet \mathbf{b})\mathbf{c} \\
(\mathbf{a} \times \mathbf{b}) \bullet (\mathbf{c} \times \mathbf{d}) &= (\mathbf{a} \bullet \mathbf{c})(\mathbf{b} \bullet \mathbf{d}) - (\mathbf{a} \bullet \mathbf{d})(\mathbf{b} \bullet \mathbf{c}) \\
(\mathbf{a} \times \mathbf{b}) \times (\mathbf{c} \times \mathbf{d}) &= (\mathbf{a} \times \mathbf{b} \bullet \mathbf{d})\mathbf{c} - (\mathbf{a} \times \mathbf{b} \bullet \mathbf{c})\mathbf{d} .
\end{aligned}$$

A.1.4 Tensors

The following summary of tensors is not general; it focuses on tensors of rank two having nine elements. Tensors arise in mathematical physics as the linear relation between two vectors (e.g., $\mathbf{a} = \mathbb{B} \bullet \mathbf{c}$ or equivalently $a_i = b_{ij}c_j$). They also can be formed from the outer product of two vectors (e.g., $\mathbb{B} = \mathbf{ac}$ or equivalently $b_{ij} = a_i c_j$); this is called a *dyad.*

Tensor Algebra

The principal algebraic operations on tensors are multiplication by vectors and other tensors. These operations are often displayed symbolically in equations, but the subscript notation is more useful in actual computations. Consider two tensors of rank two denoted by $\mathbb{B}$ and $\mathbb{C}$ and a vector denoted by $\mathbf{a}$, with components b_{ij}, c_{ij} and a_i, respectively. The product $\mathbf{a} \bullet \mathbb{B}$ is a vector having elements $a_i b_{ij}$, the product $\mathbb{B} \bullet \mathbf{a}$ is a vector having elements $b_{ij} a_j$ and the product $\mathbb{B} \bullet \mathbb{C}$ is tensor of rank two with elements $b_{ij} c_{jk}$.

Zero and Identity Tensor

The *zero tensor* is a tensor of rank two having all elements equal to zero:

$$\emptyset = [\ 0_{ij}\ ] = \begin{bmatrix} 0 & 0 & 0 \\ 0 & 0 & 0 \\ 0 & 0 & 0 \end{bmatrix} .$$

The *identity tensor* is a tensor of rank two (mathematically equivalent to the 3 by 3 identity matrix) represented in terms of the *Kronecker delta*, δ_{ij}:

$$\mathbb{I} = [\ \delta_{ij}\ ] = \begin{bmatrix} 1 & 0 & 0 \\ 0 & 1 & 0 \\ 0 & 0 & 1 \end{bmatrix} .$$

Symmetric and Antisymmetric Tensors

A general tensor $\mathbb{C}$ may be expressed as the sum of a symmetric tensor $\mathbb{S}$ (having 6 independent elements) and an antisymmetric tensor $\mathbb{A}$ (having 3). The elements of a symmetric tensor satisfy $s_{ij} = s_{ji}$, while those of an antisymmetric tensor satisfy $a_{ij} = -a_{ji}$. Symbolically this decomposition is

$$\mathbb{C} = \mathbb{S} + \mathbb{A},$$

or in display form

$$\begin{bmatrix} c_{11} & c_{12} & c_{13} \\ c_{21} & c_{22} & c_{23} \\ c_{31} & c_{32} & c_{33} \end{bmatrix} = \begin{bmatrix} s_{11} & s_{12} & s_{13} \\ s_{12} & s_{22} & s_{23} \\ s_{13} & s_{23} & s_{33} \end{bmatrix} + \begin{bmatrix} 0 & a_{12} & a_{13} \\ -a_{12} & 0 & a_{23} \\ -a_{13} & -a_{23} & 0 \end{bmatrix}.$$

Note that $\mathbb{S}^{\mathrm{T}} = \mathbb{S}$ and $\mathbb{A}^{\mathrm{T}} = -\mathbb{A}$, where T denotes the transpose; see Appendix A.1.5. Since $\det[\mathbb{C}^{\mathrm{T}}] = \det[\mathbb{C}]$, it follows that $\det[\mathbb{A}] = 0$.

Scalar Invariants

A symmetric tensor has six independent elements, and as noted in § 3.1.1, the off-diagonal elements are zero if the tensor is represented in its principal-axis system. The non-zero elements of the diagonalized tensor are its *eigenvalues*. scalar invariants The scalar invariants of symmetric tensor $\mathbb{S}$ having elements s_{ij} are the coefficients of the *characteristic polynomial* involving its eigenvalues λ, defined by

$$\begin{aligned} P(\lambda) &= \det[\mathbb{S} - \lambda \mathbb{I}] \\ &= -\lambda^3 + I_1\lambda^2 - I_2\lambda + I_3, \end{aligned}$$

where the three invariants of $\mathbb{S}$ are the trace

$$I_1 = \mathrm{trace}[\mathbb{S}] = s_{11} + s_{22} + s_{33},$$

the second invariant (with no special name)

$$I_2 = s_{22}s_{33} + s_{33}s_{11} + s_{11}s_{22} - s_{23}^2 - s_{13}^2 - s_{12}^2$$

and the determinant

$$\begin{aligned} I_3 = \det[\mathbb{S}] &= s_{11}s_{22}s_{33} + s_{12}s_{23}s_{31} + s_{13}s_{21}s_{32} \\ &\quad - s_{11}s_{23}s_{32} - s_{12}s_{21}s_{33} - s_{13}s_{22}s_{31}. \end{aligned}$$

Note that for the identity tensor $I_1 = 3$ ($\mathrm{trace}[\mathbb{I}] = 3$), $I_2 = 3$ and $I_3 = 1$ ($\det[\mathbb{I}] = 1$). An alternate definition of the second scalar invariant that is easier to represent symbolically is

$$\tilde{I}_2 = s_{ij}s_{ij} = I_1^2 - 2I_2.$$

The three values of λ that satisfy $P(\lambda) = 0$ are the eigenvalues of $\mathbb{S}$. Since $\mathbb{S}$ is symmetric, these eigenvalues are all real. Interestingly, according to the *Cayley–Hamilton theorem*, the matrix $\mathbb{S}$ satisfies its own characteristic polynomial equation:

$$\mathbb{S}^3 - I_1\mathbb{S}^2 + I_2\mathbb{S} - I_3\mathbb{I} = \mathbf{\emptyset},$$

where $\mathbb{S}^m = \mathbb{S} \bullet \mathbb{S}^{m-1}$.

Now let's find the orientation of the principal axes of a symmetric tensor denoted with superscript B, relative to an arbitrary coordinate frame. Unit vectors along the principal axes can be represented by $\mathbf{1}_i^{\mathrm{B}}$. These vectors satisfy

$$\mathbb{S} \bullet \mathbf{1}_i^{\mathrm{B}} = \lambda_i \mathbf{1}_i^{\mathrm{B}} \qquad \text{no sum on } i,$$

plus

$$b_{ij} b_{ij} = 1 \qquad \text{no sum on } i,$$

where b_{ij} is the jth component of $\mathbf{1}_i^{\mathrm{B}}$. These appear to form a set of four equations for three unknowns, but in fact only two of the three equations contained in the matrix equation are independent. Taking any two of these plus the normalizing condition gives a set of three equations for the components of each of the principal unit vectors. The algebra involved is messy in general and the solution is usually illustrated by means of specific examples.

The problem of finding the three values of λ and the principal coordinate directions is an *eigenvalue problem*; the λs are *eigenvalues* and the coordinate directions are *eigenvectors*.

A.1.5 Transpose and Inverse

The transpose of a tensor of rank two, $\mathbb{C}$, having entries c_{ij}, is denoted by $\mathbb{C}^{\mathrm{T}}$ and has entries c_{ji}. It is easily verified that $\det[\mathbb{C}] = \det[\mathbb{C}^{\mathrm{T}}]$ and that the determinant of the product of two tensors is the product of their determinants: $\det[\mathbb{C}_1 \bullet \mathbb{C}_2] = \det[\mathbb{C}_1]\det[\mathbb{C}_2]$.

The inverse of $\mathbb{C}$ is a tensor, denoted by $\mathbb{C}^{-1}$, that satisfies $\mathbb{C}^{-1} \bullet \mathbb{C} = \mathbb{I}$. Note that $\det[\mathbb{C}^{-1}] = 1/\det[\mathbb{C}]$.

A.2 Vector Calculus

Vector calculus is concerned with the quantification of spatial variations of scalar and vector fields. A field is a quantity which varies with spatial position, expressed functionally as $a(x_i)$ and $\mathbf{b}(x_i)$ or equivalently $a(\mathbf{x})$ and $\mathbf{b}(\mathbf{x})$, where $\mathbf{x} = x_i \mathbf{1}_i$ is the position vector. The fundamental agent of vector calculus is the differential operator called *del*; this is usually represented by the nabla symbol (an upside-down bold delta). In our reference Eulerian coordinate system

$$\nabla \equiv \mathbf{1}_1 \frac{\partial}{\partial x_1} + \mathbf{1}_2 \frac{\partial}{\partial x_2} + \mathbf{1}_3 \frac{\partial}{\partial x_3} = \mathbf{1}_i \frac{\partial}{\partial x_i}.$$

We will develop other representations of ∇ using general orthogonal coordinates in Appendix A.3. A similar version in our reference Lagrangian coordinate system, identified by a subscript $\mathbf{X}$, is employed in § 3.3.

Del acts both as a vector and as a differential operator. As a vector, it combines with scalars and vectors using the rules of vector algebra. Being a differential operator, it is incomplete; it needs something to act upon. It is somewhat like the Tasmanian devil of Bugs Bunny cartoons, in that it is compelled to act on whatever is placed to its right. However, it can be steered by a vector placed to its left.[4]

A.2.1 Gradient

Since del is a vector operator, when it acts on a scalar $a(\mathbf{x})$ it creates a vector:

$$\text{grad}(a) = \nabla a$$

called the *gradient*. This operation is reminiscent of multiplication of a vector and a scalar, except that the vector (del) also differentiates the scalar. The gradient points in the direction of the maximum increase of a. The magnitude of ∇a represents the rate of increase of a experienced by an observer moving in the direction of ∇a. For example, if a represents the temperature in a three-dimensional body, ∇a is the direction in which one should move to experience the maximum increase of temperature and the magnitude of ∇a is the rate which a thermometer will increase with distance traveled (in units of degrees per meter).

A scalar function of position is constant on an *iso-surface* defined by

$$a(x_i) = a_0\,,$$

where a_0 is a specified value of a. The gradient is always perpendicular to the local iso-surface. For example, if $z = f(x,y)$ denotes topographic elevation, then ∇z points directly uphill, perpendicular to elevation contours.

A.2.2 Divergence

If del combines with a vector $\mathbf{b}(\mathbf{x})$ by means of the dot product, the result is a scalar field, called the *divergence*:

$$\text{div}(\mathbf{b}) = \nabla \bullet \mathbf{b} = \mathbf{1}_i \frac{\partial}{\partial x_i} \bullet (b_j \mathbf{1}_j) = \mathbf{1}_i \bullet \mathbf{1}_j \frac{\partial b_j}{\partial x_i} = \delta_{ij} \frac{\partial b_j}{\partial x_i} = \frac{\partial b_i}{\partial x_i}\,.$$

The divergence is a measure of the tendency of vector field to diverge, i.e., it is a measure of how much the "arrows" of the vector field point out of a small enclosed surface. For example, if $\mathbf{b} = \mathbf{x}$, then

$$\nabla \bullet \mathbf{x} = \frac{\partial x_i}{\partial x_i} = 3\,.$$

[4] See the directional derivative in Appendix A.2.3.

A.2.3 Directional Derivative

The divergence results from a dot product with the vector to the right, so that the del operator on the left differentiates it. If del and the vector switch places, we get another creature called the *directional derivative*:

$$\mathbf{b} \bullet \nabla = b_i \frac{\partial}{\partial x_i}.$$

This is still an operator, in that it is looking for something to differentiate. If it acts on a scalar, a, the result, $(\mathbf{b} \bullet \nabla)a$, may be interpreted as the rate of change of a experienced by an observer constrained to move in the direction of $\mathbf{b}$.

A.2.4 Curl

If del and a vector combine using the cross product, the result is a vector field called the *curl*:

$$\operatorname{curl}(\mathbf{b}) = \nabla \times \mathbf{b} = \mathbf{1}_1 \left(\frac{\partial b_3}{\partial x_2} - \frac{\partial b_2}{\partial x_3} \right) + \mathbf{1}_2 \left(\frac{\partial b_1}{\partial x_3} - \frac{\partial b_3}{\partial x_1} \right) + \mathbf{1}_3 \left(\frac{\partial b_2}{\partial x_1} - \frac{\partial b_1}{\partial x_2} \right).$$

The curl measures the tendency for the arrows of field $\mathbf{b}$ to curl around on themselves.

The gradient has no curl and the curl has no divergence:

$$\nabla \times \nabla a = \mathbf{0} \qquad \text{and} \qquad \nabla \bullet (\nabla \times \mathbf{b}) = 0$$

for any scalar field a and vector field $\mathbf{b}$.

In three dimensional space, a vector has three degrees of freedom, while a scalar has one. The condition $\nabla \times \mathbf{b} = \mathbf{0}$ is two constraints on the vector $\mathbf{b}$, and its one remaining degree of freedom can be expressed as a scalar, with $\mathbf{b} = \nabla \psi$, where ψ is the scalar potential of the vector $\mathbf{b}$.[5]

A.2.5 Laplacian

If one del combines with another by means of the dot product, the vectorial character of both is neutralized, and the result is a second-order differential operator called the *Laplacian*, formed by the divergence of the gradient. The Laplacian of a scalar a is

$$\nabla^2 a = \nabla \bullet \nabla a = \mathbf{1}_i \frac{\partial}{\partial x_i} \bullet \mathbf{1}_j \frac{\partial a}{\partial x_j} = \mathbf{1}_i \bullet \mathbf{1}_j \frac{\partial^2 a}{\partial x_i \partial x_j} = \delta_{ij} \frac{\partial^2 a}{\partial x_i \partial x_j} = \frac{\partial^2 a}{\partial x_i \partial x_i}.$$

Written out in our reference coordinate system,

$$\nabla^2 a = \frac{\partial^2 a}{\partial x_1^2} + \frac{\partial^2 a}{\partial x_2^2} + \frac{\partial^2 a}{\partial x_3^2}.$$

The form in other coordinates systems can be found by using $\nabla^2 a = \nabla \bullet \nabla a$ and the formulas for divergence and gradient; see Appendix A.3.

[5] See Appendix A.8.

The Laplacian of a vector **b** looks the same as that of a scalar in the reference coordinate system,

$$\nabla^2 \mathbf{b} = \frac{\partial^2 \mathbf{b}}{\partial x_i \partial x_i}.$$

The form in other coordinates systems can be found by using the last formula of Appendix A.2.6 together with the formulas for curl, divergence and gradient found in Appendix A.3.

The Laplacian is encountered in many branches of mathematical sciences, and is an essential part of the description of the spatial structure of fluid flows, electric fields, magnetic fields, etc.

A.2.6 Various Vector Identities

We finish the fundament on vector calculus with a list of various vector identities. In the following a is a scalar and **b** and **c** are two general vector fields.

$$\begin{aligned}
\nabla \bullet (a\mathbf{b}) &= a\nabla \bullet \mathbf{b} + \nabla a \bullet \mathbf{b}, \\
\nabla \times (a\mathbf{b}) &= a\nabla \times \mathbf{b} + \nabla a \times \mathbf{b}, \\
\nabla \bullet (\mathbf{b} \times \mathbf{c}) &= \mathbf{c} \bullet (\nabla \times \mathbf{b}) - \mathbf{b} \bullet (\nabla \times \mathbf{c}), \\
\nabla \times (\mathbf{b} \times \mathbf{c}) &= (\nabla \bullet \mathbf{c})\mathbf{b} - (\nabla \bullet \mathbf{b})\mathbf{c} + (\mathbf{c} \bullet \nabla)\mathbf{b} - (\mathbf{b} \bullet \nabla)\mathbf{c}, \\
\nabla (\mathbf{b} \bullet \mathbf{c}) &= \mathbf{b} \times (\nabla \times \mathbf{c}) - (\nabla \times \mathbf{b}) \times \mathbf{c} + (\mathbf{c} \bullet \nabla)\mathbf{b} + (\mathbf{b} \bullet \nabla)\mathbf{c}
\end{aligned}$$

and

$$\nabla^2 \mathbf{b} = \nabla(\nabla \bullet \mathbf{b}) - \nabla \times (\nabla \times \mathbf{b}).$$

These last two formulas, with $\mathbf{b} = \mathbf{c} = \mathbf{v}$, may be used to obtain alternate forms of the Navier–Stokes equation.

A.2.7 Jacobian Determinant

The *Jacobian determinant*, denoted by W, is a product of first derivatives arrayed in the form of the determinant of a square matrix. Given a set of functions $y_i(x_j)$ where the indices i and j run from 1 to J, the Jacobian is the determinant of the set of partial derivatives:

$$W(x_j) \equiv \det \left[\frac{\partial y_i}{\partial x_j} \right].$$

For $J = 2$, the (two-dimensional) Jacobian is

$$W(x_1, x_2) = \frac{\partial y_1}{\partial x_1} \frac{\partial y_2}{\partial x_2} - \frac{\partial y_1}{\partial x_2} \frac{\partial y_2}{\partial x_1}.$$

Often the two-dimensional Jacobian is written as

$$W(x_1, x_2) = \frac{\partial (y_1, y_2)}{\partial (x_1, x_2)}.$$

A.3 Coordinate Systems

A coordinate system is a systematic means of locating a point in three-dimensional space. Each coordinate system consists of a set of three *coordinate surfaces*, with each surface characterized by the value of a coordinate. Location in space is determined by specified values of the three coordinates. The most useful coordinate systems are those in which vectors normal to the surfaces are mutually perpendicular, or *orthogonal*. If all the coordinate surfaces are planes,[6] the system is *Cartesian*; if some surfaces are not planar, the coordinate system is *curvilinear*.

A surface is a single scalar constraint on position in space. Any surface may be expressed as

$$f_j(x_1, x_2, x_3) = c_j ,$$

where f_j is a specified formula, x_i are components of the position vector using our reference Cartesian coordinate system[7] and c_j (for $j = 1$, 2 or 3) is a specified parameter, referred to as a *curvilinear coordinate*. The set of three curvilinear coordinates forms a *curvilinear coordinate system*. The commonly used curvilinear coordinate systems are cylindrical and spherical, with their own set of symbols for the curvilinear coordinates, as we shall see in Appendix A.3.4 and Appendix A.3.5. For example, a spherical surface is expressed by

$$\sqrt{x_1^2 + x_2^2 + x_3^2} = c_1 .$$

In this case the parameter c_1 is called the radius and is typically denoted by r.

In order that a given surface be a coordinate surface, it must satisfy two conditions. First, it must sweep all space as its associated coordinate is varied within specified limits. The spherical surface sweeps all space as c_1 varies from 0 to ∞. Second, it must be invertible, to give the reference coordinate positions as functions of the curvilinear coordinates:

$$\mathbf{x} = \mathbf{x}(c_1, c_2, c_3) \qquad \text{or} \qquad x_i = x_i(c_1, c_2, c_3) .$$

A.3.1 Unit Vectors, Scale Factors and Del

When the values of all three curvilinear coordinates are specified, the formula $\mathbf{x} = \mathbf{x}(c_1, c_2, c_3)$ specifies a unique point in space. If one of the coordinate values (say, c_i) changes slightly (with the other two held fixed) the point is displaced a small amount, denoted by $\mathrm{d}\mathbf{x}$ and given by[8]

$$\mathrm{d}\mathbf{x} = \left(\frac{\partial \mathbf{x}}{\partial c_i} \right) \mathrm{d}c_i .$$

[6] These planar surfaces need not be orthogonal, but in practice, they invariably are.
[7] See § 2.1.
[8] In order to ensure clarity in this presentation, we will use explicit sums and not employ the Einstein summation convention.

We may construct a unit vector, designated by $\mathbf{1}_i$, pointing in the direction of this motion by dividing by the magnitude of the displacement:

$$\mathbf{1}_i = \frac{1}{h_i}\frac{\partial \mathbf{x}}{\partial c_i}, \qquad \text{where} \qquad h_i \equiv \left\| \frac{\partial \mathbf{x}}{\partial c_i} \right\|$$

is called a *scale factor* or *Lamé coefficient* and represents the distance moved as the coordinate c_i varies. This unit vector may not be orthogonal to the c_1 coordinate surface.

If we allow all three curvilinear coordinates to vary, the displacement vector is given by

$$\mathrm{d}\mathbf{x} = \sum_{j=1}^{3} \frac{\partial \mathbf{x}}{\partial c_j} \mathrm{d}c_j = \sum_{i=j}^{3} \mathbf{1}_j h_j \mathrm{d}c_j$$

and its magnitude may be expressed as

$$\mathrm{d}x^2 = \mathrm{d}\mathbf{x} \bullet \mathrm{d}\mathbf{x} = \sum_{i,j=1}^{3} g_{ij} \mathrm{d}c_i \mathrm{d}c_j, \qquad \text{where} \qquad g_{ij} \equiv \sum_{k=1}^{3} \frac{\partial x_k}{\partial c_i}\frac{\partial x_k}{\partial c_j} = \frac{\partial \mathbf{x}}{\partial c_i} \bullet \frac{\partial \mathbf{x}}{\partial c_j}$$

are elements of the *metric tensor*, which is equivalent to a square (3 by 3) matrix or tensor of tank 2.

Note that

- if the elements g_{ij} are independent of c_i, then the coordinate system is *Cartesian*;
- if $(\partial \mathbf{x}/\partial c_i) \bullet (\partial \mathbf{x}/\partial c_j) = 0$ for all $i \neq j$, then:
  - the coordinate system is *orthogonal*;
  - the metric tensor is diagonal;
  - $\mathbf{1}_i \bullet \mathbf{1}_j = \delta_{ij}$, where δ_{ij} is the Kronecker delta;[9]
  - $\mathrm{d}x^2 = \sum_{i=1}^{3} h_i^2 \mathrm{d}c_i^2$; and
  - unit vector $\mathbf{1}_i$ (for a particular value of i) is parallel to the line of intersection of the other two surfaces;
- if the unit vectors are oriented such that $\mathbf{1}_1 = \mathbf{1}_2 \times \mathbf{1}_3$, then the coordinate system is *right-handed* and
- the set of three scale factors may be represented by $\mathbf{h}$, but this is an ordered list, not a vector.

Del and Jacobian

The change in the magnitude of a scalar function $a(\mathbf{x})$ during a displacement $\mathrm{d}\mathbf{x} = \sum_{j=1}^{3} \mathbf{1}_j h_j \mathrm{d}c_j$ is given by

$$\mathrm{d}a = \sum_{i=1}^{3} \left(\frac{\partial a}{\partial c_i} \right) \mathrm{d}c_i \qquad \text{or}$$

$$= \nabla a \bullet \mathrm{d}\mathbf{x} = \sum_{j=1}^{3} \left(\mathbf{1}_j \bullet \nabla a \right) h_j \mathrm{d}c_j .$$

[9] See Appendix A.1.4.

These two versions (on the first and second lines) are in agreement provided we define the del operator as

$$\nabla \equiv \mathbf{1}_1 \frac{\partial}{h_1 \partial c_1} + \mathbf{1}_2 \frac{\partial}{h_2 \partial c_2} + \mathbf{1}_3 \frac{\partial}{h_3 \partial c_3} = \sum_{j=1}^{3} \mathbf{1}_j \frac{\partial}{h_j \partial c_j}.$$

The similarity to the Cartesian form introduced at the beginning of Appendix A.2 is apparent if we recall that distance moved is represented by $h_1 \mathrm{d}c_1$, etc. The differential volume element is given by

$$\mathrm{d}V = W \mathrm{d}c_1 \mathrm{d}c_2 \mathrm{d}c_3, \qquad \text{where} \qquad W \equiv h_1 h_2 h_3$$

is the Jacobian determinant.

A.3.2 Vector Operators

It is important to realize that the unit vectors of a curvilinear coordinate system are functions of position. This dependence is the reason for the appearance of "extra" terms in the representations of the divergence ($\nabla \bullet \mathbf{b}$), curl ($\nabla \times \mathbf{b}$) and Laplacian ($\nabla^2$) in those systems. In this section, we derive formulas for these operations in a general orthogonal curvilinear coordinate system.[10]

First consider the divergence. Expressing an arbitrary vector $\mathbf{b}(\mathbf{x})$ as $\mathbf{b} = \sum_{i=1}^{3} b_i \mathbf{1}_i$, we have

$$\nabla \bullet \mathbf{b} = \sum_{i=1}^{3} (\mathbf{1}_i \bullet \nabla b_i + b_i \nabla \bullet \mathbf{1}_i) = \sum_{j=1}^{3} \frac{\partial b_j}{h_j \partial c_j} + \sum_{i,j=1}^{3} \frac{b_i}{h_j} G_{ij},$$

$$\text{where} \qquad G_{ij} \equiv \mathbf{1}_j \bullet \frac{\partial \mathbf{1}_i}{\partial c_j}.$$

It is readily verified that $G_{ij} = 0$ when $i = j$. Using $\mathbf{1}_i = \partial \mathbf{x} / h_i \partial c_i$ and noting that orthogonality ensures $\partial \mathbf{x}/\partial c_i \bullet \partial \mathbf{x}/\partial c_j = 0$ when $i \neq j$, we have (with no sum on i or j)

$$\begin{aligned} G_{ij} &= \frac{\partial \mathbf{x}}{\partial c_j} \bullet \frac{\partial}{h_j \partial c_j} \left(\frac{\partial \mathbf{x}}{h_i \partial c_i} \right) = \frac{1}{h_i h_j} \frac{\partial \mathbf{x}}{\partial c_j} \bullet \frac{\partial^2 \mathbf{x}}{\partial c_i \partial c_j} \\ &= \frac{1}{2 h_i h_j} \frac{\partial}{\partial c_i} \left\| \frac{\partial \mathbf{x}}{\partial c_j} \right\|^2 = \frac{1}{2 h_i h_j} \frac{\partial h_j^2}{\partial c_i} = \frac{1}{h_i} \frac{\partial h_j}{\partial c_i} \end{aligned}$$

and using $W = h_1 h_2 h_3$

$$\sum_{i,j=1}^{3} \frac{b_i}{h_j} G_{ij} = \frac{1}{W} \left(b_1 \frac{\partial (h_2 h_3)}{\partial c_1} + b_2 \frac{\partial (h_3 h_1)}{\partial c_2} + b_3 \frac{\partial (h_2 h_1)}{\partial c_3} \right).$$

[10] The mathematical development in this section is due to Prof. Paul Roberts.

Now

$$\nabla \bullet \mathbf{b} = \frac{1}{W} \sum_{i=1}^{3} \frac{\partial}{\partial c_i} \left(W \frac{b_i}{h_i} \right)$$

and it follows directly that the Laplacian is

$$\nabla^2 \equiv \nabla \bullet \nabla = \frac{1}{W} \sum_{i=1}^{3} \frac{\partial}{\partial c_i} \left(\frac{W}{h_i^2} \frac{\partial}{\partial c_i} \right).$$

Derivation of the formula for the curl proceeds in three steps. First, using the vector formulas $\mathbf{1}_1 = \mathbf{1}_2 \times \mathbf{1}_3$ and $\mathbf{a} \times (\mathbf{b} \times \mathbf{c}) = (\mathbf{a} \bullet \mathbf{c})\mathbf{b} - (\mathbf{a} \bullet \mathbf{b})\mathbf{c}$ we see that

$$\begin{aligned} \mathbf{1}_1 \times \frac{\partial \mathbf{1}_1}{\partial c_1} &= \mathbf{1}_1 \times \left(\frac{\partial \mathbf{1}_2}{\partial c_1} \times \mathbf{1}_3 - \frac{\partial \mathbf{1}_3}{\partial c_1} \times \mathbf{1}_2 \right) \\ &= \left(\mathbf{1}_1 \bullet \frac{\partial \mathbf{1}_3}{\partial c_1} \right) \mathbf{1}_2 - \left(\mathbf{1}_1 \bullet \frac{\partial \mathbf{1}_2}{\partial c_1} \right) \mathbf{1}_3 = \frac{1}{h_3} \frac{\partial h_1}{\partial c_3} \mathbf{1}_2 - \frac{1}{h_2} \frac{\partial h_1}{\partial c_2} \mathbf{1}_3. \end{aligned}$$

In step 2 we find that

$$\begin{aligned} \left(\frac{\mathbf{1}_2}{h_2} \frac{\partial}{\partial c_2} + \frac{\mathbf{1}_3}{h_3} \frac{\partial}{\partial c_3} \right) \times \mathbf{1}_1 &= \frac{1}{h_2} \frac{\partial \mathbf{1}_1}{\partial c_2} \times (\mathbf{1}_1 \times \mathbf{1}_3) - \frac{1}{h_3} \frac{\partial \mathbf{1}_1}{\partial c_3} \times (\mathbf{1}_1 \times \mathbf{1}_2) \\ &= \left(\frac{\mathbf{1}_3}{h_2} \bullet \frac{\partial \mathbf{1}_1}{\partial c_2} - \frac{\mathbf{1}_2}{h_3} \bullet \frac{\partial \mathbf{1}_1}{\partial c_3} \right) \mathbf{1}_1 = \mathbf{1}_1 \bullet \left(\frac{1}{h_3} \frac{\partial \mathbf{1}_2}{\partial c_3} - \frac{1}{h_2} \frac{\partial \mathbf{1}_3}{\partial c_2} \right) \mathbf{1}_1 \\ &= \frac{\mathbf{1}_1}{h_3} \bullet \frac{\partial}{\partial c_3} \left(\frac{1}{h_2} \frac{\partial \mathbf{x}}{\partial c_2} \right) \mathbf{1}_1 - \frac{\mathbf{1}_1}{h_2} \bullet \frac{\partial}{\partial c_2} \left(\frac{1}{h_3} \frac{\partial \mathbf{x}}{\partial c_3} \right) \mathbf{1}_1 = \mathbf{0}. \end{aligned}$$

Putting these two results together, we have

$$\begin{aligned} \nabla \times (b_1 \mathbf{1}_1) &= \left(\mathbf{1}_1 \frac{\partial}{h_1 \partial c_1} + \mathbf{1}_2 \frac{\partial}{h_2 \partial c_2} + \mathbf{1}_3 \frac{\partial}{h_3 \partial c_3} \right) \times (b_1 \mathbf{1}_1) \\ &= \frac{\mathbf{1}_2}{h_3 h_1} \frac{\partial}{\partial c_3} (h_1 b_1) - \frac{\mathbf{1}_3}{h_1 h_2} \frac{\partial}{\partial c_2} (h_1 b_1). \end{aligned}$$

Step 3 consists of repeating this result for components 2 and 3; altogether we have

$$\begin{aligned} \nabla \times \mathbf{b} &= \frac{\mathbf{1}_1}{h_2 h_3} \left(\frac{\partial}{\partial c_2} (h_3 b_3) - \frac{\partial}{\partial c_3} (h_2 b_2) \right) \\ &\quad + \frac{\mathbf{1}_2}{h_3 h_1} \left(\frac{\partial}{\partial c_3} (h_1 b_1) - \frac{\partial}{\partial c_1} (h_3 b_3) \right) + \frac{\mathbf{1}_3}{h_1 h_2} \left(\frac{\partial}{\partial c_1} (h_2 b_2) - \frac{\partial}{\partial c_2} (h_1 b_1) \right). \end{aligned}$$

A.3.3 Cartesian Coordinates

Many orthogonal coordinate systems exist, but the most common and useful are the Cartesian, cylindrical and spherical. In a Cartesian system the coordinate directions are three mutually perpendicular straight lines called *axes* and named x_1, x_2 and x_3. The coordinate surfaces are flat planes described by $x_i = c_i$ where c_i are specified constants. Each of the coordinates varies from $-\infty$ to ∞ and $\mathbf{h} = \{1, 1, 1\}$. The Cartesian system is homogeneous, with each coordinate surface equivalent to another.

In the theoretical development of the equations of motion, it is most convenient to name the Cartesian coordinates x_1, x_2 and x_3. However, in practice, these are usually named x, y and z.

A.3.4 Cylindrical Coordinates

In a cylindrical system the coordinates are oriented with respect to an axis: a straight line in space extending indefinitely in both directions. One coordinate, called the cylindrical radius and typically denoted by ϖ, is the perpendicular distance from the axis. A second coordinate, denoted by ϕ, is the angular orientation about the axis with φ increasing in the direction of $\mathbf{1}_z \times \mathbf{1}\varpi$. The third coordinate is distance parallel to the axis, commonly denoted by z. These coordinates may be related to a Cartesian system as follows:

$$\varpi = \sqrt{x_1^2 + x_2^2} \quad \phi = \arctan(x_2/x_1) \quad z = x_3$$

or in more familiar notation,

$$\varpi = \sqrt{x^2 + y^2} \quad \phi = \arctan(y/x) \qquad z = z.$$

The associated coordinate surfaces are cylinders, half planes and full planes, respectively. Each surface sweeps all space as $0 \le \varpi < \infty$, $-\pi < \phi \le \pi$ and $-\infty < z < \infty$ and $\mathbf{h} = \{1, \varpi, 1\}$.

In this coordinate system with a being an arbitrary scalar and an arbitrary vector $\mathbf{b}$ expressed as $\mathbf{b} = b_\varpi \mathbf{1}_\varpi + b_\phi \mathbf{1}_\phi + b_z \mathbf{1}_z$, where $\mathbf{1}_\varpi$, $\mathbf{1}_\phi$ and $\mathbf{1}_z$ are the cylindrical unit vectors with

$$\mathbf{1}_\varpi = \cos\phi\,\mathbf{1}_x + \sin\phi\,\mathbf{1}_y, \qquad \mathbf{1}_\phi = -\sin\phi\,\mathbf{1}_x + \cos\phi\,\mathbf{1}_y$$

$$\text{and} \qquad \frac{\partial\{\mathbf{1}_\varpi, \mathbf{1}_\phi\}}{\partial\phi} = \{\mathbf{1}_\phi, -\mathbf{1}_\varpi\},$$

the gradient, divergence, Laplacian and curl are[11]

$$\nabla a = \frac{\partial a}{\partial\varpi}\mathbf{1}_\varpi + \frac{1}{\varpi}\frac{\partial a}{\partial\phi}\mathbf{1}_\phi + \frac{\partial a}{\partial z}\mathbf{1}_z,$$

$$\nabla\bullet\mathbf{b} = \frac{1}{\varpi}\frac{\partial(\varpi b_\varpi)}{\partial\varpi} + \frac{1}{\varpi}\frac{\partial b_\phi}{\partial\phi} + \frac{\partial b_z}{\partial z},$$

$$\nabla^2 a = \nabla\bullet\nabla a = \frac{1}{\varpi}\frac{\partial}{\partial\varpi}\left(\varpi\frac{\partial a}{\partial\varpi}\right) + \frac{1}{\varpi^2}\frac{\partial^2 a}{\partial\phi^2} + \frac{\partial^2 a}{\partial z^2}$$

[11] Defined in Appendix A.2.

and

$$\nabla \times \mathbf{b} = \left(\frac{1}{\varpi} \frac{\partial b_z}{\partial \phi} - \frac{\partial b_\phi}{\partial z} \right) \mathbf{1}_\varpi + \left(\frac{\partial b_\varpi}{\partial z} - \frac{\partial b_z}{\partial \varpi} \right) \mathbf{1}_\phi + \frac{1}{\varpi} \left(\frac{\partial (\varpi b_\phi)}{\partial \varpi} - \frac{\partial b_\varpi}{\partial \phi} \right) \mathbf{1}_z .$$

A.3.5 Spherical Coordinates

In a spherical system, the coordinates are oriented with respect to a half axis: a straight line in space extending from a specified point, called the origin, to infinity in a specified direction. One coordinate, called the radius and denoted by r, is the distance from the origin. A second, denoted by θ is the angle between a ray extending from the origin and the half axis; this is the *colatitude*. A third, denoted by ϕ, is the angular orientation about the half axis with φ increasing in the direction of $\mathbf{1}_r \times \mathbf{1}_\theta$. This third coordinate is equivalent to the second of the cylindrical system. These coordinates may be related to a Cartesian system as follows:

$$r = \sqrt{x_1^2 + x_2^2 + x_3^2} \qquad \theta = \arctan\left(x_3 / \sqrt{x_1^2 + x_2^2} \right) \qquad \phi = \arctan\left(x_2 / x_1 \right)$$

or more familiarly

$$r = \sqrt{x^2 + y^2 + z^2} \qquad \theta = \arctan\left(z / \sqrt{x^2 + y^2} \right) \qquad \phi = \arctan\left(y / x \right) .$$

The associated coordinate surfaces are spheres, cones and half planes, respectively. Each surface sweeps all space as $0 \le r < \infty$, $0 \le \theta \le \pi$ and $-\pi < \phi \le \pi$ and $\mathbf{h} = \{1, r, r\sin\theta\}$.

In this coordinate system with $\mathbf{b} = b_r \mathbf{1}_r + b_\theta \mathbf{1}_\theta + b_\phi \mathbf{1}_\phi$, where $\mathbf{1}_r$, $\mathbf{1}_\theta$ and $\mathbf{1}_\phi$ are the spherical unit vectors with

$$\begin{aligned}
\mathbf{1}_r &= \sin\theta\, \cos\phi\, \mathbf{1}_x + \sin\theta\, \sin\phi\, \mathbf{1}_y + \cos\theta\, \mathbf{1}_z , \\
\mathbf{1}_\theta &= \cos\theta\, \cos\phi\, \mathbf{1}_x + \cos\theta\, \sin\phi\, \mathbf{1}_y - \sin\theta\, \mathbf{1}_z , \\
\mathbf{1}_\phi &= -\sin\phi\, \mathbf{1}_x + \cos\phi\, \mathbf{1}_y , \\
\frac{\partial \{\mathbf{1}_r, \mathbf{1}_\theta, \mathbf{1}_\phi\}}{\partial \theta} &= \{\mathbf{1}_\theta, -\mathbf{1}_r, \mathbf{0}\} \qquad \text{and} \\
\frac{\partial \{\mathbf{1}_r, \mathbf{1}_\theta, \mathbf{1}_\phi\}}{\partial \phi} &= \{\sin\theta\, \mathbf{1}_\phi, \cos\theta\, \mathbf{1}_\phi, -\sin\theta\, \mathbf{1}_r - \cos\theta\, \mathbf{1}_\theta\} ,
\end{aligned}$$

the gradient, divergence, Laplacian and curl are

$$\begin{aligned}
\nabla a &= \frac{\partial a}{\partial r} \mathbf{1}_r + \frac{1}{r} \frac{\partial a}{\partial \theta} \mathbf{1}_\theta + \frac{1}{r\sin\theta} \frac{\partial a}{\partial \phi} \mathbf{1}_\phi , \\
\nabla \bullet \mathbf{b} &= \frac{1}{r^2} \frac{\partial (r^2 b_r)}{\partial r} + \frac{1}{r\sin\theta} \frac{\partial (b_\theta \sin\theta)}{\partial \theta} + \frac{1}{r\sin\theta} \frac{\partial b_\phi}{\partial \phi} , \\
\nabla^2 a = \nabla \bullet \nabla a &= \frac{1}{r^2} \frac{\partial}{\partial r} \left(r^2 \frac{\partial a}{\partial r} \right) + \frac{1}{r^2 \sin\theta} \frac{\partial}{\partial \theta} \left(\sin\theta \frac{\partial a}{\partial \theta} \right) + \frac{1}{r^2 \sin^2\theta} \frac{\partial^2 a}{\partial \phi^2}
\end{aligned}$$

and

$$\nabla \times \mathbf{b} = \frac{1}{r\sin\theta}\left(\frac{\partial(b_\phi \sin\theta)}{\partial\theta} - \frac{\partial b_\theta}{\partial\phi}\right)\mathbf{1}_r$$
$$+ \frac{1}{r}\left(\frac{1}{\sin\theta}\frac{\partial b_r}{\partial\phi} - \frac{\partial(rb_\phi)}{\partial r}\right)\mathbf{1}_\theta + \frac{1}{r}\left(\frac{\partial(rb_\theta)}{\partial r} - \frac{\partial b_r}{\partial\theta}\right)\mathbf{1}_\phi .$$

It is straightforward to obtain additional expressions such as for $\nabla^2\mathbf{b}$, $\nabla \times (\mathbf{b} \times \mathbf{c})$ and $(\mathbf{b} \bullet \nabla)\,\mathbf{c}$ in both cylindrical and spherical coordinates.

A.4 Taylor Series

A Taylor series is a systematic means of approximating a smooth (analytic) function in the vicinity of a specified point. The simplest form of Taylor series involves a scalar function of a single variable:

$$y = f(x) .$$

The Taylor series approximation of this function about a specified point,[12] $x = x_0$, is

$$y = f(x_0) + \frac{\mathrm{d}f}{\mathrm{d}x}(x_0)(x - x_0) + \frac{1}{2}\frac{\mathrm{d}^2 f}{\mathrm{d}x^2}(x_0)(x - x_0)^2$$
$$+ \frac{1}{6}\frac{\mathrm{d}^3 f}{\mathrm{d}x^3}(x_0)(x - x_0)^3 + O\left((x - x_0)^4\right),$$

where the symbol O (say "big oh") means that the order of magnitude of the terms not explicitly shown vary in magnitude in proportion to the term in the parenthesis.

The degree of approximation is at the user's discretion. For many applications it is sufficient to replace a function by its linear approximation and write

$$y = y_0 + \left.\frac{\mathrm{d}f}{\mathrm{d}x}\right|_0 (x - x_0) ,$$

where $y_0 = f(x_0)$ and $|_0$ denotes evaluation at $x = x_0$.

If a scalar is a function of position in space, i.e., $y = f(x_1, x_2, x_3)$, the linear Taylor series about point $\mathbf{x}_0 = x_{pj}\mathbf{1}_j$ is[13]

$$y = y_0 + \left.\frac{\partial f}{\partial x_i}\right|_0 \left(x_i - x_{pi}\right) .$$

This may be expressed in vector form as

$$y = y_0 + (\nabla f)_0 \bullet (\mathbf{x} - \mathbf{x}_0) .$$

[12] If the specified point is $x = 0$, the series may be called a *Maclaurin series*.
[13] Reinstating the Einstein summation convention.

Table A.1. *Leading coefficients of the Taylor series expansions of some elementary transcendental functions. n! is the factorial function.*

Function	1	θ	θ^2	θ^3	θ^4	θ^5	θ^6	θ^7	θ^8
e^θ	1	1	1/2!	1/3!	1/4!	1/5!	1/6!	1/7!	1/8!
$e^{-\theta}$	1	−1	1/2!	−1/3!	1/4!	−1/5!	1/6!	−1/7!	1/8!
$\cosh\theta$	1	0	1/2!	0	1/4!	0	1/6!	0	1/8!
$\sinh\theta$	0	1	0	1/3!	0	1/5!	0	1/7!	0
$\sin\theta$	0	1	0	−1/3!	0	1/5!	0	−1/7!	0
$\cos\theta$	1	0	−1/2!	0	1/4!	0	−1/6!	0	1/8!
$\tan\theta$	0	1	0	1/2	0	2/15	0	17/315	0
$\ln(1+\theta)$	1	1	−1/2	1/3	−1/4	1/5	−1/6	1/7	−1/8

If a vector, $\mathbf{b} = b_j \mathbf{1}_j$, is a function of position, each component may be expanded as above:

$$b_j = b_{0j} + \left.\frac{\partial b_j}{\partial x_i}\right|_0 (x_i - x_{0i}) \,.$$

If $\mathbf{b}$ is the displacement, then the quantities $\partial b_j/\partial x_i$ are the elements of the tensor of deformation gradients and if $\mathbf{b}$ is the velocity, then these quantities are elements of the tensor of velocity gradients (see Chapter 3).

The first few terms of the Taylor series of some elementary transcendental functions are given in Table A.1.

A.5 Fourier Series and Integral

A function $f(x)$ defined on a finite interval $-\pi < x < \pi$ may be expressed in terms of a *Fourier series*:

$$f(x) = \tfrac{1}{2}a_0 + \sum_{k=1}^{\infty} a_k \cos(kx) + \sum_{k=1}^{\infty} b_k \sin(kx)\,,$$

where the coefficients a_k and b_k are related to $f(x)$ by the following integrals:

$$a_k = \frac{1}{\pi}\int_{-\pi}^{\pi} f(x)\cos(kx)\mathrm{d}x \quad \text{and} \quad b_k = \frac{1}{\pi}\int_{-\pi}^{\pi} f(x)\sin(kx)\mathrm{d}x\,.$$

If $f(x)$ is even on the interval, $b_k = 0$ for all k and the series is called a *cosine series*. If $f(x)$ is odd on the interval, $a_k = 0$ for all k and the series is called a *sine series*.

If $f(x)$ is defined on the infinite interval $-\infty < x < \infty$, then the series are replaced by integrals; now

$$f(x) = \tfrac{1}{2}a_0 + \int_0^{\infty} a(k)\cos(kx)\mathrm{d}k + \int_0^{\infty} b(k)\sin(kx)\mathrm{d}k\,,$$

where

$$a(k) = \frac{1}{\pi}\int_{-\infty}^{\infty} f(x)\cos(kx)\mathrm{d}x \quad \text{and} \quad b(k) = \frac{1}{\pi}\int_{-\infty}^{\infty} f(x)\sin(kx)\mathrm{d}x.$$

If $f(x)$ is even on the interval, $b(k) = 0$ and the representation is called a *cosine integral*. If $f(x)$ is odd on the interval, $a(k) = 0$ and the representation is called a *sine integral*.

A.5.1 Complex Representation

Complex Fourier representation is based on *Euler's formula* which relates the sines and cosines to imaginary exponentials:

$$e^{\pm \mathrm{i}q} = \cos q \pm \mathrm{i}\sin q$$

with inverse

$$\cos q = \tfrac{1}{2}\left(e^{\mathrm{i}q} + e^{-\mathrm{i}q}\right) \qquad \text{and} \qquad \sin q = \tfrac{1}{2}\mathrm{i}\left(-e^{\mathrm{i}q} + e^{-\mathrm{i}q}\right)$$

or equivalently

$$\cos q = \tfrac{1}{2}e^{\mathrm{i}q} + \text{c.c.} \qquad \text{and} \qquad \sin q = -\tfrac{1}{2}\mathrm{i}e^{\mathrm{i}q} + \text{c.c.},$$

where "c. c." means the complex conjugate of the preceding term. These last equations are often employed in the literature with the "c. c." absent, but implied.

The complex Fourier series is[14]

$$f(x) = \frac{1}{\sqrt{2\pi}}\sum_{k=-\infty}^{\infty} c_k e^{\mathrm{i}kx}, \qquad \text{with inverse} \qquad c_k = \frac{1}{\sqrt{2\pi}}\int_{-\pi}^{\pi} f(x)e^{-\mathrm{i}kx}\mathrm{d}x.$$

Similarly the complex Fourier integral is

$$f(x) = \frac{1}{\sqrt{2\pi}}\int_{-\infty}^{\infty} c(k)e^{\mathrm{i}kx}\mathrm{d}k, \qquad \text{where} \qquad c(k) = \frac{1}{\sqrt{2\pi}}\int_{-\infty}^{\infty} f(x)e^{-\mathrm{i}kx}\mathrm{d}x.$$

One way to think of Fourier representation is that information regarding a function $f(x)$ is codified in the coefficients $a(k)$ and $b(k)$ or equivalently $c(k)$, with the function being reconstituted by use of the appropriate Fourier formula.

[14] where

$$c_k = \frac{1}{2}\begin{cases} a_k - \mathrm{i}b_k & \text{for} \quad 0 < k \\ a_k & \text{for} \quad k = 0 \\ a_k + \mathrm{i}b_k & \text{for} \quad k < 0 \end{cases}.$$

Since k ranges through both negative and positive values, it is not necessary to append "c. c." to these complex expressions.

A.5.2 Fourier Representation in Three Dimensions

A function of position in three-dimensional space may be expressed as

$$f(\mathbf{x}) = \frac{1}{\sqrt{2\pi}} \int_{\mathbf{k}} c(\mathbf{k}) e^{\mathrm{i}\mathbf{k}\bullet\mathbf{x}} \mathrm{d}\mathbf{k}, \qquad \text{with inverse} \qquad c(\mathbf{k}) = \frac{1}{\sqrt{2\pi}} \int_{\mathbf{x}} f(\mathbf{x}) e^{-\mathrm{i}\mathbf{k}\bullet\mathbf{x}} \mathrm{d}\mathbf{x}.$$

A.5.3 Complex Representation and Differentiation

One big advantage of complex Fourier representation is that it directly turns calculus into algebra. That is, differentiation of the complex Fourier integral representation of $f(x)$ gives

$$\frac{\mathrm{d}f(x)}{\mathrm{d}x} = \frac{1}{\sqrt{2\pi}} \int_{-\infty}^{\infty} \mathrm{i}kc(k) e^{\mathrm{i}kx} \mathrm{d}k,$$

or symbolically

$$\frac{\mathrm{d}}{\mathrm{d}x} \to \mathrm{i}k \qquad \text{and} \qquad \frac{\mathrm{d}^2}{\mathrm{d}x^2} \to -k^2.$$

Similarly,

$$\nabla f(\mathbf{x}) = \frac{1}{\sqrt{2\pi}} \int_{\mathbf{k}} \mathrm{i}\mathbf{k} c(\mathbf{k}) e^{\mathrm{i}\mathbf{k}\bullet\mathbf{x}} \mathrm{d}\mathbf{k}$$

or symbolically

$$\nabla \to \mathrm{i}\mathbf{k} \qquad \text{and} \qquad \nabla^2 = \nabla \bullet \nabla \to -k^2.$$

A word of caution when dealing with Fourier series and Fourier integral representations of functions. These representations can be integrated with impunity, but care must be taken when differentiating, because this operation introduces factors k in the numerator of the expressions, making the derivative less convergent than its parent.

A.6 Partial Differential Equations

A *partial differential equation* (PDE) is an equation involving partial derivatives of the dependent variables, which are functions of two or more independent variables. In mathematical physics the independent variables commonly are position and time. PDEs are classified primarily by order (that is, the greatest number of differentiations). Equations governing waves and flows are often second order. PDEs are also categorized as either linear (each term contains one dependent variable in linear form, or none) or nonlinear. The equations of elasticity are linear, but the Navier–Stokes equation is nonlinear. Linear equations are further classified as having constant (not containing the independent variable) or variable coefficients.

The PDEs that are most commonly encountered in mathematical physics are linear, second order with constant coefficients. The general form of this equation[15] involving

[15] Following the notation in chapter XIII of Forsythe (1959).

Table A.2. *Classification of linear second-order PDEs. Subscripts denote differentiation.*

Discriminant value	Classification	Canonical form	Equation name
Negative	Hyperbolic	$z_{xx} - z_{yy} = 0$	Wave equation
Zero	Parabolic	$z_{xx} - z_y = 0$	Heat equation
Negative	Elliptic	$z_{xx} + z_{yy} = 0$	Laplacian

$z(x,y)$ is

$$R\frac{\partial^2 z}{\partial x^2} + S\frac{\partial^2 z}{\partial x \partial y} + T\frac{\partial^2 z}{\partial y^2} + P\frac{\partial z}{\partial x} + Q\frac{\partial z}{\partial y} + Zz = U(x,y)\,,$$

where the coefficients are constants and $U(x,y)$ is a non-homogeneous forcing term. This equation is classified as hyperbolic, parabolic and elliptic depending on the value of the discriminant[16]

$$\Delta = S^2 - 4RT\,.$$

The classifications and canonical forms of these equations are shown in Table A.2.

The time-dependent momentum and heat equations, having a single time derivative and two spatial derivatives, are classified as parabolic, while the steady versions of these equations are elliptic. Invariably these equations are valid on an open interval of time and space and are complemented by initial and/or boundary conditions. If the problem under consideration is time-dependent, we need to specify an initial condition. In either the time-dependent or steady case, we need to apply one condition on velocity (or its normal derivative) and one on temperature (or its normal derivative) at each point on the spatial boundary.

A.7 Greek Symbols

It is common practice in mathematics to employ many of the letters of the Greek alphabet, which are presented in Table A.3. Note that many upper-case Greek symbols are quite similar to Roman symbols and so are less useful as mathematical symbols. The two may be distinguished in equations by displaying the upper-case Greek symbols in upright text and the upper-case Roman symbols in italics. However, this subtle distinction is often not sufficient to avoid confusion. It is best if the Greek symbols A, B, E, Z, H, I, K, M, N, O, o, P, T, Y and X are not used in mathematical representations. The letter xi is the most difficult for English speakers to enunciate; it is usually spoken as if it were spelled “kxi”.

[16] Note the similarity to the classification of conic sections.

Table A.3. *Greek upper- and lower-case symbols, plus six variants of Greek lower-case symbols.*

Upper-case symbol	Lower-case symbol	Name	Upper-case symbol	Lower-case symbol	Name
A	α	Alpha	N	ν	Nu
B	β	Beta	Ξ	ξ	Xi
Γ	γ	Gamma	O	o	Omicron
Δ	δ	Delta	Π	π	Pi
E	ϵ	Epsilon	P	ρ	Rho
Z	ζ	Zeta	Σ	σ	Sigma
H	η	Eta	T	τ	Tau
Θ	θ	Theta	Y	υ	Upsilon
I	ι	Iota	Φ	ϕ	Phi
K	κ	Kappa	X	χ	Chi
Λ	λ	Lambda	Ψ	ψ	Psi
M	μ	Mu	Ω	ω	Omega
	Variant symbol	**Name**		**Variant symbol**	**Name**
	ε	Curly epsilon		ς	Curly sigma
	ϑ	Curly theta		φ	Curly phi
	ϱ	Curly rho		ϖ	Pomega

Mathematical usage includes several variants of Greek letters; these are listed in Table A.3.

A.8 Scalar and Vector Potentials

On occasion it is useful to express a vector in terms of a scalar potential, a vector potential or both. Suppose a vector **b** has zero curl:

$$\nabla \times \mathbf{b} = \mathbf{0}.$$

This condition is automatically satisfied by expressing **b** as the gradient of a scalar:

$$\mathbf{b} = \nabla \phi ,$$

where ϕ is the *scalar potential*. Alternatively, suppose the vector **b** has zero divergence:

$$\nabla \cdot \mathbf{b} = 0.$$

This condition can be automatically satisfied by expressing **b** as the curl of a vector:

$$\mathbf{b} = \nabla \times \mathbf{A},$$

where **A** is the *vector potential*. Typically, and without loss of generality, **A** is required to be non-divergent:

$$\nabla \times \mathbf{A} = \mathbf{0}.$$

A general vector **b** may be expressed as in terms of the scalar and vector potentials:

$$\mathbf{b} = \nabla\phi + \nabla \times \mathbf{A},$$

with **A** being non-divergent.

The scalar- and vector-potential formalism may helpful in simplifying vector partial differential equations, in both elasticity (See Appendix A.8.1) and fluid dynamics. There are two types of simple flows that often occur in modeling waves and flows: anelastic and irrotational; these afford significant simplification of the governing equations, with *irrotational flow* leading to the *velocity potential* (See Appendix A.8.2) and *anelastic flow* leading to the *stream function* (See Appendix C.4).

A.8.1 Simplification of the Navier Equation

Consider the perturbation Navier equation presented in § 7.3.1:

$$\rho_0 \frac{\partial^2 \mathbf{u}}{\partial t^2} = \left(K + \frac{\mu}{3}\right) \nabla(\nabla \bullet \mathbf{u}) + \mu \nabla^2 \mathbf{u}.$$

Expressing the displacement vector **u** in terms of the scalar and vector potentials ($\mathbf{u} = \nabla\phi + \nabla \times \mathbf{A}$), this separates into two simpler equations:

$$\rho_0 \frac{\partial^2 \phi}{\partial t^2} = \left(K + \frac{4\mu}{3}\right) \nabla^2 \phi \qquad \text{and} \qquad \rho_0 \frac{\partial^2 \mathbf{A}}{\partial t^2} = \mu \nabla^2 \mathbf{A}.$$

Longitudinal seismic waves are curl-free and the associated displacement vector may be expressed in terms of the scalar potential. Transverse seismic waves are non-divergent and may be expressed in terms of the vector potential.

A.8.2 Potential Flow

The flow of fluids (such as water and air) having small viscosity is often *irrotational*, with the velocity satisfying the vorticity equation

$$\nabla \times \mathbf{v} = \mathbf{0}.$$

Flow satisfying this equation is called *potential flow*. This equation is a first integral of the momentum equation and is, in effect, two scalar constraints on the vector function **v**. It can be "solved" by introduction of a scalar potential, denoted by ϕ and called the *velocity potential*:

$$\mathbf{v} = \nabla\phi\,.$$

The velocity potential is determined by the continuity equation.

Two-Dimensional Potential Flow

If the flow is two-dimensional (in the $x - z$ plane)[17] the velocity components are given by

$$u = \frac{\partial\phi}{\partial x} \qquad \text{and} \qquad w = \frac{\partial\phi}{\partial z}\,.$$

[17] See Appendix A.3.

Appendix B

Dimensions and Units

The fundaments of kinematics presented include:

- B.1: an introduction to dimensional analysis;
- B.2: an introduction to the international system of units; and
- B.3: a table of parameters relevant to waves and flows on Earth and their estimated magnitudes.

B.1 Dimensional Analysis

This fundament provides an orientation to the process of dimensional analysis, whereby a problem is simplified by analyzing the dimensions of its variables and parameters. The theoretical basis for this process is provided by the Buckingham Pi theorem.[1] An essential part of the process is the formation of a minimal number of *dimensionless parameters*, typically denoted by Π, each of which is an algebraic combination of the dimensional parameters in the problem. One advantage of dimensional analysis is that we do not need a set of governing equations (though that is, of course, quite helpful). All that is needed is a "grocery list" of relevant parameters.

A related procedure is *non-dimensionalization*, whereby a set of governing equations and the parameters within it are made dimensionless. Often the equations are significantly simplified by this procedure, making them more amenable to analysis and solution.

Dimensional analysis employs the seven base dimensions (length, mass, time, temperature, electric current, amount of substance and luminous intensity) of the *SI system*.[2] Mechanical systems involve only the first three (length, mass and time), denoted by L, M and T:[3]

$$\mathrm{L} = \text{length} \quad \mathrm{M} = \text{mass} \quad \mathrm{T} = \text{time}.$$

[1] This is not a proper mathematical theorem, and so is rather unfortunately named.
[2] More formally called the International System of Units; see the fundament on International System of Units found in Appendix B.2.
[3] Letters denoting dimensions are displayed in Roman typeface, while letters denoting variables are in italic.

Table B.1. *Base and supplementary SI dimensions, units and symbols.*

Dimension	Unit	Symbol
Length	Meter	m
Mass	Kilogram	kg
Time	Second	s
Temperature	Kelvin	K
Electric current	Ampere	A
Amount of substance	Mole	mol
Luminous intensity	Candela	cd
Plane angle	Radian	rad
Solid angle	Steradian	sr

Table B.2. *SI mechanical units. m = meter, kg = kilogram, s = second, rad = radian. J = N·m, W = J·s^{-1}, Pa = N·m^{-2}.*

			–	Exponents	–	–
Quantity	Unit	Symbol	m	kg	s	rad
Force	Newton	N	1	1	−2	0
Energy	Joule	J	2	1	−2	0
Power	Watt	W	2	1	−3	0
Pressure	Pascal	Pa	−1	1	−2	0

Commonly the dimensions of a variable are identified using square brackets; for example, $[g] = \mathrm{L{\cdot}T^{-2}}$ means that the dimensions of acceleration are length divided by time squared.

In addition to the seven base dimensions, the SI system contains two supplemental units: plane and solid angle. These do not play a direct role in dimensional analysis and non-dimensionalization, but they do affect the numerical relationships among the dimensionless variables.

The number of dimensionless parameters involved in a physical problem can easily be determined by use of the Buckingham Pi theorem which states that

$$N_\Pi = N_p - N_d,$$

where N_Π is the number of dimensionless parameters, N_p is the number of dimensional parameters and N_d is the number of base dimensions contained in all parameters.

Table B.3. *SI units of mechanical quantities. m = meter, kg = kilogram, s = second, rad = radian, N = newton, J = joule, Pa = pascal.*

Quantity	Unit	–	Exponents	–	–
		m	kg	s	rad
Acceleration	$\mathrm{m \cdot s^{-2}}$	1	0	−2	0
Angular acceleration	$\mathrm{rad \cdot s^{-2}}$	0	0	−2	1
Angular momentum	$\mathrm{m^2 \cdot kg \cdot rad \cdot s^{-1}}$	2	1	−1	1
Angular speed	$\mathrm{rad \cdot s^{-1}}$	0	0	−1	1
Area	$\mathrm{m^2}$	2	0	0	0
Bulk modulus	Pa	−1	1	−2	0
Compressibility	$\mathrm{Pa^{-1}}$	1	−1	2	0
Coefficient of friction	–	0	0	0	0
Density	$\mathrm{kg \cdot m^{-3}}$	−3	1	0	0
Diffusivity	$\mathrm{m^2 \cdot s^{-1}}$	2	0	−1	0
Dynamic viscosity	$\mathrm{Pa \cdot s}$	−1	1	−1	0
Force	N	1	1	−2	0
Frequency	$\mathrm{rad \cdot s^{-1}}$	0	0	−1	1
Gravitational potential	$\mathrm{m^2 \cdot s^{-2}}$	2	0	−2	0
Gravity	$\mathrm{m \cdot s^{-2}}$	1	0	−2	0
Head	m	1	0	0	0
Incompressibility	Pa	−1	1	−2	0
Kinematic viscosity	$\mathrm{m^2 \cdot s^{-1}}$	2	0	−1	0
Momentum	$\mathrm{kg \cdot m \cdot s^{-1}}$	1	1	−1	0
Poisson's ratio	–	0	0	0	0
Pressure	Pa	−1	1	−2	0
Rate of strain	$\mathrm{s^{-1}}$	0	0	−1	0
Rotation rate	$\mathrm{rad \cdot s^{-1}}$	0	0	−1	1
Shear modulus	Pa	−1	1	−2	0
Specific volume	$\mathrm{m^3 \cdot kg^{-1}}$	3	−1	0	0
Speed	$\mathrm{m \cdot s^{-1}}$	1	0	−1	0
Stress	Pa	−1	1	−2	0
Strain	–	0	0	0	0
Surface tension	$\mathrm{N \cdot m^{-1}}$	0	1	−2	0
Time	s	0	0	1	0
Torque	$\mathrm{N \cdot m}$	2	1	−2	0
Velocity	$\mathrm{m \cdot s^{-1}}$	1	0	−1	0
Volume	$\mathrm{m^3}$	3	0	0	0
Work	J	2	1	−2	0
Young's modulus	Pa	−1	1	−2	0

Table B.4. *SI thermal quantities. m = meter, kg = kilogram, s = second, K = kelvin, kmol = kilogram-mole, J = joule, W = watt. Note that 1 $J{\cdot}kg^{-1} = 1\ m^2{\cdot}s^{-2}$.*

		–	SI	exponents	–	–
Quantity	Unit	m	kg	s	K	kmol
Amount of substance	kmol	0	0	0	0	1
Boltzmann's constant	$J{\cdot}K^{-1}$	2	1	−2	−1	0
Chemical potential	$J{\cdot}kmol^{-1}$	2	1	−2	0	−1
Energy	J	2	1	−2	0	0
Entropy	$J{\cdot}K^{-1}$	2	1	−2	−1	0
Heat	J	2	1	−2	0	0
Ideal gas constant	$J{\cdot}kmol^{-1}{\cdot}K^{-1}$	2	1	−2	−1	−1
Latent heat	$J{\cdot}kg^{-1}$	2	0	−2	0	0
Specific energy	$J{\cdot}kg^{-1}$	2	0	−2	0	0
Specific entropy	$J{\cdot}kg^{-1}{\cdot}K^{-1}$	2	0	−2	−1	0
Specific gas constant	$J{\cdot}kg^{-1}{\cdot}K^{-1}$	2	0	−2	−1	0
Specific heat	$J{\cdot}kg^{-1}{\cdot}K^{-1}$	2	0	−2	−1	0
Temperature	K	0	0	0	1	0
Thermal conductivity	$W{\cdot}m^{-1}{\cdot}K^{-1}$	1	1	−3	−1	0
Thermal diffusivity	$m^2{\cdot}s^{-1}$	2	0	−1	0	0
Thermal expansion coefficient	K^{-1}	0	0	0	−1	0

As an example of dimensional analysis, suppose we wish to determine how the frequency ω of a guitar string varies with the string length L, density ρ and tension σ: $\omega = f(L,\rho,\sigma)$. At first glance, this is a formidable task, involving a large number of experiments in which L, ρ and σ are systematically varied. However, dimensional analysis shows that the task is far simpler than this.

In our example, we have four physical parameters ($[L] = \mathrm{L}$, $[\rho] = \mathrm{M{\cdot}L^{-1}}$, $[\sigma] = \mathrm{M{\cdot}L{\cdot}T^{-2}}$, and[4] $[\omega] = \mathrm{T^{-1}}$) and three dimensions (L, M and T), so that one dimensionless parameter may be constructed. Since our objective function is ω, we may choose to make the single dimensionless parameter linear in ω. This parameter is $L\omega\sqrt{\rho/\sigma}$. Now our functional relationship simplifies to

$$\omega = \frac{c}{L}\sqrt{\frac{\sigma}{\rho}},$$

where c is a dimensionless constant.[5] Now we need only conduct one experiment to determine the value of c and our job is done.

[4] The true dimensions of frequency are angle/time, but the plane angle is a supplementary dimension and so plays no role in the Buckingham Pi theorem.

[5] But having SI units of radians.

Table B.5. *Values of parameters. Values given as integers are approximations. STP means standard temperature and pressure: 273.15 K and 10^5 Pa. "Mantle" means mantle silicates at low pressure.*

Quantity	Numerical	value	Units	Symbol
Angular speed of Earth	7.2921	$\times 10^{-5}$	$rad \cdot s^{-1}$	Ω
Atmospheric pressure	1.01325	$\times 10^{5}$	Pa	p_a
Avogadro's constant	6.02214	$\times 10^{26}$	$atoms \cdot kmol^{-1}$	N_A
Boltzmann's constant	1.380649	$\times 10^{-23}$	$J \cdot K^{-1}$	k
Density				ρ
— of air at STP	1.27		$kg \cdot m^{-3}$	
— of water vapor at STP	0.804		$kg \cdot m^{-3}$	
— of liquid water	1.0	$\times 10^{3}$	$kg \cdot m^{-3}$	
— of sea water	1.029	$\times 10^{3}$	$kg \cdot m^{-3}$	
— of mantle	3.3	$\times 10^{3}$	$kg \cdot m^{-3}$	
Gravity (mean)	9.80665		$m \cdot s^{-2}$	g
— at poles	9.832		$m \cdot s^{-2}$	
— at equator	9.78		$m \cdot s^{-2}$	
Ideal gas constant	8314.4598		$J \cdot kmol^{-1} \cdot K^{-1}$	R
Incompressibility				K
— of air at STP	1.4	$\times 10^{5}$	Pa	
— of water	2.2	$\times 10^{9}$	Pa	
— of mantle	1.7	$\times 10^{11}$	Pa	
Kinematic viscosity				ν
— of air	2	$\times 10^{-5}$	$m^2 \cdot s^{-1}$	
— of water	1	$\times 10^{-6}$	$m^2 \cdot s^{-1}$	
Lapse rate of atmosphere				
— dry	9.8	$\times 10^{-3}$	$K \cdot m^{-1}$	
— mean	6.5	$\times 10^{-3}$	$K \cdot m^{-1}$	
Latent heat for water				
— of vaporization	2.5	$\times 10^{6}$	$J \cdot kg^{-1}$	L
— of melting	0.33	$\times 10^{6}$	$J \cdot kg^{-1}$	
Mass of mantle	4	$\times 10^{24}$	kg	M
Molecular weight				M_A
— of air	28.97		$kg \cdot kmol^{-1}$	
— of water	18.01528		$kg \cdot kmol^{-1}$	
Radius of Earth (mean)	6.371001	$\times 10^{6}$	m	R_E
— at poles	6.356752	$\times 10^{6}$	m	
— at equator	6.378137	$\times 10^{6}$	m	
Radius of Earth's core	3.485	$\times 10^{6}$	m	
Specific gas constant				R_s

Table B.5. *Continued*

Quantity	Numerical	value	Units	Symbol
— of dry air	287.1		$J{\cdot}kg^{-1}{\cdot}K^{-1}$	
— of water vapor	461.5		$J{\cdot}kg^{-1}{\cdot}K^{-1}$	
Specific heat				c_p
— of air	1005		$J{\cdot}kg^{-1}{\cdot}K^{-1}$	
— of water	4184		$J{\cdot}kg^{-1}{\cdot}K^{-1}$	
— of mantle	1250		$J{\cdot}kg^{-1}{\cdot}K^{-1}$	
Temperature at sea level	288.15		K	
Thermal conductivity				k
— air	0.025		$W{\cdot}m^{-1}{\cdot}K^{-1}$	
— water	0.6		$W{\cdot}m^{-1}{\cdot}K^{-1}$	
— of crust	3.0		$W{\cdot}m^{-1}{\cdot}K^{-1}$	
— of mantle	4.5		$W{\cdot}m^{-1}{\cdot}K^{-1}$	
Thermal diffusivity				κ
— of air	1.9	$\times 10^{-5}$	$m^2{\cdot}s^{-1}$	
— of water	1.4	$\times 10^{-7}$	$m^2{\cdot}s^{-1}$	
— of mantle	1	$\times 10^{-6}$	$m^2{\cdot}s^{-1}$	
Thermal expansion				α
— of air	T^{-1}		K^{-1}	
— of water[4]	2	$\times 10^{-4}$	K^{-1}	
— of mantle	3	$\times 10^{-5}$	K^{-1}	

[4]At $T = 20$ °C. Thermal expansion of water decreases to zero at 4 °C.

B.2 International System of Units

The International System of Units, called SI, establishes a set of nine fundamental quantities or *dimensions*, and associated *units* of measurement, with each unit represented by a standard symbol. These consist of seven base dimensions and units plus two supplementary dimensions and units as presented in Table B.1.

The definitions of these fundamental dimensions and units are set by international agreements and change from time to time. For example, at one time the meter was defined as the distance between two scratches on a bar of platinum kept in Paris. Since 1983 it has been defined as the length traveled by light in a vacuum in 1/299,792,458 of a second. Since 1967 a second has been defined as the duration of 9,192,631,770 periods of the radiation corresponding to the transition between the two hyperfine levels of the ground state of ^{133}cesium. But the kilogram is still (since 1889 and as of 2017) defined the old way; it is the mass of the International Prototype Kilogram, copies of which are kept at various national laboratories.

The radian is a somewhat slippery unit (as is its counterpart the steradian). It is manifest when measuring angles, angular speed or angular acceleration, but gets lost when combined with other quantities. For example, we well know that the speed v of a point on a rigidly rotating body is given by $v = \Omega r$, where r is the distance from the axis of rotation and Ω is the rate of rotation (angular speed) of the body. Now $\{r, v, \Omega\}$ separately have SI units $\{\mathrm{m}, \mathrm{m/s}, \mathrm{rad/s}\}$. But when combined into the formula $v = \Omega r$, "rad" mysteriously disappears. What is happening? In this formula, r is the distance along the circumference of the circle *per unit angle*. That is, r should have units m/rad or the like, such as m/deg. But we don't normally express distances in this manner, which requires the "rad" to disappear from the formula. This sloppy handling of the angular unit gives a correct numerical result provided the magnitude of the distance along the perimeter per angular unit is equal to the magnitude of the radius of the circle. This is true if and only if the angular unit is the radian.

In addition to the fundamental dimensions and units listed in Table B.1, in practice we encounter many derived units that have special names, a few of which are given in Table B.2. Other quantities encountered in various physical systems may be expressed in terms of the base units, as shown in Table B.3. Quantities involving thermal effects are presented in Table B.4.

A note about temperature units. The official SI temperature unit is the kelvin, denoted by K, measuring absolute temperature. Often we encounter a closely related unit: degrees Celsius, denoted by °C. The numerical value of temperature in degrees Celsius is smaller than that in kelvin by 273.15. That is, a temperature of 0°C is equal to 273.15 K. The obsolete unit centigrade is essentially identical to Celsius.

For completeness, let's note that the unit of energy *calorie* (denoted by "cal") is defined as the energy required to heat one gram of water one kelvin at a pressure of one atmosphere; 1 cal = 4.184 J. A related unit, the *Calorie*, is equal to 1000 calories; the Calorie is sometimes called the *food calorie*.

B.3 Parameter Values

Table B.5 contains values of a number of parameters relevant to the structure of the atmosphere, oceans and Earth's interior, as well as behavior of waves and flows within these regions.

Appendix C
Kinematics

The fundaments of kinematics include analysis and discussion of:

- C.1: non-inertial frames of reference and virtual forces;
- C.2: material derivative;
- C.3: finite deformation;
- C.4: the stream function; and
- C.5: flow lines and points.

C.1 Non-Inertial Frames of Reference and Virtual Forces

Virtual forces arise when the motions of objects are viewed from a non-inertial coordinate system (or frame of reference), i.e., one which is either accelerating linearly, rotating or both. It is difficult to define linear acceleration, as it is indistinguishable from gravitational acceleration. This duality is the basis of the general theory of relativity. On the other hand, we have a unique and well-defined means of defining a non-rotating coordinate system; it is one in which the distant stars appear fixed.

Let us consider two observers, called S and M. Observer S is using an inertial or stationary coordinate system (with origin S and axes $\mathbf{1}_i^{\mathrm{S}}$) to observe motions, while observer M uses a moving (i.e., translating and rotating) coordinate system (with origin M and axes $\mathbf{1}_i^{\mathrm{M}}$). Both observe a particle located at point P as shown in Figure 2.1. Let's write

$$\mathbf{x}_{SP}(t) = \overrightarrow{\mathrm{SP}}, \quad \mathbf{x}_{SM}(t) = \overrightarrow{\mathrm{SM}} \quad \text{and} \quad \mathbf{x}_{MP}(t) = \overrightarrow{\mathrm{MP}}$$

and note that

$$\mathbf{x}_{SP}(t) = \mathbf{x}_{SM}(t) + \mathbf{x}_{MP}(t)\,.$$

C.1.1 Relative Velocities

Differentiating the equation above with respect to time,[1] we have

$$\frac{\mathrm{d}\mathbf{x}_{SP}}{\mathrm{d}t} = \frac{\mathrm{d}\mathbf{x}_{SM}}{\mathrm{d}t} + \frac{\mathrm{d}\mathbf{x}_{MP}}{\mathrm{d}t}\,.$$

[1] Ignoring relativistic effects, so that S and M use the same clock.

This expression is ambiguous if the observer doing the differentiating is not specified. To remove this ambiguity, we shall define the time derivatives of vectors as measured by the two observers as follows. Consider a vector $\mathbf{b}(t)$ which is observed by both S and M:

$$\mathbf{b}(t) = b_i^S(t)\mathbf{1}_i^S = b_i^M(t)\mathbf{1}_i^M$$

(sum over i, but no sum over S or M). Both observers agree on the functions $b_i^S(t)$ and $b_i^M(t)$. This permits us to define

$$\left.\frac{\mathrm{d}\mathbf{b}}{\mathrm{d}t}\right|_S = \frac{\mathrm{d}b_i^S}{\mathrm{d}t}\mathbf{1}_i^S \qquad \text{and} \qquad \left.\frac{\mathrm{d}\mathbf{b}}{\mathrm{d}t}\right|_M = \frac{\mathrm{d}b_i^M}{\mathrm{d}t}\mathbf{1}_i^M .$$

Also,

$$\left.\frac{\mathrm{d}}{\mathrm{d}t}\right|_S (b_i^M\mathbf{1}_i^M) = \frac{\mathrm{d}b_i^M}{\mathrm{d}t}\mathbf{1}_i^M + b_i^M\left.\frac{\mathrm{d}\mathbf{1}_i^M}{\mathrm{d}t}\right|_S .$$

What is $(\mathrm{d}\mathbf{1}_i^M/\mathrm{d}t)_S$? The unit vectors of M's coordinate system translate and rotate, as observed by S. The translation doesn't affect the orientation of the unit vectors, but the rotation does. So it is sufficient to consider the time rate of change of $\mathbf{1}_i^{\mathrm{M}}$ about an axis through the origin of S's coordinate system. Using the rigid-body formula, $\mathrm{d}\mathbf{x}/\mathrm{d}t = \mathbf{\Omega} \times \mathbf{x}$, we have

$$\left.\frac{\mathrm{d}\mathbf{1}_i^M}{\mathrm{d}t}\right|_S = \mathbf{\Omega} \times \mathbf{1}_i^M ,$$

where $\mathbf{\Omega}$ is the rotation vector of M's system as observed by S.[2] Now

$$\left.\frac{\mathrm{d}\mathbf{b}}{\mathrm{d}t}\right|_S = \left.\frac{\mathrm{d}\mathbf{b}}{\mathrm{d}t}\right|_M + \mathbf{\Omega} \times \mathbf{b}$$

for any vector $\mathbf{b}$. Note that $(\mathrm{d}\mathbf{\Omega}/\mathrm{d}t)_S = (\mathrm{d}\mathbf{\Omega}/\mathrm{d}t)_M$; both observers agree on the time rate of change of rotation of M's coordinate frame, relative to S's.

We may now write

$$\mathbf{v}_S = \mathbf{v}_T + \mathbf{v}_M + \mathbf{\Omega} \times \mathbf{x}_{MP} ,$$

where $\mathbf{v}_S = (\mathrm{d}\mathbf{x}_{SP}/\mathrm{d}t)_S$ is the velocity of P measured by S, $\mathbf{v}_T = (\mathrm{d}\mathbf{x}_{SM}/\mathrm{d}t)_S$ is the translational velocity of M measured by S, $\mathbf{v}_M = (\mathrm{d}\mathbf{x}_{MP}/\mathrm{d}t)_M$ is the velocity of P measured by M and $\mathbf{\Omega} \times \mathbf{x}_{MP}$ is the velocity of P (as measured by S) due to the rotation of M's coordinate system.

[2] This equation is dimensionally consistent provided $\mathbf{\Omega}$ has dimensions of inverse time. But rotation rate has dimensions of angle per unit time; see Appendix B.2.

C.1.2 Relative and Absolute Acceleration

The absolute acceleration $\mathbf{a}_S$ of a particle is defined in the inertial frame of reference, i.e., in the frame labeled S. Using the last two equations in the previous subsection and noting that $(\mathrm{d}\mathbf{x}_{MP}/\mathrm{d}t)_M = \mathbf{v}_M$, we have

$$\begin{aligned}
\mathbf{a}_S &= \left.\frac{\mathrm{d}\mathbf{v}_S}{\mathrm{d}t}\right|_S \\
&= \left.\frac{\mathrm{d}\mathbf{v}_T}{\mathrm{d}t}\right|_S + \left.\frac{\mathrm{d}\mathbf{v}_M}{\mathrm{d}t}\right|_S + \left.\frac{\mathrm{d}\boldsymbol{\Omega}}{\mathrm{d}t}\right|_S \times \mathbf{x}_{MP} + \boldsymbol{\Omega} \times \left.\frac{\mathrm{d}\mathbf{x}_{MP}}{\mathrm{d}t}\right|_S \\
&= \left.\frac{\mathrm{d}\mathbf{v}_T}{\mathrm{d}t}\right|_S + \left.\frac{\mathrm{d}\mathbf{v}_M}{\mathrm{d}t}\right|_M + \boldsymbol{\Omega} \times \mathbf{v}_M + \left.\frac{\mathrm{d}\boldsymbol{\Omega}}{\mathrm{d}t}\right|_S \times \mathbf{x}_{MP} \\
&\quad + \boldsymbol{\Omega} \times \left(\left.\frac{\mathrm{d}\mathbf{x}_{MP}}{\mathrm{d}t}\right|_M + \boldsymbol{\Omega} \times \mathbf{x}_{MP} \right).
\end{aligned}$$

Writing $\mathbf{a}_T = (\mathrm{d}\mathbf{v}_T/\mathrm{d}t)_S$, $\mathbf{a}_M = (\mathrm{d}\mathbf{v}_M/\mathrm{d}t)_M$ and $\dot{\boldsymbol{\Omega}} = (\mathrm{d}\boldsymbol{\Omega}/\mathrm{d}t)_M$, recalling that $\mathbf{v}_M = (\mathrm{d}\mathbf{x}_{MP}/\,\mathrm{d}t)_M$ and dropping the subscripts M and MP, this becomes

$$\mathbf{a}_S = \mathbf{a}_T + \mathbf{a}_M + \boldsymbol{\Omega} \times (\boldsymbol{\Omega} \times \mathbf{x}) + 2\boldsymbol{\Omega} \times \mathbf{v} + \dot{\boldsymbol{\Omega}} \times \mathbf{x}.$$

The term on the left-hand side is the absolute acceleration and the five terms on the right-hand side are the frame, relative, centripetal, Coriolis and Poincaré accelerations, respectively. The frame acceleration is measured by observer S, but the other four terms on the right-hand side are measured by observer M.

C.1.3 Virtual Forces

The per-unit-mass version of Newton's second law[3] is simply $\mathbf{f} = \mathbf{a}$, where $\mathbf{f}$ is the body force per unit mass. Our inertial observer S uses this formula in the form $\mathbf{f}_T = \mathbf{a}_S$, where $\mathbf{f}_T$ is the "true" body force per unit mass acting on particles within a continuous body and $\mathbf{a}_S$ is the true acceleration of those particles. Our moving observer M uses $\mathbf{f}_M = \mathbf{a}_M$, where $\mathbf{a}_M$ is the acceleration of particles as observed by M. Since $\mathbf{a}_S$ and $\mathbf{a}_M$ are quite different, it follows that $\mathbf{f}_T$ and $\mathbf{f}_M$ must be as well. The difference between these forces are *virtual forces* arising from the difference in accelerations; that is,

$$\mathbf{f}_M = \mathbf{f}_T + \mathbf{f}_V,$$

where $\mathbf{f}_V$ is the set of virtual forces per unit mass.

Using the acceleration formula developed in the previous section, we can distinguish four virtual forces:

$$\mathbf{f}_V = \mathbf{f}_T + \mathbf{f}_{\text{cen}} + \mathbf{f}\mathrm{Po},$$

[3] See § 4.1 and § 4.6.

where

$$\mathbf{f}_T = -\mathbf{a}_T$$

is the acceleration force, which results from the translational acceleration of observer M,

$$\mathbf{f}_{\text{cen}} = -\boldsymbol{\Omega} \times (\boldsymbol{\Omega} \times \mathbf{x})$$

(with $\mathbf{x}$ being the position vector according to observer M) is the *centrifugal force*, which results from the rotation of observer M's frame of reference,

$$\mathbf{f}_{\text{cor}} = -2\boldsymbol{\Omega} \times \mathbf{v}$$

is the *Coriolis force*, experienced by a body moving in observer M's rotating coordinate frame, and

$$\mathbf{f}_{\text{Po}} = -\dot{\boldsymbol{\Omega}} \times \mathbf{x}$$

is the *Poincaré force*, due to the angular acceleration of observer M's frame of reference.

The centrifugal force is experienced by observer M while sitting on a merry-go-round. Observer S sees M being accelerated toward the origin (that is, M experiences a centripetal acceleration) and notes that M is subject to a centripetal force. On the other hand, M does not note the acceleration but does need to provide a centripetal force. That is, M must hold onto something in order to avoid being flung outward, along an inertial trajectory. M attributes this tendency to be flung outward to a virtual centrifugal force.

The Coriolis force is experienced by observer M moving about on a merry-go-round. Suppose the merry-go-round is spinning counterclockwise (when viewed from above) and further suppose M is moving counterclockwise, a constant (radial) distance from the axis of the merry-go-round. M is in fact spinning faster than the merry-go-round and experiences an enhanced centripetal acceleration. M needs to provide an extra centripetal force to avoid being flung to the right. M attributes this tendency to be flung to the right (that is, outward) to a virtual Coriolis force. M will experience this same tendency to be flung to the right no matter what the direction of motion. Weather systems are like observer M, with the Earth being our merry-go-round. A low pressure system is rotating faster than Earth and the extra force is provided by pressure increasing with distance from its center. In the northern hemisphere, if you stand facing the wind, the center of a storm will be to your right.

We will use a coordinate system fixed to the Earth, which is in orbital motion around the Sun and is rotating on its axis. The orbital motion involves an acceleration of magnitude $a_T = v^2/R$ where v is the orbital speed, $v = \Omega_S R$, $R \approx 1.5 \times 10^{11}$ m is the distance from Earth to Sun and $\Omega_S = 2\pi/\text{year}$ is Earth's orbital spin rate. Altogether the acceleration of our Earth-fixed coordinate system is about $a_T = 0.006\,\text{m}\cdot\text{s}^{-2}$. This small acceleration will be ignored. Also, Earth's spin rate is very uniform and the Poincaré acceleration term, $\dot{\boldsymbol{\Omega}} \times \mathbf{x}$, is negligibly small. This leaves us with

$$\mathbf{f}_V = -\boldsymbol{\Omega} \times (\boldsymbol{\Omega} \times \mathbf{x}) - 2\,\boldsymbol{\Omega} \times \mathbf{v}$$

for fluid bodies.

C.2 Material Derivative and Inertial Term

As noted in the introduction to Part II, there are two alternate ways to observe the displacement of a body: Lagrangian and Eulerian. The main difference between these two views of reality is in the treatment of time derivatives. To illustrate and quantify this difference, consider a scalar function, f, of position and time. This may be represented as either

$$f = f_L(\mathbf{X},t) \qquad \text{or} \qquad f = f_E(\mathbf{x},t),$$

where $\mathbf{X}$ and $\mathbf{x}$ are the initial and current positions of a particle of the body, respectively. The first formula is the Lagrangian description and the second is the Eulerian. We want to relate the time rates of change as experienced by two observers one moving with the particle (with $\mathbf{X}$ held constant) and one stationary (with $\mathbf{x}$ held constant). These two rates of change will be expressed as follows

$$\left.\frac{\partial f_L}{\partial t}\right|_{\mathbf{X}} = \frac{\mathrm{D}f}{\mathrm{D}t} \qquad \text{and} \qquad \left.\frac{\partial f_E}{\partial t}\right|_{\mathbf{x}} = \frac{\partial f}{\partial t}.$$

The current position and velocity of a particle are given in terms of its initial position by

$$\mathbf{x} = \mathbf{x}_L(\mathbf{X},t) \qquad \text{and} \qquad \mathbf{v} = \frac{\mathrm{D}\mathbf{x}}{\mathrm{D}t} = \left.\frac{\partial \mathbf{x}_L}{\partial t}\right|_{\mathbf{X}}.$$

Using the Eulerian description of f,

$$\frac{\mathrm{D}f}{\mathrm{D}t} = \frac{\mathrm{D}}{\mathrm{D}t} f_E\left(\mathbf{x}_L(\mathbf{X},t),t\right).$$

By the chain rule of differentiation and noting that $\partial \mathbf{x}_L/\partial t|_{\mathbf{X}} = \mathbf{v}$, this is

$$\frac{\mathrm{D}f}{\mathrm{D}t} = \left.\frac{\partial f_E}{\partial t}\right|_{\mathbf{x}_L} + \left.\frac{\partial f_E}{\partial \mathbf{x}_L}\right|_t \left.\frac{\partial \mathbf{x}_L}{\partial t}\right|_{\mathbf{X}} \qquad \text{or simply} \qquad \frac{\mathrm{D}f}{\mathrm{D}t} = \left(\left.\frac{\partial}{\partial t}\right|_{\mathbf{x}} + \mathbf{v}\left.\frac{\partial}{\partial \mathbf{x}}\right|_t\right) f.$$

In the second version we have dropped the subscript E because it has become redundant. In fact, we can omit the function f and define the *material derivative* as

$$\frac{\mathrm{D}}{\mathrm{D}t} \equiv \frac{\partial}{\partial t} + \mathbf{v}\bullet\nabla,$$

where the time derivative is measured by a fixed observer ($\mathbf{x}$ = constant is implied) and $\mathbf{v}$ is the velocity of the particle as measured by that observer. If the moving body is deforming, then $\mathbf{v}$ itself is a function of position and time.

The inertial term of the momentum equation is written as

$$\frac{\mathrm{D}\mathbf{v}}{\mathrm{D}t} = \frac{\partial \mathbf{v}}{\partial t} + (\mathbf{v}\bullet\nabla)\mathbf{v},$$

or equivalently[4]

$$\frac{\mathrm{D}\mathbf{v}}{\mathrm{D}t} = \frac{\partial \mathbf{v}}{\partial t} + \nabla\left(\frac{1}{2}v^2\right) + \boldsymbol{\omega} \times \mathbf{v},$$

where $v^2 = \mathbf{v}\bullet\mathbf{v}$ and $\boldsymbol{\omega} = \nabla \times \mathbf{v}$ is the vorticity in the rotating frame.

C.2.1 Material Derivative and Inertial Term in Cylindrical Coordinates

Consider cylindrical coordinates $\{\varpi,\phi,z\}$ as defined in Appendix A.3.4. Writing the velocity as $\mathbf{v} = v_\varpi\,\mathbf{1}_\varpi + v_\phi\,\mathbf{1}_\phi + v_z\mathbf{1}_z$ and using $\partial\{\mathbf{1}_\varpi,\mathbf{1}_\phi\}/\partial\phi = \{\mathbf{1}_\phi,-\mathbf{1}_\varpi\}$, the material derivative in cylindrical coordinates is

$$\frac{\mathrm{D}}{\mathrm{D}t} = \frac{\partial}{\partial t} + v_\varpi\frac{\partial}{\partial\varpi} + v_\phi\frac{\partial}{\varpi\,\partial\phi} + v_z\frac{\partial}{\partial z}$$

and the inertial term may be expressed as

$$\frac{\mathrm{D}\mathbf{v}}{\mathrm{D}t} = \left(\frac{\mathrm{D}v_\varpi}{\mathrm{D}t} - \frac{v_\phi^2}{\varpi}\right)\mathbf{1}_\varpi + \left(\frac{\mathrm{D}v_\phi}{\mathrm{D}t} + \frac{v_\varpi v_\phi}{\varpi}\right)\mathbf{1}_\phi + \frac{\mathrm{D}v_z}{\mathrm{D}t}\mathbf{1}_z.$$

C.2.2 Material Derivative and Inertial Term in Spherical Coordinates

Consider spherical coordinates $\{r,\theta,\phi\}$ as defined in Appendix A.3.5. Writing the velocity as $\mathbf{v} = v_r\,\mathbf{1}_r + v_\theta\,\mathbf{1}_\theta + v_\phi\,\mathbf{1}_\phi$ and using $\partial\{\mathbf{1}_r,\mathbf{1}_\theta,\mathbf{1}_\phi\}/\partial\theta = \{\mathbf{1}_\theta,-\mathbf{1}_r,\mathbf{0}\}$ and $\partial\{\mathbf{1}_r,\mathbf{1}_\theta,\mathbf{1}_\phi\}/\partial\phi = \{\sin\theta\,\mathbf{1}_\phi,\cos\theta\,\mathbf{1}_\phi,-\sin\theta\,\mathbf{1}_r - \cos\theta\,\mathbf{1}_\theta\}$, the material derivative in spherical coordinates is

$$\frac{\mathrm{D}}{\mathrm{D}t} = \frac{\partial}{\partial t} + v_r\frac{\partial}{\partial r} + \frac{v_\theta}{r}\frac{\partial}{\partial\theta} + \frac{v_\phi}{r\sin\theta}\frac{\partial}{\partial\phi}$$

and the inertial term may be expressed as

$$\frac{\mathrm{D}\mathbf{v}}{\mathrm{D}t} = \left(\frac{\mathrm{D}v_r}{\mathrm{D}t} - \frac{v_\theta^2 + v_\phi^2}{r}\right)\mathbf{1}_r + \left(\frac{\mathrm{D}v_\theta}{\mathrm{D}t} + \frac{v_r v_\theta - v_\phi^2\cot\theta}{r}\right)\mathbf{1}_\theta + \left(\frac{\mathrm{D}v_\phi}{\mathrm{D}t} + \frac{v_r v_\phi - v_\theta v_\phi\cot\theta}{r}\right)\mathbf{1}_\phi.$$

C.3 Finite Deformation

We have seen in § 3.2 that homogeneous deformation is characterized by

$$\mathbf{x} = \mathbf{x}_\mathrm{O} + \mathbf{X} + \mathbb{K}\bullet\mathbf{X},$$

[4] Using the fifth formula from Appendix A.2.6 with $\mathbf{b} = \mathbf{c} = \mathbf{v}$.

where $\mathbf{x}$ is the position of a general particle P, $\mathbf{x}_O$ is the position of particle O, $\mathbf{X}$ is the initial position of particle P and the elements of the tensor $\mathbb{K}$ are given by

$$k_{ij} = \frac{\partial x_i}{\partial X_j} - \delta_{ij}\,.$$

In § 3.3 this tensor is separated into rotation and strain assuming that the elements of $\mathbb{K}$ are small. This assumption is relaxed in this fundament.

We again seek to express $\mathbb{K}$ as

$$\mathbb{K} = \mathbb{R} + \mathbb{E}\,,$$

where $\mathbb{R}$ and $\mathbb{E}$ are the rotation and strain tensors, respectively. But first, we need to learn to recognize these two players. A body experiences deformation during displacement if the distance between two particles within the body changes in magnitude. The change δ of distance between particles P and O is given by

$$\delta = x_i x_i - X_i X_i \qquad \text{or} \qquad \delta = 2\epsilon_{ij} X_i X_j\,,$$

where

$$\epsilon_{ij} = \frac{1}{2}\left(\frac{\partial x_k}{\partial X_i}\frac{\partial x_k}{\partial X_j} - \delta_{ij}\right)$$

is an element of the *Lagrangian finite strain tensor*.

C.3.1 Finite Rotation

If $\epsilon_{ij} = 0_{ij}$; that is, if

$$\frac{\partial x_k}{\partial X_i}\frac{\partial x_k}{\partial X_j} = \delta_{ij}\,,$$

then $\mathbb{E} = \emptyset$ and $\mathbb{K} = \mathbb{R}$, with

$$\mathbb{R}\bullet\mathbb{R}^{\mathrm{T}} = \mathbb{I} \qquad \text{or} \qquad r_{ki} r_{kj} = \delta_{ij}\,.$$

This states that the inverse of $\mathbb{R}$ is equal to its transpose. This is a set of nine scalar equations, but only six of them provide constraints on the elements of $\mathbb{R}$, leaving three free to describe rotation. There are many ways to represent finite rotation. We shall choose to represent it as rotation through a finite angle ϕ about an axis $\mathbf{1}_\Omega = a_i \mathbf{1}_i$ (with $a_i a_i = 1$). This may be expressed as

$$\mathbb{R} = \mathbb{I} + \sin\phi\,\mathbb{A} + (1 - \cos\phi)\mathbb{S}\,,$$

where

$$\mathbb{A} = \begin{bmatrix} 0 & -a_3 & a_2 \\ a_3 & 0 & -a_1 \\ -a_2 & a_1 & 0 \end{bmatrix}$$

and

$$\mathbb{S} = \begin{bmatrix} a_1^2 - 1 & a_1 a_2 & a_1 a_3 \\ a_1 a_2 & a_2^2 - 1 & a_2 a_3 \\ a_1 a_3 & a_2 a_3 & a_3^2 - 1 \end{bmatrix}.$$

It is straightforward to verify that

- if $\phi \ll 1$, $\mathbb{R} \approx \mathbb{I} + \phi\mathbb{A}$;[5]
-

$$\begin{aligned} \mathbf{x} = {} & [X_1 + \sin\phi\,(a_2 X_3 - a_3 X_2) + C(a_i X_i - X_1)a_1]\,\mathbf{1}_1 \\ & + [X_2 + \sin\phi\,(a_3 X_1 - a_1 X_3) + C(a_i X_i - X_2)a_2]\,\mathbf{1}_2 \\ & + [X_3 + \sin\phi\,(a_1 X_2 - a_2 X_1) + C(a_i X_i - X_3)a_3]\,\mathbf{1}_3\,, \end{aligned}$$

  where $C = 1 - \cos\phi$;
- the determinant of $\mathbb{R}$ equals unity; and
- the transpose of $\mathbb{R}$ is equal to the inverse.

Rotation Is Not a Vector

Although rotation is determined by specifying three pieces of information and is often represented as a vector, finite rotation is not a vector, because it does not obey the commutative law (it is *non-Abelian*). That is, if body is subject to two finite rotations, its final orientation depends on the order of the rotations. To illustrate this consider two rotations, labeled 1 and 2, each with $\phi = \pi/2$, so that $\mathbb{R}_k = \mathbb{I} + \mathbb{A}_k + \mathbb{S}_k$ for k = 1 and 2. The first rotation is about the x_1 axis ($a_1 = 1, a_2 = a_3 = 0$) and the second is about the x_2 axis ($a_2 = 1, a_1 = a_3 = 0$). The rotation tensors for these are

$$\mathbb{R}_1 = \begin{bmatrix} 1 & 0 & 0 \\ 0 & 0 & -1 \\ 0 & 1 & 0 \end{bmatrix} \quad \text{and} \quad \mathbb{R}_2 = \begin{bmatrix} 0 & 0 & 1 \\ 0 & 1 & 0 \\ -1 & 0 & 0 \end{bmatrix}.$$

Suppose that these rotations act on a translucent cube having a plus, minus and a circle on the faces having normals initially oriented in the positive coordinate directions, as illustrated in Figure C.1. The result of rotations about the 1 and 2 axes, in turn, are shown on the right of the upper row and the result of rotations about the 2 and 1 axes, in turn, are shown on the right of the lower row. The orientations are clearly different. This exercise is readily visualized by taking any book and rotating it as described. The non-commutative property of finite rotation is the basis of the intriguing and challenging Rubik's cube.

[5] As given in § 3.3.

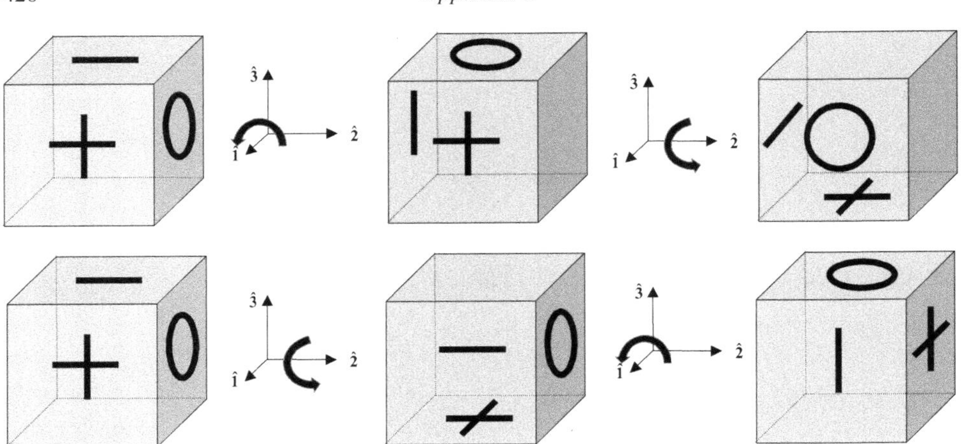

Figure C.1 A translucent cube having a plus, minus and a circle on the forward faces is rotated by $\pi/2$ radians successively about the 1 and 2 axes (top row) and the 2 and 1 axes (bottom row). The final orientations differ.

This result arises due to the nonlinear tensor $\mathbb{S}$. If this factor is ignored (as in § 3.3), then rotation does behave as a vector.

C.3.2 Finite Strain

Strain is usually represented by a symmetric tensor having six free elements. There is a fairly large number of ways to represent finite strain. Of course, we want a form that reduces to the linear version presented in § 3.3. Perhaps the most direct is to use the Lagrangian finite strain tensor introduced earlier. In terms of the local displacement $\mathbf{u} = \mathbf{x} - \mathbf{X}$, the elements of the Lagrangian strain tensor are

$$\epsilon_{ij} = \frac{1}{2}\left(\frac{\partial u_i}{\partial X_j} + \frac{\partial u_j}{\partial X_i} + \frac{\partial u_k}{\partial X_i}\frac{\partial u_k}{\partial X_j}\right).$$

If the deformation is small, the nonlinear term may be neglected and we have the Lagrangian infinitesimal strain tensor, denoted by $\mathbb{E}$, with elements given by

$$\epsilon_{ij} = \frac{1}{2}\left(\frac{\partial u_i}{\partial X_j} + \frac{\partial u_j}{\partial X_i}\right) \qquad \text{or} \qquad \epsilon_{ij} = \frac{1}{2}\left(\frac{\partial x_i}{\partial X_j} + \frac{\partial x_j}{\partial X_i}\right) - \delta_{ij}.$$

C.3.3 Volume Change

Consider a parcel of a deforming body having initial volume V_0 and volume V after deformation. The relative change of volume is given by the determinant of the deformation tensor:[6]

$$\frac{V}{V_0} = \det[\mathbb{K}] = \det\left[\frac{\partial x_i}{\partial X_j}\right] = \det\left[\delta_{ij} + \frac{\partial u_i}{\partial X_j}\right].$$

If the displacement is small ($|\partial u_i/\partial X_j| \ll 1$), this may be expressed as

$$\frac{V}{V_0} \approx 1 + \frac{\partial u_i}{\partial X_i} = 1 + \nabla_{\mathbf{X}} \bullet \mathbf{u}.$$

C.4 Stream Function

The stream function is a way to 'solve' the continuity equation. The term $\partial\rho/\partial t$ in the fluid continuity equation[7] is important in the study of wave motions that involve the compressibility of the medium. In many other situations, the anelastic approximation is valid and this term is negligibly small.[8] In this case the continuity equation simplifies to

$$\nabla \bullet (\rho \mathbf{v}) = 0.$$

This equation is a single scalar constraint on the vector function $\mathbf{v}$. This equation can be satisfied by introduction of a stream-function vector, denoted by $\boldsymbol{\Psi}$, writing

$$\rho \mathbf{v} = \nabla \times \boldsymbol{\Psi}.$$

We appear not to have gained much by this move because we have replaced one vector function, $\mathbf{v}$, by a second, $\boldsymbol{\Psi}$, each of which involves three scalar functions of position and time. However, we are free to specify a secondary constraint on $\boldsymbol{\Psi}$ since $\nabla \times \nabla s = 0$, where s is an arbitrary scalar function that we are free to specify. One possible constraint is

$$\nabla \bullet \boldsymbol{\Psi} = 0.$$

The stream function may be defined with a sign opposite to that used here; whenever using a stream function we need to be mindful of the relation between derivatives of $\boldsymbol{\Psi}$ and components of $\mathbf{v}$.

C.4.1 Two-Dimensional Flow

The advantage of using a stream function is readily apparent if the flow is two-dimensional, with $\mathbf{v}(x,y,z,t) = u(x,y,t)\mathbf{1}_x + v(x,y,t)\mathbf{1}_y$. In this case, the continuity equation simplifies to

$$\frac{\partial(\rho u)}{\partial x} + \frac{\partial(\rho v)}{\partial y} = 0.$$

[6] The determinant, defined in Appendix A.1.4, is one of the three scalar invariants. $\det\left[\partial x_i/\partial X_j\right]$ is the Jacobian determinant of the transformation of $\mathbf{X}$ to $\mathbf{x}$; this is often called "the Jacobian".
[7] Presented in § 4.7.
[8] See § 3.4.1.

This equation is automatically satisfied by the introduction of the scalar stream function, $\boldsymbol{\Psi} = \Psi \mathbf{1}_z$, so that

$$\rho \mathbf{v} = \nabla \times (\Psi \mathbf{1}_z) = -\mathbf{1}_z \times \nabla \Psi .$$

In component form

$$\rho u = \frac{\partial \Psi}{\partial y} \qquad \text{and} \qquad \rho v = -\frac{\partial \Psi}{\partial x} .$$

Use of the stream function removes the continuity equation from the set of governing equations and replaces two dependent variables, u and v, by one, Ψ.

The two-dimensional stream function, $\Psi(x,y,t)$, has two interesting physical interpretations:

- at any instant of time, the flow is along lines of constant streamline; see Appendix C.5.1; and
- if you stand facing the direction of increasing Ψ, the wind blows in your left ear and out your right.

Non-Divergent Two-Dimensional Potential Flow

If a two-dimensional potential flow is also non-divergent, then both velocity potential and stream function satisfy *Laplace's equation*:

$$\nabla^2 \phi = \nabla^2 \Psi = 0,$$

with

$$u = \frac{\partial \phi}{\partial x} = \frac{\partial \Psi}{\partial y} \qquad \text{and} \qquad v = \frac{\partial \phi}{\partial y} = -\frac{\partial \Psi}{\partial x} .$$

These equations for u and v are identical to the *Cauchy-Riemann equations*. This implies that ϕ and Ψ are *conjugate harmonic functions* of the complex variable $z = x + \mathrm{i}y$, where $\mathrm{i} = \sqrt{-1}$; that is, $\phi(x,y) + \mathrm{i}\Psi(x,y) = f(z)$. Further this means that lines of constant ϕ intersect lines of constant Ψ at right angles.

C.4.2 Other Forms

The stream function Ψ has dimensions of mass/length·time (SI units kg·m^{-1}·s^{-1}). Alternate forms of the stream function may be used. For example, the density may be expressed in terms of the scale height, in which case the dimensions and units of the stream function are the same as velocity; see § 8.4.1. If the fluid has constant density, it is common practice to use a stream function having units of m^2·s^{-1}, as is done, for example, in § 15.3, § 27.1 and Appendix G.1.

Stokes Stream Function

If the flow is axisymmetric; that is, independent of the azimuthal angle ϕ when expressed in cylindrical (ϖ,ϕ,z) or spherical (r,θ,ϕ) coordinates, the stream function vector is

$\boldsymbol{\Psi} = \Psi \mathbf{1}_\phi$ and the velocity vector is

$$\rho \mathbf{v} = \nabla \times (\Psi \mathbf{1}_\phi) = \Psi \nabla \times \mathbf{1}_\phi - \mathbf{1}_\phi \times \nabla \Psi$$

or

$$\rho \mathbf{v} = -\frac{\partial \Psi}{\varpi \, \partial z} \mathbf{1}_\varpi + \frac{\partial \Psi}{\varpi \, \partial \varpi} \mathbf{1}_z$$

in cylindrical coordinates and

$$\rho \mathbf{v} = \frac{\partial \Psi}{r^2 \sin\theta \, \partial \theta} \mathbf{1}_r - \frac{\partial \Psi}{r \sin\theta \, \partial r} \mathbf{1}_\theta$$

in spherical coordinates. These forms satisfy the continuity equation $\nabla \bullet (\rho \mathbf{v}) = 0$ in cylindrical and spherical coordinates; see Appendix A.3.4 and Appendix A.3.5.

C.5 Flow Lines and Points

Fluid flows may be characterized by various types of flow lines, many of which are described in the following.

C.5.1 Streamlines

Consider a general velocity field $\mathbf{v}(\mathbf{x},t)$. At an instant of time $t = t_0$, a *streamline* is a curve whose tangent at position $\mathbf{x} = \mathbf{x}_0$ has the same direction as $\mathbf{v}(\mathbf{x}_0, t_0)$. One and only one streamline passes through each point at any instant. The streamlines are found by integration of

$$\frac{\mathrm{d}x_1}{v_1(\mathbf{x}, t_0)} = \frac{\mathrm{d}x_2}{v_2(\mathbf{x}, t_0)} = \frac{\mathrm{d}x_3}{v_3(\mathbf{x}, t_0)},$$

where $\mathbf{v} = v_i \mathbf{1}_i$. We may write this parametrically as

$$\frac{\mathrm{d}x_i}{\mathrm{d}s} = \frac{v_i}{v},$$

where $v = \|\mathbf{v}\|$ is the speed and s is the arc length. Solutions of this equation are of the form

$$\mathbf{x} = \mathbf{f}(s; \mathbf{x}_0, t_0) \qquad \text{or} \qquad x_i = f_i(s; \mathbf{x}_0, t_0).$$

Commonly $\mathbf{f}$ is chosen such that

$$\mathbf{x}_0 = \mathbf{f}(0; \mathbf{x}_0, t_0).$$

This represents a family of streamlines. At any instant of time, this is a two-parameter family of streamlines.[9] Time is a third parameter in this representation. The streamlines are everywhere parallel to the flow at a specific instant of time.

[9] One of x_{i0} is redundant; any value of $\mathbf{x}_0$ lying on a given streamline provides a means of describing that particular line.

In two dimensions, with $\{x_1, x_2, x_3\} \to \{x, 0, z\}$ and $\{v_1, v_2, v_3\} \to \{u, 0, w\}$, the streamline may be described by

$$\frac{dz}{dx} = \frac{w(x,z,t_0)}{u(x,z,t_0)}.$$

This integrates to

$$\Psi(x,z,t_0) = \Psi_0,$$

where Ψ_0 is a constant of integration, giving, at any instant t_0, a one-parameter family of curves. The function Ψ is the *stream function*, which has a constant value on a specific streamline. To see this, take the differential of Ψ and eliminate dz using $dz = (w/u)dx$:

$$\begin{aligned} d\Psi &= \frac{\partial \Psi}{\partial x}dx + \frac{\partial \Psi}{\partial z}dz \\ &= \left(\frac{\partial \Psi}{\partial x} + \frac{\partial \Psi}{\partial z}\frac{w}{u}\right)dx = 0. \end{aligned}$$

This is satisfied by the set of equations defining the stream function presented in Appendix C.4:

$$u = \frac{\partial \Psi}{\partial z} \quad \text{and} \quad w = -\frac{\partial \Psi}{\partial x}.$$

C.5.2 Particle Trajectories

A *particle trajectory* or *path line* is the trace produced by a particle moving with the flow. If the flow is steady, a particle trajectory follows a streamline, but if the flow is unsteady, these are distinct. Particle trajectories are found by integrating

$$\frac{dx_i}{dt} = v_i(\mathbf{x}, t).$$

The solution may be expressed as

$$\mathbf{x} = \mathbf{f}(t; \mathbf{x}_0, t_0),$$

where $\mathbf{f}$ satisfies $\mathbf{x}_0 = \mathbf{f}(t_0; \mathbf{x}_0, t_0)$. As with a streamline, three of the set $(\mathbf{x}_0, t_0)$ are independent parameters. (We could set $t_0 = 0$.)

On occasion we may observe a *streak line*, which is the locus of all particles that have passed through a specified point in space; e.g., the plume of a cigarette, a contrail, a comet tail. etc. A streak line is described by the equation for particle trajectories, but with t and $\mathbf{x}_0$ held fixed and with t_0 being a free parameter.

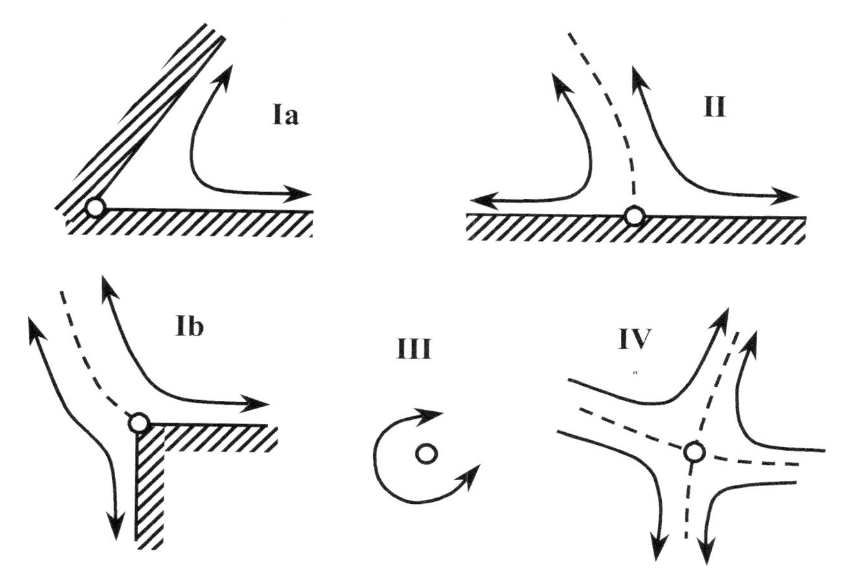

Figure C.2 Schematic of the four types of stagnation points in two-dimensional flow. The open circles represent the stagnation points, crosshatched lines indicate rigid boundaries, arrows indicate possible flow directions and dashed lines are streamlines that intersect the stagnation point.

C.5.3 Stagnation Points

A *stagnation point* is a point in a two- or three-dimensional flow field where the velocity (relative to a frame of reference fixed to a body or the ground) is zero. A stagnation point can be located either on the boundary of the fluid or in the interior. Four types of two-dimensional stagnation points can occur, as depicted schematically in Figure C.2: two on the boundary and two in the interior. In detail:

Type I. A corner point on a boundary is always a stagnation point. If the fluid occupies an angular domain less than π radians (Type Ia), there is no flow separation and flow can be in either direction, as indicated by the double-headed arrow. If the fluid occupies an angular domain greater than π radians (Type Ib), there is flow separation and flow can again be in either direction, as indicated by the double-headed arrow. The dashed stagnation line is a streamline separating the two flows on either side of the stagnation point.[10] If flow is directed into the fluid domain, the stagnation point is a separation point, with the flow on one side being much stronger. (At the corner of a building often one side is windy and the other is calm.)

Type II. A point on a smooth boundary may be a stagnation point, with flow in either direction, as indicated by the double-headed arrows. The dashed line is a streamline

[10] Assuming steady flow.

separating the two flows on either side of the stagnation point. If flow on the stagnation line is directed away from the stagnation point, it is also called a *separation point*.

Type III. An interior point may be a stagnation point with the fluid rotating about the point as if rigid. If the flow is axisymmetric, then a stagnation point in a meridional plane is in fact a stagnation line encircling the axis of symmetry.

Type IV. An interior point may be a stagnation point with the fluid in simple change-of-shape deformation.

Stagnation points are related to the topology of the flow, with the streamlines emanating from these points dividing the (steady) flow into separate domains.

Appendix D
Dynamics

The fundaments of dynamics include discussion and analysis of:

- D.1: viscoelastic behavior;
- D.2: silicate rheology;
- D.3: the constants of elasticity;
- D.4: surface tension;
- D.5: the general conservation law;
- D.6: the Euler and Bernoulli equations;
- D.7: kinetic and internal energies;
- D.8: alternate thin-layer variables;
- D.9: the Proudman–Taylor theorem; and
- D.10: the vectorial form of the vorticity equation.

D.1 Viscoelastic Behavior

There are three principal rheological classifications, with materials being categorized as elastic, fluid or viscoelastic. The first two of these have been briefly considered in §§ 6.1 and 6.2, respectively. The purpose of this fundament is to provide a short survey of viscoelastic behavior. Viscoelasticity encompasses a variety of deformational behaviors and many geophysical materials fit in this broad classification, particularly when placed under hydrostatic pressure which inhibits the formation of open cracks.

Our understanding of the viscoelastic behavior of materials is facilitated by the use of simple mechanical models of elastic and fluid behavior. We begin by introducing in Appendix D.1.1 the two prototypical rheological elements: a spring for elastic behavior and a dashpot for fluid behavior, and show how they may be combined to produce the simplest possible viscoelastic models: the Maxwell and Kelvin-Voigt models. These models are investigated in Appendix D.1.2. In each of these characterizations we imagine a simple deformational situation so that the stress, strain and rate of strain lose their tensorial character and act as scalars.

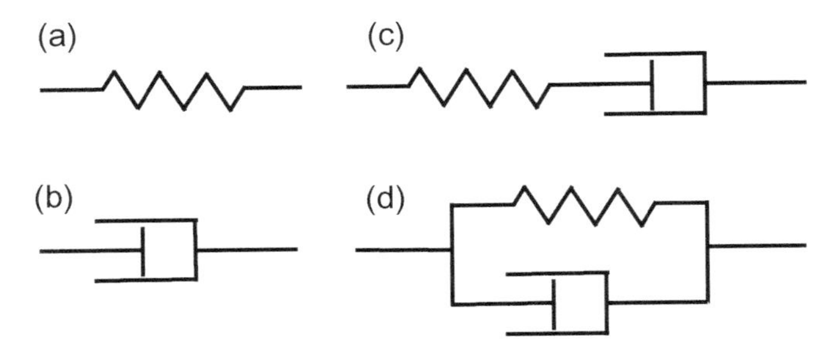

Figure D.1 Schematics of simple rheological models. (a) Spring. (b) Dashpot. (c) Maxwell material. (d) Kelvin–Voigt material.

Table D.1. *Asymptotic behavior of simple visocelastic models.*

	Maxwell	Kelvin
Small time	*Elastic*	*Fluid*
Large time	*Fluid*	*Elastic*

D.1.1 Simple Viscoelastic Models

The basic elements of viscoelastic models are a spring, representing an elastic material satisfying Hooke's law, and a dashpot (or coffee plunger), representing a viscous Newtonian fluid. These elements are represented schematically in parts (a) and (b) of Figure D.1. The rheology of a spring is characterized by an elastic modulus, called the elasticity and denoted by μ, having dimensions of stress, and the rheology of the dashpot is characterized by a viscosity, η, having dimensions of stress times time. Their ratio defines a characteristic time:[1]

$$t_r = \eta/\mu\,.$$

These two basic elements may be combined in series or parallel to make two simple models of viscoelastic materials; in series the elements form a *Maxwell material* and in parallel they form a *Kelvin–Voigt material.* As we shall see in the following section, for short time intervals ($t \ll t_r$) the Maxwell material behaves elastically and the Kevin–Voigt material behaves fluid-like, while for long time intervals the situation is reversed: the Maxwell material is fluid-like and the Kelvin–Voigt material is elastic; see Table D.1.

D.1.2 Simple Viscoelastic Rheology

To investigate the Maxwell and Kelvin–Voigt models, we begin with the stress–strain behavior of the individual elements. In this simple example, stress and strain are replaced

[1] Introduced in § 29.1.

by force, denoted by τ, and displacement, denoted by ϵ. The spring has a linear relation between force and displacement, while the dashpot has a linear relation between force and speed, which is the time rate of change of displacement:

$$\tau_S = 2\mu\epsilon_S \qquad \text{and} \qquad \tau_D = 2\eta\dot{\epsilon}_D\,.$$

Here a subscript S denotes spring and D denotes dashpot. The 2s occur here by convention, in order to avoid the occurrence of factors of 1/2 when considering more complicated situations; e.g., see § 3.3.3 and § 3.4.3.

Linear Maxwell Material

To construct a mathematical model of the Maxwell material, we note that both elements experience the same force, while the total displacement is the sum of the displacements of the two elements:

$$\tau = \tau_S = \tau_D\,, \qquad \epsilon = \epsilon_S + \epsilon_D \qquad \text{and} \qquad \dot{\epsilon} = \dot{\epsilon}_S + \dot{\epsilon}_D\,.$$

Eliminating the terms on the right-hand side of this last equation using $\dot{\tau}_S = 2\mu\dot{\epsilon}_S$ and $\tau_D = 2\eta\dot{\epsilon}_D$, we obtain a linear, first-order ordinary differential equation relating the time histories of the force and displacement for the Maxwell material:

$$\dot{\epsilon} = \frac{\dot{\tau}}{2\mu} + \frac{\tau}{2\eta}\,.$$

The nature of a Maxwell material can be illustrated by considering how the force varies if the material is subject to a sudden change of displacement. Suppose that the displacement is given by

$$\epsilon(t) = \epsilon_0 H(t)\,,$$

where ϵ_0 is a specified constant force and H is the *Heaviside step function*,[2] defined by

$$H(t) = \begin{cases} 0 & \text{if } t < 0 \\ 1/2 & \text{if } t = 0 \\ 1 & \text{if } 0 < t \end{cases}$$

The force consists of an immediate elastic response, followed by a relaxation to a force-free state; for $t > 0$

$$\tau(t) = 2\mu\epsilon_0\left(1 - e^{-t/t_r}\right).$$

On short timescales ($t \ll t_r$), a Maxwell material behaves elastically and on long timescales ($t_r \ll t$), it behaves like a fluid. Earth's silicate mantle exhibits such behavior; it can transmit elastic waves but also flows on geological timescales. However, the rheology of mantle

[2] The derivative of the Heaviside function is the *Dirac delta function*, $\delta(t)$, having the following properties: $\delta(t) = 0$ if $t \neq 0$ and $\int \delta(t)\mathrm{d}t = 1$ if the interval of integration spans the point $t = 0$.

material is more complex than this simple model, with deformation being a nonlinear function of stress. This behavior is simulated by the nonlinear Maxwell model described in the following subsection.

Nonlinear Maxwell Model

A nonlinear Maxwell model has a linear spring but a nonlinear dashpot, with the linear formula $\tau_D = 2\eta\dot{\epsilon}_D$ replaced by $(\tau_D/\mu)^n = 2\eta\dot{\epsilon}_D/\mu$, where n is a positive constant. With this modification, the Maxwell model becomes

$$\dot{\epsilon} = \frac{\dot{\tau}}{2\mu} + \frac{\mu}{2\eta}\left(\frac{\tau}{\mu}\right)^n .$$

The long-term behavior of the nonlinear Maxwell model is given by

$$\dot{\epsilon} = \frac{\mu}{2\eta}\left(\frac{\tau}{\mu}\right)^n .$$

This is a simplified version of Glen's law developed in § 6.2.3.

Kelvin–Voigt Material

In a Kelvin–Voigt material both rheological elements have the same displacement, while they share the force:

$$\tau = \tau_S + \tau_D , \qquad \epsilon = \epsilon_S = \epsilon_D \qquad \text{and} \qquad \dot{\epsilon} = \dot{\epsilon}_S = \dot{\epsilon}_D .$$

Again using the force-displacement relations for the elements we have the equation governing a Kelvin–Voigt material:

$$t_r\dot{\epsilon} + \epsilon = \tau/2\mu .$$

The nature of a Kelvin–Voigt material can be illustrated by considering how the displacement varies if the material is subject to a constant force, suddenly applied at $t = 0$, say: $\tau = \tau_0 H(t)$, where τ_0 is a specified constant force. The displacement is

$$\epsilon = \frac{\tau_0}{2\mu}\left(1 - e^{-t/t_r}\right).$$

The Kelvin–Voigt material behaves rigidly for very short times as the dashpot resists the applied force. As the dashpot slowly deforms, resistance to the applied force is taken up by the spring and the displacement relaxes to a constant asymptotic value.

D.1.3 Discussion

In addition to the simple models presented in the previous sections, viscoelastic materials can exhibit more complex behaviors, such as *work hardening*. This behavior is a slight generalization of elastic-plastic; as the material deforms ductilely, its internal structure is modified such that resistance to deformation increases. This is a permanent change; once the force has been released, the material is stronger.[3] This is a more realistic model of rigid, ductile materials, such as metals. The material first gets stronger due to strain hardening, but then cracks form, leading to failure.

Beyond the simple models investigated here, a number of more complicated models have been introduced to simulate various viscoelastic behaviors, including the *Bingham model* and the *Burgers model*.

D.2 Silicate Rheology

Silicates when molten deform as a Newtonian fluid and when solid deform plastically, through a process called sub-solidus *creep*.[4] In this section we investigate their sub-solidus rheology, focusing on viscosity as a thermodynamic function of pressure, temperature and composition: $\eta(p, T, \boldsymbol{\xi})$, where $\boldsymbol{\xi}$ represents the constituents; see Appendix E.5. This study is a complement to § 7.2.5, in which we investigate the variation of viscosity with position with Earth's mantle: $\eta(\mathbf{x})$.

Before delving into the thermodynamics of silicate rheology, it will be helpful to get oriented to silicate phase behavior and how this relates to deformation within the mantle.

D.2.1 Silicate Phase Behavior

Silicates are complex assemblages of minerals characterized by the presence of the orthosilicate ion SiO_4^{4-}. In this summary, we will consider a simple "cartoon" silicate consisting of two chemical components that has a simple eutectic equilibrium phase diagram; see Appendix E.10.2. The mass fractions of the two components are ξ and $1-\xi$. There are five temperatures that govern the phase and dynamic behavior of the system, as explained in the following summary:

- if $T_l < T$, the silicate is liquid[5] and has Newtonian rheology;
- if $T_a < T < T_l$, the silicate is primarily liquid, with isolated crystals in suspension; flow does not require crystal deformation and is still Newtonian;
- if $T_b < T < T_a$, the silicate consists of interconnected liquid and solid phases; flow requires crystal deformation and is non-Newtonian. (However, the interconnected liquid can percolate and collect in magma chambers);
- if $T_c < T < T_b$, the silicate consists of solid with isolated pockets of liquid that are dynamically impotent (Flow occurs entirely via creep of the solid with no percolation);
- if $T_d < T < T_c$ the silicate is entirely solid (Flow occurs entirely via sub-solidus creep); and
- if $T < T_d$, the silicate behave as an elastic solid; flow occurs via cracking (that is, earthquakes), rather than creep.

[3] Has a greater yield strength.
[4] Actually, several processes including dislocation creep and diffusion creep. In simple models, these processes are all modeled as power-law creep.
[5] A molten silicate is commonly called a *melt*.

In naturally-occurring silicates, T_b is very close to the eutectic temperature, T_e, and the distinction between these two temperatures can be ignored. These regimes of flow are discussed further in Appendix D.2.4.

To briefly summarize, the rheological behavior of mantle silicates is a combination of elastic and fluid behavior. Like many other materials, at sufficiently high temperature silicates are liquid and behave as Newtonian fluids and at sufficiently low temperature are solid and exhibit linear elastic behavior. But in between there are a number of complications depending on the temperature, pressure and composition. These complications, exemplified by the behavior of the viscosity: $\eta(p,T,\boldsymbol{\xi})$, are considered below, beginning in the following subsection with an analysis and discussion of the temperature dependence of viscosity.

D.2.2 Temperature Dependence

Viscous deformation is a thermally activated reaction process and the variation of η with T may be quantified by the *Arrhenius equation*,[6] which may be expressed as

$$\eta(T) = \eta_\infty e^{\beta T_m/T} = \eta_\infty e^{\beta/\upsilon_H}\,,$$

where η_∞ is the value of the viscosity at very high temperature, T_m is the melting temperature, $\upsilon_H = T/T_m$ is the *homologous temperature* and β is a dimensionless activation parameter having a magnitude considerably greater than unity. For example, Borch and Green (1987) estimate that $\beta = 29$ for olivine. The large value of β tells us that the viscosity of silicates varies strongly with temperature. Specifically, as T decreases, the factor within the exponent increases in magnitude; consequently η increases; cold silicates are sluggish.

The sign in the exponent of an Arrhenius equation for a chemical reaction rate is negative, indicating that the reaction rate increases with increasing temperature. Here the sign is positive, indicating that the viscosity decreases with increasing temperature.

D.2.3 Pressure Dependence

In developing the Arrhenius equation in the previous section, we ignored pressure effects; however, both the activation parameter, β, and the eutectic temperature, T_e, are functions of pressure. As it happens, the variation of β with p is weak and this parameter may be considered constant. It follows that a reasonable approximation is to express the viscosity as

$$\eta(p,T) = \eta_\infty e^{\beta/\upsilon_H(p)}\,, \qquad \text{with} \qquad \upsilon_H(p) = T/T_m(p)\,.$$

[6] Virtually all solids obey this equation – or a similar one; they become soft as the temperature increases to the melting temperature. A notable exception to this general rule is the most familiar solid, water ice, which is brittle right up to the melting point. Alternate Arrhenius-type equations for silicate melts are described in Ni et al. (2015).

Rheology of the mantle is governed in large part by the variation of the homologous temperature with pressure: $\mathrm{d}\upsilon_H/\mathrm{d}p$. To determine this variation, let's recall that the adiabatic temperature gradient is[7]

$$\frac{\mathrm{d}T}{\mathrm{d}p} = \frac{\alpha T}{\rho c_p} = \frac{\gamma_G T}{K},$$

where $\gamma_G = \rho c_p/\alpha K$,[8] while the variation of the eutectic temperature with pressure is quantified by the Clausius–Clapeyron equation[9]

$$\frac{\mathrm{d}T_e}{\mathrm{d}p} = \beta_e T_e, \qquad \text{where} \qquad \beta_e = \frac{\Delta v}{L}$$

is in effect the compressibility of phase change, L is the latent heat of melting and Δv is the associated change of specific volume. Using these we have

$$\frac{\mathrm{d}\upsilon_H}{\mathrm{d}p} = (\gamma_G \beta_s - \beta_e)\frac{T}{T_e},$$

where $\beta_s = 1/K$ is the adiabatic compressibility.

The sign and magnitude of $\mathrm{d}\upsilon_H/\mathrm{d}p$ are difficult to quantify, but the general consensus in the literature is that $\mathrm{d}\upsilon_H/\mathrm{d}p < 0$. This may be reasoned as follows. The compressibility of a mixture of two phases that do not undergo a change of phase is volume-weighted average of the compressibilities of the two phases, with each phase being compressed without re-arrangement of its crystalline structure. The ability of a crystalline lattice to deform in response to an increase in pressure decreases as the pressure becomes greater. This is reflected in a roughly linear increase in K with p. During a change of phase, the crystalline structure of one phase is transformed to that of the other, with the more dense phase being preferred at higher pressure, because it has a lower value of Gibbs function; see Appendix E.10. It is reasonable to suppose that the decrease in specific volume achieved by a re-arrangement of the crystal structure is greater than the amount provided by deformation of the structure. That is, β_e is greater than β_s, and sufficiently greater such that $\gamma_G \beta_s < \beta_e$.

If $\mathrm{d}\upsilon_H/\mathrm{d}p < 0$, then as depth within the mantle increases, υ_H decreases, and the exponential term in the Arrhenius equation increases. That is, mantle material becomes more viscous with depth.

D.2.4 Phase and Compositional Dependence

We now turn our attention to the variation of silicate rheology with phase and composition.

[7] From § 5.5.
[8] See Appendix E.5.1.
[9] See Appendix E.10.1.

Phase Dependence

Let's consider a silicate melt initially at a temperature above the liquidus. As it is cooled (at a rate sufficiently slow that the silicate does not transition to a glassy state), solid crystals form within the melt once the temperature drops below the liquidus. If these crystals remain in suspension within the liquid,[10] they act to impede deformation of the melt and the effective viscosity, η_e increases, as quantified by the *Einstein–Roscoe equation*:

$$\eta_\Phi = \eta \left(\frac{\Phi_{max}}{\Phi_{max} - \Phi} \right)^{5/2},$$

where η is the viscosity of the melt, Φ is the volume fraction of crystals and Φ_{max} is the value of Φ at which the crystals begin to touch and coalesce; this occurs when $T = T_a$.

Once $T < T_a$ and $\Phi > \Phi_{max}$, the rheological character of the silicate changes because flow now involves deformation of the crystals, which requires significantly greater stress. For $T_b < T < T_a$, the silicate acts as a dual continuum consisting of interconnected liquid and solid phases. Shear flow is dominated by the resistance provided by deformation (via sub-solidus creep) of the solid phase. A novel feature of this regime is that each phase is now compressible, even though the assemblage is incompressible. For example, the liquid phase can be divergent (flowing upward, say) and the solid phase convergent (settling downward), in a process called *compaction*. This process permits the assembly of liquid silicates in magma chambers beneath mid-ocean ridges and volcanoes.

Compositional Dependence

Silicate rheology is very sensitive to the presence of volatile components, particularly H_2O. These components are *incompatible*, tending to remain in the melt as silicates crystals form and grow. It follows that their influence on the liquid phase increases as the temperature is cooled toward the eutectic. Also, volatile components permit the liquid phase to wet the crystal surfaces, so that the liquid phase remains interconnected even when the mass fraction of liquid phase is very small. Volatile components also tend to reduce the density of the liquid, making it buoyant and driving compaction and fractionation.

Another (complicating and under-appreciated) aspect of mantle rheology is the coexistence of two or more sub-solidus phases having differing rheologies. For example, the lower mantle is believed to be a mixture of 20–30% of a weaker (that is, more easily deformed) phase (composed of MgO and FeO) and a 70–80% of a stronger phase ($MgSiO_3$ and $FeSiO_3$). The overall rheology of such a mixture depends on the geometry of the two phases; if the stronger phase collects in isolated pockets, strain becomes localized within the interconnected phase and the assemblage becomes more easily deformed, in a process akin to shear thinning.

[10] A reasonable assumption, given the large value of η.

D.3 Parameters of Elasticity

The rheological properties of homogeneous elastic materials are characterized by two parameters, with differing pairs employed in various fields of science and engineering. The pair used most frequently in geophysics are the bulk modulus K and the shear modulus μ. Engineers prefer to use Young's modulus, denoted by E, and Poisson's ratio, denoted by ν, while continuum mechanicians prefer the Lamé parameters λ and μ. These five parameters are related as follows:

$$E = \frac{9K\mu}{3K+\mu}, \quad \nu = \frac{3K-2\mu}{2(3K+\mu)} \quad \text{and} \quad \lambda = K - \frac{2}{3}\mu.$$

Note that $E/(1+\nu) = 2\mu$ and $E/(3-6\nu) = K$.

Ignoring the reference-state pressure,[11] the stress tensor, $\mathbb{S}$, is related to the strain tensor, $\mathbb{E}$, by

$$\mathbb{S} = \left(K - \tfrac{2}{3}\mu\right)(\boldsymbol{\nabla}_{\mathbf{X}} \bullet \mathbf{u})\mathbb{I} + 2\mu\mathbb{E},$$

where $\mathbb{I}$ is the identity tensor and $\boldsymbol{\nabla}_{\mathbf{X}} \bullet \mathbf{u}$ is the change of volume,[12] or equivalently, with $\mathbb{E} = (\boldsymbol{\nabla}_{\mathbf{X}} \bullet \mathbf{u})\mathbb{I}/3 + \mathbb{E}'$ and $\mathbb{S} = -p\mathbb{I} + \mathbb{S}'$, where a prime denotes the deviatoric portion,

$$p = -K(\boldsymbol{\nabla}_{\mathbf{X}} \bullet \mathbf{u}) \qquad \text{and} \qquad \mathbb{S}' = 2\mu\mathbb{E}'.$$

It is readily seen that the bulk modulus modulates the change of volume induced by a change of isotropic stress (that is, a change of pressure) and the shear modulus modulates changes of shape induced by deviatoric stresses.

The stress–strain relation using E and ν is written as

$$\mathbb{S} = \frac{E\nu}{(1-2\nu)(1+\nu)}(\boldsymbol{\nabla}_{\mathbf{X}} \bullet \mathbf{u})\mathbb{I} + \frac{E}{1+\nu}\mathbb{E}$$

or equivalently

$$p = -\frac{E}{3(1-2\nu)}(\boldsymbol{\nabla}_{\mathbf{X}} \bullet \mathbf{u}) \qquad \text{and} \qquad \mathbb{S}' = \frac{E}{1+\nu}\mathbb{E}'.$$

In terms of the Lamé constants these are

$$\mathbb{S} = \lambda(\boldsymbol{\nabla}_{\mathbf{X}} \bullet \mathbf{u})\mathbb{I} + 2\mu\mathbb{E}$$

or equivalently

$$p = -\left(\lambda + \tfrac{2}{3}\mu\right)(\boldsymbol{\nabla}_{\mathbf{X}} \bullet \mathbf{u}) \qquad \text{and} \qquad \mathbb{S}' = 2\mu\mathbb{E}'.$$

[11] See § 6.1.

[12] $\boldsymbol{\nabla}_{\mathbf{X}} \bullet \mathbf{u} = (V - V_0)/V_0 = (\rho_0 - \rho)/\rho$; see § 3.3.1.

D.3.1 Poisson's Ratio

Poisson's ratio, ν has a simple geometric meaning. As an elastic material is stretched in the longitudinal direction, it normally shrinks in the transverse direction. Poisson's ratio quantifies this change:

$$\nu = -\frac{d\epsilon_{\text{transverse}}}{d\epsilon_{\text{longitudinal}}}.$$

Thermodynamics constrains ν to the range $-1 \leq \nu \leq 1/2$. If the material is incompressible $\nu = 1/2$. For most materials $\nu > 0$, but some plastics have $\nu < 0$. Cork has ν close to zero, which makes it useful as a bottle stopper.

D.4 Surface Tension

Surface tension arises from the fact that the atoms or molecules of a liquid are attracted to each other.[13] An atom or molecule of a liquid is in a lower energy state when surrounded by its own kind than when it is alone or at the edge of the crowd. This attraction is balanced in the interior of the liquid, but an atom or molecule at or near the surface experiences a net force of attraction perpendicular to the surface, directed inward, called *surface tension*. Water has a particularly large surface tension because its molecules are *dipolar*; though a water molecule is electrically neutral, its positive and negative charges are not uniformly distributed. As a consequence, the more positive portion of a molecule is attracted to the more-negative portions of neighboring molecules.

Surface tension is manifest as a force per unit length, denoted by γ_s. To understand this force, consider a thin layer of fluid having width L and (variable) length x, stretched on a movable wire frame (think of a soap bubble) as shown in part (a) of Figure D.2. To keep the surface from contracting, we must exert a force $F = 2\gamma_s L$ on the movable portion of the frame having length L. The factor "2" appears because the thin layer of fluid has two surfaces. This force is constant, independent of the magnitude of x. The work done in creating one of the two surfaces is $W = Fx/2 = \gamma_s Lx = \gamma_s A$, where $A = Lx$ is the area of the surface. The (recoverable) energy of the surface is equal to the work done to create it.

Now let's consider a spherical bubble of radius r, filled with fluid having surface tension γ_s. Actually, let's consider just half the bubble, as illustrated in part (b) of Figure D.2. In order for the surface to remain in equilibrium, we must apply a downward surface-tension force equal to the surface tension times the length of the perimeter: $F = 2\pi r\gamma_s$. In order that the bubble remain in equilibrium, we must exert an upward force of equal magnitude on the exposed horizontal surface, providing a replacement for the excess pressure within the bubble of magnitude Δp: $F_p = \Delta p \pi r^2$. The two forces balance provided

$$\Delta p = \frac{2\gamma_s}{r} = \gamma_s \left(\frac{1}{r} + \frac{1}{r} \right).$$

[13] If the atoms or molecules were not attracted to each other, or if the attraction were weak, they would fly apart and form a gas, rather than a liquid.

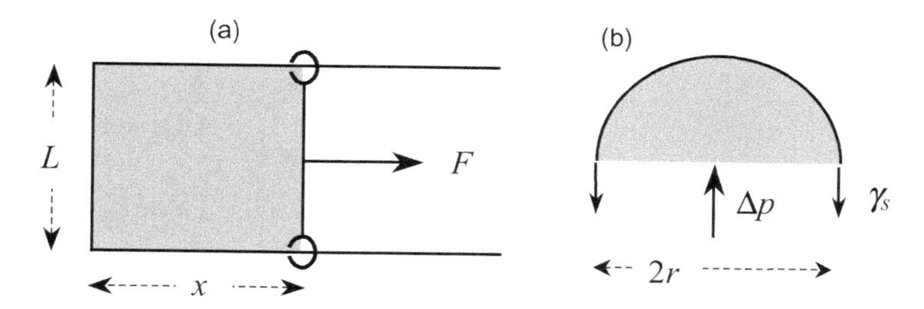

Figure D.2 (a) A thin layer of fluid (the gray rectangle) stretched on a wire frame exerts a force F on the movable portion of the frame. An equal and opposite force must be exerted on the movable portion, as shown, in order to maintain equilibrium. (b) The upper half of a spherical bubble of fluid (gray semi-circle). Surface tension exerts a downward force that must be balanced by an internal pressure Δp.

The pressure within a bubble is higher by $2\gamma_s/r$ than the outside pressure.[14] In general

$$\Delta p = \gamma_s \left(\frac{1}{r_1} + \frac{1}{r_2} \right),$$

where r_1 and r_2 are the principal radii of curvature of the surface.

D.5 General Conservation Law

The differential form of the general conservation equation is of the form

$$\frac{\partial q}{\partial t} + \nabla \bullet \mathbf{F} = s_q,$$

where q is an amount of a conserved quantity, such as chemical components of a system, $\mathbf{F}$ is the flux of that quantity and s_q is the source of that quantity.[15] A chemical component is transported from place to place through the motions of the atoms and molecules comprising the system. These motions may be macroscopic and organized or microscopic and disorganized. The flux due to macroscopic motions is called *advection* and is given by $\mathbf{F} = q\mathbf{v}$, where $\mathbf{v}$ is the (macroscopic) fluid velocity. Diffusion due to disorganized motions is called *Fickian diffusion*, and is a flux down a gradient: $\mathbf{F} = -\kappa_q \nabla q$, where κ_q is a diffusion coefficient. Putting these two together, $\mathbf{F} = q\mathbf{v} - \kappa_q \nabla q$ and the differential conservation law becomes

$$\frac{\partial q}{\partial t} + \nabla \bullet (q\mathbf{v}) = s_q + \nabla \bullet \left(\kappa_q \nabla q\right).$$

For example, if the conserved quantity is mass, then $q = \rho$, $s_q = 0$ and $\kappa_q = 0$.

[14] The pressure within a soap bubble, filled with air, is $4\gamma_s/r$, because there are two surfaces.
[15] This may be derived from the integral form of the general conservation equation, with the aid of *Gauss' theorem*.

If the source is zero and the quantities depend only on one space dimension x and time t, then the conservation equation is of the form

$$\frac{\partial q}{\partial t} + \frac{\partial F}{\partial x} = 0,$$

where $F = \mathbf{F} \bullet \mathbf{1}_x$ is the x component of $\mathbf{F}$. We encounter several versions of this equation in the study of shallow-water flow: $\{q, F\} = \{h, Q\}$ for conservation of water, $\{q, F\} = \{u, gH\}$ for conservation of x momentum and $\{q, F\} = \{uh, gM\}$ for conservation of energy; see § 19.1. With the speed u equal to zero and the diffusivity constant (so that $\mathbf{F} = \kappa_q(\partial q/\partial x)\mathbf{1}_x$), this one-dimensional conservation equation may be expressed as

$$\frac{\partial q}{\partial t} = \kappa_q \frac{\partial^2 q}{\partial x^2}.$$

This is commonly known as the *heat equation.*

D.6 Euler and Bernoulli Equations

The *Euler equation* is the inviscid form of the Navier–Stokes equation. Using the third vector identity in Appendix A.2.6, Euler's equation may be expressed as

$$\frac{\partial \mathbf{v}}{\partial t} + \nabla\left(\tfrac{1}{2}v^2\right) + (2\boldsymbol{\Omega} + \boldsymbol{\omega}) \times \mathbf{v} = \mathbf{g} - \frac{1}{\rho}\nabla p,$$

where v is the magnitude of the velocity $\mathbf{v}$ and $\boldsymbol{\omega} = \nabla \times \mathbf{v}$ is the vorticity, $\boldsymbol{\Omega}$ is the rotation rate of the coordinate system, p is pressure and t is time.

The *Bernoulli equation* is a first integral of Euler's equation, governing conservation of mechanical energy during the flow of an inviscid fluid. There are several versions of the equation, depending whether the whether the flow is steady or irrotational and whether the fluid is incompressible.

D.6.1 Irrotational Euler and Bernoulli

If the coordinate system is not rotating and if the flow is irrotational (that is, $\boldsymbol{\omega} = \boldsymbol{\Omega} = \mathbf{0}$), the fluid velocity may be expressed as the gradient of a scalar:

$$\mathbf{v} = \nabla\phi$$

and Euler's equation becomes

$$\nabla\left(\frac{\partial \phi}{\partial t} + \tfrac{1}{2}v^2 + \psi + P(p)\right) = \mathbf{0},$$

where ψ is the gravitational potential[16] and P is the pressure function satisfying $\mathrm{d}P = (1/\rho)\mathrm{d}p$.[17]

[16] $\mathbf{g} = -\nabla\psi$.
[17] P exists if the flow is barotropic: $\nabla\rho \times \nabla p = \mathbf{0}$.

This is readily integrated to yield the equation of conservation of mechanical energy, called the Bernoulli equation:

$$\frac{\partial \phi}{\partial t} + \tfrac{1}{2}v^2 + \psi + P(p) = gH\,,$$

where H is a constant of integration called the *head*. Strictly speaking, this should be a function of integration $H(t)$, but the time dependence can be incorporated into the velocity potential, so there is no loss of generality in assuming that H does not vary with time. This statement holds throughout the fluid.

The Bernoulli equation above is a statement of conservation of mechanical energy. Three of the four terms on the left-hand side are familiar: $v^2/2$, ψ and $P(p)$ are the kinetic, gravitational and pressure energies per unit mass, respectively. The unfamiliar term is $\partial\phi/\partial t$, which represents the integrated effect of fluid acceleration.

If we limit our attention to constant density, so that $P = p/\rho$, and note that $\psi = gz$ on or near Earth's surface, then the Bernoulli equation may be solved explicitly for the pressure in terms of the potentials and the speed:

$$p = p_a - \rho\left(gz + \frac{\partial \phi}{\partial t} + \tfrac{1}{2}v^2\right),$$

where p_a is a constant. If we further split pressure and density into static part and perturbation, we have

$$p_0 = p_a - \rho_0 gz \qquad \text{and} \qquad p' = -\rho' gz - \rho_0\left(\frac{\partial \phi}{\partial t} + \tfrac{1}{2}v^2\right).$$

D.6.2 Steady Rotational Euler and Bernoulli

Steady, rotational, barotropic, inviscid flow is governed by

$$\nabla\left(\tfrac{1}{2}v^2 + \psi + P(p)\right) = \mathbf{v} \times (2\boldsymbol{\Omega} + \boldsymbol{\omega})\,.$$

The term on the right-hand side is perpendicular to $\mathbf{v}$ and may be removed by taking the dot product of this equation with $\mathbf{v}$:

$$(\mathbf{v}\bullet\nabla)\left(\tfrac{1}{2}v^2 + \psi + P(p)\right) = 0\,.$$

The operator $(\mathbf{v}\bullet\nabla)$ is a directional derivative in the direction of the fluid motion;[18] that is, along *streamlines*.[19] Now the quantity $v^2/2 + \psi + P(p)$ is constant along streamlines. If we limit our attention to constant density, so that $P = p/\rho$ and write the gravitational potential as $\psi = gz$, where z is the upward coordinate, then we may write the familiar form

[18] See Appendix A.2.3.
[19] See the fundament on flow lines found in Appendix C.5.

of conservation of mechanical energy:

$$\frac{v^2}{2g} + z + \frac{p - p_a}{\rho g} = H .$$

The three terms on the left-hand side are the velocity head, elevation head and pressure head. Head has the dimensions of length and measures the height of static column of fluid within a tube.

D.7 Kinetic and Internal Energies

The purpose of this fundament is to investigate the separation the energy associated with the motions of the atoms comprising a massive body into macroscopic kinetic energy and microscopic internal energy. The total kinetic energy of a massive body, composed of N particles of identical masses m (e.g., a monatomic gas), is given by

$$E_K = \tfrac{1}{2} m \sum_{n=1}^{N} v_n^2 = \tfrac{1}{2} m \sum_{n=1}^{N} \mathbf{v}_n \bullet \mathbf{v}_n ,$$

where $\mathbf{v}_n = \mathrm{d}\mathbf{x}_n/\mathrm{d}t$ is the velocity of the nth particle, $v_n = ||\mathbf{v}_n||$ is its speed and $\mathbf{x}_n$ is its position, measured with respect to a specified coordinate system. The center of mass and mean velocity of the body, as seen by an observer fixed in the coordinate frame, are given by

$$\mathbf{x}_0 = \frac{1}{N} \sum_{n=1}^{N} \mathbf{x}_n \qquad \text{and} \qquad \mathbf{v}_0 = \frac{1}{N} \sum_{n=1}^{N} \mathbf{v}_n .$$

The particle velocities are the sum of translation and internal motions relative to the center of mass: $\mathbf{v}_n = \mathbf{v}_0 + \mathbf{v}'_n$. Now we may express the total kinetic energy as

$$E_K = \tfrac{1}{2} M v_0^2 + \tfrac{1}{2} m \sum_{n=1}^{N} \left(v'_n\right)^2 ,$$

where $M = mN$ is the total mass of the body. The first term on the right-hand side is the macroscopic translational kinetic energy while the second includes the rotational, deformational and internal kinetic energies.

Now we must endeavor to split $\mathbf{v}'_n$ into its three components: rotation, deformation and (disorganized) internal motions. Let's begin by noting that rotation is expressed as a cross product and write $\mathbf{v}'_n = \mathbf{w}_n + \mathbf{v}''_n$, where $\mathbf{w}_n = \boldsymbol{\Omega} \times (\mathbf{x}_n - \mathbf{x}_0) = \boldsymbol{\Omega} \times \mathbf{x}'_n$. The rotation $\boldsymbol{\Omega}$ is determined by crossing this equation with $\mathbf{x}'_n$ and summing over all particles. Using a standard vector identity, this is

$$\sum_{n=1}^{N} \mathbf{x}'_n \times \mathbf{w}_n = \boldsymbol{\Omega} \sum_{n=1}^{N} (x'_n)^2 - \sum_{n=1}^{N} (\boldsymbol{\Omega} \bullet \mathbf{x}'_n) \mathbf{x}'_n .$$

Noting that $\mathbf{w}_n$ and $\mathbf{v}_n''$ are uncorrelated (i.e., the sum of their product over all particles is zero), we have

$$E_K = \tfrac{1}{2}Mv_0^2 + \tfrac{1}{2}m\sum_{n=1}^{N} w_n^2 + \tfrac{1}{2}m\sum_{n=1}^{N}(v_n'')^2 .$$

The first two terms on the right-hand side represent the translational and rotational kinetic energies, respectively. Note that $\mathbf{v}_0$ is independent of position while $\mathbf{w}_n$ varies linearly with position, and that the first two parts of the kinetic energy are those experienced by a rigid body. The magnitudes of the translational and rotational kinetic energies depend on the choice of reference coordinate system and may be made identically zero by having the coordinate system move with the center of mass of the system and rotate with angular velocity $\boldsymbol{\Omega}$. However, the choice of reference system does not affect the third contribution to the kinetic energy; the magnitudes of the deformational and internal energies are independent of the observer.

If we adopt a coordinate system translating and rotating with the body, we have

$$E_K = \tfrac{1}{2}M\bar{v}^2, \qquad \text{where} \qquad \bar{v} = \left(\frac{1}{N}\sum_{n=1}^{N}(v_n'')^2\right)^{1/2}$$

is the average speed of internal motions. The velocity in this expression may have some large-scale structure if the body is being deformed. Deformation may be split into two parts: change of volume without change of shape and change of shape without change of volume. For a monatomic gas, the first of these is thermodynamically important while the second is not.

For simple systems, the kinetic energy may be equated with the internal energy: $E_K = U$ and the specific internal energy $u = U/M$ is given by $u = \bar{v}^2/2$. The specific internal energy is related to the temperature in Appendix E.8.4.

D.8 Alternate Thin-Layer Variables

This fundament contains two simplifications of the thin-layer equations developed in § 8.2.1 that hide the complications of variable density.

D.8.1 Isobaric Surface

The total pressure, $p_r(z) + p'(\mathbf{x}_H, z, t)$, may vary with horizontal position. This means that the surface on which the total pressure is a specified value, say, $p = p_0$, may not be flat. This provides us with an alternate vertical coordinate: the elevation, denoted by Z, of a surface on which the pressure is constant; this is called an *isobaric surface*. Note that Z is a function of horizontal position and pressure, while pressure is a function of position and time. In order to relate p' and Z, let's begin by considering the surface $z = Z(\mathbf{x}_H, t; p_0)$. As we move an infinitesimal distance on this surface (with t and p_0 constant), z changes by an amount $\mathrm{d}z = \nabla_H Z \bullet \mathrm{d}\mathbf{x}_H$. The pressure $p = p_r(z) + p'(\mathbf{x}_H, z, t)$ remains constant as we move along the surface, so that, using $|\partial p'/\partial z| \ll \rho_r g$, we have $\mathrm{d}p = \nabla_H p' \bullet \mathrm{d}\mathbf{x}_H + (\partial p_r/\partial z)\mathrm{d}z = 0$ or, using the differential for z and the hydrostatic relation ($\mathrm{d}p_r/\mathrm{d}z = -\rho_r g$), we have $\mathrm{d}p = \nabla_H p' \bullet \mathrm{d}\mathbf{x}_H - \rho_r g \nabla_H Z \bullet \mathrm{d}\mathbf{x}_H = 0$. This equation is satisfied for all $\mathrm{d}\mathbf{x}_H$ provided[20]

$$\nabla_H p' = \rho_r g \nabla_H Z .$$

Using this to eliminate the pressure from the horizontal momentum equation, the troublesome factor ρ_r disappears:

$$\frac{\mathrm{D}\mathbf{v}_H}{\mathrm{D}t} + f\mathbf{1}_z \times \mathbf{v}_H = -g\nabla_H Z + \mathcal{V}(\mathbf{v}_H) .$$

In our studies of waves and flows in the oceans, a constant-pressure surface of particular interest is the air–sea interface, the elevation of which is denoted by h rather than Z.

We have eliminated the variable coefficient ρ_r from the horizontal momentum equation by replacing the perturbation pressure, p', with the isobaric-surface elevation, Z. In the following section, we will eliminate ρ_r from the continuity equation by replacing the vertical coordinate z with the static pressure, p_r.

D.8.2 Pressure Coordinate

The transformation from z to p_r as the vertical independent coordinate is accomplished with the differential hydrostatic relation, $\mathrm{d}p_r = -\rho_r g \mathrm{d}z$. Using this, the continuity equation may be expressed as

$$\nabla_H \bullet \mathbf{v}_H + \frac{\partial w_p}{\partial p_r} = 0, \qquad \text{where} \qquad w_p = -g\rho_r w$$

is a scaled vertical speed (positive downward). Commonly the distinction between the reference-state pressure p_r and the actual pressure p is ignored and the subscript r is dropped.

Next consider the hydrostatic equation. In static equilibrium, this equation is $\mathrm{d}p_r = -\rho_r g \mathrm{d}z$, where z is the upward coordinate. When the flow is not quite hydrostatic, p and Z are related by a very similar expression:

$$\mathrm{d}p = -\rho g \,\mathrm{d}Z$$

with, of course, $\mathbf{x}_H$ held fixed.

[20] It is important to realize that this relation applies only to the horizontal plane; in the vertical plane $\mathrm{d}p' = -\rho_r g \mathrm{d}z$. Note the opposite sign.

D.9 Proudman–Taylor Theorem

An interesting and important feature of rapidly rotating flows is the tendency for flow structures to align with the axis of rotation. This tendency is quantified by the *Proudman–Taylor theorem*. Assuming that the fluid is barotropic and inviscid, that the system is rapidly rotating (with $\|\nabla \times \mathbf{v}\| \ll \|\boldsymbol{\Omega}\|$) and that flow is steady, the momentum equation[21] may be expressed as[22]

$$2\boldsymbol{\Omega} \times \mathbf{v} = -\nabla\left(\tfrac{1}{2}v^2 + \psi + p/\rho\right),$$

where ψ is the gravitational potential[23] and v is the magnitude of $\mathbf{v}$. The curl of this equation yields an important dynamic constraint:

$$(\boldsymbol{\Omega} \bullet \nabla)\mathbf{v} = \mathbf{0}.$$

This states that the flow is invariant in the direction of the axis of rotation. Of course this is not an absolute statement, because a number of approximations and simplifications have been made to get to this result. But it does describe a strong tendency flow in rotating fluids to align with the axis of rotation. The Proudman–Taylor constraint on flow can be severe. For example, it states that flow within a rapidly rotating sphere or spherical shell is confined to the zonal direction, apparently precluding convective motions.

On Earth's surface, only the vertical component of rotation is dynamically important, and the Proudman–Taylor theorem implies that motions in the atmosphere and oceans have a strong tendency to develop vertically-uniform structures. That is, *rotation provides structure to flow* in an otherwise homogeneous fluid. This explains – in part – why rotating systems exist and persist in the atmosphere and oceans.

D.10 Vector Vorticity Equation

A useful diagnostic equation for rotating flows is the vorticity equation, obtained by taking the curl of the Navier–Stokes equation. Noting that gravity may be expressed as a potential, so that its curl is zero, the vorticity equation is[24]

$$\frac{\mathrm{D}\boldsymbol{\omega}}{\mathrm{D}t} + 2\boldsymbol{\Omega}(\nabla \bullet \mathbf{v}) - 2(\boldsymbol{\Omega} \bullet \nabla)\mathbf{v} = \frac{1}{\rho^2}\nabla\rho \times \nabla p + \nabla \times [\nabla \bullet (\nu\nabla)]\,\mathbf{v}.$$

The density is a thermodynamic function of p and T, so that the vorticity equation may be expressed as

$$\frac{\mathrm{D}\boldsymbol{\omega}}{\mathrm{D}t} = 2(\boldsymbol{\Omega} \bullet \nabla)\mathbf{v} - 2\boldsymbol{\Omega}(\nabla \bullet \mathbf{v}) + \frac{1}{\rho^2}\left.\frac{\partial\rho}{\partial T}\right|_p \nabla T \times \nabla p + \nabla \times [\nabla \bullet (\nu\nabla)]\,\mathbf{v}.$$

[21] From § 4.7.
[22] Using formula $\nabla(v^2) = 2\mathbf{v} \times (\nabla \times \mathbf{v}) + 2(\mathbf{v} \bullet \nabla)\mathbf{v}$ from Appendix A.2.6.
[23] $\mathbf{g} = -\nabla\psi$.
[24] Using formulas from Appendix A.2.6.

This equation shows that the vorticity of a parcel of fluid can be changed by four mechanisms that are quantified by the terms on the right-hand side of the equation above.

- The first mechanism, called *vortex stretching*, involves a divergence of flow that concentrates the background vorticity; this is the dominant mechanism in the formation and maintenance of vortices.[25]
- The second mechanism concentrates the background vorticity through compression of the fluid; this mechanism is negligibly small in geophysical flows.
- The third mechanism is a baroclinic torque applied by the varying density of the fluid; this can be quite important in buoyancy-driven flows.
- The fourth mechanism requires the action of viscosity; this mechanism is important only in boundary layers or in regions of strong turbulence.

If the fluid is inviscid ($\nu = 0$), not rotating ($\boldsymbol{\Omega} = \mathbf{0}$) and barotropic ($\partial\rho/\partial T|_p = 0$) then the vorticity equation reduces to $\mathrm{D}\boldsymbol{\omega}/\mathrm{D}t = \mathbf{0}$; the vorticity of a parcel of fluid is invariant.

The scalar vorticity equation encountered in our investigation of rotating fluids in Chapter 8 has a somewhat different form than the vector equation.

[25] See Chapter 28.

Appendix E
Thermodynamics

Thermodynamics is concerned with the amounts of energy stored in – and transferred to and from – a *thermodynamic system* (i.e., a specified portion of the Universe). The behavior of energy is codified in three laws:

- The *zeroth law* defines the concept of temperature. Temperature is an *intensive thermodynamic state variable.*[1]
- The *first law* quantifies conservation of energy; it is entirely bookkeeping – keeping track of the types and amounts of energy stored in a system and transferred into or from it.
- The *second law* restricts the possible changes of forms of energy. In a nutshell, changes of energy from organized to disorganized are unconstrained, but transfers of energy from disorganized to organized are severely limited[2] and are quantified by the *thermodynamic efficiency*; see Appendix E.11. An unavoidable consequence of the second law is the degradation of the kinetic energy of a flowing fluid to heat due to the action of viscous forces.

A thermodynamic system consists of any clearly defined portion of the universe. A system is assumed to be composed of a vast number[3] of massive particles (i.e., particles having a rest mass). These particles may be atoms, electrons, ions, molecules and/or larger aggregations of matter, provided that the system consists of a vast number of them. (It is simplest to think of the particles as atoms.) A system may be categorized as *closed* or *open*; transfers of energy and matter into or out of a closed system are not permitted, while these transfers are permitted if the system is open.

The following fundaments of thermodynamics include:

- E.1: a discussion of storage and transfers of energy;

[1] Whose value depends on the state of a thermodynamic system, independent of the processes leading to the establishment of that state.

[2] This constraint can be understood in terms of probabilities. An assemblage of atoms is far more likely to be in a disorganized configuration than in an organized one. The state of disorganization of a system is characterized by the *entropy*, with an organized state having low entropy and disorganized having high. The entropy of the universe is increasing inexorably with time. We can reduce the entropy of a given system, but only at the expense of increasing the entropy of the remainder of the universe.

[3] The problem of counting vast numbers of atoms is circumvented by using the mole; see Appendix E.2.

- E.2: an introduction to the mole, which is a way to count atoms or molecules;
- E.3: an introduction to the first law of thermodynamics;
- E.4: a discussion and categorization of thermodynamic potentials, variables and parameters;
- E.5: the equations of state for density and entropy, including an alternate form for the first law and the equation of state for sea water;
- E.6: an introduction to ideal mixtures;
- E.7: development of the energy equation;
- E.8: a brief introduction to the thermodynamic behavior of ideal gases;
- E.9: a summary of the thermodynamics of the atmosphere;
- E.10: a discussion of phase equilibrium; and
- E.11: quantification of thermodynamic efficiency.

E.1 Energy Storage and Transfer

Energy can be stored within a system in two fundamental forms: *potential* and *kinetic*. The potential energy of a system is that stored in the positions of its particles (relative to force fields), while kinetic energy is that stored in the motion of its particles (relative to a specified frame of reference).[4] Forms of potential energy include chemical energy, which is due to the relative positions of electrons and nuclei affected by electromagnetic forces, and nuclear energy, which is due to the relative positions of protons and neutrons affected by nuclear forces. The kinetic energy of a body consists of the organized (macroscopic) motions of translation, rotation and deformation and its disorganized (microscopic) internal motions.[5]

The amount of energy that can be stored within a particle (that is, an atom or molecule) is related to its internal structure and is controlled by the number of its degrees of freedom. The simplest type of particle is an atom, having three degrees of freedom: the ability to move in three-dimensional space, as explained in Appendix E.8. If the particle is a molecule, it has additional degrees of freedom in the relative motions of its constituent atoms. The *equipartition theorem* states that in equilibrium the amounts of energy stored in each degree of freedom are the same. The amount of energy stored in one degree of freedom is given by $kT/2$ where T is the temperature and k is *Boltzmann's constant*: $k \approx 1.38 \times 10^{-23}$ J·K^{-1}.[6]

The categorization of stored energy as potential or kinetic is not the most practical, as the two forms are essentially equivalent, in that (at least in theory) one can be changed into the other without restriction. A more practical categorization is whether the energy is organized, and hence entirely useful, or disorganized. Energy stored in the positions or motions of macroscopic collections of particles is organized, while energy stored in the

[4] An exception to this categorization is electromagnetic energy stored in photons having no rest mass and in electric and magnetic fields.
[5] See Appendix D.7.
[6] See Appendix E.8.1.

positions and motions of the individual particles is disorganized. The latter is referred to as *internal energy*. It is common to refer to the internal energy of a system as its heat or heat content, but this is not strictly correct; as noted below, heat is energy transferred via disorganized motions.

Energy in transit can take two forms, depending whether it involves organized or disorganized motions. Transfers involving organized motion are called *work* while those involving disorganized motions are called *heat*. Work, denoted by W, is defined as the exertion of a force through a distance:

$$W = \int_{\text{path}} \mathbf{F}(\mathbf{x}) \bullet \mathrm{d}\mathbf{x},$$

where $\mathbf{F}$ is a force vector, $\mathbf{x}$ is the position vector and "path" denotes a specified path in space. This simple equation is of particular interest because it relates a thermodynamic quantity, W, to a dynamic one, $\mathbf{F}$; see, for example, § 24.3 and § 28.5.1.

Interestingly, work can involve the collective action of the disorganized motions of the particles comprising the system. To illustrate this concept, consider a system consisting of a uniform distribution of gas particles in disorganized motion, confined within a cylindrical chamber by a movable piston. As the gas particles bounce off the face of the piston, they exert a force per unit area (i.e., a *pressure*) on the piston.[7] As the piston is withdrawn from the cylinder, the gas particles move – in a collective manner – to fill the vacated space. In doing so, they exert a force (pressure times piston area) through a distance, thereby doing work. The energy for this work comes from mean kinetic energy of the gas particles. As the particles bounce off the receding piston face, they rebound with diminished speed, which is seen as a decrease of temperature.

E.2 Mole

The mole of chemistry and thermodynamics is in some respects similar to the animal of the same name: mostly hidden from view, a bit mysterious and often not too pleasant to deal with. Like the angle, the mole has units but not dimensions. That is, a mole is an amount of something, expressed as a number, with the magnitude of the number somewhat arbitrary. The mole was introduced to avoid the problem of counting individual atoms or molecules,[8] as we shall now explain.

As noted in the previous section, the energy per atom is quantified by Boltzmann's constant, which has obvious dimensions of energy per unit temperature, plus a hidden dimension of "per atom". The small numerical magnitude of Boltzmann's constant is balanced by the large number of atoms in a laboratory-sized chunk of matter. Counting 10^{23} or more atoms can be a tedious business and people found it preferable to count "egg cartons" full of atoms, rather than individual atoms. These egg cartons are called moles, with a fixed number of atoms in each carton. These cartons can have various capacities.

[7] See Appendix E.8.4.

[8] In this section, "atom" means both atom and/or molecule.

The standard carton is the *gram-mole* or simply the *mole*. It contains a number of atoms such that the total mass (in grams) of the atoms in the carton is equal to the atomic number of the atoms in the carton. The number of atoms in a mole is called *Avogadro's number* and is approximately equal to 6.02214×10^{23}.

An attractive alternate to the gram-mole is the *kilogram-mole*, or simply the *kmole*, which contains approximately 6.02214×10^{26} atoms. The adoption of the gram-mole as the standard mole is a bit unfortunate because in the SI system, mass is measured in kilograms, rather than grams. This presents us with a problem of conversion between (gram-) moles and kmoles and between grams and kilograms, with the factor 1000 popping up when trying to perform numerical calculations. To avoid this issue and attendant numerical errors, in the following we will employ the kilogram-mole (having unit "kmol") exclusively.

One confusing aspect of this topic is that the amount of stuff in a mole depends on what the stuff is. Think of eggs; a dozen ostrich's eggs makes a much larger omelet than does a dozen hen's eggs. Similarly, a kmole of helium atoms contains four times as much matter as a kmole of hydrogen, but each contains the same number of atoms. Another confusing aspect is that different people can use cartons of differing capacities. For example, the kmole contains 1000 times more atoms (or molecules) than the mole. To clarify this latter aspect, Avogadro's number has been replaced by *Avogadro's constant*, which has the dimensions of reciprocal amount of substance: $N_A \approx 6.02214 \times 10^{26}\,\mathrm{kmol}^{-1}$. Now the total number of atoms is simply the number of cartonsful n (in kmoles) times N_A (in reciprocal kmoles):

$$N = nN_A\,.$$

The mass M_A of one kmole of atoms (in kg/kmol), each of which has mass m (in kg), is

$$M_A = mN_A$$

and the mass, M, of an assembly of atoms is variously

$$M = mN = nmN_A = nM_A\,.$$

E.3 First Law of Thermodynamics

The first law of thermodynamics, which quantifies changes of the internal energy of a thermodynamic system, is initially presented in extensive, differential form in Appendix E.3.1, then is integrated in Appendix E.3.2 and finally presented in intensive form in Appendix E.3.3.

E.3.1 Differential Form

This presentation begins with a differential form of the first law involving *irreversible* changes – limited by work and heat inequalities. Later, we will limit our attention to reversible changes that can proceed in either direction. In writing the differential form of the first law, we need to distinguish between changes in quantities that are state variables and those that are not by using the usual symbol "d" for the former and the differential symbol đ for the latter.[9] Curiously, heat – a central quantity of interest – is not a thermodynamic state variable.

The internal energy of a system, denoted by U, can be increased by the transfer of heat (denoted by Q) or material (denoted by Φ) into the system and can be decreased as the system does work, W, on its surroundings. These changes are codified in the first law of thermodynamics, written as

$$\mathrm{d}U = \text{đ}Q - \text{đ}W + \mathrm{d}\Phi, \qquad \text{where} \qquad \mathrm{d}\Phi = \sum_{i}^{I} \mu_i \, \mathrm{d}N_i$$

quantifies effects of added mass; $\mathrm{d}U$ is the infinitesimal increase of internal energy of the system, $\text{đ}Q$ is an infinitesimal amount of heat added to the system, $\text{đ}W$ is an infinitesimal amount of work done by the system, $\mathrm{d}N_i$ is the number of moles of constituent i added to the system and μ_i is the energy per atom or molecule (also called the *chemical potential*) of that constituent. The integer I equals the total number of constituents within the system, with different phases counting as distinct constituents. Note that the chemical potentials μ_i have dimensions of energy per kmole.

Work and Heat Inequalities

The variables Q and W are not functions of the state of our thermodynamic system, but depend on the nature of the processes causing the changes. The small changes of these variables are subject to inequality constraints as follows:

$$\text{đ}W \le \mathrm{d}W \equiv p\,\mathrm{d}V \qquad \text{and} \qquad \text{đ}Q \le \mathrm{d}Q \equiv T\,\mathrm{d}S,$$

where p, T, $\mathrm{d}V$ and $\mathrm{d}S$ are the pressure, temperature, small increase of volume and small increase of entropy of the system, respectively. The differential $\mathrm{d}W$ is the maximum amount of work that a system can do during an expansion (assuming $\mathrm{d}V$ is positive), while $\mathrm{d}Q$ is the maximum amount of heat that could be added to our system (assuming $\mathrm{d}S$ is positive). Let's take a minute to understand these inequalities. As we saw in Appendix E.1, work is done by the system as it exerts a force through a distance, or equivalently a pressure over a volume change. If the expansion of the system (think of a piston being withdrawn from a cylinder containing an ideal gas) is infinitesimally slow, then $\text{đ}W = p\,\mathrm{d}V$, with the work being done by the system as the gas atoms bounce off the receding piston. If the expansion is rapid, the gas atoms having speed less than the piston are unable to bounce off the piston and the work done by the system is less than $p\,\mathrm{d}V$. Entropy S is a measure of disorder within the system and the second law states that a thermodynamic system cannot spontaneously become ordered; the entropy of a thermally isolated system cannot

[9] Say "d bar". Sometimes the symbol δ is used instead of đ.

decrease.[10] It follows that $\mathrm{d}Q$ is the maximum amount of heat that could be added to a system that has no internal processes which act to increase its entropy.

Equality between $đW$ and $đQ$ on the one hand and $\mathrm{d}W$ and $\mathrm{d}W$ on the other holds only for idealized reversible processes that occur infinitesimally slowly. For any real process, equality cannot be achieved; degradation of energy and disorganization are inevitable. Having said that, it is worth noting that often these losses are negligibly small and we can set $đW = \mathrm{d}W$ and $đQ = \mathrm{d}Q$.

Added Mass

The number of moles added is related to the added mass by the differential form of the last equation in Appendix E.2:

$$\mathrm{d}M_i = m_i \mathrm{d}N_i,$$

where m_i is the mass of constituent i per mole.

Note that $\sum_i^I M_i = M$ and $\sum_i^I \mathrm{d}M_i = \mathrm{d}M$, where M is the total mass of the system. If the total mass of the system is held constant ($\mathrm{d}M = 0$), then one of the differentials $\mathrm{d}M_i$ is redundant and the i sum is over all but one of the constituents. Typically, one constituent of a system is identified as the primary constituent (say, the Ith; denoted by the subscript p – for primary) and the sum is over the remaining constituents. That is, $\mathrm{d}M_p = -\sum_i^{I-1} \mathrm{d}M_i$ and

$$\mathrm{d}\Phi = \sum_i^{I-1} \tilde{\mu}_i \mathrm{d}M_i, \qquad \text{where} \qquad \tilde{\mu}_i = \frac{\mu_i}{m_i} - \frac{\mu_p}{m_p}$$

is the difference of chemical potentials, measured in units of energy/mass (= speed^2).

Discussion of the Chemical Potential

One form of potential energy that is not immediately obvious is the energy of attraction among similar molecules within a crystalline lattice. An atom in a crystal is in a lower energy state than if it were in isolation, far apart, and we must do work to separate it from its buddies; that is, we must exert energy to break a solid in two. This energy is similar to that which gives rise to surface tension; see Appendix D.4.

If an atom were added to a crystalline lattice of identical atoms, the force of attraction would do work on that atom and it would end up in a lower energy state, just as gravity does work on a falling body and a body on Earth is in a lower energy state than it would be far from Earth. The amount of energy released by the addition of a mole of atoms to the lattice is quantified by the *chemical potential*. The chemical potential μ_i is negative.

Now suppose that a particle (particle 1) of constituent 1 within a crystalline lattice of its own kind were replaced by a particle (particle 2) of constituent 2. With particle 2 being

[10] This implies that the Universe was in a highly ordered state (all matter localized and hot) immediately following the Big Bang and has been going downhill ever since.

bigger or smaller than particle 1 the lattice would need to distort to accommodate it. It takes energy to distort the lattice. The amount of energy needed to replace particle 1 by particle 2 is represented by $\tilde{\mu}_i$. Since the attraction of particle 2 to the lattice is weaker than the attraction of particle 1, the chemical potential difference $\tilde{\mu}_i$ is positive.

Reversible Form of the First Law

Assuming reversible processes (so that $đW = p\,\mathrm{d}V$ and $đQ = T\mathrm{d}S$),[11] the differential form of the first law is

$$\mathrm{d}U = T\,\mathrm{d}S - p\,\mathrm{d}V + \sum_i^{I-1} \tilde{\mu}_i\,\mathrm{d}M_i\,.$$

E.3.2 Integration of the First Law

The differential terms on the right-hand side of the reversible form of the first law ($\mathrm{d}S$, $\mathrm{d}V$ and $\mathrm{d}M_i$) are each extensive – linearly proportional to the total mass of the system, while their coefficients (T, p and $\tilde{\mu}_i$) are each intensive – independent of total mass of the system.[12] This property permits the reversible form of the differential of the first law to be integrated:

$$U = TS - pV + \sum_i^{I-1} \tilde{\mu}_i M_i\,.$$

Note that by integrating the first law, we have transitioned from a focus on processes to a focus on states; this is a statement about the thermostatic state of a system.

E.3.3 Specific Form of the First Law

The variables p, T and $\tilde{\mu}_i$ are intensive, whereas M_i, S, U and V are extensive; as a consequence, the integrated form of the first law presented above is extensive, with each term being proportional to the mass of the system. This equation may be written in specific form by dividing through by the total mass M of the system:

$$u = Ts - pv + \sum_i^{I-1} \tilde{\mu}_i \xi_i\,,$$

where $u = U/M$ is the specific internal energy, $s = S/M$ is the specific entropy, $v = V/M$ is the specific volume, and $\xi_i = M_i/M$ is the mass fraction of the ith constituent. The summation over the chemical constituents may be expressed in vector form, with the vector

[11] With this limitation, the subject would more accurately be called thermostatics.

[12] The differentials of the other thermodynamic potentials (enthalpy, Helmholtz energy and Gibbs energy) do not have this property.

$\boldsymbol{\xi}$ representing the set of mass fractions ξ_i for $i = 1$ to $I - 1$ and the vector $\tilde{\boldsymbol{\mu}}$ representing the set of mass fractions $\tilde{\mu}_i$ for $i = 1$ to $I - 1$. That is,

$$u = Ts - pv + \tilde{\boldsymbol{\mu}} \bullet \boldsymbol{\xi}, \qquad \text{where} \qquad \tilde{\boldsymbol{\mu}} \bullet \boldsymbol{\xi} = \sum_{i}^{I-1} \tilde{\mu}_i \xi_i .$$

The differential of this equation may be divided into two: the specific differential form of the first law

$$\mathrm{d}u = T\,\mathrm{d}s - p\,\mathrm{d}v + \tilde{\boldsymbol{\mu}} \bullet \mathrm{d}\boldsymbol{\xi}$$

and an auxiliary differential, known as known as the *Gibbs–Duhem relation*,

$$s\,\mathrm{d}T - v\,\mathrm{d}p + \boldsymbol{\xi} \bullet \mathrm{d}\tilde{\boldsymbol{\mu}} = 0.$$

Similarly, the work and heat inequalities may be written in specific form:

$$đw \le \mathrm{d}w = p\,\mathrm{d}v \qquad \text{and} \qquad đq \le \mathrm{d}q = T\,\mathrm{d}s,$$

where w is the work per unit mass and q is the heat per unit mass.

E.4 Thermodynamic Potentials, Variables and Parameters

Thermodynamics is plagued by a bewildering array of variables and manipulation of these variables can at times seem like a game of three-card monte. We try to make sense of all these in this fundament, by categorizing the various thermodynamic quantities that we will be encountering. These may be categorized by differentiation: a *variable* is the derivative of a *potential* and a *parameter* is a derivative of a variable. These categories of thermodynamic quantities are introduced and discussed in the following two sections.

E.4.1 Thermodynamic Potentials and Variables

The first thermodynamic potential we encountered is the internal energy, U. We will identify three others quite soon. Also, we have encountered two versions of internal energy, the extensive U and the intensive u. So before delving into potentials and variables, it might be prudent to introduce a second categorization; thermodynamic potentials and variables can be categorized as either *extensive* or *intensive* depending whether they vary linearly with the amount of material in a system or are independent of the amount. For example, mass M and volume V are extensive, while density is intensive. Also, U and S are extensive variables, while p, T and $\tilde{\mu}_i$ are intensive variables. As a general rule, it is preferable to work with intensive variables. An extensive variable may be made intensive by dividing by the total mass of the system. Variables that have been made intensive in this manner are called *specific*. Often an extensive variable is denoted by a capital letter, and its specific counter part by the corresponding lower case letter. For example, S is the entropy of a system, while s is the specific entropy.[13] This same categorization of extensive or intensive applies to equations. The equation for the internal energy has been cast in specific form in Appendix E.3.3. Each term of an extensive equation contains one and only one extensive variable, while each variable in an intensive equation is intensive.

Another way to categorize (intensive) thermodynamic variables is either dependent or independent, just as in calculus.[14] For example it is apparent from the differential for the specific internal energy that s, v and $\boldsymbol{\xi}$ are independent variables while u, T, p and $\tilde{\boldsymbol{\mu}}$ are dependent. Actually, u plays a different role from the other dependent variables in the mathematical statement of the first law; u is a *potential function* of the system, with the remaining dependent variables being defined as the first partial derivatives of $u(s, v, \boldsymbol{\xi})$:

$$p \equiv \left.\frac{\partial u}{\partial v}\right|_{s,\boldsymbol{\xi}}, \qquad T \equiv \left.\frac{\partial u}{\partial s}\right|_{v,\boldsymbol{\xi}} \qquad \text{and} \qquad \tilde{\mu}_i \equiv \left.\frac{\partial u}{\partial \xi_i}\right|_{v,s,\boldsymbol{\xi}_-},$$

where $\boldsymbol{\xi}_-$ is the set of mass fractions minus the ith entry.

A final way to categorize thermodynamic variables is to note that certain of them form *conjugate pairs*. These pairs are identified from the energy differential developed in the previous subsection; the three conjugate pairs are T & s, v & p and $\tilde{\boldsymbol{\mu}}$ & $\boldsymbol{\xi}$. One of a conjugate pair of variables is independent and the other is dependent. We have some flexibility in our choice of independent thermodynamic variables, and this choice can depend on circumstances. In deciding which variables to be independent, we need to ask ourselves: what variables can we – or do we need to – control? In many situations the answer is p and T.

It is apparent that the "natural" independent thermodynamic variables for u are s, v and $\boldsymbol{\xi}$. Often it is advantageous to use alternate sets of independent variables by swapping among p, s, T and v.[15] When changing the independent variables, we are constrained to swapping members of conjugate pairs. That is, s & v are a valid pair of independent variables, as we have seen above, as are T & v, but s & T are not a valid set. There are four permissible sets of independent variables: s & v, T & v, p & s, and p & T. Each of the four sets has an associated *thermodynamic potential* as shown in Table E.1.

In many situations in engineering and geophysics the most convenient independent variables are pressure p and temperature T. We will focus on the specific form of the first law in terms of the specific Gibbs function, which is remarkably simply:

$$\mu = \tilde{\boldsymbol{\mu}} \bullet \boldsymbol{\xi}.$$

This equation states that the chemical potential of a system is the sum of the chemical potentials of the constituents. If the system is a mixture, then each $\tilde{\mu}_i$ depends only on the

[13] Following this naming convention temperature, which is intensive, should be denoted by t, but that variable is already used for time, so we are stuck with denoting temperature by T.

[14] In the development of the equations of thermodynamics, we assume the system is spatially homogeneous and in steady state, so that all variables are independent of position and time. Later, once the proper equations have been established, we can relax this constraint and treat the thermodynamic variables as functions of $\mathbf{x}$ and t.

[15] It is never advantageous to swap $\tilde{\boldsymbol{\mu}}$ and $\boldsymbol{\xi}$.

Table E.1. *The four possible pairs of independent thermodynamic variables and their associated potentials (in specific form). The factor* $\tilde{\boldsymbol{\mu}} \cdot \mathrm{d}\boldsymbol{\xi} = \sum_i^{I-1} \tilde{\mu}_i \mathrm{d}\xi_i$ *should be appended to each of the differentials, as needed.*

Variables	Potential	Symbol	Differential
s & v	Internal energy	u	$\mathrm{d}u = T\mathrm{d}s - p\mathrm{d}v$
T & v	Helmholtz free energy	$f = u - Ts$	$\mathrm{d}f = -s\mathrm{d}T - p\mathrm{d}v$
p & s	Enthalpy	$h = u + pv$	$\mathrm{d}h = T\mathrm{d}s + v\mathrm{d}p$
p & T	Gibbs free energy	$\mu = u - Ts + pv$	$\mathrm{d}\mu = -s\mathrm{d}T + v\mathrm{d}p$

corresponding mass fraction ξ_i, but if it is a solution then each $\tilde{\mu}_i$ can be a function of all mass fractions. Often, it is convenient to employ the density ρ rather than the specific volume v; using the identity $v = 1/\rho$, the differential of the chemical potential is

$$\mathrm{d}\mu = -s\,\mathrm{d}T + (1/\rho)\,\mathrm{d}p + \tilde{\boldsymbol{\mu}} \bullet \mathrm{d}\boldsymbol{\xi}\,,$$

plus the Gibbs–Duhem relation

$$\boldsymbol{\xi} \bullet \mathrm{d}\tilde{\boldsymbol{\mu}} = -s\,\mathrm{d}T + (1/\rho)\,\mathrm{d}p\,.$$

The differential of the chemical potential gives us definitions of specific entropy and density, as dependent variables:

$$s \equiv -\left.\frac{\partial \mu}{\partial T}\right|_{p,\xi} = -\left.\frac{\partial \tilde{\boldsymbol{\mu}}}{\partial T}\right|_{p,\xi} \bullet \boldsymbol{\xi} \qquad \text{and} \qquad \frac{1}{\rho} \equiv \left.\frac{\partial \mu}{\partial p}\right|_{T,\xi} = \left.\frac{\partial \tilde{\boldsymbol{\mu}}}{\partial p}\right|_{T,\xi} \bullet \boldsymbol{\xi}\,.$$

Note that s and ρ are functions of p, T and $\boldsymbol{\xi}$. The differentials of these functions, termed the *equations of state*, are investigated in Appendix E.5. But before doing that, let's introduce and discuss the thermodynamic parameters that appear in those equations.

E.4.2 Thermodynamic Parameters

We have seen in the previous section that ρ and s are defined as first derivatives of the thermodynamic potential $\mu(p,T,\boldsymbol{\xi})$. Additional useful thermodynamic parameters naturally arise as the second partial derivatives of this potential; these are the coefficients of thermal expansion

$$\alpha \equiv \rho \left.\frac{\partial^2 \mu}{\partial T \partial p}\right|_{\xi} = -\rho \left.\frac{\partial s}{\partial p}\right|_{T,\xi} = -\frac{1}{\rho}\left.\frac{\partial \rho}{\partial T}\right|_{p,\xi},$$

the isothermal compressibility

$$\beta_T \equiv -\rho \frac{\partial^2 \mu}{\partial p^2}\bigg|_{T,\boldsymbol{\xi}} = \frac{1}{\rho}\frac{\partial \rho}{\partial p}\bigg|_{T,\boldsymbol{\xi}}$$

and the specific heat at constant pressure

$$c_p \equiv -T \frac{\partial^2 \mu}{\partial T^2}\bigg|_{p,\boldsymbol{\xi}} = T\frac{\partial s}{\partial T}\bigg|_{p,\boldsymbol{\xi}},$$

plus

$$\bar{s}_i \equiv -\frac{\partial^2 \mu}{\partial T \partial \xi_i}\bigg|_{p,\boldsymbol{\xi}_-} = -\frac{\partial \tilde{\mu}}{\partial T}\bigg|_{p,\boldsymbol{\xi}_-} = \frac{\partial s}{\partial \xi_i}\bigg|_{p,T,\boldsymbol{\xi}_-},$$

which represents the change of specific entropy with composition, and

$$\rho_{\xi_i} \equiv -\rho^2 \frac{\partial^2 \mu}{\partial p \partial \xi_i}\bigg|_{T,\boldsymbol{\xi}_-} = \frac{\partial \rho}{\partial \xi_i}\bigg|_{p,T,\boldsymbol{\xi}_-},$$

which represents the change of density with composition, where $\boldsymbol{\xi}_-$ is the set of compositions minus the ith entry.

Specific Heats

Specific heats measure the ability of a material to store heat; it is the amount of energy per unit mass needed to raise the temperature of this material by a specified amount. The ability of a body to absorb energy depends on how it is constrained, whether the volume or the pressure is held constant. As a consequence, there are two flavors of specific heat: the *specific heat at constant pressure*, denoted by c_p, and the *specific heat at constant volume*, denoted by c_v.[16] We previously encountered c_p in terms of the second partial derivative of μ with respect to T, or as the first partial derivative of s with respect to T, which doesn't seem to be in agreement with the first sentence in this paragraph. We can bring the words and math into agreement by noting that, with p and $\boldsymbol{\xi}$ constant,[17] $T\mathrm{d}s = \mathrm{d}h$ and

$$c_p = \frac{\partial h}{\partial T}\bigg|_{p,\boldsymbol{\xi}}.$$

Similarly, we can define

$$c_v \equiv \frac{\partial u}{\partial T}\bigg|_{v,\boldsymbol{\xi}} = T\frac{\partial s}{\partial T}\bigg|_{v,\boldsymbol{\xi}}.$$

These equations may be understood as follows. Suppose we reversibly add a small amount of heat per unit mass (that is, *specific heat*) $\mathrm{d}q$ to a system. Noting that $\mathrm{d}q = T\mathrm{d}s$,

[16] Since we will be using density rather than specific volume, this could just as well be called the specific heat at constant density, but nobody does.

[17] See Table E.1.

the temperature of the system increases by an amount $\mathrm{d}T = \mathrm{d}q/c_v$ if the system is held at constant volume and $\mathrm{d}T = \mathrm{d}q/c_p$ if the system is held at constant pressure. If the specific volume (or density) is held constant, all the added energy goes into speeding up the particles comprising the material (and increasing the temperature), while if the pressure is held constant, extra energy is needed to push back the surroundings as the material expands. Consequently $c_v < c_p$ and the increase of temperature will be less if the system is held at constant pressure.

The *Dulong–Petit model* of heat capacities gives

$$c_v = 3R_s, \qquad \text{where} \qquad R_s = R/M_A$$

is the specific gas constant, R is the ideal gas constant[18] and M_A is the molecular weight of the material. This model is valid for solids at high temperatures, such as in Earth's interior. In Earth's mantle, $M_A \approx 22$ and $c_v \approx 1100\ \mathrm{J{\cdot}K^{-1}{\cdot}kg^{-1}}$. Peeking ahead, we shall see in Appendix E.5.1 that $c_p = (1 + \gamma_G \alpha T)c_v$. According to Stacey (2010), the factor $(1 + \gamma_G \alpha T)$ varies between 1.03 and 1.1 in the mantle, and a good approximation is $c_p \approx 1250\ \mathrm{J{\cdot}K^{-1}{\cdot}kg^{-1}}$.

Latent Heat

The specific heats are also called sensible heats because they can be sensed by a change of temperature as heat is added to a system. This is in contrast to *latent heats* for which there is no change in temperature as heat is added to a system. As a portion of a system changes phase (at constant temperature), the system must absorb or release a certain amount of heat; the latent heat per unit mass is commonly denoted by L. The addition of latent heat to a system increases the specific entropy:

$$\Delta s = L/T,$$

where Δs is the increase of specific entropy. Note that L is the heat *added* to the system, so that the newly-formed phase is the higher-temperature (and more disorganized) phase, characterized by a greater specific entropy.

Adiabatic Compressibility

Just as there are two types of specific heats, there are two types of compressibilities. We have already encountered the isothermal incompressibility as the second partial derivative of μ or the first partial derivative of ρ with respect to p, with T held fixed. Its companion is the *adiabatic compressibility*, defined by

$$\beta_s \equiv \frac{1}{\rho}\frac{\partial \rho}{\partial p}\bigg|_{s,\xi}.$$

[18] See § 5.6.

Note that the adiabatic compressibility is the inverse of the bulk modulus:[19]

$$\beta_s = 1/K\,.$$

The bulk modulus is also called the *adiabatic incompressibility*. This is a potential source of confusion since it is the inverse of – and sounds a lot like – the adiabatic compressibility. Incompressibility has dimensions of pressure, while compressibility has dimensions of inverse pressure.

It is readily verified that

$$\beta_T = \beta_s + \frac{\alpha^2 T}{\rho c_p}$$

and $\beta_s < \beta_T$; a system held at constant temperature is more compressible than one held at constant entropy. This may be understood as follows. When compressed with s held fixed, a system heats up. The added heat translates to an added pressure[20] and an added resistance to compression.

We can readily verify that

$$c_v \beta_T = c_p \beta_s$$

and it is a useful exercise to verify the interesting identity

$$\frac{\partial(V,p)}{\partial(S,T)} = 1 \qquad \text{or equivalently} \qquad \frac{\partial(p,\rho)}{\partial(T,s)} = \rho^2\,,$$

where the symbol on the left-hand sides of these equations is the Jacobian; see Appendix A.2.7.

All the thermodynamic parameters (such as α, β_s, β_T and c_p) are themselves functions of the independent variables p, T and ξ_i. If the variations from a uniform reference state[21] are small, these may be treated as constants.

E.5 Equations of State

Using the chemical-potential differential as the fundamental form of the energy equation, s and ρ are functions of p, T and $\boldsymbol{\xi}$ with differentials (i.e., equations of state):

$$\mathrm{d}s = \frac{c_p}{T}\,\mathrm{d}T - \frac{\alpha}{\rho}\,\mathrm{d}p + \bar{\mathbf{s}}\bullet\mathrm{d}\boldsymbol{\xi} \qquad \text{and} \qquad \mathrm{d}\rho = -\alpha\rho\,\mathrm{d}T + \rho\beta_T\,\mathrm{d}p + \bar{\boldsymbol{\rho}}\bullet\mathrm{d}\boldsymbol{\xi}\,,$$

where the vector $\bar{\mathbf{s}}$ represents the set of coefficients $\bar{s}_i$ for $i = 1$ to $I-1$ and the vector $\bar{\boldsymbol{\rho}}$ represents the set of coefficients ρ_{ξ_i} for $i = 1$ to $I-1$. Commonly changes in the mass

[19] Introduced in § 5.5 and discussed in Appendix D.3.

[20] See Appendix E.8.4.

[21] The concept of a reference state is introduced in § 2.1.1. If the reference state is not uniform, these parameters may be known functions.

fraction of constituents are driven by changes of phase, with the difference in specific entropy being related to the latent heat.

E.5.1 Alternate Equations of State

In most situations, p & T are the best choice of independent thermodynamic variables and the equations of state are differentials for $\rho(p,T,\boldsymbol{\xi})$ and $s(p,T,\boldsymbol{\xi})$. But in some cases, we may want equations of state with other variables, such as $\rho(p,s,\boldsymbol{\xi})$ and $s(T,v,\boldsymbol{\xi})$. Eliminating $\mathrm{d}T$ between the differentials for $\rho(p,T,\boldsymbol{\xi})$ and $s(p,T,\boldsymbol{\xi})$ we have the differential equation of state for $\rho(p,s,\boldsymbol{\xi})$:

$$\mathrm{d}\rho = -\frac{\alpha\rho T}{c_p}\,\mathrm{d}s + \beta_s\rho\,\mathrm{d}p + \left(\frac{\alpha T\rho}{c_p}\bar{\mathbf{s}} + \bar{\boldsymbol{\rho}}\right)\bullet\mathrm{d}\boldsymbol{\xi}\,.$$

If a process proceeds isentropically and without change of composition or phase, $\mathrm{d}s = 0$, $\mathrm{d}\boldsymbol{\xi} = \mathbf{0}$ and the equation of state for density simplifies to

$$\mathrm{d}\rho = \beta_s \mathrm{d}p = (\rho/K)\,\mathrm{d}p$$

On the other hand, eliminating $\mathrm{d}p$ between the differentials for $\rho(p,T,\boldsymbol{\xi})$ and $s(p,T,\boldsymbol{\xi})$ we have the differential equation of state for $s(T,v,\boldsymbol{\xi})$:

$$\mathrm{d}s = \frac{c_v}{T}\mathrm{d}T - \frac{c_v}{\rho}\gamma_G\,\mathrm{d}\rho + \left(\bar{\mathbf{s}} + \frac{\alpha}{\beta_T\rho^2}\bar{\boldsymbol{\rho}}\right)\bullet\mathrm{d}\boldsymbol{\xi}$$

where

$$\gamma_G = \frac{\alpha}{\rho\beta_T c_v} = \frac{\alpha}{\rho\beta_s c_p} = \frac{\alpha K}{\rho c_p}$$

is the *Grüneisen parameter*. The subscript G has been added to distinguish γ_G from its close relative $\gamma = \gamma_G + 1 = c_p/c_v$ for an ideal gas.[22]

Atmospheric gases are particularly expansive in response to changes of temperature (having $\alpha = 1/T$), so that γ_G is of unit order and c_p is appreciably greater than c_v in the atmosphere. However, water and silicates are much less expansive, so that $c_p \approx c_v$ and $\beta_s \approx \beta_T$ for these materials.

Note that

$$c_v = \frac{c_p}{1 + \gamma_G \alpha T}$$

and

$$\left.\frac{\partial T}{\partial \rho}\right|_{s,\xi} = \frac{\gamma_G T}{\rho}, \qquad \text{or equivalently} \qquad \left.\frac{\partial \ln T}{\partial \ln \rho}\right|_{s,\xi} = \gamma_G\,.$$

[22] See Appendix E.8.5.

E.5.2 Alternate Form of the First Law

The specific, reversible form of the first law, $\mathrm{d}q = T\mathrm{d}s$, can be combined with the differential equation of state for entropy,[23] $\mathrm{d}s = (c_p/T)\,\mathrm{d}T - (\alpha/\rho)\,\mathrm{d}p + \bar{\mathbf{s}}\bullet\mathrm{d}\boldsymbol{\xi}$, to obtain an alternate form of the first law:

$$\mathrm{d}q = T\mathrm{d}s = c_p\,\mathrm{d}T - (\alpha T/\rho)\,\mathrm{d}p + T\bar{\mathbf{s}}\bullet\mathrm{d}\boldsymbol{\xi}\,.$$

Recall that $\mathrm{d}q$ is positive when heat is added to the system. This equation states that if the pressure of a system is increased at constant temperature and composition, then $\mathrm{d}q < 0$; as the system is compressed, it must cast off heat in order to remain at a constant temperature.

The term $\mathrm{d}p/\rho$ appearing in this equation also occurs in the differential form of the momentum equation presented at the end of § 4.6.3. Eliminating this factor between the two we have

$$\mathrm{d}q = T\mathrm{d}s = c_p\,\mathrm{d}T + \alpha T\mathrm{d}\left(\tfrac{1}{2}v^2 + \psi\right) - \mathbf{f}_D\bullet\mathrm{d}\mathbf{l} + T\bar{\mathbf{s}}\bullet\mathrm{d}\boldsymbol{\xi}\,,$$

where v is the flow speed and ψ is the gravitational potential. (Normally $\psi = gz$, where z is the upward coordinate.) To understand this equation, suppose we move a parcel upward ($\mathrm{d}\psi > 0$) reversibly, slowly and with constant composition; this equation simplifies to $\mathrm{d}q = c_p\,\mathrm{d}T + \mathrm{d}\psi$. The increase in ψ is balanced by either the addition of heat ($\mathrm{d}q > 0$) and/or cooling ($\mathrm{d}T < 0$).

E.5.3 Equation of State for Sea Water

The oceans consist of liquid water (the primary constituent) plus dissolved elements and ions (electrically charged atoms and molecules). Water has been called the universal solvent, and traces of all elements can be found in sea water (including 20 million tons of gold). The most important of these elements are those comprising salts (particularly ions of sodium and chlorine),[24] and it is common to treat the oceans as consisting of water and (a single) salt, with an equation of state given in differential form by

$$\mathrm{d}\rho = -\alpha\rho\,\mathrm{d}T + \rho\beta_T\,\mathrm{d}p + \rho_h\mathrm{d}\xi_h$$

where ρ is the density of sea water, ξ_h is the mass fraction of salts, called the *salinity* (commonly denoted by S in the oceanographic literature) and $\rho_h \equiv \partial\rho/\partial\xi_h|_{p,T}$ represents the increase of density with increasing salt content.[25] The salinity is the number of grams of salt in a kilogram of sea water.[26] The mean value of ξ_h in sea water is about 0.035 (35 grams per kilogram) and the mean density of sea water is about 1.025×10^3 kg·m^{-3}, with the 2.5% increase over the fresh-water value being due to the salt content.

[23] From Appendix E.5.
[24] 55% of sea-salt ions are chloride (Cl^-) and 30.6% are the sodium ion Na^+.
[25] See § 5.2. An integrated form is the international equation of state of seawater, 1980.
[26] The mass fraction of salt is $S/1000$.

The coefficients α, β_T and ρ_h commonly are treated as constants, with approximate values being

- $\alpha \approx 2 \times 10^{-4}\ \mathrm{K}^{-1}$;
- $\beta_T \approx 5 \times 10^{-10}$ Pa; and
- $\rho_h \approx 714\ \mathrm{kg{\cdot}m}^{-3}$.

The compressibility of water is sufficiently small that the increase in density at the bottom of the ocean due to compression is about 2%. Since variations of density with pressure have no dynamic consequences, the compression of water can be safely ignored, and the gross static structure of the oceans is particularly simple: $\rho = \rho_w$, a constant. But the small departures from this simple state, due to variations of temperature and salinity are dynamically important.

E.6 Ideal Mixture

An *ideal mixture* is a material composed of two or more constituents (usually different phases), each of which obeys its own equation of state. In particular, the equation of state for density of an ideal mixture obeys *Amagat's law of additive volumes*, which states that the volume of the mixture is the sum of volumes of each constituent.[27] Expressed in terms of densities, Amagat's law is

$$\frac{1}{\rho} = \sum_{i=1}^{I} \frac{\xi_i}{\rho_i},$$

where ρ_i is the density of the ith constituent by itself, ξ_i is the mass fraction of the ith constituent and I is the total number of constituents. Amagat's law was developed for mixtures of ideal gases, but it can apply to a mixture of liquid and gas, such as liquid water and air or exsolving gas and magma in a volcanic conduit (see Chapter 34).

E.7 The Energy Equation

Up to now, we have been investigating the behavior of homogeneous thermodynamic systems, without concern for their variation in space or time. In this section, we remedy this deficiency, by investigating the spatial and temporal variations of various forms of energy, as represented in the *energy equation*, which controls the variation of temperature of a parcel of fluid. The basic procedure is to replace differentials with material derivatives in the appropriate thermodynamic relation. By rights, the energy equation should be called the entropy equation, because it is based on the entropy differential $T\mathrm{d}s = \mathrm{d}q$, which represents the increase in the specific entropy of a parcel due to the addition of heat. The

[27] Amagat's law is more general than Dalton's law of partial pressures, which applies only to mixtures of gases.

corresponding material derivative is

$$\rho T \frac{\mathrm{D}s}{\mathrm{D}t} = \rho \frac{\mathrm{D}q}{\mathrm{D}t},$$

where q represents the sources of heat added to the parcel in question. (By multiplying through by ρ, the equation now is on a per-unit-volume basis, rather than per-unit-mass.) These consist of

- conduction of heat across the parcel boundary, represented by $\nabla \bullet (k \nabla T)$, where k is the thermal conductivity;
- viscous dissipation of mechanical energy,[28] represented by $\dot{w} = \dot{\epsilon}'_{ij} \tau_{ij}$; and
- radioactive heating, represented by $\rho \Psi_R$, where Ψ_R is the rate radioactive heating per unit mass; $[\Psi]$=W·kg^{-1}.

Altogether we have a preliminary version of the energy equation

$$\rho T \frac{\mathrm{D}s}{\mathrm{D}t} = \nabla \bullet (k \nabla T) + \dot{\epsilon}'_{ij} \tau_{ij} + \rho \Psi_R .$$

Using the entropy differential ($T\mathrm{d}s = c_p \mathrm{d}T + \alpha T \mathrm{d}\left(\frac{1}{2}v^2 + \psi\right) + T\bar{\mathbf{s}} \bullet \mathrm{d}\boldsymbol{\xi}$)[29] and noting that

$$\frac{\mathrm{D}\psi}{\mathrm{D}t} = g \frac{\mathrm{D}z}{\mathrm{D}t} = -\mathbf{g} \bullet \mathbf{v}$$

we have the energy equation in all its glory:

$$\rho c_p \frac{\mathrm{D}T}{\mathrm{D}t} = \nabla \bullet (k \nabla T) + \alpha T \rho \left(\mathbf{g} \bullet \mathbf{v} - \frac{1}{2} \frac{\mathrm{D}v^2}{\mathrm{D}t} \right) - \rho T \bar{\mathbf{s}} \bullet \frac{\mathrm{D}\boldsymbol{\xi}}{\mathrm{D}t} + \dot{\epsilon}'_{ij} \tau_{ij} + \rho \Psi_R .$$

This equation states that the change of temperature experienced by a parcel is due to six effects quantified by the six terms on the right-hand side:

- diffusion of heat (this is the most familiar term);
- change of (hydrostatic) pressure induced by vertical motion;
- acceleration of the material (this term is usually ignored);
- change of composition or phase;
- viscous dissipation of mechanical energy; and
- radioactive or radiative heating.

The terms Ψ_R and $\dot{\epsilon}'_{ij} \tau_{ij}$ cannot be negative, but the other four terms on the right-hand side can be either positive or negative.

[28] First introduced in § 4.5.1. This term may include work done by electromagnetic forces (that is, Ohmic dissipation), as well.

[29] From Appendix E.5.2, omitting the drag term, since we already have included that as the viscous dissipation term.

If radioactive heating, viscous dissipation and adiabatic compression are negligible and composition is constant, the energy equation simplifies to

$$\rho c_p \frac{\mathrm{D}T}{\mathrm{D}t} = \nabla \bullet (k \nabla T) .$$

Further, if the thermal conductivity is constant, this equation simplifies to the *heat equation*:

$$\frac{\mathrm{D}T}{\mathrm{D}t} = \kappa \nabla^2 T ,$$

where $\kappa = k/\rho c_p$ is the *thermal diffusivity*.

E.8 Ideal Gas

An *ideal gas*, also called a *perfect gas*, has a relatively simple equation of state, called the *ideal gas law*:

$$pV = NkT = nRT ,$$

where V is the volume, p is the pressure, T is the temperature, N is the number of atoms, k is Boltzmann's constant, n is the amount of substance (in moles) and R is the ideal gas constant. These two constants are discussed in Appendix E.8.1.

E.8.1 Boltzmann's Constant and the Gas Constants

In essence, Boltzmann's constant (or the Boltzmann constant) is the ratio of the product of pressure and volume to the product of the number of atoms present and temperature, as codified in the ideal gas law given above. It takes on a number of numerical forms depending on the system of units employed. It is clear from the ideal gas law and the definition of the mole that

$$R = N_A k ,$$

where $N_A \approx 6.02214 \times 10^{26}\,\mathrm{kmol}^{-1}$ is Avogadro's constant. The standard (SI) value of k is approximately[30] 1.38×10^{-23} J·K^{-1} and the standard (SI) value of R is approximately[31] 8.314 J·mol^{-1}·K^{-1} or equivalently, 8314 J·kmol^{-1}·K^{-1}. We will use the latter value, based on the kilogram-mole.

[30] More precisely, the numerical coefficient is 1.38064852. Alternate representations in other units include $k \approx 1.38 \times 10^{-23}$ m^2·kg·s^{-2}·K^{-1}, $k \approx 1.38 \times 10^{-16}$ erg·K^{-1} and $k \approx 8.617 \times 10^{-5}$ev·K^{-1}.

[31] More precisely, the numerical coefficient is 8314.4598. Alternate representations in other units include $R \approx$ 8314 m^2·s^{-2}·kmol^{-1}·K^{-1}, $R \approx$ 8.314 m^2·s^{-2}·mol^{-1}·K^{-1}, $R \approx 5.189 \times 10^{19}$ ev·mol^{-1}·K^{-1}, $R \approx$ 8314 m^3·Pa·kmol^{-1}·K^{-1} and $R \approx 8314 \times 10^7$ erg·kmol^{-1}·K^{-1}.

Actually there are two gas constants in use, the ideal gas constant denoted by R, and a specific gas constant, denoted by R_s. These are related by

$$R_s = nR/M \qquad \text{or equivalently}^{32} \qquad R_s = R/M_A\,.$$

The value of R_s depends on the nature of the material.

E.8.2 Maxwell–Boltzmann Distribution

The probability distribution, ψ, of ideal-gas speeds is given by the *Maxwell–Boltzmann distribution*:

$$\psi(v) = \sqrt{\frac{2}{\pi}}\left(\frac{m}{kT}\right)^{3/2} v^2 e^{-mv^2/2kT}\,,$$

where v is the particle speed, m is its mass, k is Boltzmann's constant and T is the temperature. The integral of this gives the probability function; the probability that a particle has speed less than v is given by

$$\begin{aligned}\Phi(v) &= \int_0^v \psi(\grave{v})\mathrm{d}\grave{v} \\ &= \operatorname{erf}\left(\sqrt{\frac{m}{2kT}}v\right) - \sqrt{\frac{2}{\pi}}\left(\frac{m}{kT}\right)^{1/2} e^{-mv^2/2kT}\,,\end{aligned}$$

where the error function, erf, was introduced in § 7.2.3. The mean speed (for which $\Phi = 1/2$) is $\bar{v} \approx 1.538173\sqrt{kT/m}$.

E.8.3 Specific Form of the Ideal Gas Law

Often it is useful to express the ideal gas law in intensive or specific form, rather in extensive form and use the density rather than the specific volume. Noting that $V = M/\rho$, where M is the total mass of the gas, the specific ideal gas law may be expressed as

$$pv = R_sT \qquad \text{or equivalently} \qquad p = \rho R_s T\,,$$

where $v = V/M = 1/\rho$ is the *specific volume*.

E.8.4 Relation to Kinetics

The perfect gas law can be derived by considering the behavior of an ideal gas confined within a cubic box having sides of length l, with the individual particles (atoms or molecules) having mean speed[33] $\bar{v}$; see Appendix D.7. We shall assume the particle

[32] See Appendix E.2.

[33] More precisely, this is the root-mean square speed: $\bar{v} = \sqrt{\sum_{n=1}^{N} v^2/N}$.

motions to be isotropic, so that the mean speed in any direction, say x, is given by $\bar{v}_x = \bar{v}/\sqrt{3}$.

The pressure p within the box is due to the change of momentum as atoms bounce off the walls; this is quantified in the next subsection, while the temperature T is a macroscopic measure of the kinetic energy of the atoms, as investigated in the following subsection. As it happens, both p and T are proportional to the square of the mean speed, $\bar{v}^2$. Eliminating $\bar{v}^2$ between the equations for p and T yields the specific form of the ideal gas law: $p = \rho R_s T$.

Pressure

The particles are confined to the box. As a particle reaches a wall of the box, it bounces off and in doing so, imparts an impulse of momentum to that wall, in a direction normal to the wall. The pressure is the macroscopic sum of all these microscopic impulses. The average particle moves toward or away from a wall with speed $\bar{v}_x$ and strikes that wall periodically with time interval $2l/\bar{v}_x$ between strikes on that wall. Each strike changes the x component of momentum of the particle by $2m\bar{v}_x$, where m is the mass of a particle. The average force exerted normal to the wall by this particle is the momentum change divided by the time interval:

$$\bar{F}_x = \frac{m\bar{v}_x^2}{l} = \frac{m\bar{v}^2}{3l}.$$

The pressure is this force multiplied by the number of particles, N, and divided by the area of the wall, l^2. Noting that $\rho = mN/V$ and $l^3 = V$, this gives

$$p = \tfrac{1}{3}\rho\bar{v}^2.$$

Temperature

The temperature is a macroscopic measure of the specific internal energy, u, of a system,[34] which is the sum of the kinetic and potential energies of its particles. Kinetic energy consists of translational energy, rotational energy and energies of internal vibrations. A monatomic ideal gas has no rotational energy, internal motions or internal potential energies; its only energy is the kinetic energy of its particles.[35] It follows that the total internal energy of a mass M of an ideal gas is equal to $M\bar{v}^2/2$ and the specific internal energy, u, is simply[36]

$$u = \tfrac{1}{2}\bar{v}^2.$$

By definition, T is proportional to u, with the constant of proportionality (for a monatomic gas) being $3R_s/2$, where R_s is *specific gas constant*:

$$u = \tfrac{3}{2}R_s T.$$

[34] The internal energy per unit mass, having SI units $\mathrm{J{\cdot}kg^{-1}}$.
[35] A diatomic ideal gas does have rotational energy; see Appendix E.8.5.
[36] See Appendix D.7.

Eliminating u between these last two equations, we have that

$$\bar{v}^2 = 3R_sT\,.$$

This expression can be used to estimate the speed of the gas particles. For example, suppose the particles in question are atoms of ^{40}Ar, having masses $m = 6.6 \times 10^{-26}$ kg or equivalently $M_A = 40$ kg·kmol^{-1}. At an absolute temperature of 300 K, the atoms have an average thermal speed of 433 m·s^{-1}. Much higher speeds than these are encountered in explosively propelled objects (bullets, etc.) and in very high speed flight (e.g., orbital flight). Typical orbital speeds are about 8 km·s^{-1}, which translates into temperatures in excess of 100,000 K.

E.8.5 Thermodynamics of an Ideal Gas

It is easy to verify that the coefficient of thermal expansion, the isothermal compressibility and the adiabatic compressibility for an ideal gas are given by

$$\alpha = \frac{1}{T}, \qquad \beta_T = \frac{1}{p} \qquad \text{and} \qquad \beta_s = \frac{1}{\gamma p}\,.$$

The last of these is equivalent to

$$K = \gamma p, \qquad \text{where} \qquad \gamma = c_p/c_v$$

is the ratio of specific heats.

We have seen in Appendix E.8.4 that the specific internal energy of a monatomic ideal gas is given by $u = 3R_s/2$ and the enthalpy is[37]

$$h = u + pv = \tfrac{3}{2}R_sT + R_sT = \tfrac{5}{2}R_sT\,.$$

Also, using the formulas for the specific heats developed in Appendix E.4.2, we have[38]

$$c_v = \tfrac{3}{2}R_s \qquad \text{and} \qquad c_p = \tfrac{5}{2}R_s = c_v + R_s$$

and $\gamma = 5/3$ for a monatomic perfect gas.

Diatomic Ideal Gas

We saw in Appendix E.8.4 that the internal energy of a monatomic ideal gas, due to translational motions only, is given by $u = 3R_sT/2$. Now translation involves three degrees of freedom, with each degree of freedom accounting for $R_sT/2$ of the internal energy. If the gas is diatomic, it has two additional degrees of freedom due to rotation.[39] In equilibrium,

[37] Introduced in Table E.1.

[38] The latter form for c_p is known as *Mayer's formula*.

[39] A diatomic gas has in addition a vibrational mode, but this mode is dynamically decoupled from the other five.

there is equipartition of energy among modes. In the case of a diatomic ideal gas, with five degrees of freedom, the internal energy is given by

$$u = 5R_sT/2 .$$

The specific heats of an ideal diatomic gas are

$$c_v = \tfrac{5}{2}R_s \qquad \text{and} \qquad c_p = \tfrac{7}{2}R_s$$

and the ratio of specific heats is $\gamma = 7/5 = 1.4$.

E.8.6 Speed of Sound

We saw in § 9.3 that the speed of sound is given by $U_s = \sqrt{K/\rho}$. Noting that $K = \gamma p$, $p = \rho\bar{v}^2/3$ and $\bar{v} = \sqrt{3R_sT}$ for an ideal gas, we have

$$U_s = \sqrt{\gamma p/\rho} = \sqrt{\gamma/3}\bar{v} = \sqrt{\gamma R_sT} .$$

The speed of sound is of the same order as, but a bit smaller than, the average speed of the atoms.

E.9 Thermodynamics of the Atmosphere

The atmosphere is an open thermodynamic system consisting of dry air, water vapor and condensed water (liquid and solid). The system is open principally because water enters via evaporation and leaves by falling to the ground as rain, snow, hail, etc.

E.9.1 Atmosphere Energy Equation

In the atmosphere, heat sources (conduction of heat, radioactive heating and viscous dissipation) are negligible and the only constituent of importance is the vapor content, ξ_v. With these simplifications and noting that $\alpha T = 1$ for an ideal gas,[40] the energy equation[41] becomes

$$c_p\frac{\mathrm{D}T}{\mathrm{D}t} = \mathbf{g}\bullet\mathbf{v} - v\frac{\mathrm{D}v}{\mathrm{D}t} - L\frac{\mathrm{D}\xi_v}{\mathrm{D}t} ,$$

where $L = T\bar{s}_v$ is the *latent heat of vaporization* of water: the energy per unit mass of water vapor, compared to the mean energy per unit mass of the atmosphere. The atmosphere gains water vapor due to evaporation near the ground and loses it due to condensation and precipitation. It follows that $\mathrm{D}\xi_v/\mathrm{D}t$ is (almost always) negative. As water vapor condenses, it essentially vanishes from the atmosphere, but not quite entirely. Like the Cheshire cat's smile, it leaves behind a virtual presence in the form of released latent

[40] See Appendix E.8.5.
[41] From Appendix E.7.

heat. This occurs as follows. Molecules of water vapor have a range of speeds and those that condense from the vapor have slower speeds than average. As these slow molecules leave the system, the remaining molecules have, on average, greater speeds. The amount of energy left behind is quantified by the latent heat, L. This heat is seen as an increase in temperature over that which would have prevailed in the absence of condensation. The differential form of this equation is

$$c_p \mathrm{d}T = \mathbf{g} \bullet \mathrm{d}\mathbf{x} - v\mathrm{d}v - L\mathrm{d}\xi_v \,,$$

where $\mathrm{d}\mathbf{x} = \mathbf{v}\mathrm{d}t$ is the change of position. Writing $\mathbf{g} = -g\mathbf{1}_z$ with z being the upward coordinate, this becomes

$$c_p \mathrm{d}T = -g\mathrm{d}z - v\mathrm{d}v - L\mathrm{d}\xi_v \,.$$

Normally the energy equation is presented without the kinetic energy term $v\mathrm{d}v$. Since $c_p \approx 1000\ \mathrm{J{\cdot}kg^{-1}{\cdot}K^{-1}}$, a change of 1 K corresponds to a speed (starting with calm conditions) of $\approx 45\ \mathrm{m{\cdot}s^{-1}}$, which corresponds to a category 2 hurricane; see Table 28.1. Since these speeds are rare, it is reasonable to omit the kinetic energy term under normal circumstances.

In the absence of condensation, $\mathrm{d}\xi_v = 0$ and the energy equation for the static atmosphere ($\mathrm{d}v = 0$) simplifies to the dry adiabat:[42]

$$\frac{\mathrm{d}T}{\mathrm{d}z} = -\frac{g}{c_p} \approx 9.8\ \mathrm{K \cdot km^{-1}} \,.$$

E.9.2 Dry Atmosphere

Earth's dry atmosphere is comprised of molecules that behave, to a good approximation, as a diatomic ideal gas,[43] obeying the equation of state

$$pv = R_d T \qquad \text{or equivalently} \qquad p = \rho R_d T \,,$$

where p is the pressure, $v = 1/\rho$ is the specific volume, T is the temperature and

$$R_d = R/M_A$$

is the specific gas constant, with $R = 8314\ \mathrm{m^2{\cdot}s^{-2}{\cdot}kmol^{-1}{\cdot}K^{-1}}$ being the ideal gas constant and M_A being the mass of one kilogram-mole of atoms (having units $\mathrm{kg{\cdot}kmol^{-1}}$).[44] The numerical value of M_A is equal to the atomic number of the gas. Earth's dry atmosphere has a mean molecular weight $M_A = 28.97\ \mathrm{kg{\cdot}kmol^{-1}}$ and the specific gas constant for the atmosphere is $R_d = 287\ \mathrm{J{\cdot}kg^{-1}{\cdot}K^{-1}}$.

[42] Given in § 5.5.
[43] See Appendix E.8.5.
[44] See the fundament on ideal gas found in Appendix E.8.

E.9.3 Moist Atmosphere

The total density (ρ) of the atmosphere is the sum of the reduced densities of dry air ($\tilde{\rho}_d$) and water ($\tilde{\rho}_w$) with the water consisting of vapor ($\tilde{\rho}_v$) and condensate (liquid or solid) ($\tilde{\rho}_c$):[45]

$$\rho = \tilde{\rho}_d + \tilde{\rho}_w \qquad \text{and} \qquad \tilde{\rho}_w = \tilde{\rho}_v + \tilde{\rho}_c \,.$$

It is convenient to divide these equations by ρ and write them as

$$1 = \xi_d + \xi_w \qquad \text{and} \qquad \xi_w = \xi_v + \xi_c \,,$$

where $\xi_i = \tilde{\rho}_i/\rho$ is the mass fraction of constituent i.[46] Under normal atmospheric conditions, $\xi_w \ll 1$ and $\xi_d \approx 1$. The ratios ξ_d and ξ_w are relatively constant, but the ratios ξ_v and ξ_c vary as water evaporates or condenses. The factor ξ_v is called the *specific humidity*.

The condensed water does not contribute to the pressure, so that *Dalton's law*, is simply

$$p = p_d + p_v \,,$$

where p_d is the pressure of the dry air and p_v is the partial pressure of the water vapor. Under normal circumstances, $p_v \ll p_d$ and $p \approx p_d$. However, this does not mean that p_v is unimportant, as its temperature dependence controls the rate of change of $\tilde{\rho}_c$. The dry air and water vapor obey the ideal gas law:

$$p_d = \tilde{\rho}_d R_d T \qquad \text{and} \qquad p_v = \tilde{\rho}_v R_v T \,,$$

where $R_d = 287\ \mathrm{J{\cdot}kg^{-1}{\cdot}K^{-1}}$ and $R_v = 461.5\ \mathrm{J{\cdot}kg^{-1}{\cdot}K^{-1}}$ are the specific gas constants for dry air and water vapor, respectively.

Combining the equations for partial pressures, the equation of state for moist air may be expressed as

$$p = \rho R_m T \qquad \text{or equivalently} \qquad p = \rho R_d T_v \,,$$

where

$$R_m = R_d(1 - \xi_v - \xi_c) + R_v \xi_v$$

is the effective gas constant and

$$T_v = T\left(1 + \frac{R_v - R_d}{R_d}\xi_v - \xi_c\right)$$

is the *virtual temperature* – the temperature of a parcel of dry air having the same density and pressure as the moist parcel. Commonly $\xi_c \ll 1$ and this factor can be neglected.[47] Note that $R_d/R_v \approx 0.622$, $(R_v - R_d)/R_d \approx 0.606$ and $T_v \approx (1 + 0.606\xi_v)T$. The mean

[45] See § 6.4.1.
[46] Commonly in the literature the mass fraction of vapor is denoted by q: $\xi_v = q$.
[47] Condensed water tends to fall from the sky.

mass concentration of water vapor in the atmosphere is roughly 2% and the effective gas constant is $R_m \approx 290\ \mathrm{J{\cdot}kg^{-1}K^{-1}}$. Since $R_d < R_v$, R_m increases with increasing water vapor and at constant p and T the density decreases with increasing water vapor.

Entropy Equation of State

The general equation of state for entropy is $T\mathrm{d}s = c_p\,\mathrm{d}T - (\alpha T/\rho)\,\mathrm{d}p + T\bar{\mathbf{s}}\bullet\mathrm{d}\boldsymbol{\xi}$.[48] Since the atmosphere behaves – to a good approximation – as a perfect gas, we can set $\alpha T = 1$. The constituent of interest in the atmosphere is water vapor, ξ_v, and the associated coefficient is the latent heat, $T\bar{s} = L$; see Appendix E.4.2. Altogether, the differential equation of state for entropy in the atmosphere is

$$T\mathrm{d}s = c_p\,\mathrm{d}T - (1/\rho)\,\mathrm{d}p + L\mathrm{d}\xi_v\,.$$

E.9.4 Saturation

The amount of water vapor that the atmosphere can contain is given by the *saturation vapor pressure*, denoted by p^*, which is highly sensitive to temperature. As a parcel of air containing water vapor ascends without condensation, the vapor partial pressure p_v remains constant, but p^* decreases because the temperature is decreasing. As long as $p^* > p_v$, all the water remains in vapor form. However, p_v cannot exceed p^*; as p^* continues to decrease, so $p_v = p^*$ must decrease. This is accomplished by converting some of the vapor to liquid, forming a cloud. Saturation variables are denoted by an asterisk; within the cloud

$$p_v = p^*\,, \qquad \tilde{\rho}_v = \rho^* \qquad \text{and} \qquad \xi_v = \xi^*\,.$$

If there is a single phase of water, according to *Gibbs's phase rule*[49] the system has three independent thermodynamic variables: e.g., p, T and p_v. When a second phase of water is present, the system has one fewer independent variables, or equivalently, there is a constraint on the permissible values of T and $p_v = p^*$. This constraint is quantified by the *Clausius–Clapeyron equation*,[50] written as

$$\frac{\mathrm{d}p^*}{\mathrm{d}T} = \frac{L}{T\Delta v}\,.$$

where L is the latent heat of vaporization of water and $\Delta v = v_v - v_l$ is the change of specific volume of water upon vaporization. Since $v_l \ll v_v$, $\Delta v \approx v_v$. Using this and the ideal gas law ($v_v = R_v T/p^*$),

$$\frac{\mathrm{d}p^*}{\mathrm{d}T} = \frac{T_L p^*}{T^2}\,, \qquad \text{where} \qquad T_L \equiv \frac{L}{R_v}\,.$$

[48] From Appendix E.5.
[49] See Appendix E.10.
[50] See Appendix E.10.1.

With $L \approx 2.5 \times 10^6$ J·kg^{-1} and $R_v \approx 461.5$ J·kg^{-1}·K^{-1}, $T_L \approx 5400$ K. Treating T_L as constant, this equation is readily integrated to yield the saturation pressure as a function of temperature:[51]

$$p^*(T) = p_0 \exp\left(\frac{T_L}{T_B} - \frac{T_L}{T}\right),$$

where p_0 is atmospheric pressure[52] and $T_B = 373\,\mathrm{K}$ is the boiling temperature of water at 1 atmosphere.

The amount of vapor in saturated air is given by the ideal-gas equation

$$\rho^* = \frac{p^*}{R_v T} = \frac{p_0}{R_v T} \exp\left(\frac{T_L}{T_B} - \frac{T_L}{T}\right)$$

or equivalently

$$\xi^* = \frac{p^*}{R_v \rho T} = \frac{R_m p^*}{R_v p},$$

where now $R_m = R_d(1-\xi^*) + R_v \xi^*$. Solving explicitly for ξ^* we have

$$\xi^* = \frac{R_d p^*}{R_d p^* + R_v(p - p^*)}.$$

This states that the mass fraction of water vapor in the atmosphere increases with p^* (and hence temperature), reaching unity when $p^* = p$.

Commonly $p^* \ll p$ in the atmosphere and

$$\xi^*(p,T) \approx \frac{R_d p^*(T)}{R_v p}.$$

Note that the differential of this expression is[53]

$$\begin{aligned} \mathrm{d}\xi^* &\approx \xi^* \left(\frac{1}{p^*}\mathrm{d}p^* - \frac{1}{p}\mathrm{d}p\right) \\ &\approx \frac{T_L \xi^*}{T^2}\mathrm{d}T + g\frac{\xi^*}{R_v T}\mathrm{d}z. \end{aligned}$$

Moist Adiabat

Substituting this differential for $\mathrm{d}\xi^*$ into the static energy differential gives an expression for the moist adiabatic lapse rate:

$$\frac{\mathrm{d}T}{\mathrm{d}z} = -\frac{g}{c_p}\zeta,$$

[51] The inverse of this is the boiling temperature as a function of pressure.

[52] 1 atmosphere = 101,325 Pa.

[53] The latter form is obtained using the differential for p^*, the hydrostatic relation $\mathrm{d}p = -g\rho\mathrm{d}z$ and the approximate ideal gas equation $p \approx \rho R_v T$.

where

$$\zeta(p,T) = \left(1 + \frac{T_c T_L \xi^*}{T^2}\right)^{-1} \left(1 + \frac{T_L \xi^*}{T}\right),$$
$$T_c = L/c_p \approx 2500\,\mathrm{K} \qquad \text{and} \qquad T_L = L/R_v \approx 5400\,\mathrm{K}.$$

The factor ζ represents the ratio of the moist to dry lapse rate. It is readily verified that $\zeta < 1$, $\mathrm{d}\zeta/\mathrm{d}p > 0$ and $\mathrm{d}\zeta/\mathrm{d}T < 0$. The moist adiabatic lapse rate is less than the dry adiabatic lapse rate and the fractional decrease becomes greater as the temperature increases and the pressure decreases. If the air is sufficiently moist,

$$\zeta \approx \frac{T}{T_c} \qquad \text{and} \qquad \frac{\mathrm{d}T}{\mathrm{d}z} \approx -\frac{gT}{L}.$$

This gradient has a maximum magnitude at the ground and decreases with altitude as the temperature decreases.

E.10 Phase Equilibrium

A simple thermodynamic system (composed of a single component or type of material) in equilibrium may be in one of three phases: solid, liquid or gas. The phase that occurs is the one having the lowest value of the Gibbs free energy.[54] Two phases are in equilibrium when the Gibbs free energies of the two phases have equal magnitude. This equality places a constraint on the permissible values of p and T. That is, a liquid composed of a single material has definite temperatures at which freezing and boiling (at constant pressure) occur: $T_F(p)$ and $T_B(p)$.

The relation between the number of phases present and the number of independent variables is given by Gibbs's phase rule:

$$f = c - p + 2,$$

where f is the number of thermodynamic degrees of freedom, c is the number of components and p is the number of phases present.

The functions $T_F(p)$ and $T_B(p)$ are determined by the Clausius–Clapeyron equation, which is developed in the next section.

E.10.1 Clausius–Clapeyron Equation

The *Clausius–Clapeyron equation* governs the variation of a phase-change temperature with pressure. In order to develop this equation, let's begin by considering a system of a single component in a single phase with the Gibbs differential[55]

$$\mathrm{d}\mu = -s\,\mathrm{d}T + v\,\mathrm{d}p.$$

[54] Introduced in Appendix E.4.1.
[55] See Table E.1.

A change of phase occurs spontaneously when the two phases have identical chemical potentials. Suppose we have two phases (labeled 1 and 2) present and in phase-change equilibrium. Let's vary the pressure and stipulate that the phases remain in equilibrium. This requires the temperature to change such that the changes in chemical potentials μ_1 and μ_2 are identical; that is $\mathrm{d}\mu_1 = \mathrm{d}\mu_2$ or equivalently

$$-s_1\,\mathrm{d}T + v_1\,\mathrm{d}p = -s_2\,\mathrm{d}T + v_2\,\mathrm{d}p\,.$$

It is a simple matter to use the relation between latent heat and entropy change developed at the end of the previous subsection and arrive at the Clausius–Clapeyron equation:

$$\frac{\mathrm{d}T}{\mathrm{d}p} = \frac{v_2 - v_1}{s_2 - s_1} = \frac{\Delta v}{\Delta s} = \frac{T\Delta v}{L}\,,$$

where Δv is the change of specific volume during the change of phase. Commonly, Δs and Δv have the same sign, so that $\mathrm{d}T/\mathrm{d}p$ is positive.

E.10.2 Eutectic Diagram

While a system composed of a single component (type of material) has a definite temperature at which melting or freezing (at constant pressure) occurs, when the system is composed of more than one component, the behavior is more complex, with solidification occurring over a range of temperatures. This behavior is illustrated most clearly in a system consisting of only two components, but this simple model can be generalized to systems composed of many components, because typically solidification involves only one type of crystal at a time.

Let's consider a homogeneous thermodynamic system consisting of two components, A and B, having a uniform temperature T. If the system were composed of only A, it would have a melting temperature T_a, and be entirely solid if $T < T_a$ and entirely liquid if $T > T_a$. The same holds true for B with melting temperature T_b. Melting occurs when the disruptive effect of thermal motions, quantified by the magnitude of T,[56] is sufficiently strong to break the chemical bonds that keep A or B in a solid state. The stronger those bonds, the greater the melting temperature.

Now let's suppose the system consists of a mass fraction $1 - \xi$ of A and ξ of B. Commonly, atoms of B do not fit comfortably in the solid phase of A and vice versa. A solid consisting of both A and B is in a higher energy state than for A or B alone and the bonds holding the solid together are weaker. This means that thermal agitation can disrupt the solid state at a lower temperature, as illustrated in Figure E.1, which is called a *eutectic phase diagram*.[57] This diagram shows the phase states of the system as a function of temperature T and mass fraction ξ. In region 1, the temperature is sufficiently high that the system is entirely molten. Similarly, in region 4, the temperature is sufficiently low that the

[56] See Appendix E.8.4.
[57] Sometimes it is called a *rabbit-ears diagram*, for an obvious reason.

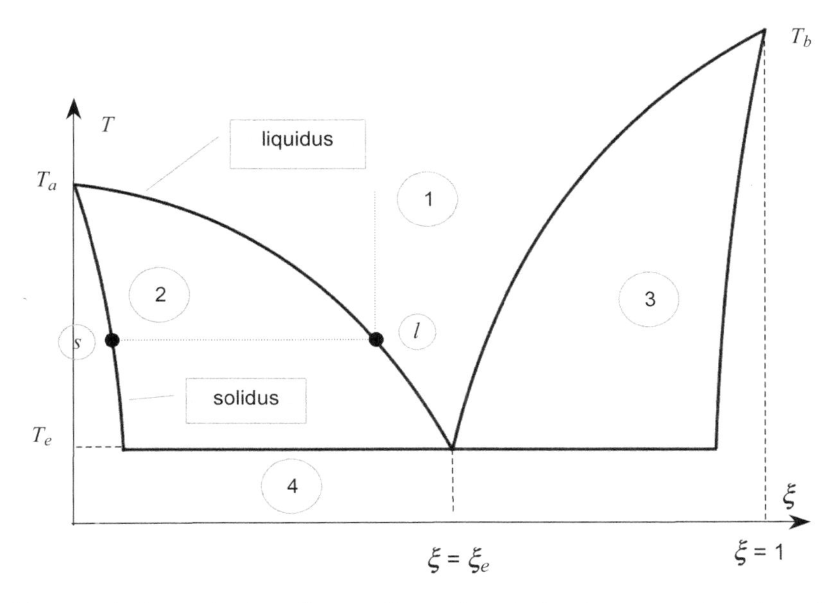

Figure E.1 A typical eutectic phase diagram. Depending whether the composition of a cooling melt is less than or greater than the eutectic composition ξ_e, the solid formed will contain predominantly material A or B. As solidification proceeds, the equilibrium composition of the liquid and solid evolve along the liquidus and solidus curves, respectively. When the system cools to $T = T_e$, the last remaining liquid has composition ξ_e.

system is entirely solid. In regions 2 and 3, the system is partially solid and partially molten. The horizontal boundary between region 4 and regions 2 and 3 is at the *eutectic temperature*, T_e. This is the lowest temperature at which liquid can exist, and this liquid must have eutectic composition ξ_e. The important curves on this diagram are the two *liquidus* curves, which separate region 1 from regions 2 and 3, and the two *solidus* curves, which separate regions 2 and 3 from region 4. In the eutectic diagram shown in Figure E.1, the melting temperature is composed of the set of curves separating regions 4 from regions 2 and 3.

Now let's consider how the system having composition ξ_0 evolves as it is gradually cooled from a high temperature. This cooling is depicted in Figure E.1 by the vertical dotted line. When the system is cooled to the liquidus temperature, solid begins to form having composition dictated by the solidus at that temperature. As cooling proceeds, sufficient solid forms to keep the composition of the liquid phase on the liquidus. At any temperature, the relative amounts of solid and liquid having compositions indicated by points l and s in Figure E.1 are determined by the *lever rule*, such that the overall composition of the system remains equal to ξ_0. When the eutectic temperature is reached, all remaining liquid solidifies. The important messages are:

- in a system containing more than one type of material, freezing takes place over a range of temperatures, rather than at a single temperature as occurs in a pure system; and

- the composition of the solid phase differs – often dramatically – from that of the parent liquid. For example, sea ice is much fresher (less salty) than sea water.

Of course, this description is highly idealized. The mantle is composed of a number of chemical components and the actual phase diagram is far more complicated than that described here. In the process of solidification, a solid phase will preferentially select atoms of those elements that are most compatible.[58] Those elements whose atoms are much smaller or much larger than those comprising the solid are *incompatible* and tend to remain in the liquid phase. So the last remaining liquid is a hodgepodge of large, small and electrochemically odd atoms that bond together very weakly, if at all. In particular, the radioactive elements (having large atoms) are quite incompatible and are concentrated in the last-to-freeze component of magma: granite. Certain metals (gold, silver, copper, etc.) are so incompatible that they do not fit into any crystal and solidify alone, forming *native metals*. Trace amounts of volatile components (having small atoms, such as water or carbon dioxide) act to lower the eutectic temperature significantly.

E.11 Thermodynamic Efficiency

As we previously noted, transfers of work and heat are quantified by the first law of thermodynamics and constrained by the second law. In this fundament we quantify the constraint that the second law places on the possible transformation of heat to work. The transformation consists of a specified system acquiring an amount Q_H of heat at a high temperature T_H, converting a portion to work W and casting off the remainder $Q_C = Q_H - W$ to a cold reservoir at temperature T_C (with $T_C < T_H$). The *thermodynamic efficiency*, η, of this process is the fraction of heat transformed to work:

$$\eta \equiv W/Q_H, \qquad \text{where} \qquad W = Q_H - Q_C.$$

The most efficient process for converting heat to work is the *Carnot cycle*, which consists of four steps that return our system to its original state, determined by its pressure and temperature: (p_H, T_H), with volume and entropy being V_H and S_H, respectively. The four steps of the Carnot cycle are as follows:

1. *Isothermal expansion* from state a to b. With the temperature fixed at T_H, the system adds heat Q_H, and expands isothermally from volume V_a to volume V_b. During this step, it gains an amount of entropy $S_H = Q_H/T_H$ and the pressure drops from p_a to $p_b (= RT_H/V_b$ if the working fluid is an ideal gas).
2. *Isentropic expansion* from state b to c. Barring exchanges of heat, the system expands isentropically from volume V_b to volume V_c. Its entropy remains constant and the temperature drops from T_H to $T_C = p_c V_c/R$.
3. *Isothermal contraction* from state c to d. With the temperature fixed at T_C, the system casts off heat Q_C, and contracts isothermally from volume V_c to volume V_d. During

[58] This principally means most close is size, but electrical affinity (valence) also plays a role.

Table E.2. *Changes of thermodynamic variables for each of the four steps of the Carnot cycle.*

Step	p	T	V	Q	ΔS
1	$p_a \to p_b$	T_H	$V_a \to V_b$	Q_H	Q_H/T_H
2	$p_b \to p_c$	$T_H \to T_C$	$V_b \to V_c$	0	0
3	$p_c \to p_d$	T_C	$V_c \to V_d$	$-Q_C$	$-Q_C/T_C$
4	$p_d \to p_a$	$T_C \to T_H$	$V_d \to V_a$	0	0

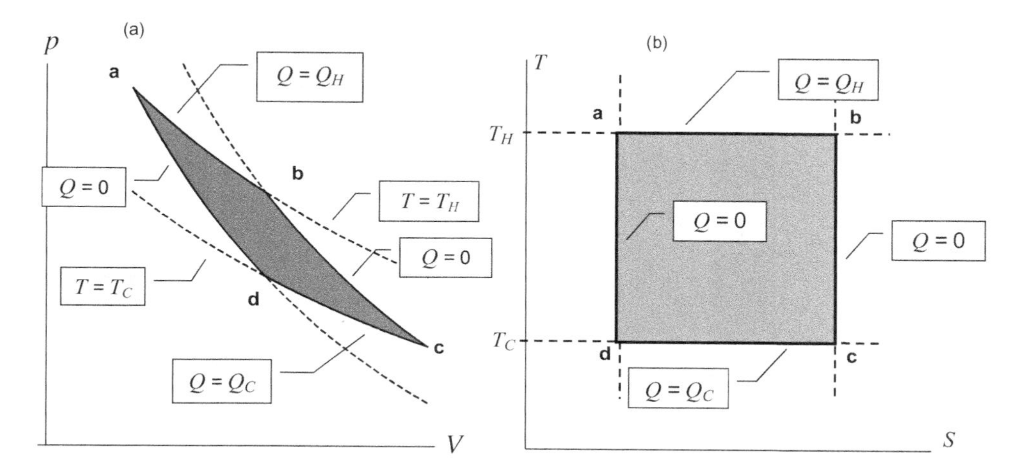

Figure E.2 (a) The Carnot cycle on the $p - V$ plane. The four steps of the cycle trace the boundary of the shaded shape in a clockwise direction. The work done is equal to the area of the shaded shape. (b) The Carnot cycle on the $T - S$ plane. The four steps trace the boundary of the shaded shape in a clockwise direction.

this step, it loses an amount of entropy $S_C = Q_C/T_C$ and its pressure drops from p_c to $p_d = RT_C/V_d$.

4. *Isentropic contraction* from state d to a. Barring exchanges of heat, the system contracts isentropically from volume V_d to volume V_a. Its entropy remains constant and the temperature increases from T_C to T_H.

The changes of variables during each of the four steps are shown in Table E.2.

As the system expands during steps 1 and 2, it does work on its surroundings represented by the area under the upper curves running from point a to point c in part (a) of Figure E.2. As the system contracts during steps 3 and 4, it absorbs work from its surroundings represented by the area under the lower curves running from point c to point a. The net work done by the system during the cycle is denoted schematically by the shaded are in this $p - V$ plot. It follows from conservation of energy that $W = Q_H - Q_C$, as previously stated. The Carnot cycle is illustrated in the $T - S$ plane in part (b) of Figure E.2.

The cycle operates at maximum efficiency provided there is no net change of entropy. This requires

$$S_C = S_H \qquad \text{or} \qquad Q_C/T_C = Q_H/T_H\,.$$

Substituting these results into the expression for η, we have the *Carnot efficiency*

$$\eta = (T_H - T_C)/T_H\,.$$

The fact that a thermodynamic system can undergo a cycle, returning to its original state, in which it absorbs a net amount of heat and performs work demonstrates that neither Q nor W is a state variable.

Appendix F
Waves

The fundaments related to wave dynamics include:

- F.1: a presentation and discussion of the behavior of waves in three dimensions;
- F.2: a summary of the Fourier representation of waves;
- F.3: an introduction to Stokes waves, which are evanescent transverse waves in a viscous fluid;
- F.4: an introductory analysis of Kelvin's ship waves;
- F.5: quantification of the energy of deep water waves;
- F.6: a brief presentation of Laplace's tidal equations; and
- F.7: analysis and discussion of inertial waves in a rotating fluid.

F.1 Wave Theory

In the main text waves are assumed to move in one direction, labeled x. This representation works for planar waves traveling through a homogeneous medium, but if the wave-front is not planar or if the material is not homogeneous, then a traveling wave needs to be treated in a three-dimensional manner, as described in this fundament.

F.1.1 Wave Phase

In § 9.2.2 we introduced the phase of a plane wave traveling in the x direction: $\theta = kx - \omega t$, where k is the wavenumber and ω is the frequency of the wave. If the wave is traveling in an arbitrary direction, the phase may be expressed as

$$\theta = k_i x_i - \omega t,$$

where x_i (for i = 1, 2, 3) are Cartesian coordinates[1] and k_i is the wavenumber in the x_i direction. The phase may be expressed in vector notation:

$$\theta = \mathbf{k} \bullet \mathbf{x} - \omega t,$$

[1] See § 2.1.

where $\mathbf{k} = k_i \mathbf{1}_i$ and $\mathbf{x} = x_i \mathbf{1}_i$. We can readily see that

$$\mathbf{k} = \nabla\theta \qquad \text{and} \qquad \omega = -\frac{\partial\theta}{\partial t}.$$

In the main text we treated k and ω as constants, independent of x and t. If $\mathbf{k}$ and ω are independent of $\mathbf{x}$ and t, $\theta =$ constant defines a set of flat plane in three-dimensional space having normal in the $\mathbf{k}$ direction and separated by a distance $\Lambda = 2\pi/k$, where $k = \|\mathbf{k}\|$. However, as a wave moves through a complicated medium, $\mathbf{k}$ and ω can vary with position. It follows from the rules of calculus that

$$\nabla\left(\frac{\partial\theta}{\partial t}\right) = \frac{\partial}{\partial t}(\nabla\theta).$$

Using the definitions of $\mathbf{k}$ and ω in terms of θ, this identity becomes the equation of *conservation of phase*:

$$\frac{\partial\mathbf{k}}{\partial t} + \nabla\omega = 0, \qquad \text{or equivalently} \qquad \frac{\partial k_i}{\partial t} + \frac{\partial\omega}{\partial x_i} = 0.$$

This is a set of three scalar equations constraining the spatial variation of $\mathbf{k}$ and ω.

A simple harmonic wave traveling in an arbitrary direction may be represented by a single Fourier component[2]

$$q = \tilde{q}\cos\theta$$

with $\tilde{q}$, k and ω being independent of position.[3] The physical nature of the wave is encapsulated in the dispersion relation $\omega = W(k)$. If the wave is a superposition of simple harmonic waves, then we may represent it as a sum over all relevant harmonic modes,[4] with the function $\tilde{q}(k)$ being a determined by the initial condition. This mathematical formalism carries over to three-dimensional waves, with the dispersion relation now being a function of $\mathbf{k}$:

$$\omega = W(\mathbf{k}).$$

F.1.2 Phase and Group Velocities

The phase and group velocities of the wave are defined as

$$\mathbf{U}_p = \frac{\omega}{\|\mathbf{k}\|}\mathbf{1}_k \qquad \text{and} \qquad \mathbf{U}_g = \frac{\partial W}{\partial k_i}\mathbf{1}_i,$$

[2] See § 9.2.
[3] Here q is some measurable property of the wave, such as displacement or speed.
[4] As in § 9.2.

where $\mathbf{1}_k$ is a unit vector parallel to $\mathbf{k}$. The phase and group speeds are[5]

$$U_p = \frac{\omega}{\|\mathbf{k}\|} \qquad \text{and} \qquad U_g = \sqrt{\frac{\partial W}{\partial k_i}\frac{\partial W}{\partial k_i}}.$$

The group velocity of a wave controls the propagation of wave energy. The propagation of wave energy is not apparent when considering a uniform wavetrain, because there is no spatial variation of wave amplitude. The role of the group velocity is made apparent if we consider a *wave packet* having a slowly varying amplitude, such as

$$q = q_0 e^{-\chi^2}\cos(\mathbf{k}\bullet\mathbf{x} - \omega t), \qquad \text{where} \qquad \chi = \|\mathbf{x} - \mathbf{U}_g t\|/L,$$

q_0 is a constant amplitude and L is a spatial distance satisfying $1 \ll kL$, so that the envelope $e^{-\chi^2}$ varies on a spatial scale much greater than the wavelength. The wave energy, quantified by the envelope of the packet, moves with the group velocity, while the magnitude of q oscillates within the envelope as $\theta = \mathbf{k}\bullet\mathbf{x} - \omega t$ varies. The speed of propagation of the packet in a specified direction x_i is given by the usual vector component relation $\mathbf{U}_g \bullet \mathbf{1}_i$.

Mathematically, there are two types of vectors, distinguished by the manner in which the vector is decomposed into components. Commonly, vectors have components that are shorter than the vector itself:

$$\mathbf{b} = b\cos\theta_i \mathbf{1}_i, \qquad \text{where} \qquad \cos\theta_i = \mathbf{1}_b \bullet \mathbf{1}_i$$

and $b = \|\mathbf{b}\|$; these are called *contravariant vectors*. Vectors of the other type, called *covariant vectors*, have components longer than the vector:

$$\mathbf{b} = b\frac{\mathbf{1}_i}{\cos\theta_i}.$$

The phase velocity is a covariant vector.[6] This is readily seen when observing the speed at which lines of constant phase move along the shore. If the wave train is moving parallel to the shore (angle of approach being zero), the lines of constant phase move with the phase speed. As the angle of approach increases, the speed at which lines of constant phase move along the shore (delineated by the point at which waves break) increases. As this angle becomes close to perpendicular, the phase speed along the shore increases without bound.[7]

[5] Velocities are vectors and speeds are scalars. It is common practice to refer to the phase and group speeds as the phase and group velocities, but this is erroneous.

[6] The phase velocity vector is proportional to the gradient of phase and gradient vectors are covariant.

[7] This speed can even exceed the speed of light. But this does not make Einstein roll over in his grave, as there is no information conveyed by the phase velocity.

F.2 Fourier Representation

A general disturbance in an infinite domain is represented by a Fourier integral

$$q(\mathbf{x},t) = \int_{\mathbf{k}} \tilde{q}(\mathbf{k}) e^{\mathrm{i}\theta} \mathrm{d}\mathbf{k} + \mathrm{c.c.} \qquad \text{with} \qquad \theta = \mathbf{k} \bullet \mathbf{x} - \omega t .$$

This is a generalization of the one-dimensional representation given in § 9.2.2. If the domain is finite, the integral is replaced by a summation. Typically, the function $\tilde{q}$ is determined by the initial condition

$$\tilde{q}(\mathbf{k}) = \int_{\mathbf{x}} q(\mathbf{x},0) e^{-\mathrm{i}(\mathbf{k}\bullet\mathbf{x})} \mathrm{d}\mathbf{k} .$$

Note that $\mathbf{k}$ may be either three-dimensional (as in the case of body waves) or two-dimensional (as in the case of surface waves). If $\mathbf{k}$ is two-dimensional, we may write $\mathbf{k} = k\mathbf{1}_k$ and express the general Fourier representation as

$$q(\mathbf{x},t) = \int_0^{\infty} I_{\chi} k \,\mathrm{d}k , \qquad \text{where} \qquad I_{\chi} \equiv \int_{-\pi}^{\pi} \tilde{q}(k,\chi) e^{\mathrm{i}\theta} \mathrm{d}\chi + \mathrm{c.c.},$$

$\theta = k(\mathbf{1}_k \bullet \mathbf{x}) - \omega t$ and χ is the angle between the wavenumber and position vectors ($\cos\chi = \mathbf{1}_k \bullet \mathbf{1}_x$). An equivalent representation is

$$q(\mathbf{x},t) = \int_{-\pi}^{\pi} I_k \mathrm{d}\chi , \qquad \text{where} \qquad I_k \equiv \int_0^{\infty} \tilde{q}(k,\chi) e^{\mathrm{i}\theta} k \,\mathrm{d}k + \mathrm{c.c.}.$$

F.2.1 Steady Disturbance

Let's suppose the disturbance is produced by an object moving steadily with velocity $\mathbf{v}$ and is steady when viewed in a frame of reference moving with the object:

$$\mathbf{x} = \mathbf{x}_{\mathrm{o}} + \mathbf{v}t ,$$

where $\mathbf{x}$ is the position measured by an observer stationary in the fluid medium and $\mathbf{x}_{\mathrm{o}}$ is the position measured by an observer moving with the object. This moving observer sees a steady disturbance, $q_{\mathrm{o}}(\mathbf{x}_{\mathrm{o}}) = q(\mathbf{x}_{\mathrm{o}} + \mathbf{v}t, t)$ or

$$q_{\mathrm{o}}(\mathbf{x}_{\mathrm{o}}) = \int_{\mathbf{k}} \tilde{q}(\mathbf{k}) e^{\mathrm{i}\theta} \mathrm{d}\mathbf{k} + \mathrm{c.c.} \qquad \text{with} \qquad \theta = \mathbf{k}\bullet\mathbf{x}_{\mathrm{o}} + (\mathbf{k}\bullet\mathbf{v} - \omega)t .$$

This formulation is consistent if and only if

$$\omega = \mathbf{k}\bullet\mathbf{v} \qquad \text{or} \qquad \omega = kv\cos\chi ,$$

where $k = \|\mathbf{x}\|$, $v = \|\mathbf{v}\|$ and χ is the angle between these two vectors.

F.3 Stokes Waves

A viscous fluid is capable of sustaining an evanescent transverse (shear) wave due to the action of the viscous force. Consider a non-rotating fluid body occupying the half-space $0 < z < \infty$, bounded by a rigid plane at $z = 0$ that is oscillating harmonically in the x direction with maximum speed U and frequency ω. The viscous force, together with the no-slip condition, induces a flow within the fluid in the x direction. This flow is governed by[8]

$$\frac{\partial u}{\partial t} = \frac{\partial}{\partial z}\left(\nu \frac{\partial u}{\partial z}\right),$$

where $u(z,t)$ is the speed in the x direction and ν is the kinematic viscosity. This equation holds in the interval $0 < z < \infty$ and is subject to the no-slip condition $u(0,t) = U\cos(\omega t)$.

If the flow is laminar with constant viscosity, the Stokes-wave solution is

$$u(z,t) = e^{-mz}\cos(\omega t - mz), \qquad \text{where} \qquad m = \sqrt{\omega/2\nu}\,.$$

On the other hand, if the flow is turbulent with $\nu_T = \hat{\varepsilon}U(z+\hat{z})$,[9] where $\hat{\varepsilon} \approx 0.02$ and $\hat{z}$ represents the surface roughness, the solution is[10]

$$u(z,t) = U\frac{\mathrm{ker}(\hat{\xi})\mathrm{ker}(\xi) + \mathrm{kei}(\hat{\xi})\mathrm{kei}(\xi)}{\mathrm{ker}^2(\hat{\xi}) + \mathrm{kei}^2(\hat{\xi})}\cos(\omega t) + U\frac{\mathrm{kei}(\hat{\xi})\mathrm{ker}(\xi) - \mathrm{ker}(\hat{\xi})\mathrm{kei}(\xi)}{\mathrm{ker}^2(\hat{\xi}) + \mathrm{kei}^2(\hat{\xi})}\sin(\omega t),$$

where

$$\xi = 2\sqrt{\frac{\omega(z+\hat{z})}{\hat{\varepsilon}U}}, \qquad \hat{\xi} = 2\sqrt{\frac{\omega\hat{z}}{\hat{\varepsilon}U}}$$

and ker and kei are Kelvin functions.[11]

F.4 Kelvin's Ship Waves

An object (ship, duck, reed in a stream) moving steadily on the surface of a body of water produces a pattern of waves that appear stationary with respect to that object. These waves are commonly called *Kelvin's ship waves* in honor of Lord Kelvin, who first analyzed them.[12] This pattern requires the phase and group speeds of the waves in the direction of motion of the body to be the same. This requirement appears to be at odds with a fundamental property of deep water waves: the phase speed is twice the group speed. This restriction is evaded by realizing that waves are vector quantities. In this fundament, we will investigate Kelvin's ship waves.

[8] See the Stokes equation in § 6.2.2.
[9] See § 23.5.2.
[10] This solution is obtained following the procedure in § 24.1.
[11] See formula 10.61.2, p.268 of Olver et al. (2010).
[12] These waves are distinct and different from the equatorial Kelvin wave; see § 17.2.

To simplify our investigation, let's employ a Cartesian coordinate system with the undisturbed surface of a body of water lying in the $x - y$ plane and consider the wake created by a duck paddling steadily in the positive x direction with velocity $\mathbf{v} = u\,\mathbf{1}_x$, leaving behind a wake that extends in the negative x direction (and in both the positive and negative y directions) relative to the duck. The steady wake produced by the duck consists of those waves having phase and group speeds in the positive x direction equal to u and group speed in the y direction equal to zero. Let's write the wave vector as $\mathbf{k} = k_x\,\mathbf{1}_x + k_y\,\mathbf{1}_y = k\left(\cos\chi\,\mathbf{1}_x + \sin\chi\,\mathbf{1}_y\right)$ and the position vector (relative to an observer stationary in the water) as $\mathbf{x} = (x_o - ut)\,\mathbf{1}_x + y_o\,\mathbf{1}_y$, where the subscript o denotes position on the water surface relative to the duck.

Let's assume the water is sufficiently deep that the dispersion relation $\omega = \sqrt{gk}$ is applicable. But we also have the steady-state constraint from Appendix F.2.1: $\omega = ku\cos\chi = k_x u$. Together these require

$$k = \frac{g}{u^2\cos^2\chi}\,.$$

This formula gives the length of waves traveling steadily at an angle χ relative to the path of our duck.

With $\omega = ku\cos\chi = k_x u$, the waves that comprise the wake have phase and group velocities (relative to an observer stationary in the water) given by

$$\mathbf{U}_p = \frac{\omega}{k}\,\mathbf{1}_k = u\cos\chi\,\mathbf{1}_k \qquad \text{and} \qquad \mathbf{U}_g = \frac{\partial\omega}{\partial k_x}\,\mathbf{1}_x + \frac{\partial\omega}{\partial k_y}\,\mathbf{1}_y = u\,\mathbf{1}_x = \mathbf{v}\,.$$

Recalling that the phase velocity is a covariant vector[13] with components of greater magnitude than the vector, we see that $\mathbf{U}_p\bullet\mathbf{1}_x = u$, as required.

The waves travel in phase with the duck. Recalling that $\theta = \mathbf{k}\bullet\mathbf{x}_o$ and $k = g/u^2\cos^2\chi$, curves of constant phase (having $\theta = \theta_0$) must satisfy

$$\theta_0 = \mathbf{k}\bullet\mathbf{x}_o = k(x_o\cos\chi + y_o\sin\chi) = \frac{g}{u^2}\left(\frac{x_o + y_o\tan\chi}{\cos\chi}\right).$$

This is a first constraint on x_o and y_o. We can obtain a second constraint by realizing that as we move along a curve of constant phase, (having $\partial\theta/\partial\chi = 0$), the values of x_o and y_o change such that

$$\frac{\partial}{\partial\chi}\left(\frac{x_o\cos\chi + y_o\sin\chi}{\cos^2\chi}\right) = 0\,,$$

which is satisfied provided

$$x_o\sin\chi\cos\chi + y_o(1 + \sin^2\chi) = 0\,.$$

[13] See Appendix F.1.2.

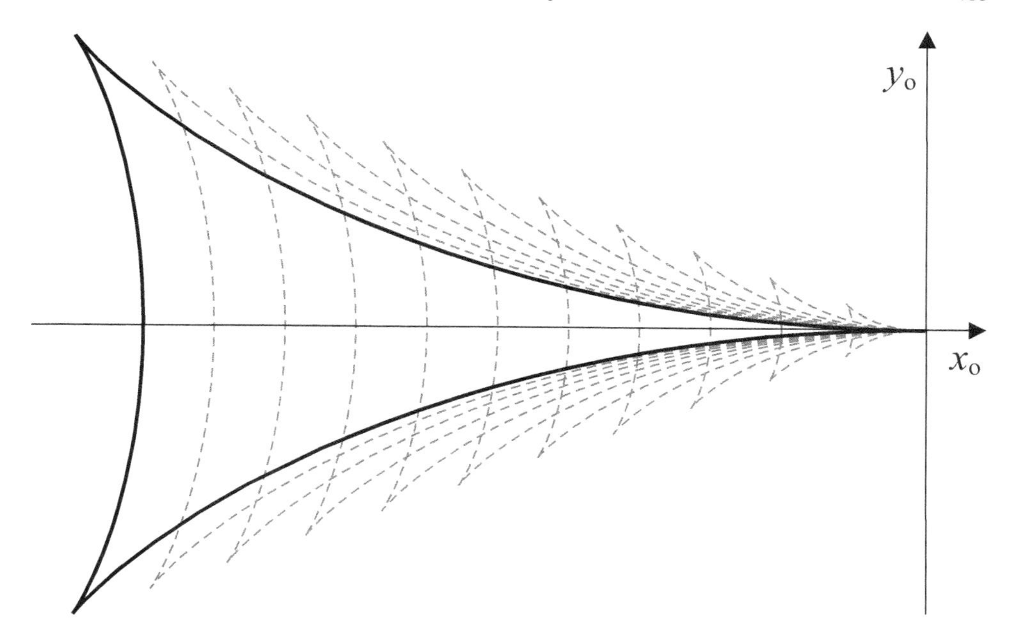

Figure F.1 Curves of constant phase within the wake produced by an object moving to the right.

These two constraints on x_o and y_o may be expressed as

$$x_o = -x_{max}(1+\sin^2\chi)\cos\chi \qquad \text{and} \qquad y_o = x_{max}\sin\chi\cos^2\chi\,.$$

where $x_{max} \equiv \theta_0 v^2/g$. This parametric representation of curves of constant phase is graphed in Figure F.1, which shows 11 curves of constant phase with successive curves separated in the x direction by a distance $2\pi u^2/g$. It is easily shown that $|y_o/x_o|$ has a maximum value of $2^{-3/2} \approx 0.35355$ when $\chi = \arcsin(1/\sqrt{3}) \approx 0.61548$ rad, corresponding to a wake lying within 0.3398 rad (19.47 degrees) of the duck's path. The full wake is a superposition of many such curves, each having a differing value of x_{max} or equivalently θ_0. The portions of the wake emanating from the origin are called the *divergent wake*, while the portion joining the two cusps is called the *transverse wake*.

The basic wake pattern produced by a moving object, illustrated in Figure F.1, is curiously independent of its size, shape or speed. This pattern is valid for deep water. When the speed of the object is comparable to or greater than the shallow-water wave speed $\sqrt{gh}$, then the pattern is affected by the finite depth of water.[14] The observed amplitude of the wake depends on several factors. The finite size of the object blurs the wave crests and troughs, while the wave amplitude as a function of angle χ depends on the shape of the object below water, with the amplitude of the divergent wake being more sensitive to the

[14] See Havelock (1908).

shape near the bow and the transverse wake being more sensitive to the shape near the stern. A bluff object (our duck) produces a wake of relatively uniform amplitude, while a flat-bottomed object (a planing speed boat) produces a wake that has large amplitude directly behind it. Whatever the object's shape, most of the divergent wake is obscured by the geometric overlap of waves of all phases. What remains in view are the crests at the cusps and the harmonic transverse wake astern of the object. Also, the spacing of successive wave crests vary as the square of the object's speed and the breaking of waves within the wake close to a speeding boat further reduces the amplitude of the wake farther behind the boat. Consequently the wake pattern behind a slowly moving object (such as a boat in a harbor) is clearly visible, whereas successive wake crests behind a planing speed boat are widely separated and rather difficult to discern.

Energy for the waves within a wake is produced by the moving object and is stored as potential energy in the deflection of the surface and as kinetic energy in the motions of the water. Barring dissipation of energy, geometric spreading dictates that the amplitude of the wake must decay as the inverse of distance from the object. The moving object causes water to pile up at the bow, forming the *bow wake*, while depth at the stern is lower because water does not immediately fill the void left by the moving object. The combination of these two effects causes the object to seem to be trying to move "uphill," or to climb onto its bow wake.[15] The energy expended by the object in climbing this hill goes into energy of the wake. If a body moving along a narrow canal is suddenly brought to a halt, the bow wake will continue on along the canal as a solitary wave; see § 13.6.

F.5 Energy of Deep-Water Surface Waves

The magnitude, v, of velocity associated with a single wave harmonic having wavenumber k is given by

$$v(x,z,t) = h_0\sqrt{gk}e^{kz},$$

where h_0 is the wave amplitude, g is the local acceleration of gravity and z is elevation from the undisturbed free surface.[16] Note that the speed does not vary with x or t. The trajectories of the water parcels are circles whose radii have maximum of h_0 at the surface and decrease exponentially with depth, as described in § 11.4.

The kinetic energy per unit area of a surface wave is given by

$$E_K = \int_{-\infty}^{0} \frac{\rho}{2} v^2 \mathrm{d}z = \frac{\rho g}{4} h_0^2 = \frac{\rho g}{16} H^2,$$

where $H = 2h_0$ is the wave height: the vertical distance between crest and trough.

The potential energy of the wave is the energy necessary to distort the water surface from flat to a wavy surface. The potential energy of a column of water is the product of mass,

[15] A planing speed boat has succeeded in climbing onto its bow wake.
[16] See § 11.3.

elevation and the local acceleration of gravity. The mass is equal to ρhA where A is the surface area, while the mean elevation is $h/2$. Since the elevation varies with down-wind distance x, we must integrate over one wavelength ($\Lambda = 2\pi/k$) to get a mean value; the potential energy per unit area is

$$E_P = \frac{1}{\Lambda}\int_0^{\Lambda} \frac{\rho g}{2} h^2 \mathrm{d}x = \frac{\rho g}{4} h_0^2 = \frac{\rho g}{16} H^2 .$$

This is identical to the expression for kinetic energy; equipartition of energy holds for surface waves. The total wave energy per unit area is

$$E = \frac{\rho g}{8} H^2 .$$

Traveling waves have kinetic and potential energy in equal amounts and the energy per unit area in a single harmonic wave is proportional to the square of the wave height.

F.6 Laplace's Tidal Equations

Oceanic tides, as well as other flows, may be modeled by a set of linearized, vertically averaged equations on a spherical shell. These equations are based on the simplifications presented in § 8.2.1, together with the assumptions that

- the nonlinear inertial term is negligibly small;
- the horizontal velocity components are independent of height;
- variations of density are unimportant; and
- deviations of the free surface from the equilibrium position are small.

Let's consider a layer of fluid on a spherical Earth, with the equilibrium level of the fluid denoted by $r = R$, where r is the spherical radial coordinate. The equilibrium depth of the fluid is $D(\theta,\phi)$, where θ is the colatitude and ϕ is longitude. The tides and other waves are visualized by the deviation of the water surface from equilibrium: $r = R + h(\theta,\phi,t)$.

The top of the ocean is a material surface; that is, at $r = R + h$

$$\frac{\mathrm{D}}{\mathrm{D}t}(r - R - h) = 0 .$$

Neglecting the nonlinear terms and applying this condition at the undeformed free surface, we have the kinematic boundary condition

$$v_r \equiv \frac{\partial r}{\partial t} = \frac{\partial h}{\partial t}$$

at $r = R$. The continuity equation for a constant-density fluid is simply $\nabla \cdot \mathbf{v}$. With $\mathbf{v} = v_r \mathbf{1}_r + v_\theta \mathbf{1}_\theta + v_\phi \mathbf{1}_\phi$ and $r \approx R$, this equation is[17]

$$\nabla \cdot \mathbf{v} = \frac{\partial v_r}{\partial r} + \frac{1}{R\sin\theta}\frac{\partial (v_\theta \sin\theta)}{\partial \theta} + \frac{1}{R\sin\theta}\frac{\partial v_\phi}{\partial \phi} = 0 .$$

[17] See Appendix A.3.5.

Integration of this equation from $r = R - D$ to $r = R$, it becomes

$$R\sin\theta\frac{\partial h}{\partial t} + \frac{\partial(D\bar{v}_\theta \sin\theta)}{\partial\theta} + \frac{\partial(D\bar{v}_\phi)}{\partial\phi} = 0,$$

where a bar denotes a variable independent of elevation.

Now consider the momentum equation. This is readily split into vertical and horizontal components, with the vertical component being the hydrostatic balance,[18] with the addition of an external gravitational force due to the Moon and Sun:

$$\frac{\partial p}{\partial r} = -\rho(g + g_{ms}),$$

where g_{ms} is the gravitational attraction due to the Moon and Sun. Note that g_{ms} is a function of horizontal position and time. Integration of this balance in the radial direction and using the dynamic boundary condition that the pressure is equal to the atmospheric pressure p_a at the surface, yields

$$p = p_a + \rho(g + g_{ms})(R + h - r).$$

Using its horizontal derivative, $\nabla_H p = \rho g \nabla_H h + \rho R \nabla_H g_{ms}$, the horizontal portion of the linearized momentum equation is[19]

$$\frac{\partial \bar{\mathbf{v}}_H}{\partial t} + f\mathbf{1}_r \times \bar{\mathbf{v}}_H + g\nabla_H h = -R\nabla_H g_{ms}.$$

In spherical coordinates, the components of this equation are

$$\frac{\partial \bar{v}_\phi}{\partial t} - f\bar{v}_\theta + \frac{g}{R\sin\theta}\frac{\partial h}{\partial\phi} = -\frac{1}{\sin\theta}\frac{\partial g_{ms}}{\partial\phi}$$

and

$$\frac{\partial \bar{v}_\theta}{\partial t} + f\bar{v}_\phi + \frac{g}{R\sin\theta}\frac{\partial(h\sin\theta)}{\partial\theta} = -\frac{1}{\sin\theta}\frac{\partial(g_{ms}\sin\theta)}{\partial\theta}.$$

F.6.1 Local Tidal Equations

Using the local Cartesian coordinates[20] introduced in § 8.2.2 and treating θ as constant except in the Coriolis parameter, the local version of Laplace's tidal equations is

$$\frac{\partial h}{\partial t} + \frac{\partial(D\bar{v})}{\partial y} + \frac{\partial(D\bar{u})}{\partial x} = 0,$$

$$\frac{\partial \bar{u}}{\partial t} + f\bar{v} + g\frac{\partial h}{\partial x} = -R\frac{\partial g_{ms}}{\partial x} \quad \text{and} \quad \frac{\partial \bar{v}}{\partial t} - f\bar{u} + g\frac{\partial h}{\partial y} = -R\frac{\partial g_{ms}}{\partial y}.$$

[18] See § 7.1.
[19] From § 8.2.1.
[20] With $\{v_r, \bar{v}_\theta, \bar{v}_\phi\} = \{w, -\bar{v}, \bar{u}\}$ and $R\{\sin\theta\,\mathrm{d}\phi, -\mathrm{d}\theta, \mathrm{d}r\} = \{\mathrm{d}x, \mathrm{d}y, \mathrm{d}z\}$.

Note that these are just the linearized quasi-geostrophic equations,[21] with an added astronomical gravitational acceleration g_{ms}. These equations are forced by the temporal and spatial variation of g_{ms}.

F.7 Inertial Waves

The purpose of this fundament is to summarize the oscillatory behavior of a rapidly rotating fluid in the absence of boundaries, density variations or other complications. Specifically, let's assume that

- the fluid is incompressible with uniform density;
- the fluid is inviscid; and
- variations from a state of rigid uniform motion are small.

With these simplifications, gravitational effects are entirely passive and can be ignored.[22] Now the continuity and momentum equations are[23]

$$\nabla \bullet \mathbf{v} = 0 \qquad \text{and} \qquad \rho \frac{\partial \mathbf{v}}{\partial t} + 2\rho \boldsymbol{\Omega} \times \mathbf{v} = -\nabla p .$$

Writing $\boldsymbol{\Omega} = \Omega \mathbf{1}_z$ and using formulas from Appendix A.2.6, the dot product and z-component of the cross product of the momentum equation become

$$\nabla^2 p = 2\rho\Omega\,\zeta \qquad \text{and} \qquad \frac{\partial \zeta}{\partial t} = 2\Omega \frac{\partial w}{\partial z},$$

where $w = \mathbf{1}_z \bullet \mathbf{v}$ and $\zeta = \mathbf{1}_z \bullet \nabla \times \mathbf{v}$ are the z components of the velocity and vorticity, respectively.

In steady state, the second of the equations above requires that

$$\frac{\partial w}{\partial z} = 0;$$

this is the well-known *Proudman–Taylor* constraint on rapidly rotating flows.[24]

In addition, the z derivative of the z component of the momentum equation is

$$\frac{\partial^2 p}{\partial z^2} = -\rho \frac{\partial^2 w}{\partial t \partial z}.$$

Eliminating w and ζ from these three equations, we have

$$\nabla^2 \frac{\partial^2 p}{\partial t^2} + 4\Omega^2 \frac{\partial^2 p}{\partial z^2} = 0.$$

This is a single linear partial differential equation with constant coefficients for a single unknown, p.

[21] See § 8.6.
[22] See § 4.7.1.
[23] From § 6.2.2.
[24] See Appendix D.9.

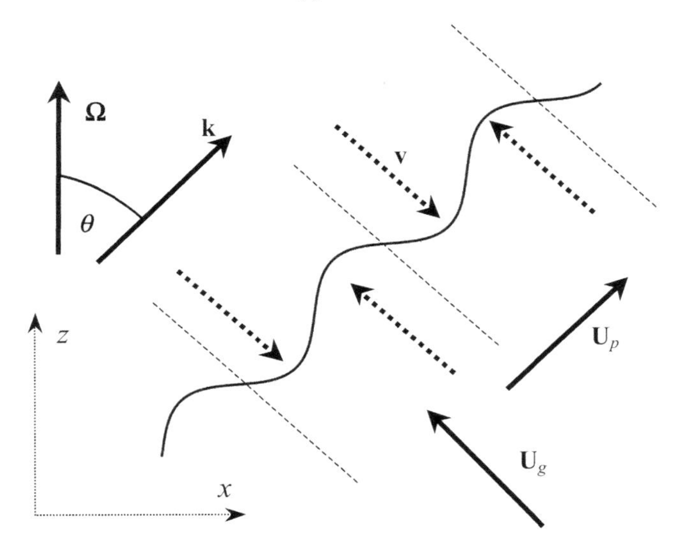

Figure F.2 Schematic of an inertial wave. Fluid motions (denoted by thick dashed arrows) are parallel to the lines of constant phase (thin dashed lines) and normal to the wave vector **k**. The phase velocity $\mathbf{U}_p$ is parallel to **k**, while the group velocity $\mathbf{U}_g$ is perpendicular to $\mathbf{U}_p$ and **k**.

Since the governing equation is isotropic in the plane normal to the rotation axis, there is no loss in generality in assuming that variables depend on x, z and t. Let's look for harmonic solutions to this partial differential equation, by assuming that p is of the form

$$\begin{aligned} p &= p_0 e^{\mathrm{i}\mathbf{k}\bullet\mathbf{x}-\mathrm{i}\omega t} + \text{c.c} \\ &= p_0 e^{\mathrm{i}kx+\mathrm{i}mz-\mathrm{i}\omega t} + \text{c.c.}, \end{aligned}$$

where $\mathbf{k} = k\mathbf{1}_x + m\mathbf{1}_z$ is the wavenumber vector, k and m are the wavenumbers in the x and z directions, respectively, $\mathbf{x} = x\mathbf{1}_x + y\mathbf{1}_y + z\mathbf{1}_z$ is the position vector in Cartesian coordinates and ω is the wave frequency.[25] Substituting this expression into the PDE for p, we obtain the dispersion relation:

$$\omega = \frac{2\Omega m}{\|\mathbf{k}\|} \qquad \text{or simply} \qquad \omega = f,$$

where $m = \sqrt{k^2+m^2}\cos\theta$, $f = 2\Omega\cos\theta$ is the Coriolis parameter introduced in § 8.1.5, and θ is the angle between **k** and **Ω** (essentially the colatitude); see Figure F.2.

The phase velocity of a wave in three dimensions, $\mathbf{U}_p$, is a vector in the direction of **k** having magnitude $\omega/\|\mathbf{k}\|$:

$$\mathbf{U}_p = \frac{\omega\mathbf{k}}{\|\mathbf{k}\|^2} = \frac{\omega}{k^2+m^2}(k\mathbf{1}_x + m\mathbf{1}_z)$$

[25] '+ c.c.' means add the complex conjugate, to achieve a real expression.

and the group velocity, $\mathbf{U}_g$, is the gradient of ω in wavenumber space:

$$\mathbf{U}_g = \nabla_{\mathbf{k}}\omega = \frac{\partial\omega}{\partial k}\mathbf{1}_x + \frac{\partial\omega}{\partial m}\mathbf{1}_z$$
$$= \frac{2\Omega k}{\left(k^2+m^2\right)^{3/2}}\left(-m\mathbf{1}_x + k\mathbf{1}_z\right).$$

The inertial wave is transverse, with fluid motions parallel to lines of constant phase. This shows us that the geostrophic wave of § 15.1 is just an inertial wave having the angle θ constrained by the geometry of a thin fluid layer.

Appendix G

Flows

The fundaments of flows consist of:

- G.1: a brief analysis and discussion of shear-flow instability;
- G.2: an introduction to boundary-layer theory; and
- G.3: some comments on models of open-channel flow.

G.1 Shear-Flow Instability

The concept of shear-flow instability was introduced in Chapter 23 to describe the onset of turbulence, but the process of flow instability was not quantified. In this fundament we rectify that deficiency by investigating the stability of a simple shear flow in a non-rotating fluid having constant density.

Let's consider an incompressible fluid confined between two planes located at $y = y_1$ and $y = y_2$ and suppose the unperturbed flow of the fluid is predominantly[1] rectilinear shear in the x direction that varies in the lateral (y) direction, but is independent of distance in the vertical (z) direction. Specifically, let $\mathbf{v}_H = u(y)\mathbf{1}_x$, where $u(y)$ is a prescribed function that satisfies the governing equations, which consist of the continuity equation and the vorticity equation developed in § 8.3.1, without the f and β terms. The continuity equation, $\nabla_H \bullet \mathbf{v}_H = 0$, is satisfied by the introduction of a stream function;[2] let

$$\mathbf{v}_H = \nabla \times (\Psi \mathbf{1}_z) = \frac{\partial \Psi}{\partial y}\mathbf{1}_x - \frac{\partial \Psi}{\partial x}\mathbf{1}_y .$$

Note that $\zeta = -\nabla_H^2 \Psi$. We will assume constant viscosity so that $\mathcal{V} = \rho \nu \nabla^2$.[3] Now the vorticity equation[4] becomes

$$\frac{\mathrm{D}\nabla_H^2 \Psi}{\mathrm{D}t} = \left(\frac{\partial}{\partial t} + \frac{\partial \Psi}{\partial y}\frac{\partial}{\partial x} - \frac{\partial \Psi}{\partial x}\frac{\partial}{\partial y}\right)\nabla_H^2 \Psi = \nu \nabla^2 \nabla_H^2 \Psi + F,$$

[1] Plus a small velocity that induces vortex stretching.
[2] See Appendix C.4.
[3] The operator $\mathcal{V}(\cdot) \equiv \partial/\partial z\,(\nu \partial(\cdot)/\partial z) + \nu \nabla_H^2(\cdot)$ was introduced in § 8.2.
[4] From Appendix D.10.

where F represents the vortex stretching necessary to maintain the steady flow against dissipative losses.

The stream function for this prescribed flow $\Psi = \int v(y)\,\mathrm{d}y$ is assumed to satisfy the vorticity equation. The big question we need to address is: *is this flow stable?* In order to answer this question, we need to assume that the basic-state solution is subject to a time-dependent perturbation of infinitesimal amplitude and arbitrary structure. If all perturbations decay with time, then the basic-state solution is stable. However if any of the perturbations grows with time, the basic-state solution is unstable; the infinitesimal perturbations will grow exponentially with time until they become sufficiently large to alter the basic-state flow. Let's suppose that

$$\Psi = \int^{y} v(\hat{y})\,\mathrm{d}\hat{y} + \Psi',$$

where a prime denotes a perturbation of infinitesimal amplitude that is a function of x, y and t. Ignoring terms in the momentum equation that are quadratic in Ψ', we have

$$\left(\frac{\partial}{\partial t} + v\frac{\partial}{\partial x}\right)\nabla_H^2\Psi' - \frac{\mathrm{d}^2 v}{\mathrm{d}y^2}\frac{\partial \Psi'}{\partial x} = \nu\nabla^2\nabla_H^2\Psi'.$$

This equation is to be solved on the interval $y_1 < y < y_2$ subject to the conditions $\Psi' = 0$ and $\partial\Psi'/\partial y = 0$ at $y = y_1$ and $y = y_2$.

This equation is linear with coefficients varying with y. We can assume the perturbations vary harmonically in x and either harmonically or exponentially in t;[5] let

$$\Psi'(x,y,t) = \tilde{\Psi}(y)e^{ikx - i\omega t} + \text{c.c.},$$

with $\tilde{\Psi}$ being complex. The goal of the following analysis will be to find the possible values of ω, with k being real. If ω can have a positive imaginary part, the shear flow is unstable. Now the governing equation becomes the *Orr–Sommerfeld equation*:

$$(kv - \omega)\left(\frac{\mathrm{d}^2}{\mathrm{d}y^2} - k^2\right)\tilde{\Psi} - k\frac{\mathrm{d}^2 v}{\mathrm{d}y^2}\tilde{\Psi} = \nu\left(\frac{\mathrm{d}^2}{\mathrm{d}y^2} - k^2\right)^2\tilde{\Psi}.$$

Typically solution of this equation requires a numerical procedure or asymptotic analysis.

G.1.1 Inviscid Instability

Commonly in the atmosphere and oceans, the laminar viscosity is so small that it plays no role in the onset of instability. Setting $\nu = 0$, the governing equation simplifies to

$$\frac{\mathrm{d}^2\tilde{\Psi}}{\mathrm{d}y^2} - k^2 - \frac{1}{v - c}\frac{\mathrm{d}^2 v}{\mathrm{d}y^2}\tilde{\Psi} = 0,$$

[5] We are ignoring variations in z; *Squire's theorem* tells us that we are missing nothing with this the simpler formulation.

where $c = \omega/k$ is the (possibly complex) phase speed of the disturbance. This is the *Rayleigh stability equation*. It is to be solved on the interval $y_1 < y < y_2$ subject to the no-flow conditions $\tilde{\Psi}(y_1) = \tilde{\Psi}(y_2) = 0$. This is an eigenvalue problem, to be solved for the complex eigenfunction $\tilde{\Psi}$ and complex eigenvalue c. Note that a positive imaginary part of ω (indicating instability) corresponds to a positive imaginary part of c. However, since the equation is satisfied if both $\tilde{\Psi}$ and c are replaced by their complex conjugates, the flow is unstable if c has a non-zero imaginary part – either positive or negative. If c is real, the flow is neutrally stable – neither growing nor decaying with time.

We may readily obtain a necessary condition for instability by multiplying the differential equation by its complex conjugate and integrating in y from y_1 to y_2. Integrating the first term by parts and using the boundary conditions, we have the integral constraint

$$\int_{y_1}^{y_2} \left(\left\| \frac{\mathrm{d}\tilde{\Psi}}{\mathrm{d}y} \right\|^2 + k^2 \|\tilde{\Psi}\|^2 + \frac{1}{v-c} \frac{\mathrm{d}^2 v}{\mathrm{d}y^2} \|\tilde{\Psi}\|^2 \right) \mathrm{d}y = 0 .$$

All the terms in this equation are real, with the possible exception of c. If c has a non-zero imaginary part, this integral constraint can be satisfied only if $\mathrm{d}^2 v/\mathrm{d}y^2$ changes sign. This result is *Rayleigh's inflection point theorem*, which states that a necessary condition for the instability of shear flow is an inflection point in the flow profile. It is not a sufficient condition because viscous forces may squelch the instability. For more detail, see Drazin and Reid (1981).

G.2 Boundary-Layer Theory

In § 7.4 we introduced the concept of boundary layers and briefly explained why they occur from a physical point of view. This fundament is an abbreviated introduction to boundary-layer theory from a mathematical point of view.

Let's illustrate boundary layers with a simple example; consider the equation

$$\varepsilon \frac{\mathrm{d}y}{\mathrm{d}x} + y = 1 ,$$

where ε is a positive adjustable parameter, to be solved for y on the interval $0 < x < 1$ subject to the condition $y(0) = 0$. The general solution of this non-homogeneous first-order linear equation is the sum of a homogeneous part $y_h = Ce^{-x/\varepsilon}$, where C is a constant of integration, and a particular part $y_p = 1$. The boundary condition is satisfied provided $C = -1$; altogether,

$$y = 1 - e^{-x/\varepsilon} .$$

This solution has a boundary-layer structure if $0 < \varepsilon \ll 1$, with the particular solution (that is, the exponential term) being very small if $\varepsilon \ll x$. Additionally, the problem becomes singular in the limit $\varepsilon \to 0$; we lose the ability to satisfy the boundary condition at $x = 0$. This example illustrates three features of boundary layers:

- the governing equation(s) contain a small parameter;
- the problem becomes singular as that parameter tends to zero; and
- the solution consists of two parts: an interior solution ($y_p = 1$ in this example) plus a boundary layer solution ($y_h = e^{-x/\varepsilon}$ in this example) that decays to zero (usually exponentially) with distance from the boundary.

Boundary-layer theory was first developed in aerodynamics,[6] but is generally applicable to a wide variety of circumstances, including many geophysical flows. The theory relies on the presence a small dimensionless parameter multiplying the highest-derivative term in the governing equation(s). If this term were absent, it would not be possible in general to find a solution of the equations that satisfies all the boundary conditions. This term is negligible in the interior of the fluid body, but within the boundary layer derivatives in the direction normal to the boundary are sufficiently large that this term becomes of dominant order, permitting satisfaction of all the boundary conditions.

Typically the thickness of the boundary layer is determined locally by the dynamic structure of the equations and the problem may be treated as a boundary-layer problem only if that thickness is small compared with the thickness of the entire fluid layer. In very simple flow problems, the boundary-layer and interior flows are decoupled, but typically there is some dynamic interaction between the two flows.

In the main text, we encounter boundary layers in various contexts. The first context is katabatic flow down a slope, with the boundary layer thickness being a result of diffusion of both momentum and heat; see § 22.2.5 and § 24.1. Katabatic-wind flow is confined entirely to the boundary layer, with no interior flow. The second context is the Ekman layer, with the thickness being determined by diffusion of momentum and rate of rotation; see Chapter 25. The Ekman layer is active, in the sense that it specifies the local interior flow speed (called Ekman pumping) normal to the boundary. The third context is surface flows in the oceans, with the boundary layer thickness involving the turbulent diffusivity of momentum and the β plane parameter; see § 27.1.1. Flow in these boundary layers is driven by a specified normal flow (the Sverdrup currents). The fourth context is the near-ground boundary layer within an atmospheric vortex, with the layer thickness being determined by a balance of turbulent diffusion of momentum and inertia; see § 28.4.3. This boundary layer is nonlinear and consequently rather difficult to tackle. When the governing equations are coupled and nonlinear, as is often the case for geophysical flows, the structure of the boundary layer is found by scale analysis.[7]

G.3 Comments On Open-Channel Flow Models

The flow of water in a channel has long been a subject of intense practical interest. Understanding of channel flow is embodied in various empirical formulas, some of which

[6] See Schlichting (1968).
[7] See § 2.4.5.

were established long ago, before the value of dimensional analysis was widely known.[8] The oldest of these is the Chézy velocity formula, published in 1775 by Antoine Chézy:

$$\bar{u} = C\sqrt{\varepsilon h},$$

where $\bar{u}$ is the mean flow speed, ε is the slope of the channel, h is the water depth and C is a constant having dimensions $\mathrm{L}^{1/2}.\mathrm{T}^{-1}$, i.e., $\sqrt{\text{acceleration}}$.[9] First, on general principles, it is undesirable to have constants of proportionality be dimensional; such constants should be dimensionless. The obvious way to non-dimensionalize the Chézy constant is to scale it with the acceleration of gravity. In fact we see from the momentum equation that g and ε appear only as a product; it follows that $\bar{u}$ must be a function of εg together, not g and ε separately. Let us write

$$C = \sqrt{g}C^*,$$

where C^* is the dimensionless Chézy coefficient, so that

$$\bar{u} = C^*\hat{u}, \qquad \text{where} \qquad \hat{u} = \sqrt{\varepsilon g h}.$$

Comparing that with the formula given in § 23.2, we see that

$$C^* = \sqrt{8/\lambda},$$

where λ is the Darcy friction factor, and the Chézy formula becomes identical to the Darcy–Weisbach formula.

Observations of river flows have long revealed that the Chézy constant is not constant, a fact we have already demonstrated. In 1888 Robert Manning proposed an "improved" flow formula, known as the *Manning formula*:

$$\bar{u} = \frac{1}{n}\varepsilon^{1/2}h^{2/3},$$

where n is the roughness coefficient. Like Chézy's constant, the roughness coefficient has the undesirable quality of being dimensional: n has dimensions $\mathrm{T}\cdot\mathrm{L}^{-1/3}$. It is not obvious how to make n dimensionless. It is clear that gravity should be involved, since water would not flow downhill in the absence of gravity. To complete the non-dimensionalization of n, we need another quantity having dimensions involving L and T, such as a length. Formula (1.11) of Sellin (1969) suggests that the appropriate dimension is the water depth, which makes the Manning formula equivalent to the Chézy formula.[10]

[8] See the fundament on dimensional analysis found in Appendix B.1.

[9] See Sellin (1969).

[10] Chow (1985) has on pages 98–99 an interesting footnote regarding Manning's formula, illustrating the extent of fuzzy thinking on this subject.

References

Allen, J. J., Shockling, M. A., Kunkel, G. J. & Smits, A. J. (2007). Turbulent flow in smooth and rough pipes. *Philos. T. Roy. Soc. A*, **365**(1852), 699–714.

Argus, D. F., Gordon, R. G. & DeMets, C. (2011). Geologically current motion of 56 plates relative to the no-net-rotation reference frame. *Geochemistry, Geophysics, Geosystems*, **12**(11), doi:10.1029/2011GC003751.

Bak, P. (1996). *How Nature Works: The Science of Self-organized Criticality*. New York: Springer.

Ball, F. K. (1956). The theory of strong katabatic winds. *Aust. J. Phys.*, **9**, 373–386.

Bascom, W. (1964). *Waves and Beaches*. Garden City, New York: Doubleday.

Batchelor, G. K. (1972). Sedimentation in a dilute dispersion of spheres. *J. Fluid Mech.*, **52**(2), 245–268.

Borch, R. S. & Green, H, W. II (1987). Dependence of creep in olivine on homologous temperature and its implications for flow in the mantle. *Nature*, **330**, 345–348.

Chow, V. T. (1985). *Open Channel Hydraulics*. New York: McGraw–Hill.

Čížková, H., van den Berg, A. P., Spakman, W. & Matyska, C. (2012). The viscosity of Earth's lower mantle inferred from sinking speed of subducted lithosphere. *Phys. Earth Planet In.*, **200–201**, 56–62; doi:10.1016/j.pepi.2012.02.010.

Clement, A. C. & Peterson, L. C. (2008). Mechanisms of abrupt climate change of the last glacial period. *Rev. Geophys*, **46**(4), doi:10.1029/2006RG000204.

Courtillot, V. & Renne, P. R. (2003). On the ages of flood basalt events. *C. R. Geosciences*, **335**(1), 113–140.

Cullen, S. (2005). Trees and wind: a practical consideration of the drag equation velocity exponent for urban tree risk management. *J. Arboriculture*, **31**(3), 101–113.[1]

Davaille, A. & Limare, A. (2009). *Laboratory studies of mantle convection*. Vol. VII of *Treatise on Geophysics*, Amsterdam: Elsevier.

Davies, G. F. & Richards, M. A. (1992). Mantle convection. *J. Geol*, **100**(2), 151–206.

Davison, A. (1956). *My Ship Is So Small*. New York: William Sloan Associates.

de Koker, N. & Stixrude, L. (2009). Self-consistent thermodynamic description of silicate liquids, with application to the shock melting of MgO periclase and $MgSiO_3$ perovskite. *Geophys. J. Int.*, **178**(1), 162–179.

de Rooy, W. C., Bechtold, P., Fröhlic, K., Hohenegger, C., Jonker, H., Mironov, D., Siebsema, A. P., Teixeira, J. & Yano, J.-I. (2013). Entrainment and detrainment in cumulus convection: an overview. *Q. J. Roy. Meteor. Soc.*, **139**(670), 1–19.

[1] This journal is now the Journal of Arboriculture & Urban Forestry Online.

Dettman, J. W. (1969). *Mathematical Methods in Physics and Engineering*. New York: McGraw–Hill.

Drazin, P. G. & Reid, W. H. (1981). *Hydrodynamic Stability*. Cambridge University Press.

Dukhovskoy, D. M. & Morey, S. L. (2010). Simulation of the Hurricane Dennis storm surge and considerations of vertical resolution. *Natural Hazards*, doi:10.1007/s11069-010-9684-5.

Dye, S. T. (2012). Geoneutrinos and the radioactive power of the Earth. *Rev. Geophys.*, **50**(3), doi:10.1029/2012RG000400.

Dyer, K. R. & Soulsby, R. L. (1988). Sand transport on the continental shelf. *Annu. Rev. Fluid Mech.*, **20**, 295–324.

Dziewonski, A. & Anderson, D. L. (1981), Preliminary reference Earth model. *Phys. Earth Planet. In.*, **25**, 297–356.

Ellison, T. H. (1956). Atmospheric turbulence. In G.K. Batchelor & R. M. Davies, eds., *Surveys in Mechanics*. Cambridge University Press, pp. 400–430.

Emanuel, K. A. (1986). An air-sea interaction theory for tropical cyclones. Part I. *J. Atmos. Sci.*, **43**, 585–604.

Emanuel, K. A. (1991). The theory of hurricanes. *Annu. Rev. Fluid Mech.*, **23**, 179–196.

Forsythe, A. R. (1959). *Theory of Differential Equations*, 6th edn. New York: Dover Publications.

Furlong, K. P. & Chapman, D. S. (2013). Heat flow, heat generation and the thermal state of the lithosphere. *Annu. Rev. Earth Pl. Sc.*, **41**, 385–410.

Griffiths, R. W. (2000). The dynamics of lava flows. *Annu. Rev. Fluid Mech.*, **32**, 477–518.

Guazzelli, É. & Hinch J. (2011). Fluctuations and instability in sedimentation. *Annu. Rev. Fluid Mech.*, **43**, 97–116.

Havelock, T. H. (1908). The propagation of groups of waves in dispersive media, with applications to waves on water produced by a travelling disturbance. *P. Roy. Soc. Lond. A Mat.*, **81**(549), 398–430.

Hellerman, S. & Rosenstein, M. (1983). Normal monthly wind stress over the world ocean with error-estimates. *J. Phys. Oceanogr.*, **13** , 1093–1104.

Hodgkinson, J. H. & Stacey, F. D. (2017). *A Practical Handbook of Earth Science*. CRC Press.

Holton, J. R. (2004). *An Introduction to Dynamic Meteorology*. Burlington, MA: Elsevier Academic Press.

Hon, K., Kauahikaua, J., Denlinger, R. & MacKay, K. (1994). Emplacement and inflation of pahoepahoe sheet flows: Observations and measurements of active lava flows on Kilauea Volcano, Hawaii. *Geol. Soc. Am. Bull.*, **106**(3), 351–370.

Jaupart C., Labrosse, S., Lucazeau, F. & Mareschal, J.-C. (2015). Temperatures, heat, and energy in the mantle of the Earth. In G. Schubert, ed., *Treatise on Geophysics*, 2nd edn, vol. 7. Oxford: Elsevier, pp. 223–270.

Jones, M. D. & Savino, J. M. (2015). *Supervolcano: The Catastrophic Event that Changed the Course of Human History*. Pronoun.

King, S. D. (2016). Reconciling laboratory and observational models of mantle rheology in geodynamic modelling. *J. Geodyn.*, **100**, 33–50.

Lansing, A. (1959). *Endurance: Shackleton's Incredible Voyage*. New York: McGraw–Hill.

Lay, T., Hernlund, J. & Buffett, B. A. (2008). Core–mantle boundary heat flow. *Nat. Geosci.*, **1**, 25–32.

Lewellen, W. S. (1993). *The tornado: its structure, dynamics, prediction and hazards*. Vol. 79 of *Geophysical Monograph Series*. American Geophysical Union.

Li, D., Katul, G. G. & Zilininkevich, S. J. (2015). Revisiting the turbulent Prandtl number in an idealized atmospheric surface layer. *J. Atmos. Sci.*, **72**, 2394–2401. doi:10.1175/JAS-D-14-0335.1.

Liu, J., Song, M., Hu, Y. & Ren, X. (2012). Changes in the strength and width of the Hadley circulation since 1871. *Clim. Past.*, **8**, 1169–1157.

Loper, D. E. (1991). The nature and consequences of thermal interactions twixt core and mantle. *J. Geomagn. Geoelectr.*, **43**(2), 79–91.

Loper, D. E. (2009). Earth's habitable loop: water, atmospheric structure, the geomagnetic field and plate tectonics. *Acta Geod. Geophys. Hu.*, **44**, 265–269. doi:10.1556/AGeod. 44.2009.3.2.

Loper, D. E. & Lay, T. (1995). The core–mantle boundary region. *J. Geophys. Res. – Sol. Ea.*, **100**(B4), 6397–6420.

Løvholt, F., Kaiser, G., Glimsdal, S., Scheele, L., Harbitz, C. B. & Pedersen, G. (2012). Modeling propagation and inundation of the 11 March 2011 Tohoku tsunami. *Nat. Hazards Earth Syst. Sci.*, **12**, 1017–28.

Madsen, O. S. (1976). A realistic model of the wind-induced Ekman boundary layer. *J. Phys. Oceanogr.*, **7**, 248–255.

Mason, B., Pyle, D. M. & Oppenheimer, C. (2004). The size and frequency of the largest explosive eruptions on Earth. *B. Volcanol.*, **66**(8), 735–748.

Mastin, L. G. & M. Ghiorso, S. (2000). A numerical program for steady-state flow of magma-gas mixtures through vertical eruptive conduits (vers. 1.05b, April 2008): U.S. Geological Survey Open-File Report 00-209. https://pubs.usgs.gov/of/2000/0209/.

Morey, S. L., Baig, S., Bourassa, M. A., Dukhovskoy, D. S. & O'Brien, J. J. (2006). Remote forcing contribution to storm-induced sea level rise during Hurricane Dennis. *Geophys. Res. Lett.*, **33**(19), L19603. doi:10.1029/2006GL027021

Murphy, G. M. (1960). *Ordinary Differential Equations and Their Solutions*. Princeton: Van Nostrand.

Ni, H., Hui, H. & Steinle-Newmann, G. (2015). Transport properties of silicate melts. *Rev. Geophys.*, doi:10.1002/2015RG000485.

Olson, P. (2016). Mantle control of the geodynamo: Consequences of top-down regulation. *Geochem. Geophys. Geosy.*, **17**, 1935–1956, doi:10.1002/ 2016GC006334.

Olver, F. J. W., Lozer, D. W., Boisvert, R. F. & Clark, C. W. (2010). *NIST Handbook of Mathematical Functions*. Cambridge University Press. http://dlmf.nist.gov.

Parish, T. R. (1988). Surface winds over the Antarctic continent: a review. *Rev. Geophys.*, **26**, 169–180.

Plank, T., Kelley, K. A., Zimmer, M. M., Hauri, E. H. & Wallace, P. J. (2013). Why do mafic arc magma contain ~ 4 wt% water on average? *Earth Planet. Sc. Lett.*, **364**, 168–179.

Pozzo, M., Davies, C., Gubbins, D. & Alfè, D. (2012). Thermal and electrical conductivity of iron at Earth's core conditions. *Nature*, **485**, 355–360.

Putirka, K. D., Perfit, M., Reynolds, F. J. & Jackson, M. G. (2007). Ambient and excess mantle temperatures, olivine thermometry, and active vs. passive upwelling. *Chem. Geol.*, **214**(3–4), 177–206; doi:10.1016/j.chemgeo.2007.01.014.

Rahmstorf, S. (2003). Timing of abrupt climate change: a precise clock. *Geophys. Res. Lett.*, **30**(10). doi:10.1029/2003GL017115.

Rahmstorf, S. (2006). Thermohaline ocean circulation. In S. A. Elias, ed., *Encyclopedia of Quaternary Sciences*. Amsterdam: Elsevier.

Rahmstorf, S., Box, J. E., Feulner, G., Mann, M. E., Robinson, A., Rutherford, S. & Schaffernicht, E. J. (2015). Exceptional twentieth-century slowdown in atlantic ocean overturning circulation. *Nature Climate Change*, **5**(5), 475–480. doi:http://dx.doi.org/10.1038/nclimate2554.

Richards, M. A., Duncan, R. A. & Courtillot, V. E. (1989). Flood basalts and hot-spot tracks: plume heads and tails. *Science*, **246**, 103–107.

Roberts, G. O. (1979). Fast viscous Bénard convection. *Geophys. Astro. Fluid*, **12**(1), 235–272.

Robock, A., (2000). Volcanic eruptions and climate. *Rev. Geophys.*, **38**(2), 191–219.

Schlichting, H. (1968). *Boundary Layer Theory*, 6th edn. New York: McGraw–Hill.

Schmittner, A., Chiang, J. C. H. & Hemming S. R. (eds.) (2007). Ocean Circulation: Mechanisms and Impacts – Past and Future Changes of Meridional Overturning. Vol. 173 of *Geophysical Monograph Series*. doi:10.1029/GM173.

Shockling, M. A., Allen, J. J. & Smits, A. J. (2006). Roughness effects in turbulent pipe flow. *J. Fluid Mech.*, **564**, 267–285.

Sellin, R. J. H. (1969). *Flow in Channels*. London: Macmillan.

Shearer, P. M. (2009). *Introduction to Seismology*, 2nd edn. Cambridge University Press.

Smith, S. D. (1980). Wind stress and heat flux over the ocean in gale force winds. *J. Phys. Oceanogr.*, **10**, 709–726.

Sovilla, B., McElwaine, J. N. & Loug, M. Y. (2015). The structure of powder snow avalanches. *Comptes Rendus Physique*, **16**(1), 97–104.

Stacey, F. D. (1991). Effects on the core of structure within D″. *Geophys. Astro. Fluid*, **60**, 157–163.

Stacey, F. D. (2010). Thermodynamics of the Earth. *Rep. Prog. Phys.*, **73**(4). doi:10.1088/0034-4885/73/4/046801.

Stacey, F. D. & Davis, P. M. (2008). *Physics of the Earth*, 4th edn. Cambridge University Press.

Stacey, F. D. & Loper, D. E. (2007). A revised estimate of the conductivity of iron alloy at high pressure and implications for the core energy balance. *Phys. Earth Planet. In.*, **161**(1–2), 13–18.

Stommel, H. (1947). Entrainment of air into a cumulus cloud. *J. Meteorology*, **4**, 91–94.

Stommel, H. (1961). Thermohaline convection with two stable regimes of flow. *Tellus*, **13**(2), 224–230.

Storchak, D. A., Schweitzer, J. & Bormann, P. (2003). The IASPEI Seismic Phase List. *Seismological Res. Lett.*, **74**(6), 761–772.

Sverdrup, H. U. (1947). Wind-driven currents in a baroclinic ocean; with application to the equatorial currents of the eastern Pacific. *Proc. Natl. Acad. Sci. USA*, **33**(11), 318–326.

Tozer, D. C. (1972). The present thermal state of the terrestrial planets. *Phys. Earth Planet. In.*, **6**(1–3), 182–197.

Vilajosana, I., Khazardze, G., Surñach, E., Lied, E. & Kristensen, K. (2007). Snow avalanche speed determination using seismic methods. *Cold Reg. Sci. Technol.*, **49**(1), 2–10.

Wallace, P. J., Planck, T., Edmonds, M. & Hauri, E. H. (2015). Volatiles in Magmas. In H. Sigurdsson, B. Holton, S. McNutt, H. Rymer & J. Stix, eds., *The Encyclopedia of Volcanoes*. Academic Press, pp. 163–183.

Williams, H. & McBirney, A. R. (1979). *Volcanology*. San Francisco: Freeman, Cooper and Co.

Yalin, M. S. (1992). *River Mechanics*. Oxford: Pergamon Press.

Zhang, Y., Xu, Z., Zhu, M. & Wang, H. (2007). Silicate melt properties and volcanic eruptions. *Rev. Geophys.*, **45**(4), doi:10.1029/2006RG000216.

Zwillinger, D. (1998). *Handbook of Differential Equations*, 3rd edn. Academic Press.

Index